This book provides a comprehensive introdu
theory of electrical circuits for students in e
sciences. The methods of circuit analysis
illustrated with the aid of numerous worked e
theory relevant to the fields of electronics, tel
systems are treated throughout.

The text is suitable for first- and second-year undergraduate courses in UK universities and polytechnics. It is also appropriate for a first course in circuits for undergraduate Electrical Engineering majors in US colleges and universities.

Selected topics are treated at a more advanced level to cater for students specializing in electrical subjects in the final years of their degree course.

Electronics texts for engineers and scientists
Editors
P.J. Spreadbury, *University of Cambridge*
P.L. Jones, *University of Manchester*

Electrical circuits

Titles in this series

H. Ahmed and P.J. Spreadbury
Analogue and digital electronics for engineers

R.L. Dalglish
An introduction to control and measurement with microcomputers

W. Shepherd and L.N. Hulley
Power electronics and motor control

T.H. O'Dell
Electronic circuit design

K.C.A. Smith and R.E. Alley
Electrical circuits

A. Basak
Analogue electronic circuits and systems

Electrical circuits

AN INTRODUCTION

K.C.A. SMITH
Emeritus Reader in Electrical Engineering
University of Cambridge
Fellow of Fitzwilliam College

R.E. ALLEY
Professor, Electrical Engineering Department
United States Naval Academy

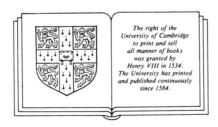

CAMBRIDGE UNIVERSITY PRESS
Cambridge
New York   Port Chester   Melbourne   Sydney

Published by the Press Syndicate of the University of Cambridge
The Pitt Building, Trumpington Street, Cambridge CB2 1RP
40 West 20th Street, New York, NY 10011–4211, USA
10 Stamford Road, Oakleigh, Victoria 3166, Australia

© Cambridge University Press, 1992

First published 1992

Printed in Great Britain at the University Press, Cambridge

British Library cataloguing in publication data

Smith, K.C.A.
  Electrical circuits.
  1. Electric equipment. Circuits. Theories
  I. Title II. Alley, R.E. III. Series
  621.319'2'01

Library of Congress cataloguing in publication data

Smith, K.C.A.
  Electrical circuits: an introduction / K.C.A. Smith and
R.E. Alley.
    p. cm. – (Electronics texts for engineers and scientists)
  Includes bibliographical references.
    ISBN 0 521 37407 3. – ISBN 0 521 37769 2 (pbk.)
    1. Electric circuits. I. Alley, R.E. II. Title. III. Series.
TK454.S585 1991
621.319'2 – dc20   89-17365   CIP

ISBN 0 521 374073 hardback
ISBN 0 521 377692 paperback

Contents

Preface   xv

1   Basic concepts, units, and laws of circuit theory

1.1	Properties of the electrical circuit	1
1.2	The lumped circuit model	3
1.3	Charge and current	5
1.4	Potential difference, energy and power	7
1.5	Ideal voltage and current sources	11
1.6	Kirchhoff's laws	13
	1.6.1 The current law	13
	1.6.2 Worked example on the current law	15
	1.6.3 The voltage law	16
	1.6.4 Worked example on the voltage law	17
1.7	Resistance	18
	1.7.1 Ohm's law	18
	1.7.2. Power dissipation in resistance	20
	1.7.3 Resistances in combination	21
1.8	Capacitance	23
	1.8.1 The voltage-current relationship for capacitance	23
	1.8.2 Energy storage in capacitance	24
	1.8.3 Capacitances in combination	24
1.9	Inductance	26
	1.9.1 The voltage-current relationship for inductance	26
	1.9.2 Energy storage in inductance	28
	1.9.3 Inductances in combination	29
1.10	Inductively coupled circuits	30
	1.10.1 Mutual inductance	31
	1.10.2 The coefficient of coupling	33

	1.10.3	The effective inductance of two series-connected coupled coils	35
1.11	Passive circuit components		36
1.12	Summary of basic circuit relations		37
1.13	Problems		37

2 Theorems and techniques of linear circuit analysis

2.1	Introduction		42
2.2	Voltage and current dividers		44
2.3	Mesh analysis		47
2.4	Worked example		50
2.5	The general mesh equations		52
2.6	The superposition and reciprocity theorems		55
	2.6.1	Superposition	55
	2.6.2	Reciprocity	58
2.7	Thévenin's theorem		58
2.8	Worked example		61
2.9	Network transformations		64
	2.9.1	The Thévenin–Norton transformation	64
	2.9.2	The star–delta transformation	66
2.10	Nodal analysis		67
2.11	Comparison of mesh and nodal analysis		71
2.12	Worked example		73
2.13	Analysis of networks containing dependent sources		75
2.14	Worked example		78
2.15	Miscellaneous theorems and techniques		81
	†2.15.1	The substitution and compensation theorems	81
	2.15.2	Circuit reduction	84
	†2.15.3	Ladder networks	87
	†2.15.4	Ring mains	88
	†2.15.5	Worked example	90
2.16	Summary		92
2.17	Problems		93

3 Alternating current circuits

3.1	Introduction	98
3.2	A.C. voltage–current relationships for the linear circuit elements	101
3.3	Representation of a.c. voltage and current by the complex exponential: Phasors	105

3.4	Voltage–current relationships for the general network branch: Impedance	109
3.5	Phasor and impedance diagrams	113
3.6	Linear circuit theorems and techniques in a.c. circuit analysis	115
3.7	Worked example	119
3.8	Admittance	123
3.9	Frequency response: transfer function	126
3.10	A.C. bridges	132
	3.10.1 The Schering bridge	133
	3.10.2 The Wien bridge	135
3.11	Worked example	136
3.12	Inductively coupled circuits	141
3.13	Resonant circuits	146
	3.13.1 Losses in inductors and capacitors	146
	3.13.2 The series resonant circuit	151
	3.13.3 The parallel resonant circuit	164
	3.13.4 Worked example	166
	†3.13.5 Definition of Q-factor in terms of stored energy	169
	†3.13.6 Multiple resonance	169
	†3.13.7 Inductively coupled resonant circuits	173
3.14	Summary	178
3.15	Problems	179

4 Power and transformers in single-phase circuits

4.1	Introduction	187
4.2	Average power	187
4.3	Reactive power and apparent power	190
4.4	Power factor	194
4.5	Worked example	196
4.6	Complex power	198
4.7	The ideal transformer	199
4.8	Worked example	201
4.9	Single-phase power transformers	201
4.10	Worked example	208
4.11	Transformer tests	209
4.12	Voltage regulation	211
4.13	Conditions for maximum efficiency	213
4.14	The autotransformer	214
4.15	Maximum power transfer	217
†4.16	The transformer bridge	221

x   Contents

| 4.17 | Summary | 224 |
| 4.18 | Problems | 225 |

5 Three-phase alternating current circuits

5.1	Introduction	231
5.2	Advantages of three-phase systems	233
5.3	Three-phase circuits	233
	5.3.1 Phase and line voltages	233
	5.3.2 Balanced load	236
	5.3.3 Worked example	237
	5.3.4 Star and delta connections	239
	5.3.5 Worked example	241
	5.3.6 Use of Y–Δ transformation	244
	5.3.7 Unbalanced load	245
	5.3.8 Worked example	245
5.4	Power, reactive power and apparent power in balanced loads	248
5.5	Worked example: power factor correction	249
5.6	Three-phase power measurement	251
	5.6.1 Alternating current meters	251
	5.6.2 Methods of power measurement	253
	5.6.3 Worked example	257
5.7	Transformers for three-phase systems	258
	5.7.1 Applications	258
	5.7.2 Equivalent circuit parameters	261
	5.7.3 Worked example	261
	†5.7.4 Harmonic currents	269
†5.8	Phase transformation	270
†5.9	Instantaneous power to balanced load	273
5.10	Summary	275
5.11	Problems	276

6 Transient and steady-state analysis

6.1	Introduction	280
6.2	Qualitative analysis of the RL circuit	281
6.3	Mathematical analysis of the RL circuit	283
6.4	Time constant	286
6.5	Natural response of some basic series circuits	288
	6.5.1 RL circuit	288
	6.5.2 RC circuit	290
	6.5.3 RLC circuit	291
	6.5.4 Q-factor and logarithmic decrement	294

Contents   xi

6.6	Total response	295
	6.6.1 *RL* circuit with sinusoidal driving voltage	296
	6.6.2 *RC* circuit with constant voltage source	297
	6.6.3 Worked example	297
	6.6.4 *RLC* circuit with constant voltage source	299
	6.6.5 *RLC* circuit with sinusoidal driving voltage	300
	6.6.6 *RLC* circuit with sinusoidal driving voltage and $\omega_0 \simeq \omega_n$	301
6.7	The D-operator	303
	6.7.1 The operators D and D^{-1}	304
	6.7.2 Solution of differential equations by D-operator	305
	6.7.3 D-impedance	309
	6.7.4 Worked example	309
	6.7.5 Thévenin's theorem in transient analysis	313
	6.7.6 Differentiating and integrating circuits	316
6.8	The unit step and related driving functions	317
	6.8.1 Step function	318
	6.8.2 Impulse function	319
	6.8.3 Worked example	323
	6.8.4 Ramp and other singularity functions	325
	6.8.5 Delayed functions	326
6.9	The Laplace transform	327
	6.9.1 Definition of the Laplace transform	328
	6.9.2 Laplace transforms of some functions of time	328
	6.9.3 Partial fractions	334
	6.9.4 Network analysis by Laplace transform	340
	6.9.5 Worked example	345
	6.9.6 Generalized impedance, network function and impulse response	347
	6.9.7 Third and higher order networks	352
	6.9.8 Worked example	355
	6.9.9 Further Laplace transform theorems	357
6.10	Pole-zero methods	359
6.11	Worked example	372
6.12	Pulse and repeated driving functions	375
	6.12.1 Pulse response of first order circuits	375
	6.12.2 Delayed singularity functions: transforms of recurrent waveforms	380
	6.12.3 Response by the Laplace transform	385
6.13	Worked example	389
†6.14	Convolution	392
	6.14.1 Representation of a function by an impulse train	392

Contents

	6.14.2	The convolution integral	394
	6.14.3	The convolution theorem	401
	6.14.4	Worked example	403
6.15	Summary		405
6.16	Problems		408

7 Non-linear circuit analysis

7.1	Introduction: linear and non-linear elements		418
7.2	Graphical analysis		419
7.3	Small-signal models		425
	7.3.1	Non-linear resistor model	425
	7.3.2	Transistor model	426
7.4	Piecewise-linear circuits		428
	7.4.1	Piecewise-linear approximation	428
	7.4.2	The ideal diode	429
	7.4.3	Combinations of resistances and ideal diodes	430
	7.4.4	The real diode	437
	7.4.5	The Zener diode	437
	7.4.6	Analysis of piecewise-linear circuits	440
	7.4.7	Worked example	440
	7.4.8	Synthesis of piecewise-linear circuits	441
	7.4.9	Worked example	443
7.5	Analytical methods		443
7.6	Rectifier circuits		448
	7.6.1	Half-wave rectifier	448
	7.6.2	Worked example	451
	7.6.3	Full-wave rectifier	453
7.7	Thyristor circuits		455
7.8	Fourier analysis of periodic waves		460
	7.8.1	Fourier expansion	460
	7.8.2	Worked example	463
	7.8.3	Odd and even functions	466
	7.8.4	Worked example	466
	7.8.5	Fourier expansion for rectifier output	469
	7.8.6	Expansion of functions of time	471
	7.8.7	Complex exponential form of Fourier series	471
	†7.8.8	Expansions for r.m.s. values and power	474
	7.8.9	Summary of formulae	479
†7.9	Filter circuits for rectifiers		481
	7.9.1	Inductor	482
	7.9.2	L-section	484

Contents

	7.9.3 Capacitor	484
	7.9.4 π-section	486
7.10	Summary	487
7.11	Problems	488

8 Two-port networks

8.1	Introduction	501
8.2	Admittance, impedance and hybrid parameters	503
	8.2.1 Admittance parameters	503
	8.2.2 Impedance parameters	505
	8.2.3 Hybrid and inverse hybrid parameters	506
8.3	Equivalent circuits and circuit models	507
8.4	Transmission, inverse transmission and $ABCD$ parameters	511
8.5	Matrix notation	513
8.6	Worked example	514
8.7	Relationships between direct and inverse $ABCD$ parameters	515
8.8	Parameter relationships for π- and T-networks	516
8.9	Worked example	517
8.10	Cascaded two-ports and chain matrices	518
8.11	Worked example	526
†8.12	Series and parallel connections of two-ports	527
†8.13	Worked example	528
†8.14	Iterative and image impedances	530
	8.14.1 Iterative impedances	530
	8.14.2 Image impedances	531
†8.15	Attenuators	533
†8.16	Worked example	534
†8.17	Insertion loss	536
†8.18	Worked example	536
8.19	Summary	537
8.20	Problems	538

Appendices

A	Units, symbols and abbreviations	542
B	The general mesh equations and proofs of the network theorems	545
C	Computer programs	549
D	Laplace transform pairs	563

Bibliography 566
Answers to problems 567
Index 573

The program listings in Appendix C are available on IBM-PC compatible diskette from the authors. For diskette prices and ordering procedure write, enclosing a self-addressed envelope, to Dr K.C.A. Smith, 50 Selwyn Road, Cambridge, CB3 9EB.

Preface

This book provides an introduction to electrical circuits that will serve as a foundation for courses in electronics, communications and power systems at first degree level. The first three chapters will be found particularly suitable as prerequisite reading for the companion volume in this series; *Analogue and digital electronics for engineers* by H. Ahmed and P.J. Spreadbury. Engineering and science students not intending to specialise in electrical subjects will find in this book most of the circuit theory required for a first degree.

The level of presentation presupposes that students will have encountered the basic ideas of electromagnetism and electrical circuits, including the laws of Faraday, Ohm and Kirchhoff. These ideas are reviewed in chapter 1. Mathematical skills are assumed to extend to the solution of first-order differential equations, and to the elements of complex algebra. Courses in mathematics taken concurrently with those in electrical subjects during the earlier part of a degree course would be expected to fill in progressively the additional mathematical background required; the subject matter has been arranged with this in mind. Sections which may give rise to mathematical difficulties on a first reading, or which may be too specialised for the general student's requirements, are indicated by an obelus (†).

A traditional approach to the development of electrical circuit theory is adopted: the concept of linearity, and the circuit theorems and analytical techniques which stem from this concept, are all presented in chapter 2 within the context of d.c. circuits. The methods and techniques of linear circuit analysis thus established are then extended to a.c. circuits in chapter 3. Familiarity with the basic material contained in chapters 1–2, together with sections 3.1–3.8 and 4.1–4.7 of chapters 3 and 4, will allow the remainder of the book to be read on a selective basis appropriate for the particular courses being followed by the student.

Chapters 4 and 5 deal mainly with various aspects of power transmission in electrical circuits; in chapter 6 methods for the general transient and steady-state analysis of circuits are described, with emphasis on Laplace transform techniques, and chapter 7 deals with the analysis of circuits incorporating non-linear elements. Chapter 8 covers the theory of two-port networks, including the modelling of non-linear devices such as the transistor; later sections of this chapter will be of interest primarily to electrical engineering students.

A suite of simple computer programs, written in BASIC, is included as an appendix, which is designed to assist the student in working through the numerous illustrative examples and problems contained in the text.

We are indebted to Professor K.F. Sander for major contributions to chapter 8; to Mr J. Barron for supplying the basis for the formal proofs of the linear network theorems contained in Appendix B; and to Dr D.E. Roberts, Dr D.M. Holburn and Mr K.L. Chau for advice on programming, and for checking and testing the programs listed in Appendix C. We wish also to acknowledge the skill and patience of Mrs Pat Silk in the preparation and typing of the greater part of the manuscript. Questions from the examination papers of several universities have been included; their permission is gratefully acknowledged.

1

Basic concepts, units, and laws of circuit theory

1.1 Properties of the electrical circuit

An electrical circuit comprises an arrangement of elements for the conversion, transmission and storage of energy. Energy enters a circuit via one or more *sources* and leaves via one or more *sinks*. In the sources energy is converted from mechanical, thermal, chemical or electromagnetic form into electrical form; in the sinks the reverse process takes place. Sources and sinks are linked by elements capable of transmitting and storing electrical energy. The familiar battery-operated flashlamp serves as a reminder of the energy flow processes in a circuit. In this device, energy is converted from chemical to electrical form in the battery and transmitted along wires to the lamp where most of the energy is converted into heat. A small but useful portion is emitted in the form of electromagnetic radiation in the visible part of the spectrum.

In an electrical circuit energy is conveyed through the agency of electrical *charge* and through the medium of *electric* and *magnetic fields*. An essential feature of any circuit, therefore, is the provision of conducting paths for the conveyance of charge. As indicated in fig. 1.1, sources and sinks are operative only when charge flows through them. The *rate* at which charge flows is referred to as the *current*; the greater the current the greater the energy transmitted between sources and sinks.

Charge is set in motion by the action of the electric field established throughout the circuit by the sources. This field provides the *electromotive force* (e.m.f.) which drives charge round the conducting paths in the circuit. Accompanying this flow of charge is the establishment of a magnetic field. Transmission of electrical energy is, therefore, manifest in a circuit by the presence of both electric and magnetic fields in addition to the movement of charge. The establishment of a field in a circuit is accompanied by an expenditure of energy, and this energy is stored within the region of space

Basic concepts, units, and laws of circuit theory

occupied by the field. On subsequent decay of the field, energy is released to the circuit and is eventually absorbed by the sinks. Thus energy can be both stored and conveyed through the medium of a field. However, for the latter process to occur the field must vary with time. Referring again to fig. 1.1, if the sources produce a constant e.m.f., the resulting currents and fields will all be constant and, in this case, there must be a continuous conducting path between sources and sinks along which charge can flow (indicated by the dashed lines in the figure). If, on the other hand, the sources produce a time-varying e.m.f., currents and fields will be time-varying and the conducting path need not be continuous.

This distinction leads to two of the major classes of circuits dealt with in this book: (1) *direct current* (d.c.) circuits in which fields are static and currents are constant and unidirectional: (2) *alternating current* (a.c.) circuits in which the directions of currents and fields alternate in a regular, periodic fashion.

It will be apparent from the above discussion that the electrical behaviour of a circuit is characterized by the strength and distribution of the currents and fields which arise when it is connected to an electrical energy source. The electrical characteristics of a circuit may, therefore, be described generally by means of three elemental properties: *resistance*, *capacitance* and *inductance* (including *mutual inductance*). Resistance is a property associated with the current-carrying paths in a circuit. Capacitance and inductance are properties associated respectively with the parts of a circuit in which electric and magnetic fields arise. Capacitive and inductive elements are often referred to as *storage elements* because of the energy storage properties of a field. A knowledge of the three elemental properties, for a particular circuit, allows us to specify, at least in principle,

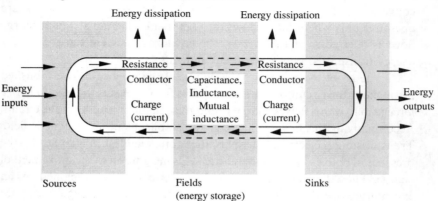

Fig. 1.1. Elements of the electrical circuit.

The lumped circuit model

the magnitudes and directions of the currents which will flow as a result of the application of a given distribution of e.m.f.

Circuits containing only the three basic elements, resistance, capacitance and inductance, are termed *passive* circuits. (*Active* circuits contain also devices such as transistors which, unlike passive elements, are capable of energy amplification.)

If the elemental properties of a passive circuit depend only on its geometry and the materials of which it is made, the circuit is described as being *linear*. If, however, these properties depend additionally on the current or e.m.f. existing in the circuit at any instant, the circuit is described as being *non-linear*. Special techniques are required for the analysis of non-linear circuits; these are dealt with in chapter 7.

Finally, it should be noted that as an inevitable consequence of the movement of charge along a conductor, electrical energy is converted into heat (we are here excluding the superconducting type of circuit), thus the circuit itself acts inherently as an energy sink.

1.2 The lumped circuit model

Practical circuits consist of interconnected assemblies of components: *resistors*, *capacitors* and *inductors*, each designed to exhibit one elemental property to the exclusion of the others.* It is, however, impossible to manufacture a component exhibiting a single property in pure form. Furthermore, all of the interconnections between components will themselves possess each of the three elemental properties to some degree. Consequently, the way in which the elemental properties are distributed in a circuit is often ill defined and, in order to render the circuit amenable to analysis, it is usually necessary to make certain simplifying assumptions and approximations. The most basic of these consists in treating the circuit as if it were composed of pure, discrete elements connected together by conductors possessing no significant properties in themselves. This approach results in the so-called *lumped circuit model*.

Consider again the flashlamp the component parts of which are depicted in fig. 1.2(*a*). Each part, comprising battery, connecting wires, and lamp, possesses resistance which is distributed in some fashion round the closed path forming the circuit. The circuit also contains distributed capacitance and inductance, but only a cursory knowledge of the principles upon which this device operates tells us that these properties can be safely neglected. The circuit model, therefore, need include only resistance as shown in fig. 1.2(*b*). In this model the battery is represented by an energy source together

---

* Note that the circuit *component* is distinguished from the circuit *property* by the terminators *-or* and *-ance* respectively.

with a concentrated or lumped resistance which accounts for all distributed resistance within the battery. The distributed resistance of the connecting wires and the resistance of the lamp are similarly represented by separate lumped resistances. These lumped elements are joined by conductors which are assumed to be *perfect*, that is, by conductors having zero resistance.

The flashlamp exemplifies the simplest possible type of modelling in which there is a close correspondence between the component parts of the real circuit and the lumped elements of the model. Most of the circuits in this book fall into this category. It should be mentioned, however, that the process of devising suitable models for the type of circuit encountered in, for example, telecommunications systems which operate at high frequencies, is

Fig. 1.2. Circuit modelling.

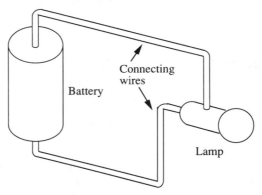

(*a*) Flashlamp: physical components

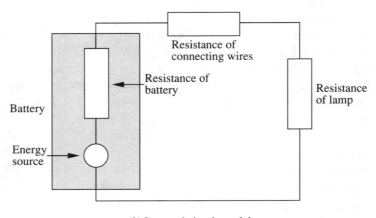

(*b*) Lumped circuit model

Charge and current

often extremely difficult. Each component and interconnection may have to be represented by a combination of elemental properties and the designer may eventually have to select for analysis one among perhaps several possible lumped models, testing each against past experience or by means of actual circuit measurement.

The lumped circuit modelling technique is directly applicable only when the dimensions of the circuit under consideration are small compared with the wavelength corresponding to the frequency of the source excitation. Circuits not falling into this category, such as high-frequency transmission lines (characterized also by a continuous distribution of elemental properties), require special methods of analysis. The lumped modelling technique provides only a starting point for the development of the theory applicable to such circuits.

1.3 Charge and current

We have stated previously that current in a conductor is equal to the rate of flow of charge. If i is the instantaneous current, and a small quantity of charge dq flows in time dt, then

$$i = \frac{dq}{dt} \tag{1.1}$$

The instantaneous current will in general vary with time (fig. 1.3(a)). We can calculate the total amount of charge q which flows during a time interval $t_1 \leq t \leq t_2$ by integrating (1.1).

$$\int_0^q dq = q = \int_{t_1}^{t_2} i \, dt$$

Fig. 1.3. Relationship between charge and current.

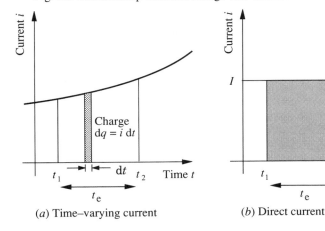

(a) Time–varying current (b) Direct current

The graphical interpretation of this integral is also shown in the figure.

If the time interval commences at the origin, $t_1=0$ and $t_2=t$, and the above integral becomes

$$q = \int_0^t i\,dt \tag{1.2}$$

For a direct current of magnitude I (fig. 1.3(b)), the charge Q which flows in a time interval $t_1 \leq t \leq t_2$ is

$$Q = I\int_{t_1}^{t_2} dt = I(t_2 - t_1) = It_e \tag{1.3}$$

where $t_e = t_2 - t_1$ is the elapsed time.

The units of charge and current are respectively the *coulomb* and the *ampere**.

Although the concept of charge is basic to our understanding of the way in which energy flows in an electrical circuit, the ampere is chosen as the fundamental electrical unit in the SI system rather than the coulomb. The reason for this is that it is easier to detect and measure charge in motion than at rest. The former gives rise to a magnetic field which in turn can be detected by utilizing forces resulting from interaction with other magnetic fields. (See definition of the ampere, appendix A.) This is discussed more fully in reference 6.

So far we have not considered the physical nature and origin of electrical charge and indeed for the purposes of the theory contained in this book it is unnecessary to do so. The established physical picture (according to the Rutherford–Bohr model of the atom) conceives of charge as being carried by atomic particles each bearing a discrete amount of charge. But, even in the smallest currents encountered in practice, the number of charge carriers involved in the transport process is very great and the discrete nature of the flow is not normally detectable. A concept of current as consisting of a smooth fluid-like flow is, therefore, adequate for nearly all practical purposes.

Detailed experimental observation reveals that charge carriers can possess two kinds of charge: positive and negative. Under the action of the same electric field, charges of different kind move in opposite directions. A given amount of positive charge moving along a conductor in one direction is indistinguishable, so far as any observable external effect is concerned,

---

* Appendix A contains information on the International System of Units (SI), and an explanation of the symbols, abbreviations and nomenclature used throughout the text.

from the same quantity of negative charge moving in the opposite direction. By an internationally accepted convention, the direction of current flow is chosen to be that of the direction of motion of positive charge.

In metallic conductors the carriers are electrons which possess negative charge and move in a direction opposite to that of the defined direction of positive current. In semiconductors and electrolytes charge of both kinds exist (carried by electrons and positively charged holes, or ions) and the current is the net result of the movement of positive and negative charge in opposite directions. It must be emphasized, however, that in circuit analysis we are not normally concerned with the nature of charge flow from this microscopic point of view, and we, therefore, talk freely about positive charge moving in metallic conductors even though the charge is in reality carried by electrons.

The reference direction of positive charge flow or current in part of a circuit is indicated diagrammatically by means of an arrow placed on or alongside the conducting path in question. The direction of current between two points A and B in a circuit may also be indicated without ambiguity by means of a double subscript notation. Thus we may write I_{AB}, which is understood to mean a current of magnitude I amperes flowing in a conventional positive sense from A to B. A positive current flowing from B to A would be written I_{BA}; it follows therefore that $I_{BA} = -I_{AB}$. This notation will be valuable in our development of techniques for circuit analysis.

1.4 Potential difference, energy and power

Consider a current of constant magnitude flowing through a section of a metallic conductor AB as shown in fig. 1.4. It is observed experimentally that the passage of current through a conductor is accompanied by the release of energy in the form of heat. It follows that the potential energy of the charge entering the conductor at A must be greater than that of the charge leaving at B since the evolution of heat implies that work is done by the charge during its passage from A to B. A potential energy difference therefore exists between the points A and B. The SI unit of potential energy difference (or simply *potential difference* (p.d.)) is the *volt*, and we say that a voltage exists between A and B. The end of the conductor at the higher potential is indicated conventionally by a $(+)$ sign and that at the lower potential by a $(-)$ sign. A double subscript notation may also be used with advantage to express the magnitude and direction (or *polarity*) of a voltage existing between two points A and B in a circuit. We may write V_{AB} which is understood to mean a p.d. of constant magnitude V volts, A being *positive* with respect to B.

Referring again to fig. 1.4, if a potential difference of one volt exists between A and B, then one coulomb of charge passing between A and B will produce one *joule* of heat energy. Generalizing this statement; if between two points on a metallic conductor there exists a constant potential difference of V volts, and a total of Q coulombs of charge passes between them, the heat output J, in joules, is given by

$$J = VQ \tag{1.4}$$

In terms of current this becomes, using (1.3),

$$J = VIt_e \tag{1.5}$$

where I is a current of constant magnitude and t_e is the elapsed time.

From (1.5) the power P (*watts*) is given by the energy dissipated in the conductor per unit time, that is,

$$P = \frac{J}{t_e} = VI \tag{1.6}$$

For the general case where both voltage and current vary with time, the energy is, at any instant of time t,

$$J = \int_0^t vi\, dt \tag{1.7}$$

and the instantaneous power is

$$p = vi \tag{1.8}$$

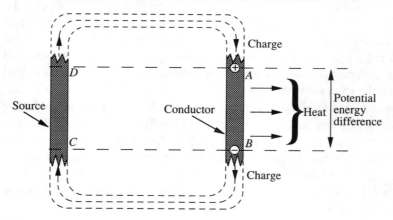

Fig. 1.4. Potential difference. The potential energy lost by the charge as it flows from A to B is recovered as it flows from C to D.

Potential difference, energy and power

It follows from the principle of conservation of energy that if heat is to be dissipated continuously in the section AB, the potential energy lost by the charge in passing from A to B must be made up by a corresponding gain in potential energy elsewhere. In the system shown in fig. 1.4 this occurs as the charge passes through a section CD of a source. The magnitude of the potential difference across CD is, of course, identical to that across AB. For obvious reasons, the latter is often referred to as a voltage *drop* (or *fall*), and the former as a voltage *rise*.

Although the relationships shown in (1.7) and (1.8) have been established by considering the particular case of a metallic conductor, they apply generally to any sink in which electrical energy is converted to some other form. Consider the circuit shown in fig. 1.5. Source and sink are joined by perfect conductors so that the p.d. across both is the same and equal to v. The polarity of this voltage is, according to our convention, indicated by the $(+)$ and $(-)$ signs. Unit positive charge, on passing through the sink from the positive terminal to the negative terminal, loses a total potential energy of v volts, and on passing through the source from the negative terminal to the positive terminal this potential energy is completely regained. The instantaneous power flow from source to sink is given by the product vi.

For circuits containing a multiplicity of elements the magnitude and direction of power flow at any particular element or in any part of the circuit may be ascertained by considering the associated directions of the voltage and current at the terminals concerned. In fig. 1.6, P is any element or part of a circuit at which the instantaneous values of voltage and current are defined. If the product vi is positive, power is being delivered to P while if the product is negative, P is supplying power to the external circuit. In terms of our double subscript notation, power is delivered to P if the product $v_{AB}i_{AB}$ is positive. (Note carefully the order of the subscripts in this product.)

Fig. 1.5. Energy flow between source and sink. Instantaneous power $p = vi$ watts.

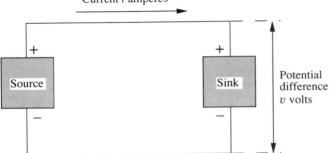

10 *Basic concepts, units, and laws of circuit theory*

If we apply this convention to fig. 1.5, we see that the direction of power flow is in accordance with the meaning which has so far been attached to the terms source and sink. That this is not always the case may be seen by comparing the two circuits shown in fig. 1.7.

In these circuits we assume that the voltage of source P is greater than that of Q and that, as a consequence, there will be a net e.m.f. acting in such a direction as to cause current to flow clockwise round the circuit as shown. Examination of the direction of this current in relation to the polarities of the two sources connected as in fig. 1.7(a), confirms that both sources are delivering power to the sink. However, if the polarity of Q is reversed, as in fig. 1.7(b), current enters its positive terminal, the product vi is positive, and we conclude that energy is being delivered to Q. In other words, what has hitherto been called a source is now effectively acting as a sink.

Many practical sources exhibit this property of reversibility. One common example is the battery which can be recharged by connecting it to

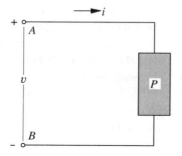

Fig. 1.6. Power in a circuit element: P receives power if product vi is positive; P delivers power if product vi is negative.

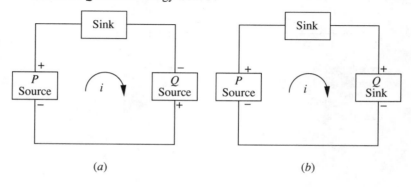

Fig. 1.7. (a) Sources P and Q deliver energy to sink. (b) Polarity of Q reversed: Q receives energy from P.

a supply capable of forcing current through it in a reverse direction. Electrical energy is thereby converted and stored in chemical form. In a resistive conductor the energy conversion is, of course, irreversible.

1.5 Ideal voltage and current sources

The function of a source is to deliver energy to the circuit to which it is connected. It does this, as we have seen, by imparting potential energy to the charges which pass through it, the energy gained by each unit of charge being equal to the p.d. across its terminals. Many practical engineering applications require a source capable of maintaining a substantially constant p.d. across its terminals irrespective of the current which flows through it. No source can be made which does this perfectly and usually, with a source such as a battery for example, the terminal voltage falls as the load current increases. This leads us to the concept of the *ideal voltage source* (also referred to as an ideal voltage generator) defined as one for which the terminal p.d. is independent of the load current. The utility of this concept lies in the fact that the electrical behaviour of a great many practical sources can be described by means of an ideal source in combination with one or more passive circuit elements. This will be discussed more fully in later chapters.

The relationship between terminal voltage and load current (called variously the *voltage–current*, *volt–ampere*, or *load characteristic*) for an ideal voltage source is shown in fig. 1.8(a). This is simply a straight line parallel to the current axis. If we are dealing with a source whose terminal voltage varies with time, then the voltage axis on this graph must be interpreted as indicating instantaneous values. Fig. 1.8(b) shows the conventional graphical symbol for the ideal voltage source.

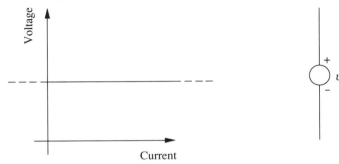

Fig. 1.8. The ideal voltage source.

(a) Voltage–current characteristic   (b) Graphical symbol

12 Basic concepts, units, and laws of circuit theory

Another type of source, of theoretical and practical importance, particularly in electronic circuits, is the *current source* (or current generator). An *ideal* current source may be defined as a source which delivers a specific current to a circuit irrespective of the voltage across its terminals. The voltage–current characteristic for such a source is shown in fig. 1.9. As with the ideal voltage source, the ideal current source can be used in combination with other circuit elements to describe the electrical characteristics of practical current sources.

While there is a universally accepted symbol for the ideal voltage source, there is no such corresponding symbol for an ideal current source. A selection of some of the more commonly used symbols is presented in fig. 1.10; in this book we shall adopt that shown in fig. 1.10(*d*). In all cases the direction of conventional positive current is specified by an arrow.

We should mention, finally, that the ideal sources which we have considered here are termed *independent* sources because the voltage or current, as the case may be, is maintained at its specified value in-

Fig. 1.9. The ideal current source: voltage–current characteristic.

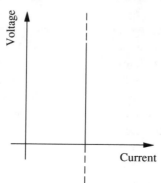

Fig. 1.10. Graphical symbols for the ideal current source.

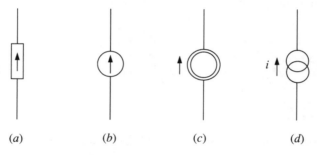

Kirchhoff's laws

dependently of any constraints imposed by the external circuit. Later (section 2.13) we shall discuss another type of ideal source the output of which is controlled by a voltage or current parameter elsewhere in the circuit. This type of source is termed a *dependent* or *controlled* source.

1.6 Kirchhoff's laws

Kirchhoff's current and voltage laws are particular expressions of two of Maxwell's general electromagnetic equations; they apply only to circuits which can be represented by a lumped model. (Their derivation from Maxwell's equations can be found in many books on electromagnetism (see for example reference 7). Within the constraints and limitations mentioned in section 1.2 concerning the lumped circuit model, these two laws provide the basis for *all* circuit analysis.

Kirchhoff's laws are basically conservation laws: the current law expresses the conservation of charge (or more explicitly the continuity of current), and the voltage law expresses the conservation of energy. Although the principles underlying Kirchhoff's laws present little conceptual difficulty, the application of the laws in circuit analysis requires a thorough appreciation of the sign conventions and rules which govern the algebraic combination of currents and voltages. In the two following sections, attention is given to these related aspects of Kirchhoff's laws as well as to the laws themselves.

1.6.1 The current law

The current law in its simplest form may be derived by considering the flow of current between two lumped elements connected together by a perfect conductor (fig. 1.11(a)). A junction such as this, between two or more elements, is called a *node*. (Kirchhoff's current law is also known as the *node law* because it relates to the currents at a node.)

Two currents are shown in fig. 1.11(a), both flowing in the conductor from left to right. Since there is no reservoir at the node in which charge can accumulate, it is obvious that the currents must be continuous through the node and we can write

$$i_1 = i_2 \tag{1.9}$$

This states simply that the currents flowing into and out of the node are equal. This relationship may, however, be expressed in another way. By transposing (1.9) we obtain

$$i_1 + (-i_2) = 0 \tag{1.10}$$

Diagrammatically this is equivalent to changing the reference direction of i_2

as shown in fig. 1.11(b). The law now states that the sum of the two currents flowing *into* the node, taking into account their algebraic signs, is zero.

Finally, a third expression of the law is obtained by reversing the direction of i_1 to give

$$i_2 + (-i_1) = 0 \tag{1.11}$$

The reference directions corresponding to this equation are shown in fig. 1.11(c), and we see that the sum of the currents flowing *out* of the node is zero.

The above arguments, based essentially on the principle of continuity of current, can be extended to include any number of elements connected together at a node, and in its general form Kirchhoff's current law can be summarized by the following relationships:

$$\sum i_{\text{in}} = \sum i_{\text{out}} \tag{1.12}$$

or

$$\sum i = 0 \tag{1.13}$$

where the sums are taken over all i.

Equation (1.12) expresses the continuity relationship in its most direct form. Equation (1.13) states that the algebraic sum of the currents flowing into (or out of) a node is zero. These equations express precisely the same relationship between the currents at a node, but they each lead to a slightly different formulation of the complete circuit equations and it is a matter of convenience which of them one chooses. We consider this further in chapter 2.

Fig. 1.11. Alternative expressions of Kirchhoff's current law at a node.

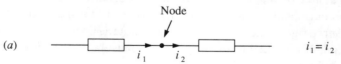

Kirchhoff's laws

Returning for a moment to the circuit of fig. 1.11(a), and to the directions assigned to the currents i_1 and i_2, it will be clear from the intervening discussion that these directions were chosen quite arbitrarily, we could equally well have shown them flowing from right to left, both flowing into or both flowing out of the node. In general, it does not matter how the currents at a node are assigned since their purpose is simply to provide a frame of reference on which to base an explicit expression of the current law. The following example will help to make this clear.

1.6.2 Worked example on the current law

Fig. 1.12(a) shows part of a circuit containing six elements. The magnitudes and directions of the currents (referred to conventional positive current) through five of the elements are indicated. Find the two currents in the conductors at A and B.

To apply Kirchhoff's current law one must first identify the nodes that are relevant to the problem. It may appear at first sight that we have to apply the law at each of the nodes a, b, c, and d separately, but if we recall that the connections between elements of our lumped circuit are perfect, and that there is no potential difference between any two points on a perfect conductor, we see immediately that the connections in the central region of this circuit can be rearranged as shown in fig. 1.12(b). There are now two clearly identifiable nodes, P and Q.

Let the two required currents be i_1 and i_2 and assign their reference directions outward from the circuit as indicated. For this problem it is immaterial which of the forms of the current law is used, so choosing (1.13),

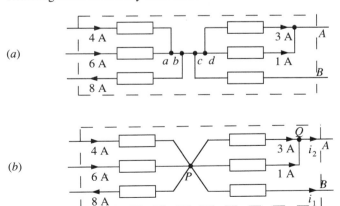

Fig. 1.12. (a) Circuit for worked example on the current law. (b) Rearrangement to identify node at P.

and equating the algebraic sum of the currents flowing *out* of the node to zero, we obtain:

At node P $\quad -4-6+8+3+1+i_1=0$
whence $\quad i_1=-2\text{A}$
At node Q $\quad -3-1+i_2=0$
whence $\quad i_2=4\text{A}$

We see that the current i_1 is negative. This simply means that our original choice of reference direction for the current was (as it transpires) opposite to the direction of conventional positive current.

This example may be used to illustrate a further consequence of the current law. If we calculate the algebraic sum of all the currents entering (or leaving) the complete circuit inside the boundary indicated by the dashed line, we find that this sum also is zero. Thus currents flowing *inwards* across the boundary are:

$$4+6-8-(-2)-4=0$$

As may be readily ascertained, a similar result is obtained for any part of the circuit defined by a closed boundary cutting through two or more of its conducting paths. Kirchhoff's current law, therefore, applies in its most general form to any region of a circuit defined by a closed boundary. This follows immediately from the fact that, within such a boundary, charge can neither be created nor destroyed, neither can it be stored. (A more complete discussion of this aspect of Kirchhoff's current law will be found in reference 4.)

1.6.3 The voltage law

The potential difference across an element has been defined in section (1.4) as the energy gained or lost by unit charge as it passes through the element. It was also shown with reference to fig. 1.5 that the potential energy gained by the charge passing through the source was equal to the potential energy lost in passing through the sink. These ideas may be expressed formally by writing $v_1=v_2$ where v_1 is the potential rise and v_2 is the potential fall.

Following the same line of argument as used in our approach to the current law, the above equation can be rearranged to give either $v_1+(-v_2)=0$ or $v_2+(-v_1)=0$, and we can interpret these as equivalent to stating that the algebraic sum of the potential differences in the circuit is zero. The two alternative forms arise because we can choose to define positive potential difference acting in either a clockwise sense or a counterclockwise sense round the circuit.

Kirchhoff's laws

The same arguments can be applied to circuits containing any number of elements and conducting paths if we take the voltage v_1 to mean the sum of all contributions to the potential rise and v_2 the sum of all contributions to the potential fall as any closed path in the circuit is traversed. Thus Kirchhoff's voltage law may be stated in either of two forms:

$$\sum v_{\text{rise}} = \sum v_{\text{fall}} \tag{1.14}$$

or

$$\sum v = 0 \tag{1.15}$$

where the sums are taken over all v and, in the latter equation, the algebraic sum is intended. The interpretation of these equations is made clear in the example given below.

1.6.4 Worked example on the voltage law

In the circuit of fig. 1.13(a), find the magnitudes and directions of the voltages across the elements P and Q.

Since there are two unknown voltages in this problem, the application of Kirchhoff's voltage law consists in choosing two paths or loops in the circuit from which two equations can be set up; these can then be solved for the two unknowns. Three possible loops are indicated in fig. 1.13(b) any two of which are sufficient to set up the required equations. For this problem we choose loops (1) and (3).

Let the magnitudes of the required voltages be V_P and V_Q, and assign

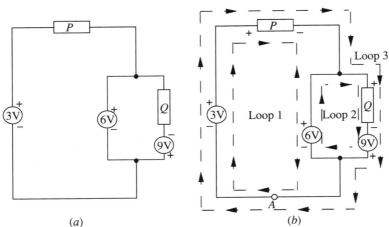

Fig. 1.13. Circuit for worked example on the voltage law.

(a)   (b)

18   *Basic concepts, units, and laws of circuit theory*

their directions (arbitrarily) as indicated by the $(+)$ and $(-)$ signs in fig. 1.13(b).

Method (1): using equation (1.14)
Starting at point A, and traversing loop (1) in a clockwise direction, we obtain a voltage rise in the 3 volt source, a fall in P, and a fall in the 6 volt source. Hence

$$3 = V_P + 6$$
$$V_P = -3\text{V}$$

Similarly for loop (3) we obtain a rise in the 3 volt source, a fall in P, a fall in Q, and a rise in the 9 volt source. Hence,

$$3 + 9 = V_P + V_Q$$

but

$$V_P = -3\text{V}$$

therefore

$$V_Q = 15\text{V}$$

Method (2): using equation (1.15).
Taking voltages acting in a clockwise sense we obtain:
for loop (1)

$$3 + (-V_P) + (-6) = 0$$

hence,

$$V_P = -3\text{V}$$

and for loop (3)

$$3 + (-V_P) + (-V_Q) + 9 = 0$$

hence,

$$V_Q = 15\text{V}$$

The application of Kirchhoff's Laws in practical circuit analysis is considered in greater detail in chapter 2.

1.7   Resistance

1.7.1   Ohm's Law

The relationship between p.d. across a conductor and the current flowing through it depends on the shape of the conductor and the materials

Resistance

of which it is made. For some materials, for instance semi-conducting compounds, this relationship may be of the general non-linear form shown in fig. 1.14(a). For metals, carbon, and many other materials the voltage–current relationship is linear, as shown in fig. 1.14(b). The ratio of voltage to current is constant, and the relationship takes the simple form known as Ohm's Law viz:

$$\frac{v}{i} = R \qquad (1.16)$$

where R is the resistance in *ohms*.

It is often convenient for the purpose of circuit analysis to express Ohm's law in the alternative form:

$$\frac{i}{v} = \frac{1}{R} = G \qquad (1.17)$$

where G is the conductance in *siemens*.

A conducting element for which Ohm's law is obeyed is called a linear resistance. Two commonly used graphical symbols for a linear resistance are shown in fig. 1.15. In this book the symbol of fig. 1.15(a) is preferred.*

Also indicated in this figure are the associated directions of voltage and current according to the sign conventions described previously. With conventional positive current flowing in the direction shown, and with v a positive number, the polarity of the voltage is as indicated by the $(+)$ and $(-)$ signs.

Materials which obey Ohm's law are called *ohmic* materials. The resistance of a bar of such material, if it is homogeneous and has a uniform

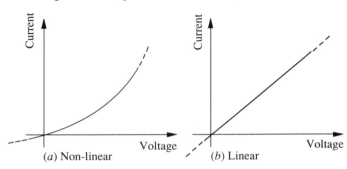

Fig. 1.14. Voltage–current relationships for resistance.

(a) Non-linear   (b) Linear

---

* The British and European standard symbol for resistance is that shown in fig. 1.15(b); in this book it is used, for pedagogical reasons, to denote immittance.

cross section, is found to be proportional to the length of the bar l and inversely proportional to its cross-sectional area A. Thus we may write

$$R = \rho \frac{l}{A} \qquad (1.18)$$

where ρ is a constant of the material known as the *resistivity*. The unit of resistivity is the *ohm metre*. The relationship (1.18) may also be written in terms of the conductance:

$$G = \sigma \frac{A}{l} \qquad (1.19)$$

where $\sigma = 1/\rho$ is a constant known as the *conductivity* of the material; this is expressed in *siemens per metre*.

1.7.2 Power dissipation in resistance

We have mentioned previously that current flowing through a resistance results in the irreversible conversion of electrical energy into heat. The power dissipated is given, according to (1.8), by the product vi, and this is always positive. This expression for the power dissipation is true whether the resistance is linear or non-linear. For the linear case, however, alternative and often more convenient expressions may be obtained by eliminating either v or i from (1.8) using Ohm's law. Thus, using $v = iR$ gives:

$$\text{power} = vi = i^2 R \qquad (1.20)$$

and using $i = v/R$ gives:

$$\text{power} = vi = \frac{v^2}{R} \qquad (1.21)$$

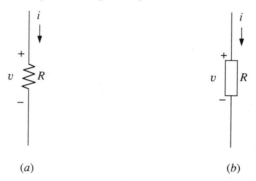

Fig. 1.15. Graphical symbols used for resistance.

(a)　　　　　　　　　　　　(b)

Resistance

The energy converted into heat in a conductor as a result of the passage of a current through it is often referred to as the 'i^2R' *loss* or, when the conductor is of copper, the *copper loss*.

1.7.3 Resistances in combination

(a) *Series connection*. With reference to fig. 1.16(a), if the single linear resistance R_s is to be equivalent to the series combination of the two linear resistances R_1 and R_2, a voltage v_{AB} applied to either circuit must cause the same current to flow. Let this current be i then, by Ohm's law, the voltage across R_1 is $v_1 = iR_1$, and the voltage across R_2 is $v_2 = iR_2$. Therefore, by Kirchhoff's voltage law,

$$v_{AB} = v_1 + v_2 = i(R_1 + R_2)$$

But $v_{AB} = iR_s$, hence

$$iR_s = i(R_1 + R_2)$$

The equivalent resistance of two resistances connected in series is, therefore, given by

$$R_s = R_1 + R_2 \tag{1.22}$$

Now consider the series combination of any number of resistances $R_1, R_2 \ldots R_n$. The equivalent resistance R can be found by repeated application of (1.22), taking resistances two at a time, to give

$$R = R_1 + R_2 + \ldots + R_n \tag{1.23}$$

In terms of conductances $G = 1/R$, $G_1 = 1/R_1$ etc., (1.23) becomes

$$\frac{1}{G} = \frac{1}{G_1} + \frac{1}{G_2} + \ldots + \frac{1}{G_n} \tag{1.24}$$

Fig. 1.16. Equivalent resistance for series and parallel combinations of resistances.

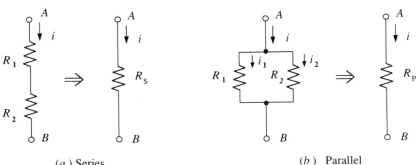

(a) Series   (b) Parallel

(b) Parallel connection. As for the series connection, the criterion for equivalence between the two circuits shown in fig. 1.16(b) is that, under the action of the same applied voltage, the resulting currents must be identical. Thus:

$$\text{Current through } R_1 = i_1 = \frac{v_{AB}}{R_1}$$

$$\text{Current through } R_2 = i_2 = \frac{v_{AB}}{R_2}$$

Therefore the total current i is, by Kirchhoff's current law,

$$i = i_1 + i_2 = v_{AB}\left(\frac{1}{R_1} + \frac{1}{R_2}\right)$$

But

$$i = v_{AB}/R_p$$

hence

$$\frac{v_{AB}}{R_p} = v_{AB}\left(\frac{1}{R_1} + \frac{1}{R_2}\right)$$

or

$$\frac{1}{R_p} = \frac{1}{R_1} + \frac{1}{R_2} \tag{1.25}$$

The parallel combination of two resistances is encountered very frequently in circuit analysis and it is often denoted symbolically by $R_1//R_2$.

Equation (1.25) is then often more conveniently expressed as

$$R_1//R_2 \equiv R_p = \frac{R_1 R_2}{R_1 + R_2} \tag{1.26}$$

We refer to this expression as the 'product over sum rule'. Note that this rule is not applicable to combinations of *more* than two resistances in parallel.

By repeated application of (1.25) we obtain the equivalent resistance of any number of resistances connected in parallel viz.:

$$\frac{1}{R} = \frac{1}{R_1} + \frac{1}{R_2} + \ldots + \frac{1}{R_n} \tag{1.27}$$

In terms of conductances $G = 1/R$, $G_1 = 1/R_1$, etc., (1.27) becomes:

$$G = G_1 + G_2 + \ldots + G_n \tag{1.28}$$

Capacitance 23

This formula is useful for calculating the resistance of many paths in parallel.

1.8 Capacitance

Capacitance is that property of the circuit which defines the distribution of the electric field within the circuit when it is energized. In the following sections we establish the voltage–current relationships for a linear capacitance and from this an expression for the energy stored in terms of the voltage across the *plates* of the capacitance. Relationships are also derived which allow us to represent any arrangement of interconnected lumped capacitances as a single lumped capacitance.

1.8.1 The voltage-current relationship for capacitance

When a source of voltage is applied to two conductors forming the plates of a capacitance, positive charge is transferred from the conductor connected to the negative terminal of the source to the conductor connected to the positive terminal. The quantity of charge transferred is found to depend on the size and shape of the conductors and on the *dielectric* medium between them. For vacuum, air and many other dielectric materials the charge transferred is proportional to the applied voltage. Thus we may write

$$q = Cv \tag{1.29}$$

where q is the instantaneous value of the charge, and C is a constant of proportionality known as the capacitance. The unit of capacitance is the *farad*, and the graphical symbol for a linear capacitance used in this book is shown in fig. 1.17(*a*).

The relationship (1.29) defines a linear capacitance, and it is true only if the applied voltage is not so high as to cause breakdown (electrical conduction) or other changes in the dielectric.

Fig. 1.17. Graphical symbols used for capacitance.

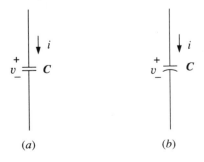

(*a*)   (*b*)

24 Basic concepts, units, and laws of circuit theory

The current i which flows as a result of the application of a voltage is found by differentiating (1.29):

$$i = \frac{dq}{dt} = \frac{d}{dt}(Cv)$$

Provided we are dealing with a conductor–dielectric system of fixed geometry, this relationship becomes:

$$i = C\frac{dv}{dt} \tag{1.30}$$

By integrating (1.30) we obtain an explicit expression for the voltage in terms of current. Thus,

$$v = \frac{1}{C}\int_0^t i\,dt + v_0 \tag{1.31}$$

in which v_0 is the voltage across the capacitance at $t=0$.

1.8.2 Energy storage in capacitance

Consider a capacitance through which a current i flows as a result of an applied voltage v. The instantaneous power flow to the capacitance is given by the product vi which becomes, using (1.30), $vC\,dv/dt$. During the interval dt the flow of energy to the capacitance is, therefore,

$$v\left(C\frac{dv}{dt}\right)dt = Cv\,dv$$

and the total energy delivered to the capacitance when its voltage is v is given by:

$$\text{Energy} = C\int_0^v v\,dv = \tfrac{1}{2}Cv^2 \tag{1.32}$$

This stored energy is released by the capacitance when the voltage is reduced to zero. (During the release of energy the product vi will be negative according to the sign convention established in section 1.4.) Because the energy of any closed system cannot change instantaneously (instantaneous change of energy implying infinite power), it follows that the voltage across a capacitance cannot change instantaneously.

1.8.3 Capacitances in combination

(a) *Series connection.* Referring to fig. 1.18(a), the two circuits between A and B are equivalent if, when the same voltage is applied to each,

Capacitance

the resulting currents are equal. Let the current be i and the voltage be v_{AB}. If we assume zero initial voltage, then the voltage v_1 across C_1 will be, from (1.31),

$$v_1 = \frac{1}{C_1} \int_0^t i\,dt$$

and the voltage across C_2 will be

$$v_2 = \frac{1}{C_2} \int_0^t i\,dt$$

Therefore, using Kirchhoff's voltage law, we have

$$v_{AB} = v_1 + v_2 = \left(\frac{1}{C_1} + \frac{1}{C_2}\right) \int_0^t i\,dt$$

but

$$v_{AB} = \frac{1}{C_s} \int_0^t i\,dt$$

hence

$$\frac{1}{C_s} = \frac{1}{C_1} + \frac{1}{C_2} \tag{1.33}$$

An expression for the equivalent capacitance of any number of capacitances in series may be found by repeated application of (1.33), viz.:

$$\frac{1}{C} = \frac{1}{C_1} + \frac{1}{C_2} + \ldots + \frac{1}{C_n} \tag{1.34}$$

Fig. 1.18. Equivalent capacitance for series and parallel combinations of capacitances.

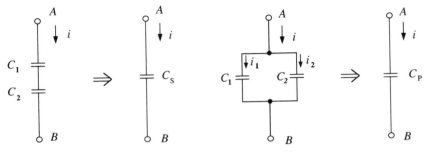

(a) Series   (b) Parallel

(b) Parallel connection. For the parallel circuit of fig. 1.18(b), the current through C_1 is, from (1.30),

$$i_1 = C_1 \frac{dv_{AB}}{dt}$$

and the current through C_2 is

$$i_2 = C_2 \frac{dv_{AB}}{dt}$$

The total current is given by

$$i = i_1 + i_2 = (C_1 + C_2) \frac{dv_{AB}}{dt}$$

But, for the equivalent capacitance C_p,

$$i = C_p \frac{dv_{AB}}{dt}$$

hence

$$C_p = C_1 + C_2 \tag{1.35}$$

By repeated application of (1.35), we may extend this result to include any number of capacitances connected in parallel, thus,

$$C = C_1 + C_2 + \ldots + C_n \tag{1.36}$$

1.9 Inductance

Inductance is that elemental property of a circuit which defines the magnetic field distribution when the circuit is energized. In the following sections we derive voltage–current and other relationships for inductance corresponding to those derived for capacitance.

1.9.1 The voltage-current relationship for inductance

Fig. 1.19(a) shows part of a current carrying circuit. The magnetic flux created by the current links with the circuit itself, the amount of flux linkage being a function of the circuit geometry. By forming the circuit into a coil, as represented schematically in fig. 1.19(b) the flux linkage is enhanced. Provided there is no iron or other magnetic material present, the flux, and therefore the flux linkage, is found to be proportional to the current. If we denote the flux linkage by ϕ, then

$$\phi = Li \tag{1.37}$$

Inductance

where L is a constant of proportionality, dependent upon the circuit geometry, which is called the *self-inductance* (or simply *inductance*). The unit of inductance is the *henry*, and the graphical symbol preferred in this book is shown in fig. 1.20(*a*).

The relationship between voltage and current is derived using Faraday's law of induced e.m.f. This states that the e.m.f. induced in a circuit is equal to the rate of change of flux linkage, that is,

$$\text{e.m.f.} = -\frac{d}{dt}(\phi) \tag{1.38}$$

In this equation the negative sign indicates that the induced e.m.f. acts in a direction such as to oppose the cause of the change of flux linkage (Lenz's law).

Fig. 1.19. Flux linkage.

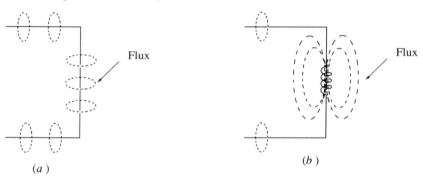

Fig. 1.20. Graphical symbols used for inductance.

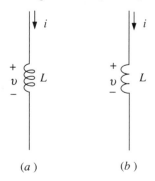

Combining (1.37) and (1.38) we have

$$\text{e.m.f.} = -\frac{\text{d}}{\text{d}t}(Li)$$

which becomes, for a conductor system of fixed geometry,

$$\text{e.m.f.} = -L\frac{\text{d}i}{\text{d}t} \tag{1.39}$$

Now consider the situation shown in fig. 1.21 in which a time varying voltage v is applied to part of a circuit which has no properties other than pure inductance. This will cause a current i to flow whose rate of change is such that the induced e.m.f. will exactly counterbalance the applied voltage, hence we may write

$$v = L\frac{\text{d}i}{\text{d}t} \tag{1.40}$$

The current i which flows as a result of applying v is found explicitly by integrating (1.40), thus,

$$i = \frac{1}{L}\int_0^t v\,\text{d}t + i_0 \tag{1.41}$$

where i_0 is the current flowing through the inductance at $t=0$.

1.9.2 Energy storage in inductance

Referring to fig. 1.21, the instantaneous power flow to the inductance is vi and the energy in the time interval $\text{d}t$ is given by

$$iL\frac{\text{d}i}{\text{d}t}\text{d}t = Li\,\text{d}i$$

Fig. 1.21. Voltage–current relationship for inductance.

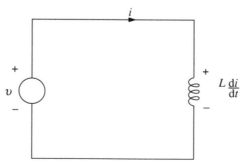

Inductance

hence, the total energy storage at any current i is:

$$\text{Energy} = L \int_0^i i\,di = \tfrac{1}{2}L i^2 \tag{1.42}$$

Remarks similar to those made in connection with the storage and release of energy in a capacitance (section 1.8.2) apply also in the case of inductance. During the acquisition of energy the product vi is positive. During the release of energy the current diminishes, the polarity of the voltage reverses and the product vi becomes negative. The energy, and therefore the current, cannot change instantaneously.

1.9.3 Inductances in combination

(a) *Series connection.* The circuits shown in fig. 1.22(a) are equivalent if, upon application of identical voltages, the resulting currents are equal. From (1.40) the voltage across L_1 will be $v_1 = L_1 di/dt$, and that across L_2 will be $v_2 = L_2 di/dt$. The total voltage will, therefore, be $(v_1 + v_2)$, and this must be equal to v_{AB}. Hence,

$$v_{AB} = (v_1 + v_2) = L_1 \frac{di}{dt} + L_2 \frac{di}{dt} = L_s \frac{di}{dt}$$

or

$$L_s = L_1 + L_2 \tag{1.43}$$

Extending this to any number of elements by repeated application we have

$$L = L_1 + L_2 + \ldots + L_n \tag{1.44}$$

Fig. 1.22. Equivalent inductance for series and parallel combinations of inductances.

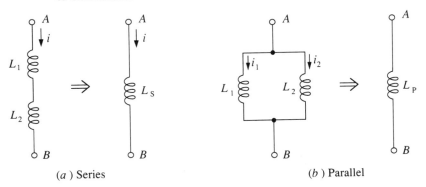

(a) Series   (b) Parallel

(b) *Parallel connection.* Adopting the same criterion for equivalence for the circuits of fig. 1.22(b), we have for the sum of the currents i_1 and i_2 flowing in L_1 and L_2 respectively (assuming zero initial currents),

$$i = i_1 + i_2 = \frac{1}{L_1} \int_0^t v_{AB} \, dt + \frac{1}{L_2} \int_0^t v_{AB} \, dt$$

But this sum must be equal to the current through L_p, hence,

$$\frac{1}{L_p} \int_0^t v_{AB} \, dt = \frac{1}{L_1} \int_0^t v_{AB} \, dt + \frac{1}{L_2} \int_0^t v_{AB} \, dt$$

or

$$\frac{1}{L_p} = \frac{1}{L_1} + \frac{1}{L_2} \tag{1.45}$$

Again, by repeated application we may extend this result to include any number of elements:

$$\frac{1}{L} = \frac{1}{L_1} + \frac{1}{L_2} + \ldots + \frac{1}{L_n} \tag{1.46}$$

1.10 Inductively coupled circuits

When two circuits are brought into close proximity, the flux produced by current flowing in one circuit can link with the other circuit as shown schematically in fig. 1.23. Faraday's law applies regardless of the source of flux linkage so that if the current varies so will the flux linkage, and an e.m.f. will be induced in the second circuit. We say that the two circuits are inductively coupled.

The inductive coupling effect can be enhanced by shaping the two circuits in the form of coils wound closely together (fig. 1.24), as in the transformer. Sometimes the inductive coupling effect is the cause of unwanted interference between adjacent circuits, and steps have to be taken to reduce the coupling by the provision of magnetic screens.

Fig. 1.23. Inductive coupling between two circuits.

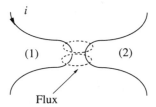

Inductively coupled circuits

In this section the concept of mutual inductance is introduced, and from this the voltage–current relationships are established for inductively coupled circuits.

1.10.1 Mutual inductance

The amount of flux linked with the second circuit in fig. 1.23 (or fig. 1.24) depends on the geometry of the two circuits and, in the absence of any non-linear magnetic materials, it will be proportional to the current in the first circuit. Let us designate the flux linkage by ϕ_{21} (read as flux linkage in circuit (2) resulting from a current in circuit (1)), and let the current be i_1. We may then write

$$\phi_{21} = M_{21} i_1 \qquad (1.47)$$

where M_{21} is a constant known as the mutual inductance of the two circuits. As for self-inductance, this is expressed in units of the henry. The mutual inductance is conventionally taken as being always positive.

By the same arguments we can derive a similar equation for the situation where circuit 2 carries a current i_2, that is,

$$\phi_{12} = M_{12} i_2 \qquad (1.48)$$

Applying Faraday's law to (1.47) and (1.48) we obtain for the voltage induced in (2), $M_{21} di_1/dt$; and for the voltage induced in (1), $M_{12} di_2/dt$. (It is assumed here that the two circuits are fixed geometrically.)

It may be shown that for *iron-free* circuits, $M_{21} = M_{12}$ (see for example reference 6). Therefore, in what follows the mutual inductance will be designated simply as M.

We can now establish the complete voltage–current relationships for mutual inductance with reference to fig. 1.25, which shows the circuit model for the inductively coupled coil arrangement of fig. 1.24. The coils possess pure self-inductances L_1 and L_2, and the mutual inductance between them

Fig. 1.24. Inductive coupling between two coils: fluxes reinforce with currents and winding directions shown.

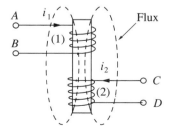

is M; resistance is assumed to be negligible. Both coils are wound in the same direction so that, with the currents i_1 and i_2 assigned as shown (both flowing clockwise when viewed from above in fig. 1.24), the fluxes produced are in the same direction through the two coils. The self-induced voltage in coil (1) will be, from (1.40), $L_1 di_1/dt$ and to this we must add the voltage induced in coil (1) due to current in coil (2), namely, $M di_2/dt$. Hence, the total voltage across AB is

$$v_1 = L_1 \frac{di_1}{dt} + M \frac{di_2}{dt} \tag{1.49}$$

The two voltages are in the same sense and therefore add because the fluxes are in the same direction.

Similarly, for the second coil, the voltage across CD is

$$v_2 = L_2 \frac{di_2}{dt} + M \frac{di_1}{dt} \tag{1.50}$$

If the direction of either current were reversed, the signs of the second terms in these equations would be negative.

It is sometimes important in circuit diagrams (and on the coils themselves) to indicate the relative directions of the windings. This is done by polarity markings. The conventional scheme is as follows: place a dot on one terminal of coil (1) and imagine the current i_1 to enter the dotted terminal. Now determine the direction of current i_2 in coil (2) that will give flux in the same direction as that produced by i_1. Place a dot on the terminal of coil (2) that i_2 enters. (If there are more than two coils, then any one of them is used as reference and the dots are placed on all others with respect to the reference coil.) According to this convention, the top terminal of each coil of fig. 1.24, and each inductance in fig. 1.25, should have a dot. (Alternatively, of course, the bottom terminal of each coil may be marked.) The ends of the coils so marked are called *corresponding ends*.

Fig. 1.25. Circuit model of the inductively coupled coil arrangement shown in fig. 1.24. Dots indicate 'corresponding ends'.

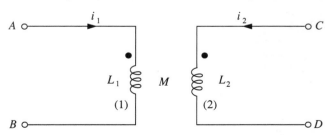

Inductively coupled circuits

Note that with assigned currents *both* entering (or both leaving) corresponding ends, the terms containing M in the circuit equations (1.49) and (1.50) are positive. If currents are assigned such that one current enters and the other leaves a corresponding end, the terms containing M are negative. If no information is provided concerning corresponding ends in a particular circuit containing inductively coupled coils, the mutual inductance terms must be written: $\pm M di/dt$.

The scheme described above implies that the corresponding ends of coils will all move in potential together when a voltage is applied to one of them. This provides a simple method of identifying, by means of a voltage measurement, the corresponding ends of a system of inductively coupled coils.

Sometimes, a circuit model, by separating out effects that cannot actually be separated physically, provides help in understanding how the circuit works. Fig. 1.26 is an alternative circuit model for the inductively coupled coil arrangement shown in fig. 1.24 in which the induced voltages due to mutual inductance have been represented by ideal sources. These sources cannot be isolated from L_1 and L_2, and so this model is not physically realizable, however, the circuit equations are the same as those for fig. 1.25 and as far as external connections are concerned, the circuits of figs. 1.25 and 1.26 exhibit identical behaviour. It will be noted that the two sources in fig. 1.26 are examples of dependent sources, that is, their values depend on currents flowing in the circuit itself.

1.10.2 The coefficient of coupling

Consider the circuit of fig. 1.26 with its terminals AB connected to a voltage source and terminals CD short circuited. The circuit for this situation is shown in fig. 1.27. Applying Kirchhoff's voltage law to the circuits (1) and (2) we obtain

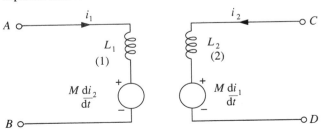

Fig. 1.26. Alternative circuit model of Fig. 1.24. L_1 and L_2 are separate inductances with no mutual inductance between them.

$$v_1 = L_1 \frac{di_1}{dt} + M \frac{di_2}{dt}$$

and

$$0 = M \frac{di_1}{dt} + L_2 \frac{di_2}{dt}$$

since $v_2 = 0$.

Upon rearrangement and elimination of di_2/dt, these equations yield

$$\frac{di_1}{dt} = \frac{v_1 L_2}{L_1 L_2 - M^2}$$

The equivalent input inductance (that is, the effective inductance between the terminals AB) is

$$L_{AB} = \frac{v_1}{di_1/dt} = \frac{L_1 L_2 - M^2}{L_2}$$

We can now calculate the energy stored in the system using (1.42), thus,

$$\text{Energy} = \tfrac{1}{2} L_{AB} i_1^2 = \tfrac{1}{2} \frac{(L_1 L_2 - M^2) i_1^2}{L_2} \tag{1.51}$$

The stored energy must be positive, otherwise, the inductance could act as a source of energy indefinitely, and this is not possible for a passive circuit element. Therefore, $(L_1 L_2 - M^2)$ must be greater than or equal to zero and so

$$M^2 \leq L_1 L_2$$

We usually write

$$M = k(L_1 L_2)^{\frac{1}{2}} \quad (0 \leq k \leq 1) \tag{1.52}$$

where k is the *coefficient of coupling*. When k has a value near unity, the

Fig. 1.27. Circuit for deriving the relationship: $M = k(L_1 L_2)^{\frac{1}{2}}$.

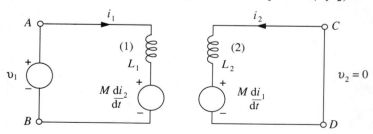

Inductively coupled circuits

coupling is said to be close, or tight. For $k=1$, all the flux produced by a current in one coil links all the turns of the other coil.

1.10.3 The effective inductance of two series-connected coupled coils

In the circuit of fig. 1.28 the two coils each have self-inductances L_1 and L_2 and mutual inductance M. Since there are two possible ways of arranging the series connection we have two possible circuit models (derived from fig. 1.26) as shown in fig. 1.29. In both arrangements the same current exists in both coils, however, in fig. 1.29(a) the coils are connected in series aiding while in fig. 1.29(b) they are connected in series opposing. The effective inductance in each case may be found by applying Kirchhoff's voltage law. For fig. 1.29(a) we have

$$v = L_1 \frac{di}{dt} + M \frac{di}{dt} + L_2 \frac{di}{dt} + M \frac{di}{dt}$$

or

$$v = (L_1 + L_2 + 2M) \frac{di}{dt}$$

and so the effective inductance for circuit (a) is

$$L_3 = L_1 + L_2 + 2M \qquad (1.53)$$

Fig. 1.28. Series connected coupled coils.

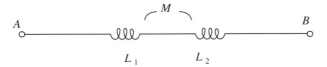

Fig. 1.29. Circuit models for series connected coupled coils.

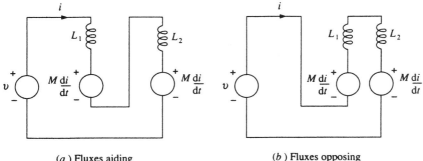

(a) Fluxes aiding   (b) Fluxes opposing

When the same procedure is applied to circuit (b) the result is

$$L_4 = L_1 + L_2 - 2M \tag{1.54}$$

1.11 Passive circuit components

Passive circuits are made up of components – resistors, capacitors, inductors and mutual inductors – each designed, ideally, to exhibit one of the elemental properties to the exclusion of the others. These components are usually, although not always, required to be time-invariant and linear, that is, their properties should not be affected by changes in operating conditions, nor should their values depend in any way on the applied currents and voltages.

Real components fall short of these ideal requirements in several respects. Generally, all three elemental properties will be associated with any given component so that it will exhibit its named property over a limited range of frequencies only. A resistor, for example, is often constructed from a length of high-resistivity wire wound upon a suitable former to form a compact coil. Such a coil will possess inductance; there will also be capacitance between the turns of the coil.* It is sometimes necessary, therefore, to devise circuit models like those shown in fig. 1.30, which will represent the characteristic behaviour of the component over a particular range of operating frequencies.

The 'purity' of a component refers to the extent to which it is free from extraneous (or stray or parasitic) elements. Most modern components possess a high degree of purity, and it is usually permissible – for example, in the case of a resistor – to neglect stray inductance and capacitance and to represent it simply by a resistance in the circuit model, but the designer must always take care to ensure that the model accurately reflects the properties of the circuit.

The properties of real components also vary, to some extent, with time

Fig. 1.30. Models for passive circuit components.

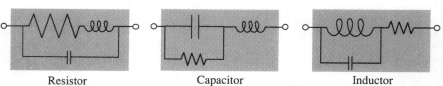

Resistor   Capacitor   Inductor

* By folding back the length of wire on itself before winding it into a coil, the self flux linkage and therefore the inductance may be greatly reduced – a technique known as *bifilar winding*.

and temperature, the latter effect being dependent both on the ambient temperature and on the power dissipated in the component itself, that is, upon applied voltages and currents. Components also suffer from long-term changes in the values of their properties – an effect known as 'ageing'.

1.12 Summary of basic circuit relations

Table 1.1 contains a summary of the basic equations and laws of circuit theory. In this table v and i denote, respectively, the voltage and current across and through the circuit element concerned at a particular instant of time.

The reader will observe that there exists a striking symmetry in these circuit relations. Kirchhoff's two laws, for instance, are of identical mathematical form with v in one replacing i in the other: the same applies to Ohm's law expressed in its resistive and conductive forms. We see also that the formula for the series combinations of resistances is of the same form as the parallel combination of conductances. The complete sets of equations applying to capacitance and inductance evince this underlying symmetry also, i interchanging with v and C interchanging with L. The equations relating to combinations of L and combinations of C reveal a reciprocal relationship of the same type as that existing between R and G.

These symmetries are a manifestation of a general principle of circuit theory which we call *duality*. Expressed in general terms this principle states that, for any linear circuit whose behaviour is described by a certain set of equations, a dual circuit can be found for which the circuit equations are of the same mathematical form. However, in the equations for the dual circuit, current and voltage are interchanged and each element is replaced by its dual element: R for G, L for C, etc. For example, a circuit comprising two resistances connected in series to an ideal voltage source would have as its dual two conductances connected in parallel to an ideal current source. This principle sometimes provides alternative and illuminating ways of approaching circuit analysis and synthesis; we refer to it at several points throughout this text. (There are certain restrictions to this general principle: for instance, it cannot be applied directly to circuits containing mutual inductance. See for example references 1 and 4.)

1.13 Problems

1. A circuit element is shown in fig. 1.31 for which the reference directions of voltage and current are defined.

(*a*) If $v = -3$ V and $i = -2$ A (both constant), is the element acting as a source or as a sink? What is the power delivered or received by the element?

(*b*) If $v = 3$ V (constant) and $i = (2t+1)$A, what is the total amount of

Table 1.1. *Summary of basic equations and laws of circuit theory*

Description	Law or relationship	Unit	Equation
Charge and current	$i = \dfrac{dq}{dt}$	ampere	1.1
	$q = \displaystyle\int_0^t i\,dt$	coulomb	1.2
Energy	$J = \displaystyle\int_0^t vi\,dt$	joule	1.7
Power	$p = vi$	watt	1.8
Kirchhoff's laws Current	$\sum i_{\text{in}} = \sum i_{\text{out}}$		1.12
	$\sum i = 0$		1.13
Voltage	$\sum v_{\text{rise}} = \sum v_{\text{fall}}$		1.14
	$\sum v = 0$		1.15
Resistance (linear) Ohm's law	$R = \dfrac{v}{i}$	ohm	1.16
	or		
	$G = \dfrac{i}{v}$	siemen	1.17
Power	$i^2 R$ or $\dfrac{v^2}{R}$	watt	1.20, 1.21
Series combination	$R = R_1 + R_2 + \ldots + R_n$	ohm	1.23
	$\dfrac{1}{G} = \dfrac{1}{G_1} + \dfrac{1}{G_2} + \ldots + \dfrac{1}{G_n}$	ohm	1.24
Parallel combination	$\dfrac{1}{R} = \dfrac{1}{R_1} + \dfrac{1}{R_2} + \ldots + \dfrac{1}{R_n}$	siemen	1.27
	$G = G_1 + G_2 + \ldots + G_n$	siemen	1.28
Capacitance (linear) Charge and voltage	$q = Cv$	coulomb	1.29
Current and voltage	$i = C\dfrac{dv}{dt}$	ampere	1.30
	$v = \dfrac{1}{C}\displaystyle\int_0^t i\,dt + v_0$	volt	1.31
Stored energy	$\tfrac{1}{2}Cv^2$	joule	1.32
Series combination	$\dfrac{1}{C} = \dfrac{1}{C_1} + \dfrac{1}{C_2} + \ldots + \dfrac{1}{C_n}$	farad^{-1}	1.34

Problems

Table 1.1. (*cont.*)

Description	Law or relationship	Unit	Equation
Parallel combination	$C = C_1 + C_2 + \ldots + C_n$	farad	1.36
Inductance (linear)			
Flux linkage and current	$\phi = Li$	weber	1.37
Voltage and current	$v = L\dfrac{di}{dt}$	volt	1.40
	$i = \dfrac{1}{L}\displaystyle\int_0^t v\,dt + i_0$	ampere	1.41
Stored energy	$\tfrac{1}{2}Li^2$	joule	1.42
Series combination	$L = L_1 + L_2 + \ldots + L_n$	henry	1.44
Parallel combination	$\dfrac{1}{L} = \dfrac{1}{L_1} + \dfrac{1}{L_2} + \ldots + \dfrac{1}{L_n}$	henry^{-1}	1.46
Mutual inductance (linear)			
Flux linkage and current	$\phi_{21} = Mi_1$	weber	1.47
Voltage and current	$v_1 = L_1\dfrac{di_1}{dt} \pm M\dfrac{di_2}{dt}$	volt	1.49
Coefficient of coupling	$k = \dfrac{M}{(L_1 L_2)^{\frac{1}{2}}}$		1.52

charge that flows during the interval $0 \leqslant t \leqslant 10$ seconds? Calculate the total energy delivered or received by the element during this interval.

2. State which of the elements A, B, C, D and E in the circuit of fig. 1.32 are sources and which are sinks. Find the total power transfer from sources to sinks.

3. What is the resistance looking into the terminals AB of the circuit shown in fig. 1.33? If the terminals AB are connected together, what is the conductance between points C and D?

4. In the circuit of fig. 1.34, find the voltages V_{AE}, V_{BE} and V_{CE}, and the currents I_{AB} and I_{CB}.

Fig. 1.31. Circuit for problem 1.

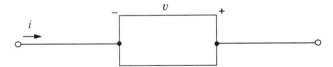

40 Basic concepts, units, and laws of circuit theory

Fig. 1.32. Circuit for problem 2.

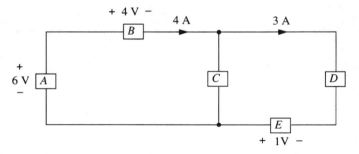

Fig. 1.33. Circuit for problem 3.

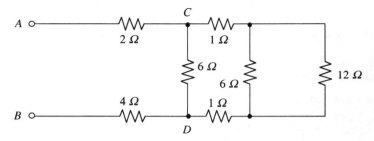

Fig. 1.34. Circuit for problem 4.

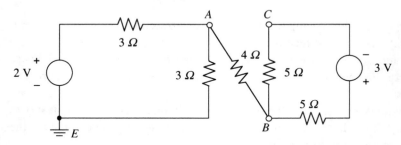

Fig. 1.35. Circuit for problem 6.

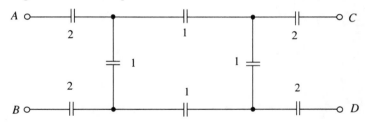

Problems

5. A 1 kΩ resistor and a 2 μF capacitor are connected in series across an ideal current source which delivers a current $i = 5e^{-2t}$ mA. The capacitor is uncharged at $t=0$. Determine the energy stored in the capacitor and the voltage across the current source at the instant $t=0.5$ s.

6. Find the change in the capacitance measured at AB in fig. 1.35 when terminals CD are connected together (capacitance in μF).

7. The mutual inductor in the circuit of fig. 1.36 has a coupling coefficient of 0.5; what is its mutual inductance? Determine the current i, the voltage v, and the total energy stored in the circuit at an instant 4 seconds after closure of the switch.

Fig. 1.36. Circuit for problem 7.

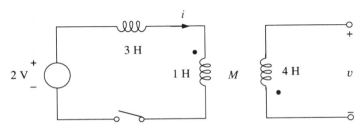

2

Theorems and techniques of linear circuit analysis

2.1 Introduction

By making the assumption that all of the elements in a circuit are linear, the analysis is greatly simplified. Although all real circuits are non-linear to some degree, in most cases a linear treatment gives sufficiently accurate results, and even for circuits containing highly non-linear elements, methods can often be devised for dealing with them on a linear basis. It is for these reasons that the study of linear circuit theory is of paramount importance in electrical engineering science.

The theorems and techniques of linear circuit analysis presented in this chapter, while being of general usefulness and validity, are developed in the context of d.c. circuits. The advantages of this approach are twofold: firstly, the theory can be developed on the simplest possible basis and in terms which will be familiar to most students. Secondly, the study of d.c. circuit theory is of great practical importance in its own right since it arises in many branches of power and electronic systems analysis.

D.C. linear circuits comprise assemblies of linear lumped resistances together with ideal direct voltage and current sources. The theory appertaining to such idealized circuits is concerned with real situations since many types of source found in practice, a battery for example, can be represented to a good approximation by an ideal source in combination with a lumped resistance.

A typical voltage–current characteristic, or load characteristic, for a *practical voltage source* is shown in fig. 2.1(*b*). The terminal voltage V falls with increasing load current I, but over a certain part of the working range, between points AB, the characteristic can be represented by a straight line. Over this region the voltage–current relationship is of the form

$$V = V_0 - IR_0 \tag{2.1}$$

where R_0 is the negative of the slope of the straight line.

Introduction

The circuit model which gives precisely this relationship is shown in fig. 2.1(c). On open circuit ($I=0$) there is no voltage drop across R_0, and the terminal voltage is V_0. When a load current is drawn, the internal voltage drop is IR_0, and the terminal voltage falls by this amount. R_0 is called the *internal resistance*, or *output resistance*, of the practical voltage source. It must be stressed that the model of fig. 2.1(c), sometimes referred to as a linear voltage source, is applicable only to practical sources that exhibit a straight-line load characteristic.

Before proceeding it is necessary to define some of the terms used in connection with the analysis described in this and subsequent chapters. A number of these have already been introduced in chapter 1 but are included here for the sake of completeness. The definitions given below are illustrated with reference to fig. 2.2.

> *Node*: An equipotential junction, formed by perfect conductors, between two or more elements. A junction between three or more elements, for example node *B*, is termed a principal node.
>
> *Branch*: A path containing one or more series-connected elements

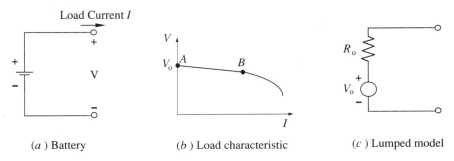

Fig. 2.1. Practical voltage source (battery). The linear lumped model is valid for region *AB* of the load characteristic.

(*a*) Battery   (*b*) Load characteristic   (*c*) Lumped model

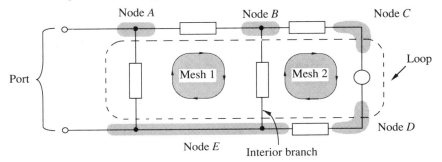

Fig. 2.2. Illustrating network terminology.

44   Theorems and techniques of circuit analysis

joining two principal nodes; for example, path *BCDE*. Path *BE* is an *interior* branch.

Loop: Any connected, closed path in a circuit. The closed path *ABCDEA* forms a loop.

Mesh: A loop which cannot be subdivided into smaller loops by interior branches. The distinction between a mesh and a loop is a rather fine one. Simply by redrawing a circuit it is possible for a mesh to become a loop and vice versa. The loop shown in the figure is divided into two meshes by interior branch *BE*. This branch and branch *BCDE* may be interchanged, in which case what was formerly mesh 1 becomes a loop.

Port: A pair of terminals, or nodes, in a network, through which connections are made to external sources or other networks.

Network or *circuit*: Used in this text interchangeably. In more advanced analysis, a distinction is sometimes made (see for example reference 2).

2.2   Voltage and current dividers

Voltage and current dividers form two of the most common building blocks of electrical circuits. The basic voltage divider (also called a potential divider) is shown in fig. 2.3(a). With voltage V across terminals PQ, voltages V_a and V_b are established across resistances R_a and R_b. These voltages may be related to V by the methods discussed in section 1.7.

The combined resistance of R_a and R_b in series is $R_a + R_b$, therefore, the current is $V/(R_a + R_b)$ and the voltage across R_a is given by

Fig. 2.3. Divider circuits.

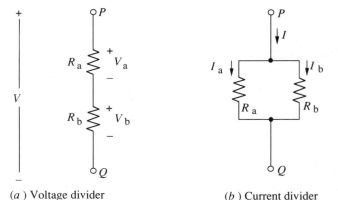

(*a*) Voltage divider   (*b*) Current divider

Voltage and current dividers

$$V_a = \frac{R_a}{R_a + R_b} V \qquad (2.2)$$

Likewise,

$$V_b = \frac{R_b}{R_a + R_b} V \qquad (2.3)$$

We see that in each of these expressions the voltage V is divided in the ratio of the particular resistance concerned to the total resistance.

The current divider is shown in fig. 2.3(b). To establish the relationships between the main current I and the branch currents I_a and I_b, we observe that the voltage across PQ is the product $I(R_a//R_b) = IR_aR_b/(R_a+R_b)$ (see section 1.7.3 for 'product-over-sum' rule). Therefore, the current through R_a is given by

$$I_a = \frac{IR_aR_b}{R_a+R_b} \cdot \frac{1}{R_a} = \frac{R_b}{R_a+R_b} I \qquad (2.4)$$

Likewise the current through R_b is

$$I_b = \frac{IR_aR_b}{R_a+R_b} \cdot \frac{1}{R_b} = \frac{R_a}{R_a+R_b} I$$

In this case each branch current is found by taking the fraction of the total current equal to the resistance in the *opposite* branch divided by the sum of the branch resistances.

The divider circuits shown in fig. 2.3, one a series circuit the other a parallel circuit, are duals (see section 1.12). This will be readily apparent if (2.4) is expressed in terms of the conductances $G_a = 1/R_a$ and $G_b = 1/R_b$:

$$I_a = \frac{1/G_b}{1/G_a + 1/G_b} I = \frac{G_a}{G_a + G_b} I \qquad (2.5)$$

Comparing (2.2) and (2.5) we see that these expressions are of similar form with voltage and current interchanged and resistance and conductance interchanged.

Voltage dividers are used extensively in electronic and power circuits. One common application, illustrated in fig. 2.4, is to provide a fixed or variable degree of voltage control (or attenuation). An input voltage V_1 is applied at the terminals AB (the input port) and a proportion of this voltage V_2 is extracted at terminals CD (the output port). We might use the circuit of fig. 2.4(a) for example, to measure a very high voltage utilizing a voltmeter capable of measuring only a relatively low voltage.

According to (2.3) the voltages at the input and output ports (fig. 2.4(a)) are related by

$$V_2 = \frac{R_2}{R_1 + R_2} V_1 \tag{2.6}$$

R_1 and R_2 may be adjusted to provide the requisite division or attenuation.

In practical circuits the resistances R_1 and R_2 forming the two 'arms' of the divider may each consist of combinations of separate series or parallel elements, in which case, before applying (2.6) the appropriate reduction formulae (section 1.7.3) must be used to find the two equivalent resistances. In cases where the two arms of the divider contain elements connected simply in parallel (see problem 3 at the end of this chapter) it is convenient to express the divider relationship (2.6) in terms of conductances rather than resistances. Let $R_1 = 1/G_1$ and $R_2 = 1/G_2$, where G_1 and G_2 are the total conductances in the two arms of the divider, then substituting in (2.6) gives

$$V_2 = \frac{1/G_2}{1/G_1 + 1/G_2} V_1 = \frac{G_1}{G_2 + G_1} V_1 \tag{2.7}$$

Note that G_1 now replaces R_2 in the numerator of the divider ratio.

It must be emphasized that (2.6) and (2.7) are true only if the terminals

Fig. 2.4. Two-port voltage divider circuits.

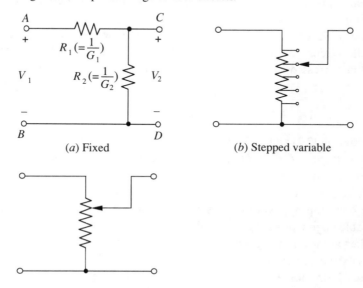

(a) Fixed

(b) Stepped variable

(c) Continuously variable (potentiometer)

Mesh analysis

CD are open circuit. Generally, the output port will be connected to some other circuit which will draw current, and this must be taken into account when calculating the attenuation. If *R* is the effective resistance presented by the external circuit to the output port, then (2.7) becomes

$$V_2 = \frac{(R_2//R)}{R_1 + (R_2//R)} V_1 \tag{2.8}$$

The circuits shown in fig. 2.4 fall into the general category known as two-port networks; these are discussed in chapter 8.

2.3 Mesh analysis

The general objective in circuit analysis is the establishment of a set of equations relating the circuit variables, voltages and currents, in terms of the circuit elements. This is achieved using Kirchhoff's two laws. The solution of these equations yields specific expressions for each of the circuit variables. In mesh analysis source voltages are specified and are treated in the equations as the independent variables; solutions are found for the currents in every branch, these being treated as the dependent variables.

The principles involved in mesh analysis may be illustrated with reference to the single-mesh circuit shown in fig. 2.5 in which source voltages V_1 and V_2 and elements R_1 and R_2 are specified.

First, the current *I* is assigned (fig. 2.5(*b*)) and then Kirchhoff's voltage law (KVL) is used in either of the forms (1.14) or (1.15) to write down the circuit equation. Choosing the latter, and traversing the circuit in a clockwise direction starting at point *A*, we have:

$$\Sigma v = 0 \tag{1.15}$$

Fig. 2.5. (*a*) Single-mesh circuit. (*b*) Circuit with assigned current and resulting voltage drops.

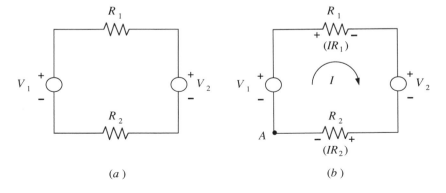

that is

$$+V_1 - IR_1 - V_2 - IR_2 = 0$$

In this equation a rise in potential is indicated by a positive sign, a fall by a negative sign. (Strict adherence to the conventions discussed in section 1.3 and 1.4 is necessary when setting up circuit equations of the above form.)

On rearrangement of this equation we obtain:

$$I = \frac{V_1 - V_2}{R_1 + R_2} \qquad (2.9)$$

If, in (2.9) V_2 is numerically greater than V_1, the current is negative; this simply means that the direction of the conventional positive current in the circuit will be in the opposite sense to that assigned.

The above procedure for the analysis of a single-mesh circuit may be readily extended to circuits containing two or more meshes. Fig. 2.6 shows an example of a two-mesh circuit in which R_3 is common, or *mutual*, to meshes (1) and (2).

Two possible ways of assigning currents are shown in figs. 2.6(b) and 2.6(c). In the first and perhaps most obvious, a current is assigned to every branch in the circuit; in the second currents are assigned to meshes.

Considering first fig. 2.6(b) and applying KVL in the form (1.15), we have for mesh (1), traversing the path $ABCD$,

$$V_1 - I_p R_1 - I_r R_3 = 0 \qquad (2.10)$$

and for mesh (2), traversing $DCEF$, we have

$$I_r R_3 - I_q R_2 - V_2 = 0 \qquad (2.11)$$

In the above equations there are three current variables but these are not independent since, by application of Kirchhoff's current law (KCL) at node C, we can see that $I_p = I_q + I_r$ or

$$I_r = I_p - I_q \qquad (2.12)$$

Substituting (2.12) in (2.10) and (2.11) we obtain

$$I_p R_1 + (I_p - I_q) R_3 = V_1 \qquad (2.13)$$
$$I_q R_2 - (I_p - I_q) R_3 = -V_2 \qquad (2.14)$$

Thus, in reality, there are only two independent variables and these may be evaluated from (2.13) and (2.14). Having found I_p and I_q, we may determine the current in the mutual resistance R_3 using (2.12).

The two paths we have chosen here, to set up the two independent equations necessary to achieve a solution, are not the only possible ones; for

Mesh analysis

example, we might have chosen instead of the mesh $DCEF$, the loop $ABEF$. (It is important to note that the paths chosen must be such that every circuit element is traversed at least once.) The method we are discussing here is, therefore, more generally termed loop analysis, but for circuits that can be drawn on a flat surface to form a series of 'windows' (as in this example) a more convenient and systematic solution is achieved by choosing meshes rather than loops.

It will be evident from the foregoing argument that it is unnecessary to specify every branch current when setting up the mesh equations; it is sufficient to assign a current to each mesh as in fig. 2.6(c). The mesh currents I_1 and I_2 are identical with the branch currents I_p and I_q in fig. 2.6(b), but in the mutual element R_3 the actual current flowing from C to D, corresponding to I_r, is $(I_1 - I_2)$. The relationship (2.12) is automatically satisfied and the mesh equations may be written down directly in terms of the mesh currents. This approach to mesh analysis was originally due to James Clerk Maxwell and is sometimes referred to as *Maxwell's cyclic current method*.

We now repeat the analysis using this method. In applying KVL in the form (1.15) to any particular mesh we traverse the mesh in the direction of the assigned current; where mesh currents meet in a mutual resistance such as R_3, the appropriate current difference is taken to calculate the voltage drop.

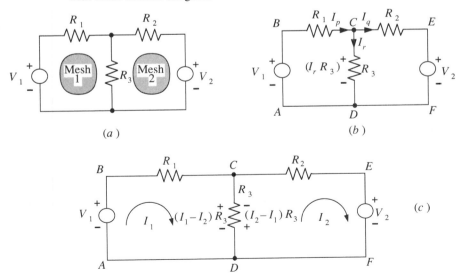

Fig. 2.6. (a) Two-mesh circuit; (b) with branch currents assigned; (c) with mesh current assigned.

Traversing mesh (1), path $ABCD$,

$$V_1 - I_1R_1 - (I_1 - I_2)R_3 = 0 \tag{2.15}$$

and traversing mesh (2), path $DCEF$,

$$-(I_2 - I_1)R_3 - I_2R_2 - V_2 = 0 \tag{2.16}$$

Note carefully the polarity of the voltage drops in fig. 2.6(c), and the order in which the mesh currents appear in the difference terms. In (2.15), relating to mesh (1), the first member of the term representing the difference current in R_3 is I_1; in (2.16), relating to mesh (2), the first member is I_2. It may be remarked also that, according to (2.16), there is no rise of potential in mesh (2); this, however, is simply a consequence of the particular choices made concerning direction of assigned current and direction of traverse.

The above procedures demonstrate the principles underlying mesh analysis using either branch currents or mesh currents. In practical circuit analysis it is useful to be able to write down the mesh equations in a consistent fashion, and in a form requiring a minimum of algebraic manipulation to reach a solution. This can be accomplished, using Maxwell's cyclic current method, by adopting two working rules:

> *Rule* 1. Place all resistive voltage drops (products of current and resistance) on one side of the mesh equation. These terms are always positive.
>
> *Rule* 2. Place all source voltages on the other side of the equation attaching the appropriate ($\pm$) sign as follows: source voltages acting in the same sense as the direction of the assigned mesh current take a ($+$) sign otherwise they take a ($-$) sign.

Using these working rules (2.15) and (2.16) are written down directly as:

$$I_1R_1 + (I_1 - I_2)R_3 = V_1$$
$$(I_2 - I_1)R_3 + I_2R_2 = -V_2$$

Equations of this form, in which voltage drops appear on one side and source voltages on the other, are sometimes referred to as *balance equations*.

2.4 Worked example

Two batteries of nominal voltage 12 V are connected to a charger as shown in fig. 2.7(a). The charger has an open-circuit voltage of 14 V and an internal resistance of 1.2 Ω. Before being placed on charge one battery (A) has an e.m.f. (open circuit voltage) of 12.1 V and an internal resistance of 0.1 Ω; the other battery (B) has an e.m.f. of 11.9 V and an internal resistance of 0.15 Ω. At the end of the charging period the e.m.f. of battery (A) rises to

Worked example

12.6 V, and the e.m.f. of battery (B) rises to 12.5 V. Assuming that the internal resistances of the batteries do not change during charging, and that batteries and charger can be represented by linear voltage sources, determine: (a) the initial charging currents; (b) the final charging currents; (c) the circulating current through the batteries when the charger is disconnected.

Solution: The linear circuit model of the practical circuit is shown in fig. 2.7(b). Mesh currents I_1 and I_2 are assigned in a clockwise direction as shown. Using the working rules presented above the mesh equations may be written immediately as:

$$1.2I_1 + 0.1(I_1 - I_2) = 14 - V_A$$

and

$$0.1(I_2 - I_1) + 0.15I_2 = V_A - V_B$$

Rearranging we obtain

$$(1.20 + 0.1)I_1 - 0.1I_2 = 14 - V_A$$
$$-0.1I_1 + (0.1 + 0.15)I_2 = V_A - V_B$$

(a) *Initial charging currents.* Substituting the initial values of V_A and V_B gives

$$1.3I_1 - 0.1I_2 = 1.9$$
$$-0.1I_1 + 0.25I_2 = 0.2$$

Fig. 2.7. Circuits for worked example. For battery (A): V_A (initial) = 12.1 V; V_A (final) = 12.6 V. For battery (B): V_B (initial) = 11.9 V; V_B (final) = 12.5 V.

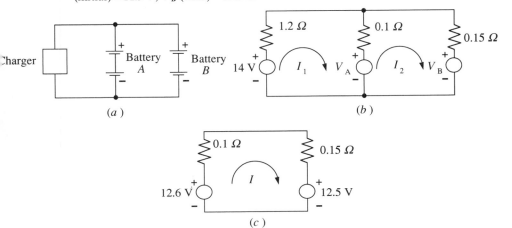

Solving these for I_1 and I_2 (for example by Gaussian elimination and back substitution) we obtain: $I_1 = 1.57$ A and $I_2 = 1.43$ A.

Therefore, the initial charging current in battery (A) is $(I_1 - I_2) = 0.14$ A, and initial charging current in battery (B) is $I_2 = 1.43$ A.

(b) *Final charging currents.* Substituting the final values of V_A and V_B we obtain

$$1.3 I_1 - 0.1 I_2 = 1.4$$
$$-0.1 I_1 + 0.25 I_2 = 0.1$$

giving: $I_1 = 1.14$ A, and $I_2 = 0.86$ A.
Therefore, the final charging current in (A) is $(I_1 - I_2) = 0.28$ A, and final charging current in (B) is $I_2 = 0.86$ A.

(c) *Circulating current.* With the charger disconnected, the circuit reduces to that shown in fig. 2.7(c) for which the mesh equation is

$$(0.1 + 0.15)I = 12.6 - 12.5 = 0.1$$

giving: $I = 0.4$ A.
Note that this last result implies that energy is being transferred from battery (A) to battery (B).

2.5 The general mesh equations

When analysing a network containing a large number of meshes, it is advantageous to adopt a systematic approach to the formulation and solution of the mesh equations. These formal procedures will now be considered with reference to the three-mesh network shown in fig. 2.8.

With mesh currents assigned as shown and using the working rules enunciated previously we obtain:

for mesh (1)

$$I_1 R_1 + (I_1 - I_3) R_3 + (I_1 - I_2) R_2 = V_1 - V_2$$

for mesh (2)

$$I_2 R_4 + (I_2 - I_1) R_2 + (I_2 - I_3) R_5 = -V_3 + V_2$$

for mesh (3)

$$(I_3 - I_2) R_5 + (I_3 - I_1) R_3 = -V_4$$

Upon rearrangement these equations become:

$$(R_1 + R_2 + R_3) I_1 - R_2 I_2 - R_3 I_3 = V_1 - V_2$$
$$-R_2 I_1 + (R_2 + R_4 + R_5) I_2 - R_5 I_3 = V_2 - V_3$$
$$-R_3 I_1 - R_5 I_2 + (R_3 + R_5) I_3 = -V_4$$

The general mesh equations

These mesh equations conform to a standard pattern and it is often convenient to use a formal notation for their description and manipulation. Let

$$R_{11} = R_1 + R_2 + R_3$$
$$R_{12} = -R_2$$
$$R_{13} = -R_3 \text{ and so forth.}$$

Then we obtain:

$$\begin{aligned} R_{11}I_1 + R_{12}I_2 + R_{13}I_3 &= V_1 - V_2 = V_{11} \\ R_{21}I_1 + R_{22}I_2 + R_{23}I_3 &= V_2 - V_3 = V_{22} \\ R_{31}I_1 + R_{32}I_2 + R_{33}I_3 &= -V_4 = V_{33} \end{aligned} \quad (2.17)$$

The coefficients R_{11}, R_{22}, R_{33}, lying along the leading diagonal of the array formed by the terms on the left-hand side of the equations, are called the *self resistances*; each is the sum of the separate resistances contained in the mesh indicated by the relevant subscripts. The coefficients $R_{12}(=R_{21})$, $R_{13}(=R_{31})$ etc., are symmetrically disposed about the leading diagonal and are called the *mutual resistances* since each is the resistance in the branch shared by the meshes indicated by the relevant subscripts. The mutual resistance terms are all negative if all cyclic currents are assigned in the same direction. Finally, notice that V_{11}, V_{22}, and V_{33}, are the net e.m.f.s acting round each of the meshes indicated by the attached subscripts. The extension of this formal notation to networks containing meshes of higher number than three will be obvious.

Various techniques are available for solving linear simultaneous equations of the form represented by (2.17). For only a small number of

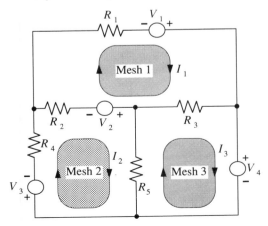

Fig. 2.8. Example of a three-mesh network.

equations, a numerical solution would generally be obtained by means of Gaussian elimination and back substitution. A large number of numerical equations would be solved by standard routines available on most digital computers.* For the present purposes the method of determinants will be used. This will enable us to deduce the solution of (2.17) in symbolic form, and at the same time will allow us to introduce the notation and methods required to develop several important circuit theorems.

The solution for I_1 may be written as the ratio of two determinants, thus,

$$I_1 = \frac{\begin{vmatrix} V_{11} & R_{12} & R_{13} \\ V_{22} & R_{22} & R_{23} \\ V_{33} & R_{32} & R_{33} \end{vmatrix}}{\begin{vmatrix} R_{11} & R_{12} & R_{13} \\ R_{21} & R_{22} & R_{23} \\ R_{31} & R_{32} & R_{33} \end{vmatrix}} \quad (2.18)$$

Putting the determinant in the denominator equal to Δ and expanding using Cramer's rule we obtain

$$I_1 = \frac{V_{11}}{\Delta}\begin{vmatrix} R_{22} & R_{23} \\ R_{32} & R_{33} \end{vmatrix} - \frac{V_{22}}{\Delta}\begin{vmatrix} R_{12} & R_{13} \\ R_{32} & R_{33} \end{vmatrix} + \frac{V_{33}}{\Delta}\begin{vmatrix} R_{12} & R_{13} \\ R_{22} & R_{23} \end{vmatrix}$$

$$= V_{11}\frac{\Delta_{11}}{\Delta} - V_{22}\frac{\Delta_{21}}{\Delta} + V_{33}\frac{\Delta_{31}}{\Delta} \quad (2.19)$$

where $\Delta_{11} = \begin{vmatrix} R_{22} & R_{23} \\ R_{32} & R_{33} \end{vmatrix}$ is the minor of Δ, that is, the determinant remaining when the first row and first column are deleted from Δ. Similar meanings may be attached to Δ_{21} and Δ_{31}.

Now if the determinants in (2.19) are expanded, it will be seen that Δ has the dimensions of (resistance)^3 whereas Δ_{11}, etc., have dimensions of (resistance)^2. Thus, we may write (2.19) as

$$I_1 = \frac{V_{11}}{r_{11}} - \frac{V_{22}}{r_{21}} + \frac{V_{33}}{r_{31}} \quad (2.20)$$

where r_{11}, r_{21}, r_{31} are coefficients having dimensions of resistance.

Solutions for currents I_2 and I_3 are found in a similar fashion. The general equations and solutions for a network containing any number of meshes are presented in Appendix B.

---

* A program, written in BASIC, for solving simultaneous equations is listed in Appendix C.

2.6 The superposition and reciprocity theorems

2.6.1 Superposition

Returning again to the single mesh circuit of fig. 2.5 and its solution (2.9) we see that the current may be written

$$I = \frac{V_1}{R_1 + R_2} - \frac{V_2}{R_1 + R_2}$$

The term $V_1/(R_1 + R_2)$ represents a current due to V_1 acting alone and which flows in a clockwise direction, while the term $V_2/(R_1 + R_2)$ represents a current due to V_2 acting alone and flowing in a counter-clockwise direction (indicated by the negative sign). The actual current I is formed by the *superposition* of these individual currents.

The same superposition principle is evident in the solution (2.20) for the current I_1 in the three-mesh network. This may be written in full as

$$I_1 = \frac{V_1}{r_{11}} - \frac{V_2}{r_{11}} - \frac{V_2}{r_{21}} + \frac{V_3}{r_{21}} - \frac{V_4}{r_{31}}$$

$$= \frac{V_1}{r_{11}} - V_2\left(\frac{1}{r_{11}} + \frac{1}{r_{21}}\right) + \frac{V_3}{r_{21}} - \frac{V_4}{r_{31}}$$

We see that I_1 is composed of four individual currents, each due to one of the voltage sources acting in the circuit alone. Each of the individual currents depends only on the value of the relevant voltage source, and is independent of the values of the other voltage sources acting in the circuit. The general proposition demonstrated by these two examples is embodied in the superposition theorem which may be stated as follows:

The total current flowing in any branch of a network containing ideal voltage sources is equal to the algebraic sum of the currents which would flow in that branch if each of the ideal voltage sources in turn acted alone, the other sources being reduced to zero. (A formal proof of this theorem is contained in Appendix B.)

It follows from this theorem and Ohm's Law that the voltage between any two nodes in a network is equal to the algebraic sum of the voltages arising between those nodes due to each of the voltage sources in the network acting alone.

The superposition theorem is of considerable importance in the theory of linear network analysis since it provides a starting point for the development of several other useful theorems and techniques. It is also sometimes used as a practical alternative to the method of mesh analysis for finding the current in a specified branch of a network. To illustrate this we consider again the circuit of fig. 2.6 (repeated in fig. 2.9(a)). To find the current in, say, the branch CD containing R_3 we determine the current in this branch due to each of the sources V_1 and V_2 acting alone. Let I_1 be the current due to V_1 with V_2 reduced to zero as in fig. 2.9(b). (Note that V_2 is reduced to zero by replacing it with a short circuit *not* by open circuiting the branch EF.) In this modified circuit we see that R_2 and R_3 form a parallel combination whose resistance is given by $R_2 R_3/(R_2 + R_3)$. The total resistance across V_1 is therefore $R_1 + R_2 R_3/(R_2 + R_3)$ and the current delivered by the source V_1 is given by

$$\frac{V_1}{R_1 + R_2 R_3/(R_2 + R_3)}$$

Now R_2 and R_3 together form a current divider hence, by (2.4), the current I_1 is given by the fraction of the total current equal to $R_2/(R_2 + R_3)$, that is

$$I_1 = \frac{V_1}{R_1 + R_2 R_3/(R_2 + R_3)} \cdot \frac{R_2}{R_2 + R_3}$$

or

$$I_1 = \frac{R_2 V_1}{R_1(R_2 + R_3) + R_2 R_3}$$

Fig. 2.9. Illustrating the superposition theorem.

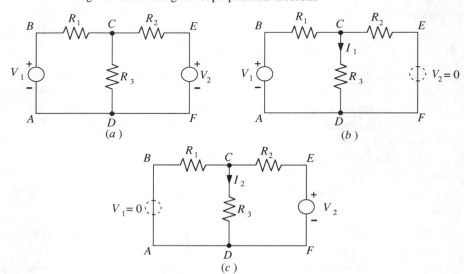

The superposition and reciprocity theorems

Similarly, with V_2 acting alone as shown in fig. 2.9(c), the current I_2 is given by

$$I_2 = \frac{R_1 V_2}{R_2(R_1 + R_3) + R_1 R_3}$$

The total current when both sources act together is, according to the superposition theorem, $(I_1 + I_2)$ directed from C to D. The same result may, of course, be obtained by solving (2.15) and (2.16) for the two mesh currents and taking the appropriate difference. Generally speaking, the use of the superposition principle in practical problems is of advantage only when it is desired to find one particular branch current in a network involving not more than two or three meshes.

The superposition principle is a direct consequence of the properties of a linear network. Consider the situation depicted in fig. 2.10 in which I_1 and I_2 are the contributions to the total current I which flow in one branch of a network as a result of sources V_1 and V_2 acting together in that network. In the linear case (fig. 2.10(a)) we see that each individual contribution is unaffected by the value of the source voltage producing the other. In other

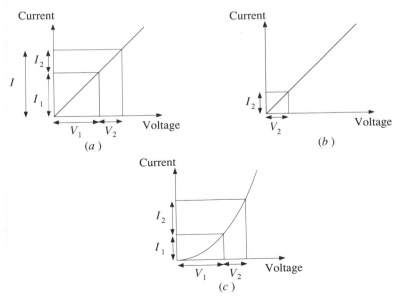

Fig. 2.10. Voltage–current characteristics of linear and non-linear circuits. (a) Linear: voltages V_1 and V_2 produce currents I_1 and I_2. (b) V_1 reduced to zero, V_2 still produces I_2. (c) Non-linear: response I_2 to V_2 depends on value of V_1.

words, V_1 could take any value including zero (fig. 2.10(b)) without affecting the contribution I_2 due to V_2. In the non-linear case, however, (fig. 2.10(c)) it is clear that the contribution I_2 to the total current is dependent not only on V_2 but also on the particular value of V_1. The current due to each source acting alone cannot, therefore, be superposed to find the total current when both act together.

2.6.2 Reciprocity

The reciprocity theorem states that:

The current produced in any one branch of a network by an e.m.f. acting in a second branch, is equal to the current which would be produced in the second branch if the e.m.f. were transferred to the first branch.

Alternatively, we may state: *the voltage produced at any one node of a network by a current source acting at another node, is equal to the voltage at the first node if the current source were transferred to the second node.* A proof of this theorem is presented in Appendix B.

Like the superposition theorem the reciprocity theorem can occasionally save work in practical problems but it is mainly of value for the theoretical insights which it can provide. An example of this will be encountered in the theory of bridge circuits contained in section 3.10.

2.7 Thévenin's theorem

According to Thévenin's theorem any network consisting of linear resistances and ideal sources, having two terminals AB (fig. 2.11) may be replaced by an equivalent network consisting of a single resistance R_T in series with a single ideal voltage source V_T; in other words the network may be replaced by a practical voltage source of the form shown in fig. 2.11(b).

The theorem asserts that: *so far as any external network connected across AB is concerned, the given network and its equivalent are indistinguishable if V_T is made equal to the e.m.f. that would appear across AB on open circuit, and*

Fig. 2.11. Thévenin's theorem.

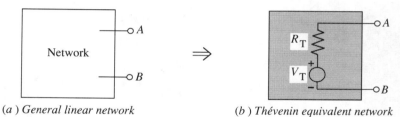

(a) General linear network   (b) Thévenin equivalent network

if R_T is made equal to the resistance that would exist between AB when all sources internal to the given network are rendered inoperative. By 'inoperative' we mean that voltage sources must be replaced by short circuits and current sources must be replaced by open circuits.

If the detailed configuration of a circuit is known, the Thévenin equivalent may be found theoretically. For example, the circuit shown in fig. 2.12(a) contains one current source and two resistances, the values of which are given. We may deduce by inspection that the voltage across AB on open circuit is 4 V, the terminal A being positive with respect to B. (No current flows through the 2 Ω resistance when AB are open circuit so that under these conditions the open circuit voltage must be identical to that across the current source.) The resistance across AB when the current source is made inoperative is, by inspection, 3 Ω. The Thévenin equivalent network is, therefore, as shown in fig. 2.12(b).

Frequently the internal details of a practical two-terminal circuit may not be known with exactitude because of limitations of the lumped modelling technique, variations associated with manufacturing tolerances of components, etc. In such cases the circuit in question can be completely characterized by its Thévenin equivalent, the elements of which can be determined by measurements made external to the circuit.

One method of finding the Thévenin equivalent of a network is illustrated in fig. 2.13(a). A variable resistor R_L is connected to the terminals AB, and the current drawn by this resistor is measured by means of an ammeter. The terminal voltage is measured by a voltmeter the resistance of which must be sufficiently high that the current flowing through it does not affect the measurement of the current through R_L.

A series of measurements of voltage and current are made for various settings of the resistor, and the results plotted in the form of a graph of voltage versus current. Provided the network under test is linear, the graph will be of straight-line form similar to that shown in fig. 2.13(b). The slope of the line gives R_T and the intercept on the voltage axis gives V_T.

Fig. 2.12. Application of Thévenin's theorem.

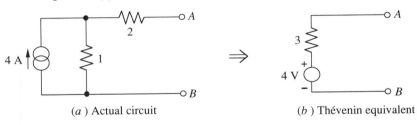

(a) Actual circuit   (b) Thévenin equivalent

60   Theorems and techniques of circuit analysis

To prove Thévenin's theorem we consider the arrangement shown in fig. 2.14. P represents the given linear network with terminals AB, and Q is an external linear network to be connected to P. Let V_T be the open circuit e.m.f. across AB, and let R_T be the resistance between AB when all sources inside P are made inoperative. We may assume, without loss of generality, that Q does not contain any sources since, by the superposition theorem, the effects of these would be independent of any currents caused by the sources in P. Let the resistance across CD be R_Q. We wish to show that when Q is connected to P, the resulting current that flows between the two networks is precisely the same as that which would flow if P were replaced by the series combination of V_T and R_T.

We now consider the situation when an ideal voltage source V is connected between A and C, as indicated by the dashed lines in fig. 2.14, the circuit being completed by joining B to D. The resulting current that flows

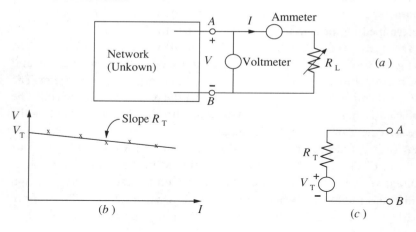

Fig. 2.13. Determination of the Thévenin equivalent circuit by measurement.

Fig. 2.14. Circuit for proof of Thévenin's theorem.

Worked example

round the path $ACDB$ can, according to the superposition theorem, be regarded as being made up of two components: (1) a current due to the combined action of all the sources in P; (2) a current due to the additional source V. Now let V be adjusted so that it is equal to V_T; with the polarities shown the current must fall to zero since there is then no net e.m.f. acting round the path $ACDB$. Thus, since the total resistance round this path is $(R_T + R_Q)$, the current (2) must be $V_T/(R_T + R_Q)$ flowing counterclockwise, and this must be equal to the current (1) flowing in the opposite direction. But $V_T/(R_T + R_Q)$ is precisely the current that would flow between P and Q if P were replaced by an ideal voltage source V_T in series with a resistance R_T. This proves the theorem. A formal, mathematical proof of this theorem is presented in Appendix B.

2.8 Worked example

A temperature sensitive resistor (thermistor) is used in a Wheatstone Bridge circuit for the measurement of the temperature of a water bath as shown in fig. 2.15(a). The temperature is indicated on the meter M, which has a sensitivity of 50 μA at full-scale deflection and a resistance of 300 Ω. A 4 V battery, of negligible internal resistance, is used to energize the bridge. The thermistor has a resistance of 1000 Ω at a water temperature of 50 °C and its resistance decreases by 5% for each degree increase in temperature.
(a) Find the value of the resistance to which R_4 must be set to give zero reading on the meter when the temperature of the bath is 50 °C.
(b) With R_4 set as in (a) above, find the temperature of the bath corresponding to full-scale deflection on the meter.

Solution. (a) We first label the nodes of the circuit $ABCD$ as shown. The required condition of zero current through the meter is obtained when the potential of node A is the same as that of B, that is, (referring potentials to the node D) when $V_{AD} = V_{BD}$. These two voltages are most easily found by recognizing that the bridge, under the given zero current condition, constitutes two separate voltage divider circuits as shown in fig. 2.15(b). To emphasize this the circuit has been split into two parts with a separate source for each part; it will be appreciated that this makes no essential difference to the operation of the circuit.

The required expressions for V_{AD} and V_{BD} may be found using (2.3). Thus, for zero meter current:

$$V_{AD} = V_{BD}$$

$$\frac{R_2}{R_1 + R_2} V_s = \frac{R_4}{R_3 + R_4} V_s$$

which upon rearrangement becomes

$$R_1 R_4 = R_2 R_3 \tag{2.21}$$

This relationship expresses the so-called *balance condition* for the Wheatstone bridge. Note that (2.21) is independent of the source voltage V_s.

At a temperature of 50 °C the thermistor has a resistance of $R_3 = 1000\,\Omega$ hence, substituting into (2.21) the given resistance values for R_1 and R_2 we obtain

$$R_4 = \frac{R_2 R_3}{R_1} = \frac{750 \times 1000}{1000} = 750\,\Omega$$

(b) We are required to find a relationship between the current flowing through the meter and the resistance R_3 of the thermistor from which the temperature for full-scale deflection may be deduced. Thévenin's theorem may be used to find such a relationship.

The general approach in applying Thévenin's theorem to a problem of this kind consists in removing from the circuit in question the branch through which it is desired to find the current. The Thévenin equivalent is then found of the remaining network that exists across the two terminals

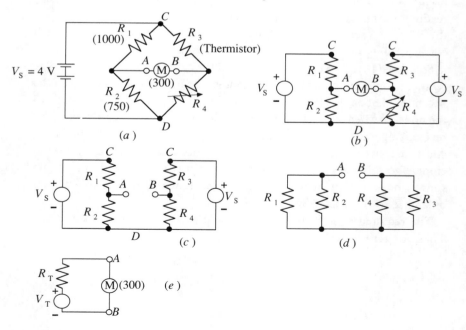

Fig. 2.15. Circuits for worked example.

Worked example

exposed as a result of removing the branch. Finally, the branch is reconnected to the equivalent circuit, thus forming a single mesh circuit from which the current is easily found.

In this example the branch of interest is that containing the meter; on its removal, terminals AB are exposed and we see that the circuit remaining is essentially that of fig. 2.15(c).

Since the Thévenin equivalent voltage V_T is, by definition, that voltage which exists across AB under open circuit conditions, we may again (as in part (a) of this example) use the voltage divider principle to determine the voltage across AB. Thus, V_T is given by

$$V_T = V_{AD} - V_{BD} = \frac{R_2}{R_1 + R_2} V_s - \frac{R_4}{R_3 + R_4} V_s$$

Substituting actual values gives:

$$V_T = \frac{750 \times 4}{1000 \times 750} - \frac{750 \times 4}{R_3 + 750}$$

To find the Thévenin equivalent resistance R_T across AB we render the internal source V_s inoperative by replacing it by a short circuit. The circuit of fig. 2.15(c) then reduces to that shown in fig. 2.15(d). It is seen that R_1 and R_2 now form a parallel combination; likewise R_3 and R_4 form a parallel combination. R_T is therefore given by (using the 'product-over-sum' rule):

$$R_T = \frac{R_1 R_2}{R_1 + R_2} + \frac{R_3 R_4}{R_3 + R_4}$$

substituting actual values:

$$R_T = \frac{1000 \times 750}{1000 + 750} + \frac{R_3 \times 750}{R_3 + 750}$$

The simplified circuit with the meter branch reconnected is shown in fig. 2.15(e). Since the current for full-scale deflection is $50 \mu A$ we obtain the following relationship for the circuit:

$$50 \times 10^{-6} = \frac{V_T}{R_T + 300}$$

Substituting the expressions for V_T and R_T into this equation and performing some algebraic manipulation we find that $R_3 = 1062 \Omega$.

Now the relationship between the temperature T and the resistance of the thermistor is $R_3 = 0.95^{(T-50)} \times 10^3$, hence the temperature corresponding to a resistance of $R_3 = 1062 \Omega$ is given by

$$T = \frac{\log \frac{1062}{1000}}{\log 0.95} + 50 = 48.8 \,°C$$

An alternative approach to this problem would be to use mesh analysis, but this involves setting up and solving three mesh equations – a somewhat tedious procedure. The power of the Thévenin approach lies in the fact that removal of one branch from a circuit often renders the remaining part of the circuit amenable to a simple form of analysis from which the equivalent circuit can be found. In this example, removing the branch containing the meter reduces the number of meshes from three to two; furthermore, by employing the artifice of the voltage divider, mesh analysis is avoided altogether.

2.9 Network transformations

If two networks have the same Thévenin equivalent circuit at corresponding pairs of terminals or ports, then, so far as any external connections are concerned, the two networks are indistinguishable. This corollary of Thévenin's theorem allows us to establish the conditions for which two or more networks are electrically equivalent. We are thus able to replace a network or part of a network with a different but electrically equivalent network, and this is of considerable practical significance in the analysis of circuits. Such a procedure is known as network transformation. Two important examples of network transformation will now be considered.

2.9.1 The Thévenin–Norton transformation

The circuit shown in fig. 2.16(a) is a practical voltage source of the type introduced in section 2.1. Another type of source, called a *practical current source*, is shown in fig. 2.16(b). This consists of an ideal current source in *parallel* with a linear resistance. According to Thévenin's theorem the two circuits are electrically equivalent if they both present the same open circuit voltage at their terminals, and if they both present the same terminal resistance when their sources are made inoperative; in other words, if they both possess the same Thévenin equivalent circuit. The circuit of fig. 2.16(a) is, of course, its own Thévenin equivalent.

The open circuit voltage of the practical current source is IR_2, and its resistance when the ideal current source is replaced by an open circuit is R_2. Therefore, the two circuits are equivalent if

$$IR_2 = V \text{ and } R_2 = R_1 \tag{2.22}$$

Network transformations

They are, of course, equivalent only so far as external connections are concerned; internally the two circuits are fundamentally dissimilar since the practical current source dissipates power continually in its own resistance even when its terminals are open circuit. Note that the equivalence holds only if the direction of I in fig. 2.16(b) is such that the same voltage polarity is produced at the terminals of the two circuits.

We conclude, therefore, that an ideal voltage source of magnitude V in series with a resistance R is equivalent to an ideal current source of magnitude V/R in parallel with a resistance R (or a conductance $G = 1/R$). The two circuits are duals of one another. Thévenin's theorem may therefore be restated in the following form:

Any network with two accessible terminals AB may, so far as external circuits are concerned, be replaced by an ideal current source I in parallel with a conductance G, where I is the current that would flow if the terminals AB were short-circuited, and G is the conductance across AB if the current source were open-circuited.

The theorem in this form was first stated by E.L. Norton, and it is consequently known as Norton's theorem, although it should be realized that it is not fundamentally different from Thévenin's theorem. The procedure of replacing the circuits of fig. 2.16, one by the other, using the relationships (2.22), is known as the Thévenin–Norton transformation.

Many practical voltage sources exhibit near ideal characteristics over part of their working range, and it is permissible to represent them by an ideal voltage source without series resistance. In this case the Thévenin–Norton transformation cannot be applied since the parameters of the Norton circuit are indeterminate.

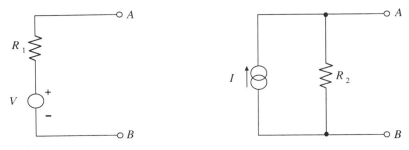

Fig. 2.16. The Thévenin–Norton transformation. By Thévenin's theorem the circuits are equivalent at AB if $I = V/R_2$ and $R_2 = R_1$.

(a) Practical voltage source (Thévenin circuit)

(b) Practical current source (Norton circuit)

2.9.2 The star-delta transformation

The use of Thévenin's theorem for establishing the relationships expressing equivalence between networks is not confined to those possessing a single accessible pair of terminals. The same principle may be applied to multi-terminal, or multi-port networks, by considering each corresponding pair of ports in turn, all other ports being open-circuited. This procedure may be illustrated with reference to the two circuits shown in fig. 2.17; one a star-connected arrangement of three resistances, the other a delta-connected arrangement. Since there are no current or voltage sources included in these circuits, we need consider only the resistances presented at each of the corresponding ports.

Considering first port(1) (terminals AC), with terminal B open-circuit, the resistance at this port is $(R_a + R_c)$. At the corresponding port(1') (terminals $A'C'$), with the terminal B' open-circuit, the resistance is $R_1 // (R_2 + R_3)$. The condition for equivalence is, therefore,

$$R_a + R_c = \frac{R_1(R_2 + R_3)}{R_1 + R_2 + R_3}$$

Similarly, by comparing the resistances at the two remaining ports, we obtain

$$R_c + R_b = \frac{R_3(R_1 + R_2)}{R_3 + R_1 + R_2}$$

and

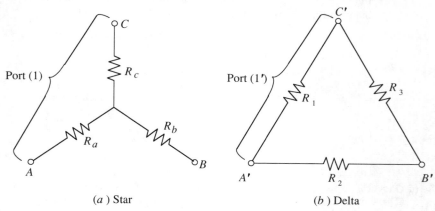

Fig. 2.17. The star–delta transformation.

(a) Star   (b) Delta

$$R_b + R_a = \frac{R_2(R_3 + R_1)}{R_2 + R_3 + R_1}$$

After algebraic manipulation the following sets of relations are established:

$$R_1 = R_a + R_c + \frac{R_a R_c}{R_b}; \quad R_a = \frac{R_1 R_2}{R_1 + R_2 + R_3}$$

$$R_2 = R_b + R_a + \frac{R_b R_a}{R_c}; \quad R_b = \frac{R_2 R_3}{R_1 + R_2 + R_3} \quad (2.23)$$

$$R_3 = R_c + R_b + \frac{R_c R_b}{R_a}; \quad R_c = \frac{R_3 R_1}{R_1 + R_2 + R_3}$$

The procedure of conversion between the two circuits shown in fig. 2.17, using the relations (2.23), has come to be variously known as the star–delta, Y–Δ, Y–mesh transformation, and it finds application particularly in the analysis of power systems. The same transformation occurs in the theory of two-port networks and is there known as the Tee–Pi(T–π) transformation, so-called because in this context the circuits of fig. 2.17 are drawn rather differently and they resemble the shapes from which the name derives (see section 8.8).

The star–delta transformation is a particular case of a more general theory relating to multi-port networks, which has become established as *Rosen's theorem* (see ref. 1).

2.10 Nodal analysis

It will be recalled that in mesh analysis, currents are assigned to each of the meshes in the circuit under consideration, and the mesh equations are formulated by applying Kirchhoff's voltage law to each mesh in turn. In nodal analysis the 'dual' of this procedure is adopted: voltages are assigned to each node and the nodal equations are formulated by applying Kirchhoff's current law.

Node voltages are specified by choosing one node in the circuit as the reference with respect to which voltages at all other nodes are defined. Thus, in fig. 2.18, node O is chosen as the reference and V_A signifies the voltage of node A with respect to that of node O; similarly V_B is the voltage of node B with respect to node O. (It is not usual to employ a double-subscript notation (such as V_{AO}, V_{BO}) for this purpose since the second subscript would merely be repeated.) Any node may, in principle, be chosen as the reference but the nodal equations take their simplest form if the node to which the greatest number of elements is attached is selected. In many practical circuits this node will be the common (ground) terminal and will constitute an obvious choice.

In the circuit of fig. 2.18, the voltage at node C is specified and has the value V; V_A and V_B are the two unknown voltages that have to be determined by setting up and solving two independent nodal equations. We say that such a circuit contains two *independent nodes*.

In applying Kirchhoff's current law at a node it is convenient to use the form (1.13) (section 1.6.1)

$$\sum i = 0 \tag{1.13}$$

where $\sum i$ is interpreted as the algebraic sum of the currents flowing *away* from the node. Thus, at the node A in fig. 2.18, we have

$$I_{AO} + I_{AB} + I_{AC} = 0$$

Now the current I_{AO} flowing through R_1 is clearly V_A/R_1; the current I_{AB} is $V_{AB}/R_3 = (V_A - V_B)/R_3$, and the current $I_{AC} = V_{AC}/R_4 = (V_A - V)/R_4$. Hence, the nodal equation at A is

$$\frac{V_A}{R_1} + \frac{V_A - V_B}{R_3} + \frac{V_A - V}{R_4} = 0 \tag{2.24}$$

Notice that the first member of each of the terms in this equation is the assigned voltage of node A itself. This is a consequence of choosing the positive direction of current as that flowing away from a node.

At node B we have a current source feeding current into the node; this may be treated as a current $(-I)$ flowing away from the node. As before, the current through each resistance attached to node B is found by taking the

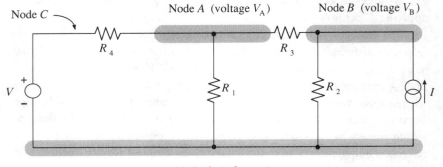

Fig. 2.18. Nodal analysis: assignment of node voltages.

Nodal analysis

difference voltage divided by the value of the resistance. Thus, at B the nodal equation is

$$-I + \frac{V_B}{R_2} + \frac{V_B - V_A}{R_3} = 0 \tag{2.25}$$

Rearranging the above equations we obtain

$$\left(\frac{1}{R_1} + \frac{1}{R_3} + \frac{1}{R_4}\right)V_A - \frac{1}{R_3}V_B = \frac{V}{R_4} \tag{2.26}$$

$$-\frac{1}{R_3}V_A + \left(\frac{1}{R_2} + \frac{1}{R_3}\right)V_B = I \tag{2.27}$$

Solving these two equations will yield specific expressions for V_A and V_B. If the currents in the various branches are required, these may be found by substitution in the appropriate difference terms in the nodal equations; for example, the current I_{BA} through R_3 is given by $(V_B - V_A)/R_3$.

In the above nodal equations we observe that the coefficient of V_A in (2.26) is simply the sum of the conductances attached to node A; this is termed the *self-conductance* at A. Likewise the self-conductance at node B appears as the coefficient of V_B in (2.27). The coefficient of V_B in (2.26) and of V_A in (2.27), namely $(1/R_3)$, is the *mutual-conductance* between nodes A and B. The mutual-conductance terms in the nodal equations are always negative.

On the right-hand side of each nodal equation we have a term representing the current injected into the node concerned from the source attached to that particular node. In the case of node A (equation 2.26) the ideal voltage source V together with R_4 constitute the effective current source, as will be readily apparent if a Thévenin–Norton transformation is carried out according to the principles discussed in section 2.9.1. When this is done, we may redraw the circuit of fig. 2.18 as shown in fig. 2.19, replacing the practical voltage source by an equivalent practical current source. It is now immediately obvious that the current injected into node A has a magnitude V/R_4.

The equations obtained in nodal analysis possess a formal similarity to those obtained in mesh analysis (equations 2.17), and a similar subscript notation is employed when it is required to express them in a general form. Thus, for a circuit such as that in fig. 2.18 in which there are two independent nodes we may write

$$\left.\begin{array}{l} G_{11}V_1 + G_{12}V_2 = I_{11} \\ G_{21}V_1 + G_{22}V_2 = I_{22} \end{array}\right\} \tag{2.28}$$

where G_{11}, G_{22} are the self-conductances at the first and second independent nodes; $G_{12} = G_{21}$ is the mutual-conductance between them; and I_{11} and I_{22} are the net currents injected into the first and second nodes from the ideal current sources attached to them.

It will be appreciated that by carrying out the transformation shown in fig. 2.19, (which may be done mentally), and by using the concept of self- and mutual-conductances, the nodal equations for the circuit of fig. 2.18 could have been written directly in the form (2.26)/(2.27). The reader unfamiliar with network analysis is, however, advised to set up the equations initially in the form (2.24)/(2.25) as in this way there is less likelihood of error.

In general a circuit containing N independent nodes will give rise to N independent nodal equations, the equation at the kth node being of the form

$$\sum_{n=1}^{N} G_{kn} V_n = I_{kk} \tag{2.29}$$

This is a balance equation which expresses the continuity of current at a node and which corresponds to Kirchhoff's current law in the form (1.12).

As a final point of general interest in connection with the nodal equations, we may observe that each nodal equation is an expression of the superposition principle. Consider, for example, the first equation in (2.28). The first term represents the current flowing away from node (1) through all the conductances attached to that node, and with the second node voltage set to zero. The second term represents the current flowing from node (2) to node (1) with the first node voltage set to zero. The superposition of these two currents gives the net current flowing away from node (1), and this equals the current injected into this node from the attached current sources.

Fig. 2.19. The Thévenin–Norton transformation applied to the circuit of fig. 2.18.

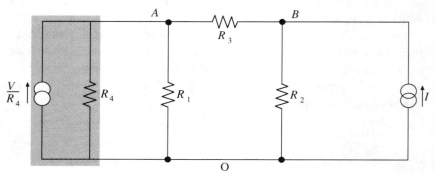

2.11 Comparison of mesh and nodal analysis

For circuits containing current sources, mesh analysis using the cyclic current method is generally less straightforward than nodal analysis. The voltage across a current source is constrained by the circuit in which it resides; it is necessary, therefore, to assign voltage drops to all current sources within the circuit before KVL can be applied to the loops in which they are contained. These unknown voltage drops must then be eliminated by combining the appropriate number of equations. Such complication can sometimes be avoided by assigning branch currents, rather than mesh currents, and by making a judicious choice of loops so as to avoid branches containing current sources. An alternative approach is to first transform practical current sources to practical voltage sources using the inverse Thévenin–Norton transformation, but rarely does this result in a more concise and labour saving solution than can be attained by other means.

Nodal analysis, on the other hand, suffers from no such constraints. It may be used freely for circuits containing both voltage and current sources, as we have seen in the case of the circuit of fig. 2.18, and it more often than not affords a method of solution involving fewer simultaneous equations than mesh analysis. Exceptions to this general rule include the symmetrical ladder type of circuit discussed in section 2.15.3.

We have seen that a network containing M independent meshes, that is, one in which there are essentially M unknown mesh currents to be found, requires the solution of M simultaneous equations. A network possessing N independent nodes leads to N nodal equations. By determining M and N for a particular circuit we are often able to make a rational choice as to which of the two methods of analysis to use. Unless we are dealing with a very large and complex circuit, it is an easy matter to determine N: count the total number of nodes N_T and the number of voltage sources N_V, then N is given by

$$N = N_T - N_V - 1 \qquad (2.30)$$

The reason why one must subtract N_V nodes from the total N_T in this expression stems from the fact that each voltage source is connected to the circuit at two nodes; the voltage of one of these is, therefore, defined with respect to the other and only one can be counted as an independent node. The total N_T also contains the reference node and this must be subtracted as well.

The determination of M often presents considerably greater difficulties, particularly if the circuit is drawn with branches crossing one another. Two ways of drawing the Wheatstone bridge circuit are illustrated in fig. 2.20. We have no difficulty in distinguishing three independent meshes in fig.

72 *Theorems and techniques of circuit analysis*

2.20(a) but these are not nearly as apparent in fig. 2.20(b). Some circuit configurations are not mappable onto a plane surface without the necessity for crossing branches, so that it is not possible to get round this difficulty (as in this case) simply be redrawing the circuit.

For such circuits it is often easier to determine M indirectly by first counting the total number of nodes N_T, and then applying the following relation (derived from a theorem of mathematical topology):

$$M = E - N_T + 1 \tag{2.31}$$

where E is the number of elements in the circuit including sources.

Thus, considering fig. 2.20(b), we see that $N_T = 5$, $E = 7$ whence from (2.31), $M = 3$. Also applying (2.30) with $N_V = 1$, we deduce that $N = 3$. For this circuit, the number of simultaneous equations required is the same; however, the reader may care to check that if the source internal resistance R_S is negligibly small, then nodal analysis confers an immediate advantage.

Fig. 2.20. The Wheatstone Bridge circuit drawn in alternative ways.

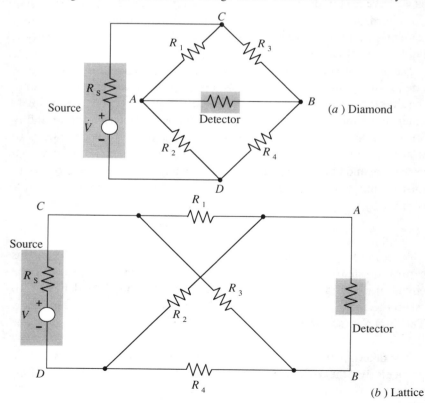

Worked example

The worked example of section 2.4 is also slightly easier using nodal rather than mesh analysis because the circuit possesses essentially a single independent node.

Judicious choice of the reference node can sometimes lead to a simpler and more direct solution using nodal analysis even when N and M are the same. For example in fig. 2.20, if we were interested only in finding the voltage across the detector, we could choose B as the reference node and solve for the voltage V_A to give the detector voltage directly. Using mesh analysis on the other hand two cyclic currents would have to be found, their differences calculated and finally Ohm's law applied.

2.12 Worked example

In the circuit of Fig. 2.21 find the magnitude and direction of the current through the 2 V source, and the magnitude and polarity of the voltage across the 6 A current source.

Solution:

The circuit contains a total of five nodes; these are identified and labelled in the figure. Since there are nine elements the number of independent meshes is, from (2.31), $M = 9 - 5 + 1 = 5$. Note, however, that one mesh contains a current source so that a solution by cyclic current mesh analysis would have to be preceded by an inverse Thévenin–Norton transformation. Alternatively, branch currents could be assigned. In either case the minimum number of simultaneous equations would be four.

There are two voltage sources in the circuit so that the number of

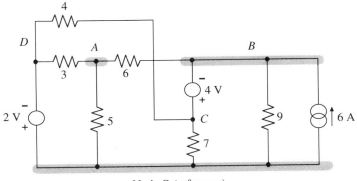

Fig. 2.21. Circuit for worked example on nodal analysis; A and B are the two independent nodes.

Node O (reference)

independent nodes is, from (2.30), $N = 5 - 2 - 1 = 2$. Thus nodal analysis involves only two simultaneous equations.

Node O is chosen as the reference because, (a) it has the greatest number of elements attached, (b) by solving for the node voltage V_B, the voltage across the current source can be found directly. Of the remaining nodes, D is specified (-2 V), and C is specified with respect to B ($+4$ V). Hence, nodes A and B are identified as the two independent nodes.

At A the nodal equation is

$$I_{AB} + I_{AO} + I_{AD} = 0$$

$$\frac{V_A - V_B}{6} + \frac{V_A}{5} + \frac{V_A - (-2)}{3} = 0$$

At B the nodal equation is

$$I_{BA} + I_{BC} + I_{BO} - 6 = 0$$

To find the current I_{BC} we note that this is equal to the current flowing away from node C through the 4 Ω and 7 Ω resistances. The voltage of C with respect to O is $(V_B + 4)$, hence,

$$I_{BC} = I_{CO} + I_{CD} = \frac{V_B + 4}{7} + \frac{(V_B + 4) - (-2)}{4}$$

The complete nodal equation at B is then

$$\frac{V_B - V_A}{6} + (V_B + 4)(\tfrac{1}{7} + \tfrac{1}{4}) + \tfrac{1}{2} + \frac{V_B}{9} - 6 = 0$$

After algebraic manipulation the two nodal equations become:

$$(\tfrac{1}{5} + \tfrac{1}{6} + \tfrac{1}{3})V_A - \tfrac{1}{6}V_B = -\tfrac{2}{3}$$
$$-\tfrac{1}{6}V_A + (\tfrac{1}{6} + \tfrac{1}{7} + \tfrac{1}{4} + \tfrac{1}{9})V_B = 6 - \tfrac{1}{2} - 1 - \tfrac{4}{7}$$

Solving we obtain $V_A = 0.47$ V and $V_B = 5.97$ V.

The voltage across the current source is therefore 5.97 V, B positive with respect to O.

The current through the 2 V source I_{DO} is given by

$$I_{DO} = I_{CD} + I_{AD} = \frac{(V_B + 4) - (-2)}{4} + \frac{V_A - (-2)}{3}$$

Substituting the values for V_A and V_B found above we obtain $I_{DO} = 3.82$ A, this is, the current flows through the source from D to O and has a magnitude of 3.82 A.

2.13 Analysis of networks containing dependent sources

The theory presented in this chapter so far has been concerned with networks containing only independent ideal current and voltage sources. If dependent sources are present (see section 1.5) then the general mesh and nodal equations take a slightly different form, but the techniques of analysis remain essentially the same. The four possible types of dependent or controlled source are depicted in fig. 2.22. In each case the value of the source, voltage or current, is proportional to the value of a current or voltage in some other part of the network.

As an example of the analysis of such a circuit we consider the configuration shown in fig. 2.23. This type of circuit arises in the theory of bipolar transistors and is there termed the hybrid-π model (ref. 5). In this circuit the current source is controlled by the voltage established across the resistance R_1. V_1 is the voltage applied to the input port, and V_2 the voltage at the output port. Apart from the presence of the dependent current source, this circuit is similar in all respects to that shown in fig. 2.18, and the nodal analysis proceeds in a similar fashion. Voltages V and V_2, measured with respect to the reference node O, are assigned to the two independent nodes A and B.

The nodal equations are:

$$\left(\frac{1}{R_1}+\frac{1}{R_3}+\frac{1}{R_4}\right)V-\frac{1}{R_3}V_2=\frac{V_1}{R_4} \qquad (2.32)$$

Fig. 2.22. Dependent (controlled) sources; α, β, γ, δ are control constants.

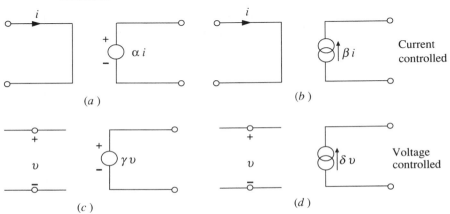

$$-\frac{1}{R_3}V + \left(\frac{1}{R_2} + \frac{1}{R_3}\right)V_2 = -gV \qquad (2.33)$$

These equations will be seen to be of similar form to (2.26) and (2.27), appertaining to the circuit of fig. 2.18. However, in this case we may transpose (2.33) to obtain

$$\left(g - \frac{1}{R_3}\right)V + \left(\frac{1}{R_2} + \frac{1}{R_3}\right)V_2 = 0 \qquad (2.34)$$

We now observe that the coefficients in (2.32) and (2.34) taken together no longer possess the symmetry about the leading diagonal, which is a characteristic of the equations for networks containing only independent sources. Solutions for V and V_2 are easily found since, from (2.34)

$$V = -V_2 \left(\frac{1}{R_1} + \frac{1}{R_2}\right) \Big/ \left(g - \frac{1}{R_3}\right)$$

Hence, substitution in (2.32) gives an explicit expression for V_2.

It is frequently of interest to determine the Thévenin equivalent of a circuit containing a dependent source; in such a case however, the techniques that have been described so far in relation to circuits containing only independent sources are inadequate. In particular, the Thévenin equivalent resistance cannot be found simply by rendering all sources within the circuit inoperative and then determining the resistance of the network remaining. This is because for a circuit containing a dependent source the value of the Thévenin resistance, as well as the e.m.f. depends critically on the dependent source and its controlling parameters.

Two methods are commonly used for finding, analytically, the Thévenin equivalents of circuits containing dependent sources. These methods are applicable also to circuits containing only independent sources, but would not normally be used in such cases since they are rather more cumbersome than the techniques already described. However, the underlying principles

Fig. 2.23. Circuit containing a voltage-controlled current source with control constant g.

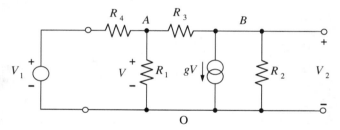

Analysis of networks containing dependent sources

of the methods may be understood by reference to the circuits and techniques previously considered.

The first method to be described is conceptually similar to the practical method of finding the Thévenin equivalent of a circuit illustrated in fig. 2.13. If in this test circuit we make the load R_L infinite, the current will be zero and the voltmeter then registers the open-circuit voltage, which is identical to the Thévenin e.m.f. Let this voltage be $V_{o.c.}$. If next the load is made zero, the ammeter will register the short-circuit current; let this be $I_{s.c.}$. Now since the short-circuit current is that which results from the application of the full Thévenin e.m.f. across the Thévenin resistance, we have the following relations:

$$V_T = V_{o.c.} \text{ (open-circuit voltage)} \tag{2.35}$$

$$R_T = \frac{V_{o.c.} \text{ (open-circuit voltage)}}{I_{s.c.} \text{ (short-circuit current)}} \tag{2.36}$$

Note that placing a short circuit across the terminals of a circuit may, from a practical standpoint, result in damage. It must be stressed that the relation (2.36) provides the basis for an *analytical* approach to the determination of the Thévenin resistance, it does not represent a practical means of measuring this parameter.

The principles underlying the second method of finding the Thévenin equivalent of a circuit containing a dependent source may also be understood with reference to fig. 2.13. Suppose the load R_L is replaced by a voltage source of magnitude V_a, and suppose all internal sources in the circuit *with the exception of dependent sources* are made inoperative, then the ammeter will register a current I_a the magnitude of which will be determined by the effective Thévenin resistance according to

$$I_a = \frac{V_a}{R_T}$$

that is

$$R_T = \frac{V_a \text{ (applied voltage)}}{I_a \text{ (resulting current)}} \tag{2.37}$$

The Thévenin e.m.f. is found by determining the open-circuit voltage, but this is not, as was the case for the first method, a prerequisite for finding R_T. The relation (2.37) is sometimes used as a basis for the practical determination of R_T; here we are concerned only with its utility as a method of analysis.

Both of the methods described are illustrated in the following example.

78 Theorems and techniques of circuit analysis

2.14 Worked example

The characteristics of an operational amplifier may be modelled by the circuit shown in fig. 2.24(a). Resistances R_i and R_o are, respectively, the input and output resistances of the amplifier, and AV is a dependent source with control constant A.*

Figure 2.24(b) shows the circuit of a common type of electronic d.c. amplifier incorporating an operational amplifier modelled in accordance with fig. 2.24(a). The input signal to the amplifier is provided by the source V_1 at terminals AB, the output signal appears at terminals CD.

Find expressions for the Thévenin equivalent e.m.f. and the Thévenin equivalent resistance at the output terminals of the amplifier. Explain how these are related to the overall gain V_2/V_1 and the output resistance of the amplifier.

Solution: method 1

First we establish by means of a nodal analysis an expression for the open-circuit voltage V_2 which, according to (2.35), represents the Thévenin e.m.f. Note that for the purposes of this analysis the signal source V_1 must be regarded as an *internal independent* source.

Let V be the voltage at the node X, then, at node X

$$\frac{V-V_1}{R_1} + \frac{V-V_2}{R_2} + \frac{V}{R_i} = 0 \tag{2.38}$$

and at node C

$$\frac{V_2-V}{R_2} + \frac{V_2-(-AV)}{R_o} = 0 \tag{2.39}$$

Eliminating V from these equations we find

$$V_2 = \left[\frac{R_o - AR_2}{AR_1 + R_oR_1/R_i + R_1R_2/R_i + R_o + R_1 + R_2} \right] V_1$$

The resistance R_i is, in practice, large in relation to the other resistances in the circuit in which case the above expression reduces to the simpler form:

$$V_2 = \left[\frac{R_o - AR_2}{R_o + (1+A)R_1 + R_2} \right] V_1 \tag{2.40}$$

---

* In the context of the theory of operational amplifiers the constant A is normally referred to as the *gain*. A full treatment of operational amplifiers and their application will be found in reference 5. The derivation of the model of fig. 2.24(a) is considered in section 8.3.

Worked example

This expression gives the Thévenin equivalent circuit e.m.f. V_T. The quantity in square brackets gives the overall gain V_2/V_1 of the amplifier. If A is very large, the gain becomes, to a good approximation, $-R_2/R_1$.

To determine the Thévenin resistance we first find the current at the output port when terminals CD are short circuited. This current will be the sum of the currents flowing in R_2 and R_o, that is,

$$I_{s.c.} = \frac{V}{R_2} + \frac{(-AV)}{R_o} = \left[\frac{R_o - AR_2}{R_2 R_o}\right] V \qquad (2.41)$$

But from (2.38) with $V_2 = 0$, and again assuming that R_i is very large, V is given by

$$V = \frac{V_1}{R_1} \cdot \frac{1}{\left(\frac{1}{R_1} + \frac{1}{R_2}\right)} = \frac{V_1 R_2}{R_1 + R_2}$$

Substituting for V in (2.41) we then obtain

$$I_{s.c.} = \frac{V_1 R_2}{(R_1 + R_2)} \cdot \frac{(R_o - AR_2)}{R_2 R_o} = \left[\frac{R_o - AR_2}{R_o(R_1 + R_2)}\right] V_1 \qquad (2.42)$$

Now by (2.36) the Thévenin resistance is given by the ratio of open-circuit voltage to short-circuit current, therefore, combining (2.40) and (2.42) gives

Fig. 2.24. Operational amplifier circuits for worked example.

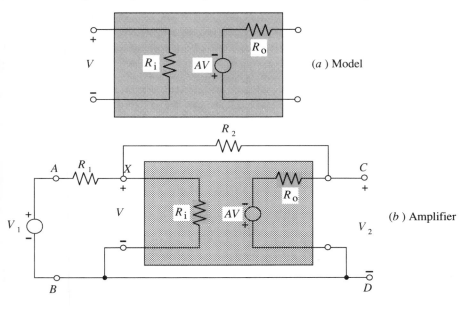

$$R_T = \frac{V_{\text{o.c.}}}{I_{\text{s.c.}}} = \frac{(R_o - AR_2)V_1}{R_o + (1+A)R_1 + R_2} \cdot \frac{R_o(R_1 + R_2)}{(R_o - AR_2)V_1}$$

or

$$R_T = \frac{R_o(R_1 + R_2)}{R_o + (1+A)R_1 + R_2} \tag{2.43}$$

The Thévenin resistance, in the context of amplifier theory, is termed the *output* resistance of the overall amplifier.

Method 2

The Thévenin e.m.f. is found as in method 1. In order to determine R_T by means of the relation (2.37), terminals AB are short-circuited, thus removing the signal source, and a voltage source V_a is connected to the terminals CD. Note, however, that the dependent source is still active since the controlling voltage V is now derived from the applied voltage V_a via resistances R_1 and R_2. Let the polarity of V_a be such that the terminal C is positive with respect to D, and let the reference direction of the resulting current I_a be into the terminal C. For the reasons given above the effect of R_i will be ignored. The nodal equation at C is then

$$\frac{V_a}{R_1 + R_2} + \frac{V_a - (-AV)}{R_o} - I_a = 0 \tag{2.44}$$

An expression for V is most easily obtained by observing that R_1 and R_2 form a voltage divider, hence we may write:

$$V = \left(\frac{R_1}{R_1 + R_2}\right) V_a \tag{2.45}$$

Substituting (2.45) in (2.44) gives

$$I_a = \frac{V_a}{R_1 + R_2} + \frac{V_a}{R_o} + \frac{A}{R_o} \cdot \frac{R_1}{(R_1 + R_2)} \cdot V_a$$

Therefore, by (2.37)

$$R_T = \frac{V_a}{I_a} = \left[\frac{1}{R_1 + R_2} + \frac{1}{R_o} + \frac{AR_1}{R_o(R_1 + R_2)}\right]^{-1}$$

which upon rearrangement becomes

$$R_T = \frac{R_o(R_1 + R_2)}{R_o + (1+A)R_1 + R_2}$$

This method for determining R_T is more direct and involves less work than method 1 because it is not necessary to first find the Thévenin e.m.f.

The amplifier circuit shown in fig. 2.24(b) contains only linear, bilateral elements and one may, as we have demonstrated, apply standard techniques of linear circuit analysis. However, the circuit contains also an active element, here represented by the dependent source AV, which gives it the properties of voltage and power gain. The consequence of this is that the reciprocity theorem does not apply to this circuit. We may easily check that this is so by considering the short-circuit currents that arise at the input and output ports as a result of the same voltage applied, in turn, at opposite ports. Consider first terminals AB short-circuited, and a voltage V_a applied at terminals CD. Making the assumption that R_i is infinite, the current in the short circuit at AB is $V_a/(R_1+R_2)$. Now consider CD short-circuited and V_a applied at AB; the current in this case is from (2.42)

$$\frac{V_a}{(R_1+R_2)} \cdot \frac{(R_o - AR_2)}{R_o}$$

Clearly, the two short-circuit currents are not the same, as would be the case if the reciprocity theorem were true. Circuits of this description, for which the reciprocity theorem does not hold, are said to be *non-reciprocal*. This is the subject of further discussion in chapter 8. It should be noted that the superposition theorem *is* applicable to such circuits.

2.15 Miscellaneous theorems and techniques

†2.15.1 The substitution and compensation theorems

We have seen that the analysis of a circuit can often be facilitated by judicious use of the appropriate linear network theorem; two useful additions to those theorems already discussed are presented below.

The *substitution theorem* is useful if it is required to change the values of the elements in one branch of a circuit, or substitute alternative kinds of elements, without changing voltages and currents elsewhere in the circuit.

Fig. 2.25(a) shows one branch of a circuit containing a resistance R and ideal voltage source V. The branch voltage is V_{AB} and the branch current I_{AB}, these being fixed values. The substitution theorem states that: *this branch may be replaced by another branch without anywhere changing voltages and currents provided the substitute branch also has voltage V_{AB} when carrying current I_{AB}*. This condition can be satisfied by various combinations of R and V which satisfy the branch equation $V_{AB} = I_{AB}R + V$. The maximum possible values are given by:

$$V_{AB} = I_{AB}R_{max} \quad (V=0); \quad V_{AB} = V_{max} \quad (R=0)$$

Thus, either a resistance R_{max} alone or a source V_{max} alone may be substituted for the original elements (figs. 2.25(b) and (c)). Two other combinations of elements that may be substituted are shown in figs. 2.25(*d*) and (*e*). With these combinations the voltage across the current source adjusts automatically to the value required to satisfy the branch equation.

The *compensation theorem* may be employed when it is required to evaluate the effect which a modification in the resistance of one branch of a network has on the currents and voltages at any part of that network. For example we may wish to know how the insertion of an ammeter, possessing some small but finite resistance, will affect the operation of a circuit.

Let AB in fig. 2.26(*a*) be the branch that is to undergo modification. The sources in the remainder of the network will drive a current I through this branch. When the resistance of the branch is changed by an amount ΔR the current will change by some increment ΔI as shown in fig. 2.26(*b*). Note that if the resistance of the branch is increased then the current will be reduced, that is, ΔI will be negative.

The compensation theorem states that: *if the current in a branch of a network before modification is I, and the resistance in that branch is changed by an amount ΔR, the incremental change of current and voltage in any part of the network is that produced by an ideal voltage source of value $I(\Delta R)$ acting in the modified branch and directed in the opposite sense to I.*

The theorem may be proved by considering the change of e.m.f. necessary in AB to reduce ΔI to zero, that is, to restore the current to its original value I. This is accomplished by introducing an additional voltage source of magnitude $I(\Delta R)$, acting in the same sense as I (fig. 2.26(*c*)); this source voltage exactly compensates for the voltage drop across ΔR thereby effectively restoring the network to its original condition. Now, if the effect

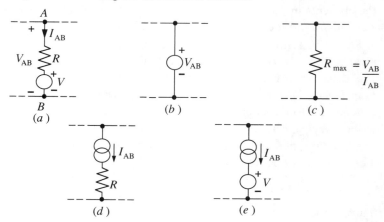

Fig. 2.25. Illustrating the substitution theorem.

of introducing this source is to reduce ΔI to zero, we can say, according to the superposition theorem, that this source acting alone in the network must produce a current ΔI in AB flowing in the opposite direction to $(I+\Delta I)$. Correspondingly, an increase in current from I to $(I+\Delta I)$ must be that effected by a source of magnitude $I\,(\Delta R)$ acting in the opposite sense to I. The incremental current in AB, and all other incremental currents and voltages in the network, may therefore be found from the circuit of fig. 2.26(d) in which all sources other than the added source are made inoperative.

The compensation theorem cannot be applied to the situation where a branch is open-circuited since under such circumstances voltages and currents are indeterminate.

An illustration of the compensation theorem is provided by the circuit shown in fig. 2.27. The elements R_Z and V_Z represent a piecewise-linear circuit model of a Zener diode. (See chapter 7 for piecewise-linear circuit theory.) The circuit is designed to supply (ideally) a constant voltage to the load R_2. What is the change in load voltage if R_2 is decreased by 10%? This problem could, of course, be solved using the standard methods of analysis already presented in this chapter, however, the nature of the circuit renders it amenable to an approximate solution that is sufficiently·accurate for most design purposes. We note that R_Z is small compared with both R_1 and R_2; the voltage V will therefore be very nearly equal to the Zener voltage V_Z. Thus, the current in R_2 (before branch modification) will be, to a good approximation, $5.6/10^3$ A. Since R_2 is reduced, ΔR_2 will be negative and the compensation voltage is $-100\times 5.6/10^3 = -0.56$ V.

Fig. 2.26. Illustrating the compensation theorem.

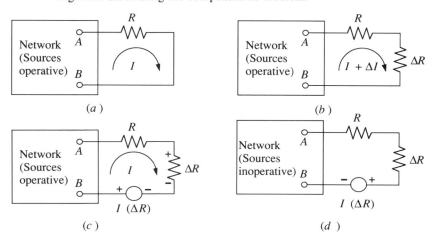

The incremental change of voltage ΔV is then found from the circuit shown in fig. 2.27(b). The modified branch resistance is now 900Ω and, because we are concerned only with the incremental change in load voltage, the other voltage sources in the circuit have been replaced by short circuits. The resulting combination of R_1 and R_Z in parallel is approximately equal to R_Z since $R_1 \gg R_Z$. We now recognize that the circuit is reduced to a simple potential divider from which

$$\Delta V = -0.56 \frac{22}{22+900} = -13.4 \,\text{mV}$$

2.15.2 Circuit reduction

Circuits can often be rendered more tractable for purposes of analysis by first reducing them to a simpler form. A typical situation is depicted in fig. 2.28, in which the effect of varying parameters in one part of a network configuration is to be investigated whilst keeping the remainder fixed. In such a situation it is often convenient to reduce the fixed part of the network to its simplest possible form, usually its Thévenin equivalent, before proceeding with the analysis proper, since the subsequent analytical

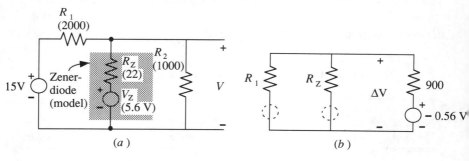

Fig. 2.27. Application of the compensation theorem to a Zener diode voltage-stabiliser circuit.

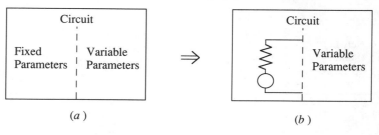

Fig. 2.28. Circuit reduction used to simplify part of a network.

relations are thereby simplified and it becomes easier to investigate the effect of parameter changes.

Circuit reduction may be effected by the direct application of Thévenin's theorem to a complete part of a circuit (as illustrated in section 2.8) or by the use of mesh or nodal analysis. Alternatively, individual nodes and meshes may be eliminated as desired by repeated application of the Thévenin–Norton transformation or the star–delta transformation. The basis of the Thévenin–Norton transformation method is indicated in fig. 2.29. We suppose that the nodes AB are connected to some other part of a network and it is desired to simplify the portion between A and B.

First, the three practical voltage sources are transformed to their equivalent current sources (fig. 2.24(b)), which results in the elimination of nodes O, P, and Q. Current sources are then added together and resistances combined in parallel to produce a single practical current source (fig. 2.29(c)). Finally, if convenient, the current source may be transformed to a voltage source (fig. 2.29(d)).

Although the above step-by-step transformation procedure, making use of diagrams, can be useful and informative, the same process can be

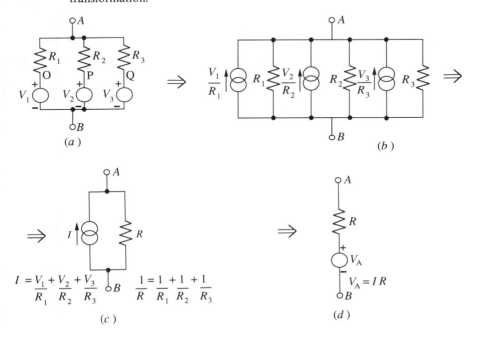

Fig. 2.29. Circuit reduction by application of the Thévenin–Norton transformation.

performed analytically by means of nodal analysis. Taking B as the reference node and solving for the node voltage V_A we obtain:

$$\frac{V_A - V_1}{R_1} + \frac{V_A - V_2}{R_2} + \frac{V_A - V_3}{R_3} = 0$$

or

$$\left(\frac{1}{R_1} + \frac{1}{R_2} + \frac{1}{R_3}\right)V_A = \frac{V_1}{R_1} + \frac{V_2}{R_2} + \frac{V_3}{R_3} \qquad (2.46)$$

Putting

$$\frac{1}{R} = \frac{1}{R_1} + \frac{1}{R_2} + \frac{1}{R_3} \quad \text{and} \quad I = \frac{V_1}{R_1} + \frac{V_2}{R_2} + \frac{V_3}{R_3}$$

$$\frac{V_A}{R} = I$$

The above relations may be interpreted in terms of figs. 2.29(c) and (d).

Equation (2.46) in its general form, relating to any number of parallel sources, is sometimes referred to as *Millman's theorem*.

The process of circuit reduction using the star–delta transformation (or its inverse) may be illustrated with reference to the circuit of fig. 2.30(a). It will be appreciated that reduction of this circuit cannot be effected by simple series and parallel additions of resistances because of the bridging resistance R_3. This difficulty is overcome by recognizing that the three resistances within the shaded box form a delta configuration. Using the relations 2.23 the delta comprising, R_1, R_2, R_3 is transformed to the star

Fig. 2.30. Circuit reduction using the inverse star–delta transformation.

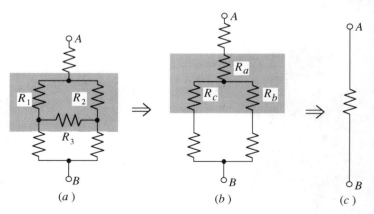

Miscellaneous theorems and techniques

comprising R_a, R_b, R_c (fig. 2.30(b)), after which series and parallel combination lead to a single resistance (fig. 2.30(c)).

†2.15.3 Ladder networks

There are certain types of network that do not lend themselves readily to the standard methods of analysis so far discussed in this chapter, and for which special techniques are used. The *ladder* network falls into this category.

Suppose we wished to find the node voltages V_{EO} and V_{CO} at the end and mid-points of the ladder network shown in fig. 2.31(a). A mesh-node count at once indicates that a standard method of analysis would involve setting up and solving four simultaneous equations. The following step-by-step procedure is rather simpler and is particularly convenient to carry out numerically using a small calculator.

We start by assuming that the voltage across the end of the ladder V_{EO} is 1 V. The current I_{EO} in the end $2\,\Omega$ resistance is then 1/2 A, and the voltage V_{DO} is 3 V (terminals EO assumed open circuit). The calculation then proceeds as follows:

$$I_{DO} = \tfrac{3}{2}\,\text{A}$$
$$I_{CD} = I_{DO} + I_{DE} = \tfrac{3}{2} + \tfrac{1}{2} = 2\,\text{A}$$

Fig. 2.31. Ladder networks.

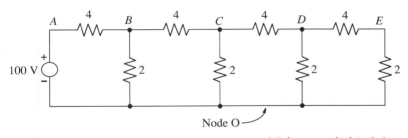

(a) Asymmetrical (unbalanced)

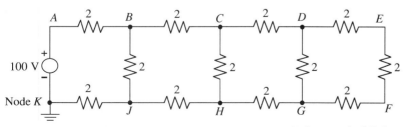

(b) Symmetrical (balanced)

$$V_{CD}=4I_{CD}=8\text{ V}$$
$$V_{CO}=V_{DO}+V_{CD}=3+8=11\text{ V}$$
$$I_{CO}=V_{CO}/2=\tfrac{11}{2}\text{ A}$$
$$I_{BC}=I_{CO}+I_{CD}=\tfrac{11}{2}+2=\tfrac{15}{2}\text{ A}$$
$$V_{BC}=4I_{BC}=30\text{ V}$$
$$V_{BO}=V_{CO}+V_{BC}=11+30=41\text{ V}$$
$$I_{BO}=V_{BO}/2=\tfrac{41}{2}\text{ A}$$
$$I_{AB}=I_{BO}+I_{BC}=\tfrac{41}{2}+\tfrac{15}{2}=\tfrac{56}{2}\text{ A}$$
$$V_{AB}=4I_{AB}=112\text{ V}$$
$$V_{AO}=V_{BO}+V_{AB}=41+112=153\text{ V}$$

This last figure is the voltage of the source assuming 1 V at the end of the ladder. But the actual value of V_{AO} is 100 V hence the true value of V_{EO} must be $1\times(100/153)$ V. Likewise all voltages and currents in the above calculation must be scaled in similar proportion to obtain true values. The voltage at the mid-point of the ladder is therefore $V_{CO}=11\times(100/153)$ V.

A similar procedure may be adopted in the case of a symmetrical ladder of the form shown in fig. 2.31(b). As far as the calculation of voltages across the rungs of a ladder network is concerned it is immaterial how the total resistance between rungs is distributed on the two sides of the ladder. It is convenient, therefore, in the analysis of a symmetrical ladder to first lump together the resistances between rungs and then proceed as for the analysis of an asymmetrical ladder. For the particular resistance values given in fig. 2.31 the voltages across each of the two forms of ladder are identical.

To determine node voltages with respect to ground (node K in fig. 2.31(b)) for the symmetrical ladder it is necessary only to apply the symmetry principle, once having found voltages across the ladder. For example, suppose the voltage V_{HK} is required. Using the procedure detailed above we first find that V_{CH} (corresponding to V_{CO} in fig. 2.31(a)) is equal to $11\times(100/153)$ V. We may then deduce, by symmetry, that V_{HK} is $\tfrac{1}{2}(100-11\times(100/153))$ V.

The ladder method described above is of value mainly for numerical calculation; extremely unwieldy expressions result if one attempts to apply the method in symbolic form. Other methods of dealing with ladder networks are discussed in chapter 8.

†2.15.4 Ring mains

Ring mains are employed extensively in power distribution systems. A number of loads are connected to a single distribution point via parallel conductors which form a closed loop or ring, as indicated schematically in fig. 2.32(a). This type of connection results in better utilization of the distribution conductors compared with a straight parallel connection. Calculations on ring mains are performed with the aim of determining voltages at loads and currents in ring conductors so that the correct conductor cross sections may be specified.

The lumped circuit model of a ring main with three loads is shown in fig. 2.32(b). R_1, R_2, R_3 and R_4 represent conductor resistances between points on the ring. By opening out the ring and treating the resulting circuit as if it were fed by identical sources, one at each end, the analysis is considerably simplified. The circuit in this form resembles somewhat the symmetrical ladder of fig. 2.31(b).

If the loads are specified as fixed resistances, and an exact analysis is required, then the ladder method described above can be used. Normally, however, the loads are specified in terms of the maximum currents to be drawn at particular points on the ring, in which case a mesh analysis is appropriate. In the following worked example both methods of analysis are illustrated.

Fig. 2.32. Ring main.

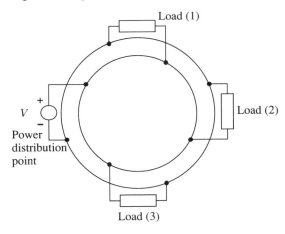

(a) Schematic diagram

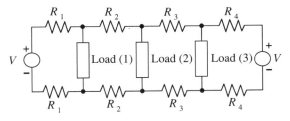

(b) Lumped circuit model

†2.15.5 Worked example

A ring main is supplied from a power point at 240 V and has three loads. The total length of the ring is 12 m and the loads and the power point are spaced at equal intervals of 3 m. Each of the ring conductors has a resistance of 0.067 Ω/m. Find the currents in the ring conductors and the voltages at each of the loads: (a) if the loads are specified as three consecutive resistances of 24 Ω, 16 Ω, and 12 Ω; (b) if the loads are specified as three consecutive currents of 10 A; 15 A and 20 A.

Solution
(a) Each of the conductors in the ring main has a resistance of 0.067 Ω/m, therefore, the total resistance of two conductors (in series) over a 3 m length is $2 \times 3 \times 0.067 = 0.402\,\Omega$. The circuit model is shown in fig. 2.33(a).

We first find the contributions to the current I_1 due to each of the two voltage sources acting alone. The total current is then found from the superposition of the two separate contributions. To find the contribution due to the left-hand source, replace the right-hand source by a short-circuit and assume a current of 1 A to flow in this short-circuit (fig. 2.33(b)). Using the ladder method the calculation proceeds as follows:

$I_{EO} = 1$ A (assumed)
$V_{DO} = 1 \times 0.402 = 0.402$ V
$I_{DO} = 0.402/12 = 0.0335$ A
$I_{CD} = I_{DO} + I_{EO} = 0.0335 + 1 = 1.0335$ A
$V_{CD} = 1.0335 \times 0.042 = 0.4155$
$V_{CO} = V_{CD} + V_{DO} = 0.4155 + 0.402 = 0.8175$
$I_{CO} = 0.8175/16 = 0.0511$
$I_{BC} = I_{CO} + I_{CD} = 0.0511 + 1.0335 = 1.0846$
$V_{BC} = 1.0846 \times 0.402 = 0.4360$
$V_{BO} = V_{BC} + V_{CO} = 0.4360 + 0.8175 = 1.2535$
$I_{BO} = 1.2535/24 = 0.05223$
$I_{AB} = I_{BO} + I_{BC} = 0.05\;223 + 1.0846 = 1.1368$
$V_{AB} = 1.1368 \times 0.402 = 0.4570$
$V_{AO} = V_{AB} + V_{BO} = 0.4570 + 1.2535 = 1.7105$

But the actual value of V_{AO} is 240 V, therefore the true value of the current $I_{AB} = 1.1368 \times (240/1.7105) = 159.50$ A and the true value of the current $I_{EO} = 1 \times (240/1.7105) = 140.31$ A.

Now the right-hand source is restored and the left-hand source replaced by a short-circuit. The contribution to I_1 due to this source acting alone must, according to the reciprocity theorem, be -140.31 A. Therefore, when

Worked example

the two sources act together $I_1 = 159.50 - 140.31 = 19.19$ A. Having found the current I_1 we now employ a step-by-step procedure, working from left to right, in the circuit of fig. 2.33(a).

$I_1 = 19.19$ A (calculated)
$V_{AB} = 19.19 \times 0.402 = 7.7144$ V
$V_{BO} = 240 - 7.7144 = 232.29$ V
$I_{BO} = 232.29/24 = 9.6785$ A
$I_2 = 19.19 - 9.6785 = 9.5115$ A
$V_{BC} = 9.5115 \times 0.402 = 3.8236$ V
$V_{CO} = 232.29 - 3.8236 = 228.46$ V
$I_{CO} = 228.46/16 = 14.279$ A
$I_3 = 9.5115 - 14.279 = -4.7676$ A
$V_{CD} = -4.7676 \times 0.402 = -1.9166$ V
$V_{DO} = 228.46 - (-1.9166) = 230.37$ V
$I_{DO} = 230.37/12 = 19.198$ A
$I_4 = -4.7676 - 19.198 = -23.965$ A
$V_{DE} = -23.965 \times 0.402 = -9.6342$ V
$V_{EO} = 230.37 - (-9.6342) = 240$ V

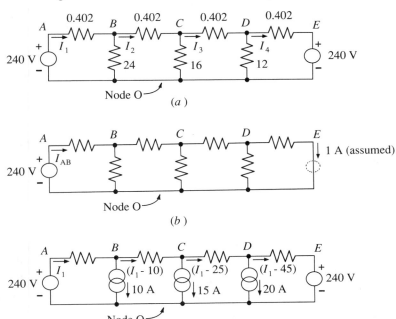

Fig. 2.33. Ladder circuits for worked example.

92   Theorems and techniques of circuit analysis

This last figure, of course, provides a check on the accuracy of the calculations.

(b) With currents of 10 A, 15 A, and 20 A specified, the circuit model becomes that shown in fig. 2.33(c). Branch current I_1 is assigned and KCL is used to write down the other currents in the ring conductors as indicated. Then, using KVL round the loop AEO, we obtain

$$0.402[I_1 + (I_1 - 10) + (I_1 - 25) + (I_1 - 45)] = 240 - 240$$

or

$$4I_1 - 80 = 0$$
$$I_1 = 20 \text{ A}$$

(This is simply the average current drawn by the three loads.) Therefore

$$V_{BO} = 240 - 20 \times 0.402 = 231.96$$
$$V_{CO} = 231.96 - (20 - 10) \times 0.402 = 227.94$$
$$V_{DO} = 227.94 - (20 - 25) \times 0.402 = 229.95$$
$$V_{EO} = 229.95 - (20 - 45) \times 0.402 = 240 \text{ (check)}$$

It should be noted that the load currents specified here are those obtained by dividing the power point voltage, 240 V, by each of the three resistances specified in part (a) of the problem. The actual load currents calculated in part (a) are the same as these to within about 5%. We may, therefore, obtain an approximate solution to the ring main problem, when load resistances are specified, by first calculating approximate load currents (using the power point voltage) and then employing the method of solution outlined in part (b). Comparing the load voltages obtained in parts (a) and (b), we see that these are in agreement to within $\frac{1}{4}$%.

2.16   Summary

Two standard methods of analyzing linear circuits are available: mesh analysis and nodal analysis.

In mesh analysis currents are assigned to every branch in the circuit or, alternatively, currents are assigned to every mesh (Maxwell's cyclic current method). Kirchhoff's voltage law is then applied to set up the requisite number of simultaneous mesh equations, there being as many equations as assigned currents.

In nodal analysis, the dual of mesh analysis, voltages are assigned to nodes, one node being chosen as a reference. Kirchhoff's current law is then applied to set up the requisite number of nodal equations. Of the two methods, nodal analysis is usually easier to apply, and often results in fewer equations. Mesh analysis is generally unsuitable for circuits containing ideal current sources whereas nodal analysis may be used for circuits containing both voltage and current sources.

The labour of circuit analysis can sometimes be reduced by employing linear circuit theorems: superposition, reciprocity and Thévenin. The last mentioned, often used in conjunction with the potential divider circuit, is particularly useful for reducing parts of a circuit to a simpler form. Other methods of circuit reduction include Thévenin–Norton and star–delta transformations. The recognition of standard circuit building blocks – potential and current dividers, bridge circuits etc. – forms an important part of the art of circuit analysis.

For some circuits, the ladder circuit for example, special step-by-step methods are available which obviate the necessity of solving a large number of simultaneous equations.

2.17 Problems

1. A d.c. power supply has an output voltage of 5 V at its terminals on open circuit. A 2 Ω resistor connected across its terminals causes the output to fall by 0.1 V. Derive a linear circuit model for the supply.

2. A certain d.c. power supply has output potential differences of 600 V and 650 V when the output current is 0.4 A and 0.2 A respectively. What simple arrangement (a) of an ideal current generator in parallel with a resistance and (b) of an ideal current generator in parallel with a conductance will give the same relation between output p.d. and current?

3. For the voltage divider network shown in fig. 2.34, determine the output voltage V_o. If the output terminals are connected to a circuit having an input resistance of 10 kΩ, what then is the output voltage?

4. A device draws a constant current through a divider network as shown in fig. 2.35. A multi-range voltmeter V_m which draws a current of 1 mA at full-scale deflection, is used to measure the voltage supplying the device. When set to its 300 V range, it reads 90 V. What is the device voltage with the voltmeter removed?

Fig. 2.34. Circuit for problem 3.

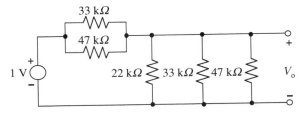

5. Show, by means of mesh analysis, that each of the mesh currents in the circuit of fig. 2.36 results from the superposition of two independent components, one proportional to V_1 the other proportional to V_2. If V_1 equals 1 V and V_2 is short-circuited, what will be the currents in the 3 Ω and 4 Ω resistances?

Apply this last result together with the reciprocity theorem to write down the new current in the 4 Ω resistance if V_1 equals 1 V and V_2 is made equal to 3 V.

Deduce, from the above results, the Thévenin equivalent circuit of the network across AB, as seen by the voltage source V_1, when V_2 equals 2 V. (Hint: to find the Thévenin e.m.f. consider the voltage V_1 required to reduce I_1 to zero; to find the equivalent resistance consider the current I_1 when AB is short circuit.)

6. In the circuit shown in fig. 2.37, additional generators are to be inserted into branches AB and BC so that the currents then flowing in the existing generators are each increased by 1 A from their original values. By means of

Fig. 2.35. Circuit for problem 4.

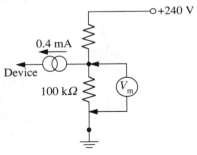

Fig. 2.36. Circuit for problem 5.

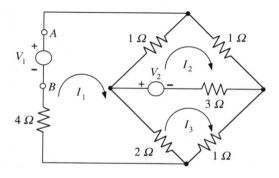

the superposition and reciprocity theorems or otherwise, find the e.m.f.s of the additional generators.
(Cambridge University: First year)

7. In the circuit of fig. 2.38 each resistance is 0.5 Ω. How many current unknowns are involved in a mesh analysis? How many voltage unknowns are involved in a nodal analysis? Find the potential of the node P with respect to the node O.

8. Find the current flowing in the 4 Ω resistor in the circuit of fig. 2.39 giving reasons for the choice of method for conducting this calculation. Describe and compare at least two other methods which might have been used instead.
(Newcastle University: First year)

9. Determine the current I shown in the network represented by fig. 2.40. Also calculate the voltage of the point C with respect to ground.
(Cambridge University: Second year)

10. The resistance of each arm and of the detector of a Wheatstone Bridge is 1 kΩ. The bridge is driven by a 10 V battery of negligible internal

Fig. 2.37. Circuit for problem 6.

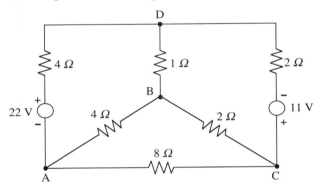

Fig. 2.38. Circuit for problem 7.

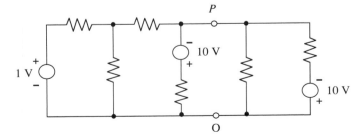

resistance. Use Thévenin's theorem to find an expression for the detector current, accurate to within 1%, if the resistance of one arm of the bridge is increased by r ohms ($r \leq 5$).

11. Figure 2.41 shows a circuit which may be used for temperature measurement. R_T is a thermistor whose resistance is 20 kΩ at 0 °C and 2 kΩ at 100 °C. M is a sensitive ammeter whose internal resistance is 5 kΩ and gives full-scale deflection at a current of 25 μA.

Calculate the ohmic values to which R_1 and R_2 must be set so that M gives zero deflection at 0 °C and full-scale deflection at 100 °C.
(Newcastle University: First year)

12. In the circuit of fig. 2.42 each resistor has the ohmic value stated. Show that when viewed from the output terminals AB, the circuit is equivalent to a generator having an e.m.f. of 20 V and an internal resistance of 3 Ω.

What are the two possible ohmic values of resistor R which when connected across the output terminals, will absorb a power of 32 W?
(Newcastle University: Second year)

13. Twelve identical pieces of wire each of resistance 1 ohm are connected together to form a skeleton cube. Find the resistance between opposite ends of a diagonal of the cube.

14. Two 240 V generators of low internal resistance are connected together

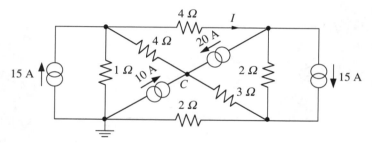

Fig. 2.39. Circuit for problem 8.

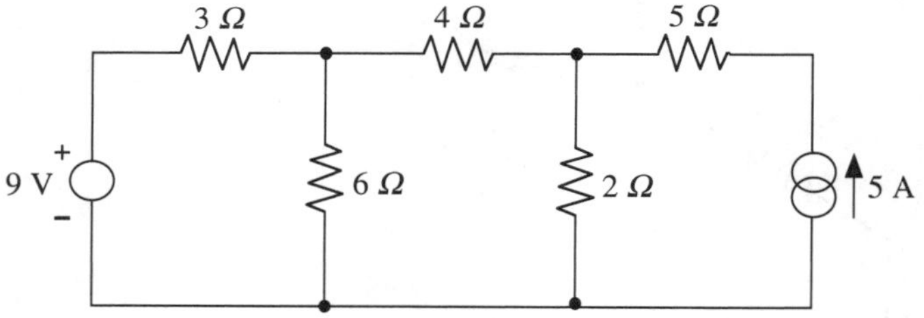

Fig. 2.40. Circuit for problem 9.

in opposition by a cable 15 m long, and two loads are connected; one of 24 Ω at a distance of 6 m from one end of the cable, and one of 16 Ω at a distance of 4 m from the other end. If the voltage across the 24 Ω load is found to be 2% below the generator voltage, what is the resistance per unit length of each conductor of the cable?

If the voltage of one generator changes by 2%, what will be the current in the part of the cable between the loads?
(Cambridge University: Second year)

15. A ring main of total length 1 km has five load points distributed as shown in the Table. Find the point at which the voltage is minimum. If the voltage drop is nowhere to exceed 1 V plus 2% of the nominal supply voltage of 240 V, calculate the minimum cross-sectional area of copper required in each cable (resistivity of copper $= 1.6 \times 10^{-8}$ Ωm).

Distance (m)	200	300	400	600	800
Load (A)	30	20	40	20	40

Fig. 2.41. Circuit for problem 11.

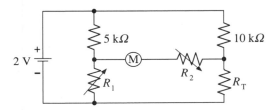

Fig. 2.42. Circuit for problem 12.

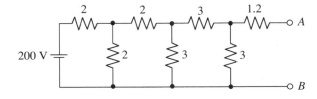

3

Alternating current circuits

3.1 Introduction

The class of circuits described as 'alternating current circuits' (abbreviated to a.c. circuits) comprises networks of linear lumped elements that may include capacitance and inductance as well as resistance. It has become a common and convenient practice to use the abbreviation, 'a.c.' as a qualifying adjective. Thus, we speak of an 'a.c. voltage', an 'a.c. current', and so on. In such circuits the sources of excitation produce time-varying voltages and currents described by sinusoidal functions of the form:

$$v = V_m \sin \omega t \quad \text{or} \quad i = I_m \sin \omega t \qquad (3.1)$$

We may regard the above expressions as functions of time t or functions of angle ωt, the latter often being the more convenient. Waveforms corresponding to (3.1) are shown in fig. 3.1, as functions of both time and angle, and the various relevant parameters are defined.

An a.c. circuit is, by definition, one in which steady-state conditions obtain; that is, any transient conditions arising in the circuit at the time of switching will have died away leaving the circuit in an equilibrium state in which the amplitudes of all currents and voltages are constant. The time origin in the above equations is therefore of no consequence so that the alternating voltages and currents in an a.c. circuit can be described equally by the cosine functions:

$$V = V_m \cos \omega t \quad \text{or} \quad i = I_m \cos \omega t \qquad (3.2)$$

For this reason the term *cisoid* is sometimes used to describe in a general way the waveforms encountered in a.c. circuits.

Although the choice of time origin is arbitrary the relative time (or angular) displacement between waveforms, which we call *phase*, is of vital

Introduction

importance in describing the electrical behaviour of an a.c. circuit. In fig. 3.2 we illustrate voltage and current waveforms displaced by phase angle ϕ. Either of the two waveforms may be regarded as the reference with respect to which the phase of the other is measured. If we select the voltage waveform as reference, then, in terms of the sine function the two waveforms are described by

$$v = V_m \sin \omega t \quad \text{(phase reference)}$$

and

$$i = I_m \sin(\omega t - \phi)$$

In this case we say that the current waveform *lags* the voltage waveform by angle ϕ since, as may be seen from fig. 3.2, the current waveform passes through zero in a positive going direction (point B) at an instant of time

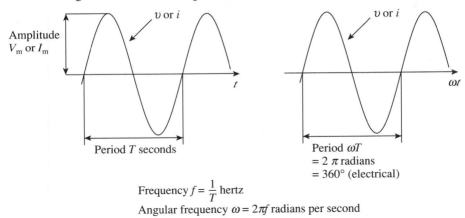

Fig. 3.1. Sinusoidal voltage and current waveforms.

Frequency $f = \dfrac{1}{T}$ hertz

Angular frequency $\omega = 2\pi f$ radians per second

Fig. 3.2. Phase displacement between two sinusoidal waveforms.

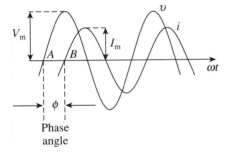

later than the corresponding point (point A) of the voltage waveform. If, on the other hand, we choose to take the current waveform as the reference, then the two waveforms are described by the functions:

$$i = I_m \sin \omega t \text{ (phase reference)}$$

and

$$v = V_m \sin(\omega t + \phi)$$

and we say that the voltage waveform *leads* the current waveform by angle ϕ. In all of the above we could have used the cosine function instead of the sine function; providing either one or the other is used consistently the method of describing phase is the same.

The following points concerning the meaning and use of phase should be noted: (1) Its definition in any circuit depends upon the choice of the phase reference waveform. (2) It is independent of waveform amplitude. (3) It has meaning only when referred to waveforms of the same frequency. (4) A negative sign attached to the phase angle signifies that the associated waveform is lagging, conversely, a positive sign signifies a leading waveform. (5) A waveform leading by a phase angle β, say, (measured in degrees) may also be described as lagging by an angle $\gamma = 360 - \beta$. It is conventional practice to choose whichever of the two possible angles is numerically less than 180°.

An important part of a.c. circuit analysis is concerned with the calculation of power. In d.c. circuits the calculation of power in resistance is effected using the expressions $I^2 R$ or V^2/R (equations (1.20) and (1.21)); in a.c. circuits these same expressions can be conveniently used by working in terms of *effective* values of current and voltage rather than with their amplitudes. The relationship between the effective value of an alternating current and its amplitude may be derived by considering the amplitude of the alternating current required to produce the same mean energy dissipation in a resistance R as that produced by a d.c. current of I amperes.

The instantaneous power in a resistance R carrying a current i is, from (1.20), $i^2 R$; therefore, the energy dissipation over one complete period T is

$$\int_0^T i^2 R \, dt$$

If I is the effective value of the current waveform, then over a period T the energy dissipation must be $I^2 RT$ hence,

$$I^2 RT = \int_0^T i^2 R \, dt$$

A.C. voltage – current relationships

or

$$\text{Effective value } I = \left[\frac{1}{T} \int_0^T i^2 \, dt \right]^{\frac{1}{2}} \tag{3.3}$$

This expression allows the effective value of any periodic current waveform to be evaluated. Because of the mathematical form of (3.3) the effective value is also known as the *root mean square* (r.m.s.) value or magnitude.

For the particular case of a sinusoidal waveform we have

$$I = \left[\frac{1}{T} \int_0^T (I_m \sin\omega t)^2 \, dt \right]^{\frac{1}{2}} = \frac{I_m}{\sqrt{2}} \tag{3.4}$$

In a similar way it can be shown that the effective value of a sinusoidal alternating voltage is $V_m/\sqrt{2}$.

Finally, we mention a parameter which is useful for purposes of comparison between various types of periodic waveform; this is the *form factor* defined as the ratio of the r.m.s. value to the half-cycle average value. (The average value of a sine-wave over a complete cycle is, of course, zero.)

It can be shown that the half-cycle average of a sine-wave is equal to $2V_m/\pi$, hence the form factor is $(V_m/\sqrt{2})/(2V_m/\pi) = 1.11$. This may be compared with a form factor of unity for a square-wave and 1.155 for a triangular wave. The form factor indicates the extent to which a wave form exhibits a 'peaked' characteristic. (See ref. 11, for a more complete discussion.)

3.2 A.C. voltage–current relationships for the linear circuit elements

We now consider the form of the voltage developed across each of the three circuit elements when a sinusoidal current passes through them (fig. 3.3). In the following analysis we choose to describe the current by the cosine function because it is mathematically slightly more convenient. Essentially the same results would be obtained using the sine function.

(a) *Resistance*. The instantaneous voltage across the resistance R is

$$v_R = Ri$$
$$= RI_m \cos\omega t$$

or

$$v_R = V_{Rm} \cos\omega t \tag{3.5}$$

where

$$V_{Rm} = RI_m \tag{3.6}$$

102 Alternating current circuits

Converting to r.m.s. magnitudes by dividing both sides of (3.6) by $\sqrt{2}$ we obtain

$$V_R = RI \tag{3.7}$$

Thus, an Ohm's law type of relationship exists between alternating current and voltage magnitudes for a resistive element. From (3.5) we see that voltage and current are in phase.

(b) *Inductance.* Using (1.40) the voltage across the inductance is given by

$$v_L = L\frac{di}{dt} = L\frac{d}{dt}(I_m \cos\omega t)$$

$$= -\omega L I_m \sin\omega t$$

$$= \omega L I_m \cos\left(\omega t + \frac{\pi}{2}\right)$$

or

$$v_L = V_{Lm} \cos\left(\omega t + \frac{\pi}{2}\right) \tag{3.8}$$

where $V_{Lm} = \omega L I_m$.

Fig. 3.3. A.C. voltage–current relationships, in terms of instantaneous values, for the basic circuit elements.

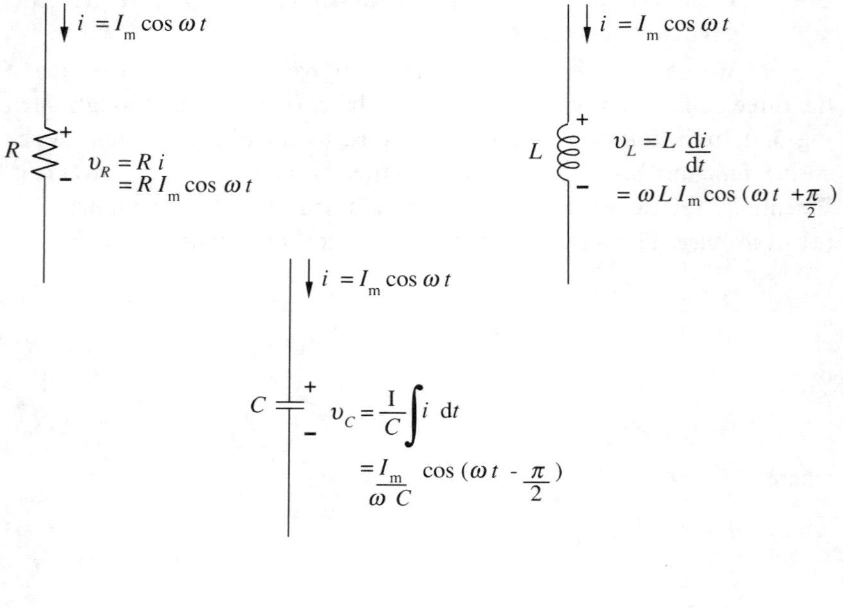

$i = I_m \cos \omega t$

R, $v_R = Ri = RI_m \cos \omega t$

$i = I_m \cos \omega t$

L, $v_L = L\frac{di}{dt} = \omega L I_m \cos(\omega t + \frac{\pi}{2})$

$i = I_m \cos \omega t$

C, $v_C = \frac{1}{C}\int i\, dt = \frac{I_m}{\omega C} \cos(\omega t - \frac{\pi}{2})$

A.C. voltage – current relationships

Converting to r.m.s. magnitudes we obtain

$$V_L = \omega L I \tag{3.9}$$

The quantity ωL is known as the *inductive reactance*, and since it is the ratio of a voltage to a current, it has dimensions of ohms. It is usually denoted by the symbol X_L hence (3.9) may be written

$$V_L = X_L I \tag{3.10}$$

Equation (3.8) shows that for an inductive element the voltage waveform *leads* the current waveform by phase angle $\pi/2$ radians.

(c) *Capacitance*. Using (1.31) the voltage across the capacitance is given by

$$v_C = \frac{1}{C}\int_0^t i\,dt + v_0 = \frac{1}{C}\int_0^t I_m \cos\omega t\,dt + v_0$$

Since we are dealing with circuits in which voltages are purely sinusoidal, there is no initial voltage across the capacitance so that we may put $v_0 = 0$. On integrating we obtain:

$$v_C = \frac{I_m}{\omega C}\sin\omega t$$

$$= \frac{I_m}{\omega C}\cos\left(\omega t - \frac{\pi}{2}\right)$$

or

$$v_C = V_{Cm}\cos\left(\omega t - \frac{\pi}{2}\right) \tag{3.11}$$

where $V_{Cm} = \dfrac{I_m}{\omega C}$

In terms of r.m.s. magnitudes we have

$$V_C = \frac{I}{\omega C} = X_C I \tag{3.12}$$

The quantity $X_C = 1/\omega C$ is called the *capacitive reactance*, measured in ohms, and we see from (3.11) that in this case the voltage *lags* the current by $\pi/2$ radians.

We conclude that, for each of the three elements, voltage is proportional to current in terms of either amplitudes or r.m.s. magnitudes. For the inductive and capacitive elements both voltage and current are sinusoidal but suffer a phase displacement of $\pi/2$ radians or 90 electrical degrees. We say that, for these elements, voltage and current are 'in quadrature'.

Alternating current circuits

It is evident from the above theory that in a.c. circuit analysis the application of Kirchhoff's laws requires the addition of voltages or currents that will differ in phase if the circuit contains two or more elements of different kinds. Because of this we cannot combine the magnitudes of the voltages given by (3.7), (3.10) and (3.12) using direct algebraic addition; however, we could proceed by adding the instantaneous values given by the trigonometric functions (3.5), (3.8) and (3.11). For instance, if the total voltage across a series combination of just two elements, say resistance and inductance, were required, we could proceed as follows:

Total voltage $v = v_R + v_L$

$$= V_{Rm}\cos\omega t + V_{Lm}\cos\left(\omega t + \frac{\pi}{2}\right)$$

$$= RI_m\cos\omega t + X_L I_m\cos\left(\omega t + \frac{\pi}{2}\right)$$

$$= RI_m\cos\omega t + X_L I_m\cos\omega t\cos\frac{\pi}{2} - X_L I_m\sin\omega t\sin\frac{\pi}{2}$$

$$= RI_m\cos\omega t - X_L I_m\sin\omega t$$

$$v = \sqrt{(R^2 + X_L^2)} I_m \cos(\omega t + \alpha)$$

where

$$\alpha = \tan^{-1}\frac{X_L}{R} \quad \left(\text{also written as } \alpha = \arctan\frac{X_L}{R}\right)$$

The result is a voltage of amplitude $\sqrt{(R^2 + X_L^2)}I_m$, and phase angle α with respect to the original current flowing through the two elements. The quantity $\sqrt{(R^2 + X_L^2)}$ has dimensions of ohms and is called the *impedance* of the series-connected elements. In principle the total voltage across any number of series-connected elements could be derived by repeated application of the above trigonometrical procedure taking voltages two at a time. We should then find that the general result was of the form:

$$\text{Instantaneous voltage} = [\text{Impedance}] \begin{bmatrix} \text{Amplitude} \\ \text{of} \\ \text{current} \end{bmatrix} \begin{bmatrix} \text{Phase displaced} \\ \text{sine or cosine} \\ \text{function} \end{bmatrix} \quad (3.13)$$

Both the impedance and the phase displacement are functions of the resistances and reactances in the circuit under consideration, and their determination by the above trigonometrical procedure for each particular circuit would be tedious in the extreme. Alternative approaches are therefore adopted based either (a) on a geometrical and graphical interpretation of the trigonometrical equations presented above or (b) on the use of the complex exponential and complex algebra. The latter is the most convenient and flexible approach, and is presented in the following sections.

3.3 Representation of a.c. voltage and current by the complex exponential: Phasors

From the discussion and results contained in the previous sections it will be clear that in a.c. circuit analysis it is only the magnitudes and relative phases of voltages and currents that are of interest. The use of the complex exponential to represent a.c. voltages and currents allows the analysis of circuits to be effected in terms of magnitude and phase only; it provides also a direct and simple means for depicting graphically the relationships among a.c. quantities. We assume in the following that the reader is familiar with the meaning of complex number and with the elements of complex algebra.

The basis of the method is provided by the Euler relation:

$$Ae^{j\theta} = A\cos\theta + jA\sin\theta \tag{3.14}$$

where $j = \sqrt{-1}$ and A and θ are respectively the modulus (or amplitude) and argument (or angle) of the complex exponential.

The relation (3.14) defines a complex number $\boldsymbol{A}$* whose real and imaginary parts are respectively $A\cos\theta$ and $A\sin\theta$, thus,

$$\boldsymbol{A} = a + jb \tag{3.15}$$

where

$$a = A\cos\theta, \quad b = A\sin\theta$$

Therefore, we may write

$$a^2 + b^2 = A^2\cos^2\theta + A^2\sin^2\theta$$

that is

$$A = \sqrt{(a^2 + b^2)}$$

and

$$\frac{b}{a} = \frac{\sin\theta}{\cos\theta}$$

---

* Complex quantities will be signified in this text by the use of bold italic type.

that is

$$\theta = \tan^{-1}\frac{b}{a}$$

We may depict the relationship (3.14) graphically by means of the Argand diagram (fig. 3.4) in which the complex exponential defines a point P in the plane with polar coordinates (A, θ). The right-hand side of (3.14) defines the same point in terms of the Cartesian coordinates $(A\cos\theta, A\sin\theta)$.

From a slightly different point of view we may regard the complex exponential $e^{j\theta}$ as an operator. With this interpretation multiplication of a real scalar quantity A by $e^{j\theta}$ simply causes A to rotate in the Argand diagram by an amount θ without change of amplitude. An important special case occurs when $\theta = \frac{\pi}{2}$ radians (or 90°); then (3.14) becomes:

$$Ae^{j\pi/2} = A\cos\frac{\pi}{2} + jA\sin\frac{\pi}{2}$$

that is,

$$Ae^{j\pi/2} = A(0) + jA(1)$$

or

$$e^{j\pi/2} = j$$

Similarly,

$$e^{j(-\frac{\pi}{2})} = -j$$

Fig. 3.4. Representation of the complex exponential $Ae^{j\theta}$ on the Argand diagram.

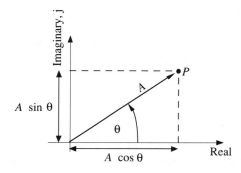

Representation by complex exponential: phasors

Thus multiplication of a real quantity by $+j$ causes a rotation of $90°$ in a positive (counter-clockwise) sense, while multiplication by $-j$ causes a rotation of $90°$ in a negative sense. It also follows that multiplication by j^2 is equivalent to a rotation of $180°$. This interpretation of the complex exponential must be borne in mind throughout the following theory.

A saving of labour, particularly in numerical work, is achieved by writing the complex exponential as

$$A\mathrm{e}^{\mathrm{j}\theta} \equiv A\underline{/\theta} \tag{3.16}$$

Now consider the argument θ in (3.14) to be a function of time such that $\theta = (\omega t + \phi)$, and let A be identically equal to the amplitude V_m of an alternating voltage wave form; (3.14) then becomes

$$V_\mathrm{m} \mathrm{e}^{\mathrm{j}(\omega t + \phi)} = V_\mathrm{m}\cos(\omega t + \phi) + \mathrm{j} V_\mathrm{m}\sin(\omega t + \phi) \tag{3.17}$$

We see that the complex exponential on the left-hand side of this expression can be used to represent mathematically either the co-sinusoidal or sinusoidal forms of the alternating voltage; the particular form being expressed by specifying either the real part (Re) or imaginary part (Im) as required, viz.

$$v = V_\mathrm{m}\cos(\omega t + \phi) = \mathrm{Re}\, V_\mathrm{m}\mathrm{e}^{\mathrm{j}(\omega t + \phi)} \tag{3.18}$$

or

$$v = V_\mathrm{m}\sin(\omega t + \phi) = \mathrm{Im}\, V_\mathrm{m}\mathrm{e}^{\mathrm{j}(\omega t + \phi)} \tag{3.19}$$

The interpretation of (3.17) on the Argand diagram is shown in fig. 3.5(a). The line OP, representing the amplitude V_m of the alternating voltage, rotates with angular speed ω and we refer to this line as a *rotating phasor*. The projection of the point P on to the real and imaginary axes defines time-varying coordinates proportional respectively to $V_\mathrm{m}\cos(\omega t + \phi)$ and $V_\mathrm{m}\sin(\omega t + \phi)$.

If the complex exponential is not specified by (3.18) or (3.19), then it is understood that either form is applicable and we can describe in general terms the instantaneous value of any alternating voltage by

$$v = V_\mathrm{m}\mathrm{e}^{\mathrm{j}(\omega t + \phi)} \tag{3.20}$$

This expression may be rewritten as

$$v = V_\mathrm{m}\mathrm{e}^{\mathrm{j}\phi}\mathrm{e}^{\mathrm{j}\omega t} = \mathbf{V_\mathrm{m}}\mathrm{e}^{\mathrm{j}\omega t} \tag{3.21}$$

where

$$\mathbf{V_\mathrm{m}} = V_\mathrm{m}\mathrm{e}^{\mathrm{j}\phi} \tag{3.22}$$

108 Alternating current circuits

The expression (3.22) defines a *complex voltage* that is *independent* of time and which contains only the amplitude and phase information concerning the alternating voltage. We interpret the complex voltage on the Argand diagram as shown in fig. 3.5(b). Since the complex voltage is time-invariant the line *OP* does not now rotate and it is consequently termed a *stationary phasor*. The complex voltage function defined by (3.22) is also referred to as a stationary phasor or, more simply, as a phasor.

In practical circuit analysis it is, for the reasons given in section (3.1), better to work in terms of r.m.s. values in which case the complex voltage is written

$$V = V e^{j\phi} \equiv V \underline{/\phi} \qquad (3.23)$$

where $V = V_m/\sqrt{2}$, and the notation of (3.16) has been used. The Argand diagram is modified accordingly as shown in fig. 3.5(c).

Fig. 3.5. Representation of an alternating voltage in complex exponential form by means of the Argand diagram. (a) Rotating phasors. (b) and (c) Stationary phasors.

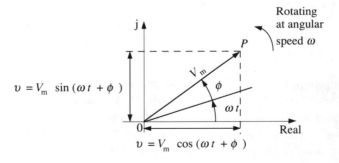

(a) Instantaneous values

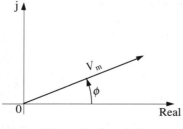

(b) Amplitude and phase

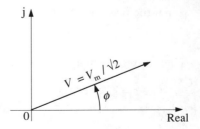

(c) Magnitude and phase

Voltage – current relationships: impedance

The above treatment is, of course, applicable to the representation of currents and we may write

$$I = Ie^{j\beta} \equiv I\underline{/\beta} \tag{3.24}$$

3.4 Voltage–current relationships for the general network branch: Impedance

The results of section (3.3) will now be used to derive the voltage–current relationship for the series-connected elements shown in fig. 3.6. This arrangement is called the *general network branch* because it is completely representative of any series-connected combination of lumped passive elements. Once the voltage–current relationship is determined for this circuit then it becomes possible to solve, at least in principle, any a.c. linear lumped network.

To make the following treatment completely general we assume that the current i passing through the branch has amplitude I_m and possesses a phase angle ϕ measured with respect to some other voltage or current elsewhere in the circuit of which the branch shown in fig. 3.6 forms part. We wish to determine the amplitude and phase angle of the total branch voltage v.

The branch voltage v is, by Kirchhoff's voltage law,

$$v = v_R + v_L + v_C$$

which may be written, using the instantaneous voltage–current relationships for the individual circuit elements derived in chapter 1 (equations (1.16), (1.31) and (1.40)):

$$v = Ri + L\frac{di}{dt} + \frac{1}{C}\int i\, dt \tag{3.25}$$

Now the instantaneous current may be represented in complex exponential form (see equation (3.18)) by

$$i = I_m\cos(\omega t + \phi) = \mathrm{Re}\, I_m e^{j(\omega t + \phi)}$$
$$= \mathrm{Re}\, \boldsymbol{I}_m e^{j\omega t}$$

therefore (3.25) becomes (taking the real part as understood)

Fig. 3.6. The general network branch.

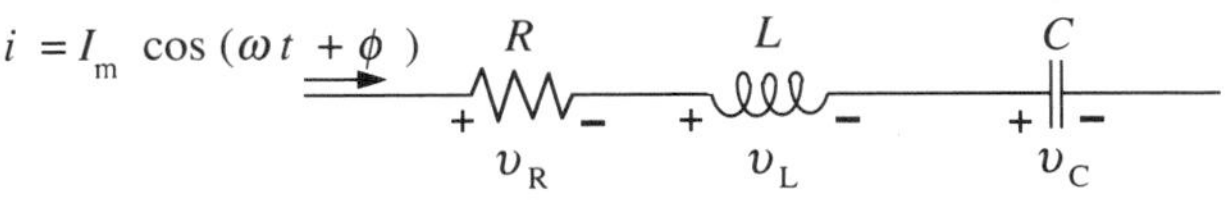

$$v = RI_m e^{j\omega t} + LI_m \frac{d}{dt} e^{j\omega t} + \frac{I_m}{C} \int e^{j\omega t}$$

$$= RI_m e^{j\omega t} + j\omega L I_m e^{j\omega t} + \frac{I_m}{j\omega C} e^{j\omega t} + (\text{const.})$$

The constant of integration is zero since the voltages and currents in an a.c. circuit are purely sinusoidal. (A finite value for this constant would imply that a direct voltage existed across the terminals of the capacitor.) We may therefore write

$$v = \left(R + j\omega L + \frac{1}{j\omega C}\right) I_m e^{j\omega t} \tag{3.26}$$

The quantity in brackets in the above equation is called the *complex impedance* and is denoted by the symbol **Z**, thus

$$\text{Complex impedance } \mathbf{Z} = R + j\omega L + \frac{1}{j\omega C} \tag{3.27}$$

$$= R + j\left(\omega L - \frac{1}{\omega C}\right)$$

or

$$\mathbf{Z} = R + jX \tag{3.28}$$

where

$$X = \left(\omega L - \frac{1}{\omega C}\right) \tag{3.29}$$

The quantity X, called the *reactance* of the branch, is the difference between the inductive and capacitive reactances.

Since **Z** is a complex number it may be converted from the Cartesian form (3.28) to polar form:

$$\mathbf{Z} = R + jX = Z e^{j\theta} = Z\underline{/\theta} \tag{3.30}$$

where

$$Z = \sqrt{(R^2 + X^2)} \tag{3.31}$$

and

$$\theta = \tan^{-1} \frac{X}{R} \tag{3.32}$$

The modulus Z of the complex impedance, measured in units of ohms, is

Voltage – current relationships: impedance   111

often referred to simply as the impedance. (Z, the complex impedance, is also often called the impedance and we have to understand from the context of the theory or argument in question which is meant.) The argument θ in (3.30) is called the *angle* of the complex impedance.

Substituting (3.27) into (3.26), the expression for the instantaneous branch voltage becomes

$$v = ZI_m e^{j\omega t} \tag{3.33}$$

Now, using (3.30) and recalling that $I_m = I_m e^{j\phi}$, this may be written

$$v = Z e^{j\theta} I_m e^{j\phi} e^{j\omega t}$$

or

$$v = ZI_m e^{j(\omega t + \phi + \theta)} \tag{3.34}$$

In all of the above expressions for v, the real part has been understood, therefore, from (3.18) the branch voltage is

$$v = ZI_m \cos(\omega t + \phi + \theta) \tag{3.35}$$

We see from this expression that the amplitude V_m of the branch voltage is given by

$$V_m = ZI_m \tag{3.36}$$

and the phase angle with respect to the current is θ, the angle of the complex impedance.

We have thus established, with the aid of complex exponential theory, the required voltage–current relationship in terms of instantaneous values. The reader should now compare (3.35) with (3.13). (The same result could have been obtained, rather more laboriously, by using the trigonometrical methods and results of section 3.2.) Of rather greater practical significance, however, is the voltage–current relationship for the general branch in terms of complex voltages and currents; this is derived as follows. In terms of the complex exponential the instantaneous branch voltage may be written

$$v = V_m e^{j(\omega t + \phi + \theta)} = \mathbf{V_m} e^{j\omega t} \tag{3.37}$$

where

$\mathbf{V_m} = V_m e^{j(\phi + \theta)}$ is the complex voltage.

Combining (3.37) and (3.33) gives

$$\mathbf{V_m} e^{j\omega t} = Z\mathbf{I_m} e^{j\omega t}$$

or

$$V_m = ZI_m \qquad (3.38)$$

Note that eliminating $e^{j\omega t}$ from both sides of the above equation has the effect of converting from a rotating phasor system to a stationary phasor system.

In terms of r.m.s. magnitudes (3.38) becomes

$$V = ZI \qquad (3.39)$$

In words:

Complex voltage = complex impedance × complex current

It will be recognized that (3.39) is of a similar mathematical form to the Ohm's law encountered in d.c. circuit theory, consequently it is often referred to as Ohm's law for a.c.

This equation allows us to write down the voltage drops across the elements of the general branch in terms of the complex impedance. From (3.27) the complex impedance of the general branch is $Z = R + j\omega L + 1/j\omega C$ therefore

$$V = \left(R + j\omega L + \frac{1}{j\omega C}\right)I$$

or

$$V = RI + j\omega LI + \frac{1}{j\omega C}I \qquad (3.40)$$

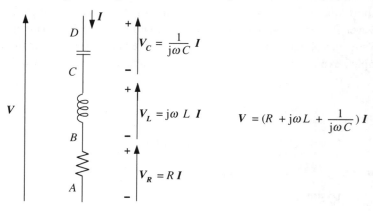

Fig. 3.7. Voltage drops across the elements of the general branch and the complex voltage–current relationship.

Phasors and impedance diagrams

The terms in this equation represent the voltage drops across each of the elements in the general branch, as shown in fig. 3.7. Practical a.c. circuit analysis is carried out in terms of complex voltage drops and currents using relationships of the form (3.40), not in terms of instantaneous quantities. It will be noted that arrows, as well as ($+$) and ($-$) signs, have been used in fig. 3.7 to indicate the polarity, or reference direction, of the a.c. voltage drops across the circuit elements. This practice has been widely adopted in British textbooks, although not in American textbooks.

3.5 Phasor and impedance diagrams

The representation of complex voltages and currents on the Argand diagram provides a valuable pictorial aid to the interpretation of the algebraic procedures used in a.c. circuit analysis, and it greatly facilitates the understanding of circuit operation; indeed, it is customary to illustrate the operation of a.c. circuits by this means, often without explicit reference to the complex exponential notation. Such diagrams are referred to as *phasor diagrams* since they illustrate the relationships between the various sinusoidal voltages and currents in a circuit interpreted in their phasor form. (The concept of phasors and phasor diagrams can also be developed on the basis of a geometrical interpretation of sinusoidal wave forms. See for example reference 1.)

The phasor diagram for the general network branch provides a basis for the construction of all other such diagrams. To derive this we use the relationship (3.40). Recalling that $1/j = -j$, this may be written as

$$V = R\mathbf{I} + j\omega L\mathbf{I} - j\frac{1}{\omega C}\mathbf{I} \tag{3.41}$$

Now let us in the first instance suppose that we have chosen the current as the reference waveform in the circuit, that is, its phase angle is chosen to be zero. Then we can write $\mathbf{I} = Ie^{j0} = I$, and (3.41) becomes

$$V = RI + j\omega LI - j\frac{1}{\omega C}I \tag{3.42}$$

This equation is represented diagrammatically in fig. 3.8. Interpreted strictly from the point of view of the Argand diagram we have the situation shown in fig. 3.8(*a*). The voltage drop across the resistance is $V_R = RI$, and this is represented by a phasor along the real or reference axis which is, of course, the direction of the current phasor. The voltage drop across the inductance is of magnitude $V_L = \omega LI$, and this is represented by a phasor along the positive imaginary axis. Similarly, the voltage drop across the capacitance is $V_C = I/\omega C$; represented by a phasor along the negative imaginary axis.

An alternative interpretation of (3.42) is shown in fig. 3.8(b). In this diagram the phasors have been drawn head-to-tail (in a manner analogous to the vector polygon used in force diagrams in the fields of mechanics and structures), and the polygon is closed by a *resultant* voltage equal to ZI, the magnitude of the total branch voltage. The angle α, which the resultant voltage makes with the reference direction, is given by (3.32), namely, $\alpha = \tan^{-1}(\omega L - 1/\omega C)/R$. Notice that there is a topological similarity between the circuit diagram of fig. 3.7 and the phasor diagram of fig. 3.8(b): the order of the phasors in the phasor diagram corresponds to the order of the voltage drops in the circuit diagram. Corresponding points have been indicated in the two diagrams following the order $ABCD$.

In developing fig. 3.8 we chose to take the current as reference; if more generally we take the branch current to have some phase angle ϕ measured with respect to another voltage or current variable elsewhere in the circuit, then (3.40) becomes

$$V = RI e^{j\phi} + j\omega L I e^{j\phi} - j\frac{1}{\omega C} I e^{j\phi}$$

$$= \left(RI + j\omega L I - j\frac{I}{\omega C}\right) e^{j\phi} \qquad (3.43)$$

We see that (3.43) is simply (3.42) multiplied by $e^{j\phi}$, which means that in the phasor diagram all phasors are rotated bodily through an angle ϕ as shown in fig. 3.9.

It is important to appreciate that the phasor diagram shows phasor voltages and currents in a fixed relationship to one another, consequently, although it is customary to draw the Argand diagram with real and

Fig. 3.8. Phasor diagrams for the general branch.

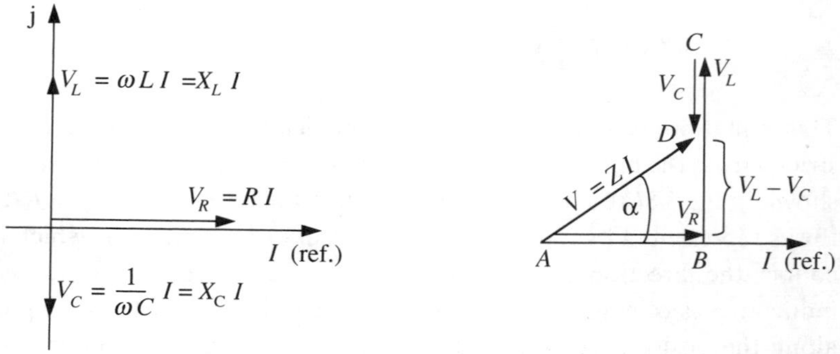

imaginary axes respectively horizontal and vertical, it is not mandatory to draw the phasor diagram with the reference phasor horizontal. Electrical power engineers, for instance, often take the system voltage as reference and traditionally draw this vertically on their phasor diagrams. Phasor diagrams may therefore be constructed in any of the forms typified by figs. 3.8 or 3.9, and with considerable freedom as to choice of reference phasor and reference direction.

Finally, we mention one other diagram (related to the phasor diagram) that shows the relationships between resistance, reactance and impedance in a circuit or part of a circuit; this is the so-called *impedance diagram*. The impedance diagram for the general network branch is shown in fig. 3.10, and it is obtained by dividing each of the voltage phasors in fig. 3.8(*b*) by the magnitude of the current *I*. The impedance diagram may be oriented in any convenient direction to suit the problem in hand.

3.6 Linear circuit theorems and techniques in a.c. circuit analysis

In chapter 2 a number of analytical techniques and circuit theorems were developed based essentially on the linear properties of the direct current circuits considered. This property of linearity depended upon the constancy of the ratio of voltage to current for each resistive element of the circuit, that is, upon Ohm's law. The same approach may be used in the case of a.c. circuits since, as we have seen in the immediately preceding

Fig. 3.9. Illustrating the effect on the phasor diagrams in fig. 3.8 of shifting the phase of the current by angle ϕ.

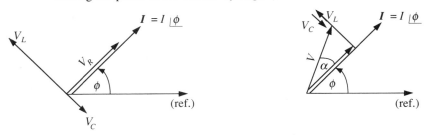

Fig. 3.10. The impedance diagram.

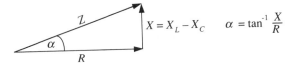

$$X = X_L - X_C \quad \alpha = \tan^{-1}\frac{X}{R}$$

Table 3.1

D.C. Formulation	Equation No.	A.C. Formulation
Ohm's law $V = RI$	(1.16)	$V = ZI$
$V = \dfrac{1}{G} I$	(1.17)	$V = \dfrac{1}{Y} I$
Elements in series $R = R_1 + R_2 + \ldots + R_n$	(1.23)	$Z = Z_1 + Z_2 + \ldots + Z_n$
Elements in parallel $\dfrac{1}{R} = \dfrac{1}{R_1} + \dfrac{1}{R_2} + \ldots + \dfrac{1}{R_n}$	(1.27)	$\dfrac{1}{Z} = \dfrac{1}{Z_1} + \dfrac{1}{Z_2} + \ldots + \dfrac{1}{Z_n}$
$G = G_1 + G_2 + \ldots + G_n$	(1.28)	$Y = Y_1 + Y_2 + \ldots + \dfrac{1}{Y_n}$
$R = \dfrac{R_1 R_2}{R_1 + R_2}$	(1.26)	$Z = \dfrac{Z_1 Z_2}{Z_1 + Z_2}$
Voltage divider $V_2 = \dfrac{R_2}{R_1 + R_2} V_1$	(2.6)	$V_2 = \dfrac{Z_2}{Z_1 + Z_2} V_1$
$V_2 = \dfrac{G_1}{G_2 + G_1} V_1$	(2.7)	$V_2 = \dfrac{Y_1}{Y_2 + Y_1} V_1$

sections, the ratio of complex voltage to complex current is constant for linear resistive, inductive and capacitive elements. Commencing, therefore, with the statement of Ohm's law in its a.c. form, (3.39), all the theorems and techniques developed for d.c. circuits can equally well be developed for a.c. circuits; the only difference being the replacement of the symbol R by the symbol Z in the linear equations. We may, therefore, immediately adopt the theory of chapter 2 in its entirety by the simple expedient of working in terms of complex voltages and currents and writing Z for R. Some analogous expressions and theorems for d.c. and a.c. circuits are presented in table 3.1. In this table the symbol Y, called the *admittance*, denotes the reciprocal of the impedance Z (see section 3.8).

As an example of the procedure we consider the Thévenin equivalent circuit for an a.c. network. This comprises an ideal a.c. voltage source in series with a complex impedance as shown in fig. 3.11(*a*). The Thévenin–Norton transformation is carried out in a way exactly analogous to that shown in fig. 2.16; with the result as shown in fig. 3.11(*b*).

Theorems and techniques in a.c. circuit analysis

The graphical symbol used in this book to indicate a complex impedance is also shown in fig. 3.11. It will be observed that the ideal a.c. voltage and current sources are distinguished by having a sine-wave symbol enclosed within the circle; this is a common but not universal practice.

The application of the phasor method will now be illustrated with reference to the circuit shown in fig. 3.12, which is analogous to the single-mesh d.c. circuit of fig. 2.5. Analysis of the circuit proceeds, as in the d.c. case, with the assignment of a current I, and this is followed by the application of Kirchhoff's voltage law to set up the circuit equation. Exactly the same conventions are applied to determine the signs of the terms in the circuit equation, and the working rules discussed in section 2.3 may be used with advantage. Although not strictly necessary for the purposes of analysis, the voltage drop across the resistance is also specified in fig. 3.12 so that the relationship between all three voltages in the mesh may be clearly seen on the phasor diagram (fig. 3.13).

Applying KVL to this circuit:

$$IR = V_1 - V_2$$

Fig. 3.11. The Thévenin–Norton transformation for an a.c. network.

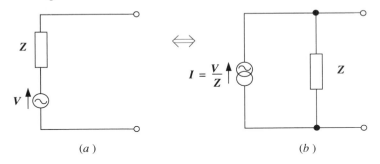

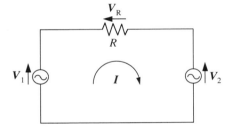

Fig. 3.12. Single-mesh a.c. circuit.

118 *Alternating current circuits*

or

$$I = \frac{V_1 - V_2}{R} = \frac{V_R}{R} \tag{3.44}$$

Let us now suppose that the two sources have equal magnitude V and that the phase angle of one can be varied with respect to the other, that is, $V_1 = V\angle 0$ (phase reference), and $V_2 = V\angle\phi$, where ϕ is variable. With both sources equal in magnitude and with $\phi = 0$ no current will flow; the phasor diagram for this situation is shown in fig. 3.13(a). If ϕ is now increased, say to 60°, (3.44) becomes

$$I = \frac{V\angle 0 - V\angle 60}{R}$$

$$= \frac{V}{R}[(\cos 0 + j\sin 0) - (\cos 60 + j\sin 60)]$$

$$= \frac{V}{R}(0.5 - j0.866)$$

$$= \frac{V}{R}\angle -60$$

This result is interpreted on the phasor diagram as shown in fig. 3.13(b). The directions of the phasors should be carefully observed; in particular, it

Fig. 3.13. Phasor diagrams for the circuit of fig. 3.12. Illustrating the effect of changing the phase of V_2 while keeping the magnitudes of V_1 and V_2 constant.

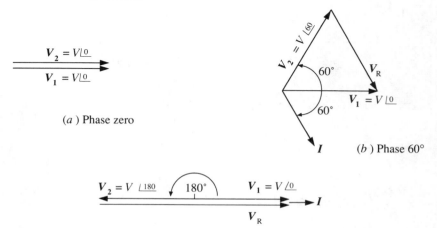

Worked example

should be noted that the direction of V_R is from the tip of V_2 towards the tip of V_1. This accords with the usual convention for vector addition, representing the relations $V_R = V_1 - V_2$ or $V_R + V_2 = V_1$. The phasor V_R lies parallel to the current phasor; a result which is to be expected since voltage and current must be in phase for a purely resistive element.

If now the phase of V_2 is further advanced until $\phi = 180°$, the phasor diagram becomes as shown in fig. 3.13(c). The current has a magnitude of $2V/R$ and a phase angle zero. We see that the effect of changing the phase of V_2 by 180° is the same as reversing the reference direction of V_2 in the circuit diagram, that is, changing the terminal polarity of the source.

3.7 Worked example

The circuit of fig. 3.14(a) is to be operated at mains power frequency (50 Hz).
(a) Find the complex impedance between the terminals AB and between the nodes D and B. Sketch an impedance diagram for the circuit.
(b) The circuit is connected to a mains power supply of 240 V; find the currents in each branch of the circuit and the voltage of node D with respect to B.
(c) Check the value for the node voltage obtained in (b) by using the method of nodal analysis.
(d) Sketch a phasor diagram for the complete circuit.

Solution
(a) First the reactances of the two inductances and the capacitance are found:

$$\omega L_1 = 2\pi \times 50 \times 15.9 \times 10^{-3} = 5$$
$$\omega L_2 = 2\pi \times 50 \times 66.8 \times 10^{-3} = 21$$
$$1/\omega C = 1/(2\pi \times 50 \times 398 \times 10^{-6}) = 8$$

These values are entered on the diagram (in brackets) as shown. Units of ohms are understood throughout the problem.

Let the complex impedances of the left- and right-hand branches between D and B be Z_1 and Z_2 respectively, then

$$Z_1 = 6 + j21; \quad Z_2 = 30 - j8$$

Expressed in polar form, using the notation of (3.16), these impedances become

$$Z_1 = 21.84 \underline{/74.1}; \quad Z_2 = 31.04 \underline{/-14.9}$$

120 *Alternating current circuits*

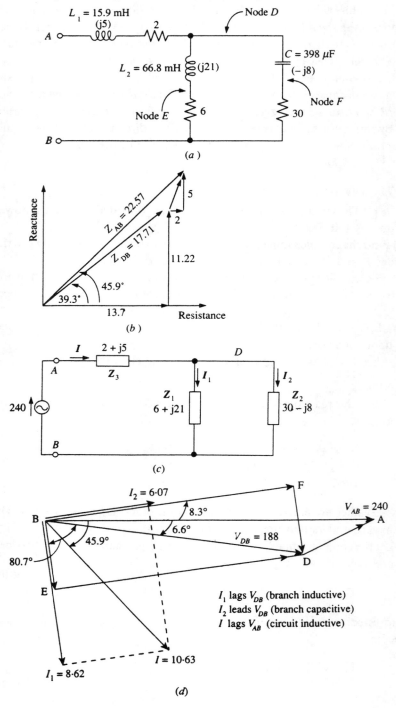

Fig. 3.14. Diagrams for the worked example of section 3.7.

Worked example

(Note: in numerical work it is customary to express the argument of a complex quantity in degrees rather than radians.)

The resultant impedance between DB is

$$Z_{DB} = Z_1 // Z_2 = \frac{Z_1 Z_2}{Z_1 + Z_2}$$

To evaluate this impedance the reader will recall that addition of complex quantities is effected by expressing them in Cartesian form, while multiplication or division is effected by expressing quantities in polar form. Thus

$$Z_1 + Z_2 = 6 + j21 + 30 - j8 = 36 + j13 = 38.27 \underline{/19.9}$$

and

$$Z_{DB} = \frac{21.84 \underline{/74.1} \times 31.04 \underline{/-14.9}}{38.27 \underline{/19.9}}$$

$$= \frac{21.84 \times 31.04 \underline{/74.1 - 14.9 - 19.9}}{38.27}$$

$$Z_{DB} = 17.71 \underline{/39.3} = 13.7 + j11.2$$

(Note that the procedure known as rationalization is generally to be avoided in numerical evaluation of expressions of the above form.)

The impedance between AB is then

$$Z_{AB} = 13.7 + j11.22 + 2 + j5 = 15.7 + j16.22$$

or

$$Z_{AB} = 22.57 \underline{/45.9}$$

The impedance diagram is illustrated in Fig. 3.14(b).

(b) Unless stated otherwise voltages are always expressed in terms of their r.m.s. magnitudes. Fig. 3.14(c) shows the 240 V supply, assumed to be an ideal source, connected to AB. The supply voltage is chosen as the reference phasor, that is, it takes zero phase angle. Since it is the branch currents which have to be evaluated, we assign currents to each branch as shown. (Assignment of mesh currents would entail an extra step in the calculation.)

The current drawn from the supply is found from the a.c. form of Ohm's law:

$$I = \frac{V}{Z_{AB}} = \frac{240 \underline{/0}}{22.57 \underline{/45.9}} = 10.63 \underline{/-45.9} \text{ A}$$

This result is interpreted as a current of magnitude 10.63 A, lagging the supply voltage by angle 45.9°.

122 *Alternating current circuits*

The voltage of node D with respect to B is given by

$$V_{DB} = IZ_{DB} = 10.63\underline{/-45.9} \times 17.71\underline{/39.3}$$

or

$$V_{DB} = 188.3\underline{/-6.6} \text{ V}$$

The currents I_1 and I_2 are found by dividing this voltage by the respective branch impedances:

$$I_1 = \frac{V_{DB}}{Z_1} = \frac{188.3\underline{/-6.6}}{21.84\underline{/74.1}} = 8.62\underline{/-80.7} \text{ A}$$

and

$$I_2 = \frac{V_{DB}}{Z_2} = \frac{188.3\underline{/-6.6}}{31.04\underline{/-14.9}} = 6.07\underline{/8.3} \text{ A}$$

Alternatively, for this part of the problem, the branch currents could be evaluated by means of the current divider principle discussed in section 2.2.

In this case

$$I_1 = \frac{Z_2 I}{Z_1 + Z_2} = \frac{31.04\underline{/-14.9} \times 10.63\underline{/-45.9}}{38.27\underline{/19.9}} = 8.62\underline{/-80.7} \text{ A}$$

and

$$I_2 = \frac{Z_1 I}{Z_1 + Z_2} = \frac{21.84\underline{/74.1} \times 10.63\underline{/-45.9}}{38.27\underline{/19.9}} = 6.07\underline{/-8.3} \text{ A}$$

(c) The nodal equation at D is

$$V_{DB} - \frac{240\underline{/0}}{Z_{AD}} + \frac{V_{DB}}{Z_1} + \frac{V_{DB}}{Z_2} = 0$$

But

$$Z_{AD} = 2 + j5 = 5.38\underline{/68.2}$$

therefore

$$\frac{V_{DB} - 240\underline{/0}}{5.38\underline{/68.2}} + \frac{V_{DB}}{21.84\underline{/74.1}} + \frac{V_{DB}}{31.04\underline{/-14.9}} = 0$$

Rearranging this equation gives

$$V_{DB}\left(\frac{1}{5.38\underline{/68.2}} + \frac{1}{21.84\underline{/74.1}} + \frac{1}{31.04\underline{/-14.9}}\right) = \frac{240\underline{/0}}{5.38\underline{/68.2}}$$

$V_{DB}(6.9 - j17.26 + 1.254 - j4.4 + 3.11 + j0.828) \times 10^{-2}$
$= 44.61\underline{/-68.2}$
$V_{DB}(11.27 - j20.83) \times 10^{-2} = 44.61\underline{/-68.2}$
$$V_{DB} = \frac{44.61\underline{/-68.2}}{0.237\underline{/-61.6}} = 188.4\underline{/-6.6} \text{ V}$$

(d) The phasor diagram for the complete circuit is shown in Fig. 3.14(d). Note that the current I is the phasor resultant of I_1 and I_2 (indicated by the dash lines). The voltage drop from A to D and the voltage drop from B to D are together equal to the source voltage. The voltage drops across the 6 Ω and 8 Ω resistances in each branch are in phase, respectively, with the currents I_1 and I_2 (lines BE and BF); the voltage drops (lines ED and FD) are in quadrature.

As an additional exercise the reader may care to evaluate the voltage V_{EF}. (Answer $188.43\underline{/-155.77}$)

The reader should note that in working through problems of this type in manuscript, complex quantities (here indicated by bold italic type) may, if so desired, be indicated by a bar above or below the quantity concerned; thus, V_{DB} may be written $\underline{V_{DB}}$ or $\overline{V_{DB}}$. Normally, however, this is unnecessary since, as will be appreciated from the above calculations, there is in numerical work little possibility of confusion between complex quantities and their moduli.

3.8 Admittance

It was found to be convenient in d.c. circuit analysis to define a quantity called conductance, the reciprocal of resistance. The formal equations of nodal analysis were expressed in terms of conductance in section 2.10. In a similar way we find it convenient in a.c. circuit analysis to define a quantity called the complex admittance, the reciprocal of complex impedance:

$$Y \equiv \frac{1}{Z} \qquad (3.45)$$

If α is the angle of the impedance and β the angle of the admittance this becomes, in polar form

$$Y\underline{/\beta} = \frac{1}{Z\underline{/\alpha}}$$

where $Y = \frac{1}{Z}$, and $\beta = -\alpha$.

Thus, there is a reciprocal relationship between the magnitudes of the admittance and the impedance.

Expressed in Cartesian form the complex admittance may be written

$$Y = G + jB \tag{3.46}$$

where G is the conductance and B the susceptance. Both G and B are expressed in units of siemens.

Some care is required in the interpretation of G and B, as defined in (3.46), since these quantities do not always bear a simple reciprocal relationship to resistance and reactance. This will be clear if we consider the complex admittance of the general network branch shown in Fig. 3.15(a). At an angular frequency ω we have

$$Y = \frac{1}{Z} = \frac{1}{R + j\omega L + \dfrac{1}{j\omega C}}$$

or

$$Y = \frac{1}{R + jX}$$

where

$$X = \left(\omega L - \frac{1}{\omega C}\right)$$

The process of rationalization (multiplying numerator and denominator by the complex conjugate of $R + jX$), allows us to deduce the conductance and susceptance.

$$Y = \frac{1}{R + jX} \frac{R - jX}{R - jX} = \frac{R - jX}{R^2 + X^2}$$

Fig. 3.15. Complex admittance of series and parallel circuits.

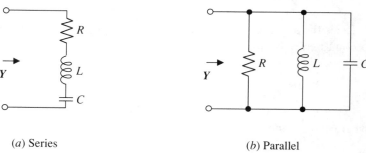

(a) Series (b) Parallel

Admittance

or

$$Y = \frac{R}{R^2 + X^2} + j\frac{-X}{R^2 + X^2} \qquad (3.47)$$

Comparing (3.47) with (3.46) we see that for the general series-connected network branch

$$\text{Conductance} = \frac{R}{R^2 + X^2}$$

$$\text{Susceptance} = \frac{-X}{R^2 + X^2}$$

Thus, conductance and susceptance for a series circuit are functions of both resistance and reactance.

The practical use of admittance lies mainly in situations where there are a number of elements connected in parallel. For instance, in the circuit of fig. 3.15(b) we have, by an expression analogous to (1.27)

$$Y = \frac{1}{Z} = \frac{1}{R} + \frac{1}{j\omega L} + j\omega C$$

or

$$Y = \frac{1}{R} + j\left(\omega C - \frac{1}{\omega L}\right)$$

In this case the conductance is identically equal to the reciprocal of the circuit resistance. It should also be noted that if the inductive reactance ωL is smaller than the capacitive reactance $1/\omega C$, the susceptance is negative, the converse of the series-connected case.

The formal equations of nodal analysis for a.c. circuits expressed in terms of admittance, yield expressions analogous to those expressed in terms of conductance for the d.c. case. For example, the a.c. formulation for a two-node circuit (corresponding to (2.28)) would be:

$$Y_{11} V_1 + Y_{12} V_2 = I_{11}$$

and $\qquad (3.48)$

$$Y_{21} V_1 + Y_{22} V_2 = I_{22}$$

Here Y_{11} and Y_{22} are the self admittances at nodes (1) and (2) respectively, and $Y_{12} = Y_{21}$ is the mutual admittance. The currents I_{11} and I_{22} represent the net current injected into nodes (1) and (2) respectively from current sources attached to those nodes.

3.9 Frequency response: transfer function

The concept of a two-port network has already been mentioned in section 2.2, and the theory of such networks is considered in more detail in chapter 8. Here we introduce the concept of the transfer function for an a.c. two-port network. The basic circuit is shown in fig. 3.16. An alternating voltage V_1, with angular frequency ω, impressed at the input port (1) gives rise to a voltage V_2 at the output port (2). In the context of electronics and communications systems these voltages would be referred to as *signals*.

The input and output voltages will be related by some linear function dependent upon the arrangement of elements in the network; in general this will be a function of frequency. The network can therefore be characterized by a *transfer function* (or *frequence response function*) defined by

$$V_2 = H(j\omega) V_1$$

or

$$H(j\omega) = \frac{V_2}{V_1} = \frac{\text{Output voltage}}{\text{Input voltage}} \tag{3.49}$$

The transfer function defined in this way, as a complex function of frequency ($j\omega$), refers to the steady-state conditions only. Other more general definitions are given in section 6.9.6.

The concept of the transfer function is not necessarily restricted to voltage ratios in a network; the relationship between any two network parameters (the ratio of output current to input voltage for example) may be expressed in a similar fashion.

In practice the transfer function is most useful when expressed in its polar form:

$$H(j\omega) = |H(j\omega)| e^{j \arg H(j\omega)} \tag{3.50}$$

Now $|H(j\omega)|$ (the modulus) and $\arg H(j\omega)$ (the angle or phase) are real functions of ω so that (3.50) may be written

$$H(j\omega) = H(\omega) e^{j\phi(\omega)} = H(\omega)\underline{/\phi(\omega)} \tag{3.51}$$

Fig. 3.16. Two-port a.c. network with transfer function $H(j\omega)$.

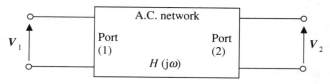

where

$$H(\omega) = |H(j\omega)| \text{ and } \underline{/\phi(\omega)} = \arg H(j\omega)$$

It may be noted that in conformity with our usual notation for complex quantities (see section 3.3) $H(j\omega)$ could also be written simply as $\mathbf{H}$, but because other definitions of the transfer function exist we have chosen to use the explicit form here.

To demonstrate the analytical procedures used to find the transfer function of a network, we consider the simple capacitance–resistance CR network shown in fig. 3.17. This circuit is employed extensively as an interstage coupling network in electronic amplifiers, its function being to prevent or block the passage of signals of zero frequency (d.c.) while allowing the passage of signals of higher frequency.

To find the transfer function of this network we may use the voltage divider principle (table 3.1: a.c. analogue of (2.6)). Assuming that no current is drawn from the output port, we may write

$$V_2 = \frac{R}{R + \dfrac{1}{j\omega C}} V_1$$

Hence

$$H(j\omega) = \frac{V_2}{V_1} = \frac{R}{R + \dfrac{1}{j\omega C}}$$

It is usually most convenient to express the transfer function in polar form; we therefore rewrite the above equation as

$$H(j\omega) = \frac{1}{1 + \dfrac{1}{j\omega CR}} \tag{3.52}$$

The denominator can now be expressed as

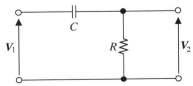

Fig. 3.17. CR coupling network.

128 *Alternating current circuits*

$$1 + \frac{1}{j\omega CR} = 1 - j\left(\frac{1}{\omega CR}\right) = \sqrt{\left[1 + \left(\frac{1}{\omega CR}\right)^2\right]}\underline{/\alpha}$$

where

$$\alpha = \tan^{-1}\left(-\frac{1}{\omega CR}\right)$$

The modulus of $H(j\omega)$ is then

$$H(\omega) = \frac{1}{\sqrt{\left[1 + \left(\frac{1}{\omega CR}\right)^2\right]}} \quad (3.53)$$

and the angle is given by

$$\phi(\omega) = -\alpha = -\tan^{-1}\left(\frac{-1}{\omega CR}\right) \quad (3.54)$$

(Note that it is not necessary to rationalize (3.52) in order to express $H(j\omega)$ in polar form.)

If we examine these expressions for the magnitude and angle of the transfer function, we see that

for $\omega \to 0$: $H(\omega) \to 0$; $\phi(\omega) \to 90°$
for $\omega \to \infty$: $H(\omega) \to 1$; $\phi(\omega) \to 0$

This result is to be expected since as far as direct voltages are concerned the capacitance acts as an open circuit; as far as very high frequencies are concerned the capacitance is effectively a short circuit.

Two methods of illustrating graphically the way in which the magnitude and phase angle of the transfer function vary with frequency are commonly used. The first method, based on the Argand diagram, is shown in fig. 3.18. In this diagram the modulus and angle are plotted in polar coordinates as a function of ω. The locus of the tip of the vector OP, traced out as ω varies,

Fig. 3.18. Locus diagram: plot of $H(j\omega) = H(\omega) \underline{/\phi(\omega)}$ on the Argand diagram.

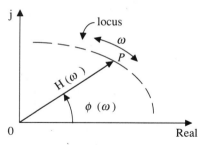

defines a line characteristic of the circuit transfer function. Diagrams of this type are called variously *locus, polar* or *Nyquist* diagrams. (Similar diagrams can also be drawn to show the effect of varying any one of the circuit parameters at a fixed frequency). It is important to appreciate that ω is not a function of time and, although based on the Argand diagram, the locus diagram is not to be confused with the rotating phasor diagram.

The locus diagram corresponding to (3.52) is shown in fig. 3.19; for the simple CR network the diagram takes the form of a semi-circle. Many other circuits exhibit a similar behaviour and give rise to locus diagrams of circular form. This graphical approach has been elaborated and extended to include many different electrical devices, particularly rotating electrical machines, and it has come to be known as the *circle diagram* method.

The second method whereby the transfer function may be depicted graphically is shown in fig. 3.20. This figure is drawn for the particular CR network of fig. 3.17. Here the modulus and phase of $H(j\omega)$ are plotted

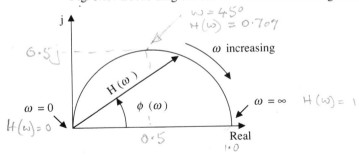

Fig. 3.19. Locus diagram for the CR network of fig. 3.17.

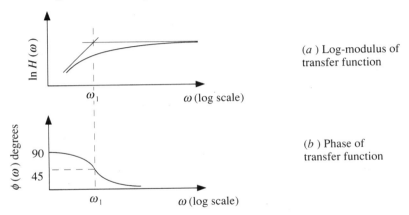

Fig. 3.20. Bode diagrams for the CR network.

(a) Log-modulus of transfer function

(b) Phase of transfer function

separately as functions of ω on a logarithmic basis. Graphs of this type are known as *Bode diagrams* or *Bode plots* after their originator.

Taking natural logarithms of (3.51) we obtain

$$\ln H(j\omega) = \ln H(\omega) + \ln e^{j\phi(\omega)} = \ln H(\omega) + j\phi(\omega) \tag{3.55}$$

In fig. 3.20(a) the real part of this expression, namely $\ln H(\omega)$, is plotted on a *linear* scale against ω which is plotted on a *logarithmic* scale. Likewise the imaginary part $\phi(\omega)$ is plotted in fig. 3.20(b) also using linear-log scales. The logarithmic basis of the Bode diagram allows large changes in the values of the parameters to be accommodated, and it provides additional practical advantages which are discussed below.

Frequently the modulus of the transfer function is expressed in logarithmic units of the decibel (abbreviation dB) (for an explanation of the decibel, see for example reference 5) and the ordinate in the Bode plot is scaled in these units in preference to the natural logarithmic scale used in fig. 3.20(a). The decibel is defined strictly in terms of the ratio of two power levels P_1 and P_2, that is,

$$\text{Power ratio (dB)} = 10 \log_{10} \frac{P_2}{P_1} \tag{3.56}$$

If we are concerned with two voltages V_1 and V_2 established across *identical* resistances, the power ratio is proportional to $(V_2/V_1)^2$ and (3.56) becomes

$$\text{Power ratio (dB)} = 10 \log_{10} \left(\frac{V_2}{V_1}\right)^2 = 20 \log_{10} \frac{V_2}{V_1} \tag{3.57}$$

In general the resistances at the input and output ports of a network are not the same, nevertheless, by convention (3.57) is applied to the modulus of the transfer function, defined by (3.49), without regard to the resistances associated with the input and output ports. According to this convention we therefore write

$$|H(j\omega)| \, (\text{dB}) = 20 \log_{10} H(\omega) \tag{3.58}$$

Fig. 3.21 shows the Bode diagram for the CR network (modulus only) with the ordinate scaled in decibels. The dB values are all negative since $V_2 = V_1$ corresponds to 0 dB, and V_2 can never be greater than V_1.

It will be observed that in fig. 3.21 (or fig. 3.20(a)) the curve is asymptotic to two straight line segments. The reason for this will be appreciated if we consider the modulus of $H(j\omega)$ for very low and very high frequencies.

At sufficiently low frequencies, such that $\omega CR \ll 1$, (3.52) becomes $H(\omega) = \omega CR$. $H(\omega)$ is therefore proportional to ω which means that the log-modulus versus ω relationship is of straight-line form provided ω is plotted on a logarithmic scale. Using (3.57) we have $|H(j\omega)|(dB) = 20\log_{10}\omega CR$. For each decade of frequency therefore the modulus changes by $20\log_{10}(10/1) = 20$ dB. This determines the asymptote for low frequencies.

At sufficiently high frequencies ($\omega CR \gg 1$) we have $H(\omega) = 1$, and $20\log_{10}H(\omega) = 0$. Thus, the curve is asymptotic to a horizontal straight line at high frequencies. The two asymptotes meet (point A in Fig. 3.21) at a frequency ω_1 such that $\omega_1 CR = 1$ or $\omega_1 = 1/CR$. At this frequency the *actual* value of $H(\omega)$ is, from (3.53), $1/\sqrt{2}$ or 0.707, in other words the true value of $H(\omega)$ is about 30% lower than the value represented by the intersection of the asymptotes. Expressed in decibels this is equal to -3.03 dB. At this frequency the phase angle is $45°$ (fig. 3.20(*b*)). The frequency ω_1 is called variously the *turnover frequency*, the *corner frequency*, the *break frequency*, or the '*3 dB point*'. Since the maximum departure of the true curve from the approximate curve formed by the two asymptotes is only 3 dB it is sufficient for many practical purposes to represent the modulus of the transfer function by its straight-line approximation. The principles of construction of the asymptotes in the Bode diagram, here discussed in relation to the CR network, may be readily extended to networks of greater complexity.

The chief advantage of the Bode diagram is that it affords a ready means of finding graphically the overall transfer function of several networks connected in cascade. Suppose we wish to find the overall transfer function $H_0(j\omega)$ of two cascaded networks with individual transfer functions $H_1(j\omega)$ and $H_2(j\omega)$. From the definition of the transfer function given in (3.49) it will be readily apparent that $H_0(j\omega) = H_1(j\omega)H_2(j\omega)$. It is a somewhat tedious procedure to multiply the moduli of the two transfer functions together over a wide frequency range, especially if one or both have been determined by experiment. However, if the results are presented on a Bode plot, advantage can be taken of the properties of logarithms, and the

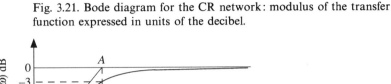

Fig. 3.21. Bode diagram for the CR network: modulus of the transfer function expressed in units of the decibel.

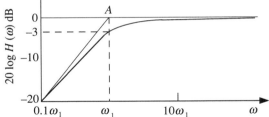

132 *Alternating current circuits*

ordinates can be simply added since

$$\ln H_0(j\omega) = \ln [H_1(j\omega)H_2(j\omega)]$$
$$= \ln H_1(\omega) + \ln H_2(\omega) + j[\phi_1(\omega) + \phi_2(\omega)]$$

This result applies equally well of course when moduli are expressed in decibels. The ordinate addition is a particularly simple operation when the straight-line approximation is used in the Bode diagram.

In the above discussion it has been assumed that individual transfer functions will not be changed by connection of the networks one to another; often in practice this will not be the case. Each network will be affected by connection to both preceding and succeeding networks to some extent, and due allowance must be made for this when determining the individual transfer functions either from theory or experiment.

3.10 A.C. bridges

A.C. bridge networks form a large and important class of circuits used for measurement purposes, for filtering (separating wanted signals from unwanted signals), and for use in certain types of oscillator. The essential characteristic common to all of the various circuits falling under this heading is that for a finite input or source signal they produce zero output signal under one particular set of conditions: the so-called *balance* conditions.

The basic circuit configuration of one common form of a.c. bridge, used mainly for measurement purposes, is shown in fig. 3.22; this will be recognized as the a.c. analogue of the Wheatstone bridge discussed in section 2.8.

The balance conditions may be derived using methods essentially similar to those of section 2.8 (equation 2.21); they occur when the impedances in the arms of the bridge are such that

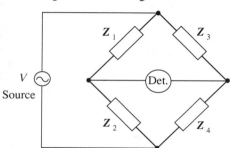

Fig. 3.22. A.C. bridge network.

A.C. bridges

$$Z_1 Z_4 = Z_2 Z_3$$

or

$$Z_1 = \frac{Z_2 Z_3}{Z_4} \tag{3.59}$$

By arranging Z_1 to be the component whose value is to be measured, and by arranging Z_2, Z_3 and Z_4 to be components of accurately known value, the value of Z_1 can be determined. We can arrange either to make Z_2 and Z_3 fixed and to vary Z_4 to achieve balance, in which case the bridge is known as a *product* bridge, or we can make Z_2 and Z_4 fixed and vary Z_3, in which case the bridge is known as a *quotient* bridge. In all bridges of this type the balance conditions are independent of the magnitude of the source voltage although the sensitivity with which the balance can be detected will suffer if the voltage is too low. The balance conditions may or may not depend on frequency.

For any given arrangement of elements forming a bridge circuit the source and detector may be interchanged without in any way affecting the signal registered by the detector. This follows directly from the reciprocity theorem. The new bridge circuit resulting from such an interchange is called the *conjugate* of the original and it has the same balance conditions.

Two examples of bridges conforming to the basic configuration shown in fig. 3.22 will serve to illustrate the techniques of analysis and some of the important characteristics of this type of circuit.

3.10.1 The Schering bridge

The circuit of this bridge is shown in fig. 3.23. Its main use is for the measurement of capacitance, particularly the capacitance of high voltage cables and insulators under working conditions. The unknown capacitance

Fig. 3.23. The Schering bridge.

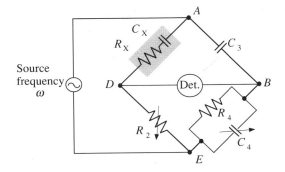

is C_x, and R_x is a series resistance representing the losses associated with this capacitance (see section 3.13.1 for an account of losses in capacitors). C_3 and C_4 are respectively fixed and variable low-loss capacitors of accurately known values. R_2 is an accurately calibrated variable resistance and R_4 is a fixed standard resistance. (When this bridge is used in high voltage applications, the node E is earthed, and the node A is at high potential.)

From (3.59) the balance conditions are given by

$$Z_1 = \frac{Z_2 Z_3}{Z_4} = Z_2 Z_3 Y_4$$

where Y_4 is the admittance of the arm BE. Therefore,

$$R_x + \frac{1}{j\omega C_x} = R_2 \frac{1}{j\omega C_3} \left(\frac{1}{R_4} + j\omega C_4 \right) = \frac{R_2}{j\omega C_3 R_4} + \frac{R_2 C_4}{C_3}$$

Equating real parts on the two sides of this equation gives

$$R_x = \frac{R_2 C_4}{C_3}$$

and equating imaginary parts gives

$$C_x = \frac{C_3 R_4}{R_2}$$

Two balance conditions are thus obtained reflecting the fact that the potentials of D and B at balance must be the same in both magnitude and phase. It will be noticed that the balance conditions are independent of frequency so that a highly stable source is not required. By adjusting R_2 and C_4 alternately the bridge can be made to converge to a balance indicated by

Fig. 3.24. Phasor diagrams for the balanced Schering bridge of fig. 3.23. (a) Left-hand arm ADE; (b) right-hand arm ABE. Currents I'_B and I''_B are the components of I_B through R_4 and C_4 respectively.

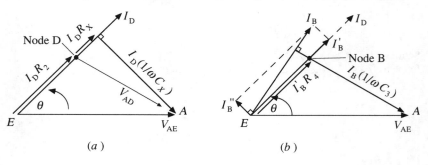

(a)   (b)

A.C. bridges

a null detector reading. The detector would normally consist of an oscilloscope or a sensitive integrated circuit amplifier with its output connected to some form of indicator. For very accurate measurements the detector may be tuned so that it is sensitive only to the bridge source frequency; by this means the effects of spurious signals arising, for example, from the power supply mains (hum pick-up) may be reduced.

Phasor diagrams for the balanced Schering bridge are shown in fig. 3.24. For the sake of clarity the diagrams for the left and right-hand arms are shown separately. Referring first to the left-hand arm fig. 3.24(a), we may understand how this diagram is constructed by considering the current I_D flowing from A to E via D (remembering that no current flows in the detector branch at balance) and its relationship to the source voltage V_{AE}, which is conveniently taken as reference. The branch ADE contains both capacitance and resistance so that the current I_D must lead V_{AE} by some angle less than 90°. The arm DE is purely resistive so that the voltage $V_{DE}(=I_D R_2)$ must be in phase with the current I_D, also the voltage drop across R_x must be in phase with I_D. These voltage phasors therefore coincide with the current phasor I_D. The voltage across $C_x(=I_D/(\omega C_x))$, lags I by 90° and this voltage phasor is therefore perpendicular to the current phasor. The phasor diagram for the right-hand arm of the bridge is similarly constructed and is shown in fig. 3.24(b). When the two diagrams are superimposed, the points marked D and B must coincide since there is no potential difference across the detector at balance.

3.10.2 The Wien bridge

This bridge (fig. 3.25) is an example of the type for which the balance conditions depend upon frequency. It finds application as a frequency determining network in certain types of oscillator (see reference 5).

Fig. 3.25. The Wien bridge.

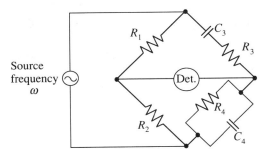

From (3.59) we have

$$Z_1 = \frac{Z_2 Z_3}{Z_4} = Z_2 Z_3 Y_4$$

that is

$$R_1 = R_2 \left(R_3 + \frac{1}{j\omega C_3} \right) \left(\frac{1}{R_4} + j\omega C_4 \right)$$

$$= \frac{R_2 R_3}{R_4} + \frac{R_2}{j\omega C_3 R_4} + j\omega C_4 R_2 R_3 + \frac{R_2 C_4}{C_3}$$

Hence equating real and imaginary parts:

$$\frac{R_1}{R_2} = \frac{R_3}{R_4} + \frac{C_4}{C_3}$$

and

$$\omega^2 = \frac{1}{R_3 C_3 R_4 C_4}$$

This second balance condition gives, for any fixed set of component values, the frequency at which there will be zero detector signal. At this frequency the phase shift between source and detector signals will be zero, and it is this property of the circuit which is important when it is incorporated in so-called phase-shift oscillator circuits.

3.11 Worked example

The output at the detector terminals of a sensitive bridge circuit using a high impedance detector contains an unwanted sinusoidal signal component at an angular frequency ω_0.

(a) Show that by interposing a twin-T filter, of the type shown in fig. 3.26, between the bridge detector terminals and the detector, the unwanted signal may be eliminated.

(b) If the unwanted signal is caused by inductively coupled pick-up from mains power lines operating at 50 Hz, suggest suitable values for R and C in fig. 3.26.

(c) Sketch the general form of the modulus of the transfer function for the twin-T filter as a function of frequency.

Solution

The twin-T filter is itself a form of bridge circuit in which the balance conditions are frequency dependent. At one particular frequency, therefore, the output V_2 falls to zero.

Worked example   137

Several of the methods discussed in Chapter 2 may be used for solving part (a) of this example; from the arguments presented in section 2.11 it will be apparent that mesh analysis is the most cumbersome. Two alternative methods are presented below: one employing nodal analysis, the other the T-π transformation.

(a) *Method 1: nodal analysis*

We may assume that no current is drawn from the output port of the filter since the detector has a high input impedance. The input voltage V_1 supplied by the bridge may, for the purposes of this problem, be regarded as a source voltage of fixed magnitude.

Choosing the reference node as shown, there remain three nodes whose voltages are undetermined. One of these, the node O, is the output node with assigned voltage V_2. The other nodes are P and Q with assigned voltages V_P and V_Q.

At O the nodal equation is

$$\frac{V_2 - V_Q}{R} + \frac{V_2 - V_P}{1/j\omega C} = 0$$

which reduces to

$$\left(\frac{1}{R} + j\omega C\right)V_2 - j\omega C\, V_P - \frac{1}{R}V_Q = 0 \qquad (3.60)$$

The nodal equations at P and Q are

$$\frac{V_P - V_1}{1/j\omega C} + \frac{V_P - V_2}{1/j\omega C} + \frac{V_P}{R/2} = 0$$

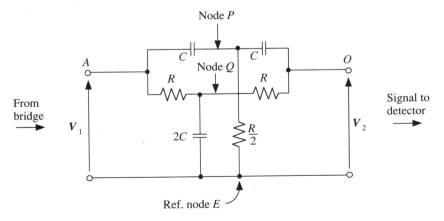

Fig. 3.26. Circuit for worked example: twin-T filter.

138 *Alternating current circuits*

and

$$\frac{V_Q - V_1}{R} + \frac{V_Q - V_2}{R} + \frac{V_Q}{1/j\omega 2C} = 0$$

which reduce to

$$-j\omega C V_2 + \left(\frac{2}{R} + 2j\omega C\right) V_P + (0) V_Q = j\omega C V_1 \qquad (3.61)$$

and

$$-\frac{1}{R} V_2 - (0) V_P + \left(\frac{2}{R} + 2j\omega C\right) V_Q = \frac{V_1}{R} \qquad (3.62)$$

Note that the coefficients of V_Q in (3.61) and of V_P in (3.62) are both zero. This is because no mutual element exists between the nodes P and Q. Solving for V_2 using the method of determinants (section 2.5) we obtain

$$V_2 = \frac{1}{\Delta} \begin{vmatrix} 0 & -j\omega C & -\frac{1}{R} \\ j\omega C V_1 & \left(\frac{2}{R} + 2j\omega C\right) & 0 \\ \frac{V_1}{R} & 0 & \left(\frac{2}{R} + 2j\omega C\right) \end{vmatrix} \qquad (3.63)$$

where Δ is the determinant formed by the array of coefficients in (3.60), (3.61) and (3.62).

Now at the frequency ω_0 of the unwanted signal $V_2 = 0$, therefore, expanding the numerator of (3.63) and equating to zero we obtain

$$-j\omega_0 C V_1 \left[-j\omega_0 C \left(\frac{2}{R} + 2j\omega_0 C\right)\right] + \frac{V_1}{R}\left[-\left(\frac{2}{R} + 2j\omega_0 C\right)\left(-\frac{1}{R}\right)\right]$$
$$= 0$$

Equating real (or imaginary) parts of this expression to zero gives, finally,

$$\omega_0 = \frac{1}{CR}$$

(a) *Method 2: T-π transformation*
Using relationships analogous to (2.23), each of the two T-connected sections of the filter may be transformed separately to give the corresponding π-configuration as shown in fig. 3.27. These two π-sections may then be combined to form the π-equivalent of the original network. When this is

Worked example

done we see that impedance ($Z_1//Z_4$) spans the input port and, since the voltage there is constant and equal to V_1, it cannot affect the output in any way. The two remaining arms of the π-section constitute a voltage divider, hence we may derive a general expression from which V_2 may be found at any frequency. However, in this problem we wish to find only the frequency at which V_2 becomes zero. This will occur when $Z_2//Z_5 = Z_2 Z_5/(Z_2 + Z_5)$ becomes infinite, that is when $(Z_2 + Z_5)$ tends to zero.

From (2.23) and fig. 2.17

$$Z_2 = \frac{1}{j\omega C} + \frac{1}{j\omega C} + \frac{(1/j\omega C)(1/j\omega C)}{(R/2)} = \frac{2}{j\omega C} - \frac{2}{\omega^2 C^2 R}$$

$$Z_5 = R + R + \frac{(R)(R)}{(1/2j\omega C)} = 2R + 2j\omega C R^2$$

Therefore V_2 is zero at the frequency of the unwanted signal ω_0 when

$$\frac{2}{j\omega_0 C} - \frac{2}{\omega_0^2 C^2 R} + 2R + 2j\omega_0 C R^2 = 0$$

which gives, on equating real or imaginary parts to zero,

$$\omega_0 = 1/CR$$

(b) At the frequency of the mains power lines, $\omega_0 = 2\pi \times 50 = 314$, therefore, the product $CR = 1/314$. The actual values of C and R are chosen on practical grounds: capacitors with large stable values of capacitance (with

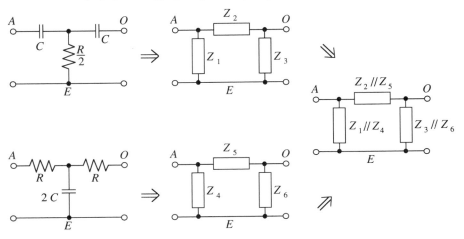

Fig. 3.27. The T-π transformation applied to the twin-T filter of fig. 3.26; by symmetry, $Z_1 = Z_3$ and $Z_4 = Z_6$.

the requisite low leakage resistance necessary for this application) are bulky and expensive; very small values of C would make the circuit too sensitive to stray capacitances between components. With these considerations in mind a suitable value for C would be $0.1\,\mu\text{F}$, giving $R = 33\,\text{k}\Omega$. In practice R would be adjusted to produce the precise CR product required by using shunt or series trimming resistors.

(c) The general form of the modulus of the transfer function $|V_2/V_1|$, may be found without calculation by considering the extremes of the frequency range; from $\omega = 0$ (d.c.) to $\omega \to \infty$. At $\omega = 0$, all the capacitances are effectively open circuit and the output node O is connected to the input node A through series resistance $2R$. On the assumption that no current is drawn from node O, there can be no voltage drop across this resistance, therefore, the input and output voltages are the same and $|V_2/V_1| = 1$. Similar considerations apply for $\omega \to \infty$: in this case the capacitances act as short circuits thus connecting input and output terminals directly together. We conclude that the plot of $|V_2/V_1|$ versus ω must take the general form shown in fig. 3.28; the curve being asymptotic to the value unity for high frequencies. Because of the characteristic shape of the transfer function, the term *notch-filter* is often used in connection with this circuit.

It will be obvious that this circuit can be effective only if the wanted bridge frequency is well separated from the notch frequency. Typically bridge circuits operate at 1 kHz or so and the notch filter frequency is arranged to be at 50 Hz. The reactance of the capacitive elements in the circuit are therefore some twenty times smaller at the wanted frequency than at the notch frequency, consequently, the attenuation produced by the filter at the bridge frequency is small. The reader may care to confirm that for the filter in this example. $|V_2/V_1| = 0.98$ at 1 kHz.

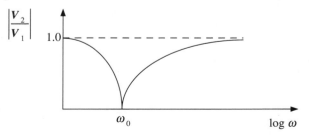

Fig. 3.28. Modulus of the transfer function, as a function of frequency, for the circuit of Fig. 3.26.

3.12 Inductively coupled circuits

The a.c. voltage–current relationships for a mutual inductance may be readily established using the methods of section 3.3. We take as our starting point the general relationship (1.49) between the instantaneous voltage v_1 and currents i_1 and i_2 indicated in fig. 3.29(a), namely,

$$v_1 = L_1 \frac{di_1}{dt} \pm M \frac{di_2}{dt} \tag{1.49}$$

It will be recalled (section 1.10) that the ($\pm$) signs in the second term of this equation arise because of the two alternative ways of connecting the coils: either with fluxes aiding ((+) sign) or fluxes opposing ((−) sign).

Since all currents and voltages are sinusoidal they may be represented in complex exponential form by:

$$v_1 = V_{1m} e^{j\omega t}; \quad i_1 = I_{1m} e^{j\omega t}; \quad i_2 = I_{2m} e^{j\omega t}$$

Substituting in (1.49) and taking the real part as understood, we obtain

$$V_{1m} e^{j\omega t} = L_1 \frac{d}{dt}(I_{1m} e^{j\omega t}) \pm M \frac{d}{dt}(I_{2m} e^{j\omega t})$$

or

$$V_{1m} e^{j\omega t} = j\omega L_1 I_{1m} e^{j\omega t} \pm j\omega M I_{2m} e^{j\omega t}$$

By cancelling out $e^{j\omega t}$ from each term in this expression, and dividing throughout by $\sqrt{2}$, we obtain the required relationship in terms of stationary phasors and r.m.s. magnitudes:

Fig. 3.29. Voltage–current relationships for mutual inductance.

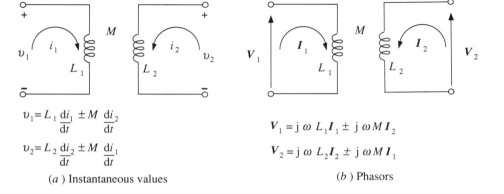

$v_1 = L_1 \dfrac{di_1}{dt} \pm M \dfrac{di_2}{dt}$

$v_2 = L_2 \dfrac{di_2}{dt} \pm M \dfrac{di_1}{dt}$

(a) Instantaneous values

$V_1 = j\omega L_1 I_1 \pm j\omega M I_2$

$V_2 = j\omega L_2 I_2 \pm j\omega M I_1$

(b) Phasors

$$V_1 = j\omega L_1 I_1 \pm j\omega M I_2 \tag{3.64}$$

A similar expression for the voltage V_2 is obtained:

$$V_2 = j\omega L_2 I_2 \pm j\omega M I_1 \tag{3.65}$$

These results are illustrated in fig. 3.29(b).

The analysis of circuits containing mutual inductance is almost invariably carried out using the method of mesh analysis rather than nodal analysis because of the difficulty of defining reference nodes and node potentials in circuits containing separate and distinct parts coupled only by mutual inductance. For a similar reason the relationship (2.31), relating numbers of nodes and elements in a circuit, does not apply directly, although it may be extended to include circuits containing separate parts (see reference 4).

The application of (3.64) and (3.65), and the general method of analysis of a.c. circuits containing mutual inductance, may be illustrated with reference to the circuit shown in fig. 3.30. This consists of two separate meshes linked only by mutual inductance; corresponding ends of the coils are indicated by the dots (see section 1.10). Currents I_1 and I_2 are assigned to meshes (1) and (2) as shown, and these have been chosen in a clockwise direction.

For mesh (1) the self-impedance is $(R_1 + j\omega L_1)$, and the voltage drop is $(R_1 + j\omega L_1)I_1$; to this we must add, according to (3.64), the mutual inductance term $j\omega M I_2$. Hence the mesh equation is

$$(R_1 + j\omega L_1)I_1 - j\omega M I_2 = V \tag{3.66}$$

The sign of the mutual inductance term follows from the dot convention discussed in section 1.10: if one assigned current *enters* a dotted end and the other *leaves* a dotted end, fluxes oppose and the mutual inductance term is negative. The voltage on the right-hand side of (3.66) takes a (+) sign

Fig. 3.30. Mesh analysis of mutually-coupled circuits.

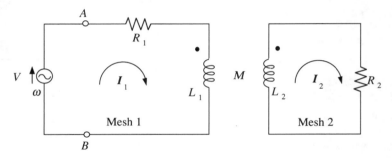

Inductively coupled circuits

because the assigned current I_1 leaves the voltage source in the direction of the reference polarity arrow (working rule 2, section 2.3).

For mesh (2) the voltage drop across the self-impedance is $(R_2 + j\omega L_2)I_2$, and the mutual inductance term is $-j\omega M I_1$. There is no voltage source in this mesh so that the equation is

$$-j\omega M I_1 + (R_2 + j\omega L_2)I_2 = 0 \qquad (3.67)$$

Notice that the mutual inductance term in this equation is treated as a voltage drop (in conformity with (3.65)) and not as a source voltage, even though it is the e.m.f. due to the mutual inductance effect that causes current to circulate in mesh (2). It is clear from the circuit model discussed in section 1.10.1, and illustrated in fig. 1.26, that it would be equally valid to write (3.67), *ab initio*, as

$$(R_2 + j\omega L_2)I_2 = j\omega M I_1$$

where the term on the right is regarded as a source of voltage. However, the formulation (3.67) is more consistent with the equations of mesh analysis developed in chapter 2.

We shall now use (3.66) and (3.67) to find the impedance of the complete circuit at the terminals AB, as seen from the voltage source. This impedance, called the *driving point impedance*, is discussed in more general terms in chapter 8.

From (3.67)

$$I_2 = \frac{j\omega M I_1}{R_2 + j\omega L_2}$$

and substituting in (3.66) we obtain

$$(R_1 + j\omega L_1)I_1 - \frac{j^2\omega^2 M^2 I_1}{R_2 + j\omega L_2} = V$$

hence

$$I_1 \left[R_1 + j\omega L_1 + \frac{\omega^2 M^2}{R_2 + j\omega L_2} \right] = V$$

The impedance at AB is then given by

$$Z_{AB} = \frac{V}{I_1} = R_1 + j\omega L_1 + \frac{\omega^2 M^2}{R_2 + j\omega L_2}$$

Rationalizing the last term in this expression we obtain

$$Z_{AB} = R_1 + j\omega L_1 + \frac{\omega^2 M^2 R_2 - j\omega(\omega^2 M^2 L_2)}{R_2^2 + (\omega L_2)^2}$$

or

$$Z_{AB} = \left[R_1 + \frac{\omega^2 M^2 R_2}{R_2^2 + (\omega L_2)^2}\right] + j\omega\left[L_1 - \frac{\omega^2 M^2 L_2}{R_2 + (\omega L_2)^2}\right] \quad (3.68)$$

Examination of (3.68) shows that if the mutual inductance between the two meshes is negligibly small the driving point impedance becomes identical to that of mesh (1) alone, that is, $(R_1 + j\omega L_1)$. With increasing M, the effective resistance, as seen at the terminals AB (first term in brackets), is increased while the effective inductance (second term in brackets) is decreased. This result is unaffected by the relative winding directions of the two coils since the final expression contains M only as a squared term; the sign associated with M in the original mesh equations is therefore immaterial.

Another frequently encountered network configuration involving mutual inductance is shown in fig. 3.31. In this the mutually coupled coils are connected in series. Taking voltage drops in sequence round the circuit, and with due regard for the dot convention, we obtain

$$j\omega L_1 I - j\omega MI + j\omega L_2 I - j\omega MI = V \quad (3.69)$$

or

$$j\omega(L_1 + L_2 - 2M)I = V$$

This shows that the effective driving point impedance at AB is

$$Z_{AB} = j\omega(L_1 + L_2 - 2M) \quad (3.70)$$

That is, the effective inductance is $(L_1 + L_2 - 2M)$. If the connections to one of the coils are reversed, the signs of the mutual inductance terms in (3.69) become positive and the effective inductance is then $(L_1 + L_2 + 2M)$.

These relationships provide the basis of a method for determining the mutual inductance between two coupled coils. Measurements are made, usually by means of an a.c. bridge method, of the effective inductance of

Fig. 3.31. Series-connected coupled coils.

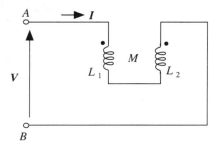

coils connected in series in the two alternative circuit configurations; first, with fluxes aiding and then with fluxes opposing. If these measurements yield effective values of, say, L_3 and L_4 ($L_3 > L_4$), then we have, according to the above theory,

$$L_3 = L_1 + L_2 + 2M$$

and

$$L_4 = L_1 + L_2 - 2M$$

Combining these two equations gives

$$4M = L_3 - L_4$$

hence, M is determined in terms of the measured values.

We complete this discussion of inductively coupled circuits by considering a useful circuit transformation that resembles in some ways the star–delta (T-π) transformation discussed previously. This is the so-called T-equivalent of two coupled coils and is illustrated in fig. 3.32. Fig. 3.32(a) shows two coupled coils, similar to those shown in fig. 3.29, but with one corresponding end of each coil joined to form a common connection (terminal B). The circuit configuration shown in fig. 3.32(a) can be arranged to be electrically equivalent to fig. 3.32(b), so far as any external connections are concerned, by choosing suitable values for the inductances L_a, L_b and L_c. The utility of this transformation lies in the fact that the inductances in fig. 3.32(b) are separate and distinct having no mutual coupling between them, and this can lead to analytical simplification in some circuits of practical interest.

To obtain the connecting relationships for the two circuits we consider the inductance that appears at corresponding terminal pairs (the remaining

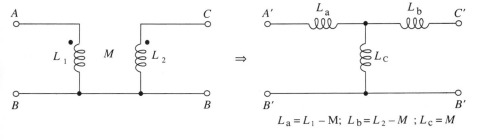

Fig. 3.32. The T-equivalent of two coupled coils.

$L_a = L_1 - M$; $L_b = L_2 - M$; $L_c = M$

(a)   (b)

terminal being open circuit). For equivalence we therefore require the same inductances at AB and $A'B'$, hence,

$$L_1 = L_a + L_c$$

Similarly at CB and $C'B'$

$$L_2 = L_b + L_c$$

Also, using similar arguments to those leading to (3.70), we have at AC and $A'C'$

$$L_1 + L_2 - 2M = L_a + L_b$$

Combining and rearranging these equations we obtain:

$$L_a = L_1 - M; \qquad L_b = L_2 - M; \qquad L_c = M \tag{3.71}$$

If non-corresponding ends are joined at B, the connecting equations become:

$$L_a = L_1 + M; \qquad L_b = L_2 + M; \qquad L_c = -M \tag{3.72}$$

A final point to note is that although the circuit transformation illustrated in fig. 3.32 has here arisen in connection with a.c. circuits, it is valid for source excitations of any form.

3.13 Resonant circuits

3.13.1 Losses in inductors and capacitors

(a) *Inductors*

A practical inductor consists typically of a length of wire wound into the form of a coil to enhance the self flux linking effect and therefore the inductance. The wire will possess some resistance so that it is natural to think of an inductor as having a lumped linear circuit model of the form shown in fig. 3.33(*a*). It might be supposed that the resistance r could be regarded as a fixed characteristic parameter, but it is found that to describe satisfactorily the electrical behaviour of an inductor, it is necessary to adjust the value of r according to the frequency at which the inductor is to operate. The reason for this is that the power loss in an inductor arises with increasing frequency because of eddy currents induced in the wire itself. The net effect of this is to cause the current in the wire to concentrate near the surface, producing the so-called 'skin effect' and thereby increasing the conduction loss (see reference 6).

Eddy currents induced in any conducting material in proximity to the inductor, for example a screening can, will also contribute to the losses, as will hysteresis losses in any magnetic materials that may be present. All these sources of loss affect the value of r to be ascribed to the circuit model.

The effective inductance will also vary with the frequency because of stray capacitances between turns of the coil. We must conclude therefore that the constants L_s and r in the circuit model of fig. 3.33(a) are true only for one particular frequency, or at best for a narrow band of frequencies.

We shall see later in our study of resonant circuits that there are good reasons for making inductors as loss free as possible; in terms of the circuit model, this implies that for a given value of L_s, r should be as small as possible. We find it convenient therefore to take as a measure of the excellence of an inductor the dimensionless ratio $\omega L_s/r$, where L_s and r are values appropriate to the particular angular frequency ω at which the inductor is to be operated. This ratio is termed the quality factor or Q-factor of the inductor. Thus, by definition,

$$Q = \frac{\text{Reactance}}{\text{Series resistance}} \equiv \frac{\omega L_s}{r} \tag{3.73}$$

At radio frequencies the Q-factor of an inductor may typically be of the order of 100, and when Q is this large, certain approximations may be made in the theory of resonant circuits which greatly simplifies their analysis. Furthermore, we are generally concerned with only narrow bands of frequency which means that Q can be regarded as being substantially constant.

Although the series circuit model shown in fig. 3.33(a) is the most natural representation for an inductor, it is often analytically convenient to use the parallel representation shown in fig. 3.33(b). The series and parallel representations will both uniquely describe the electrical behaviour of a linear inductor at one particular frequency if they both present the same complex impedance at their terminals at that frequency. The transformation relations may therefore be derived by equating complex impedances.

Fig. 3.33. Lumped circuit models for an inductor.

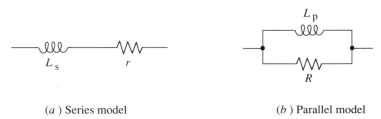

(a) Series model   (b) Parallel model

148 Alternating current circuits

For fig. 3.33(a) the impedance is $r+j\omega L_s$; for fig. 3.33(b) the impedance is $j\omega L_p // R$, thus,

$$r+j\omega L_s = \frac{j\omega L_p R}{R+j\omega L_p}$$

Rationalizing the right-hand side of this expression gives

$$r+j\omega L_s = \frac{j\omega L_p R(R-j\omega L_p)}{R^2+(\omega L_p)^2}$$

$$= \frac{\omega^2 L_p^2 R}{R^2+(\omega L_p)^2} + j\omega \frac{L_p R^2}{R^2+(\omega L_p)^2}$$

and by equating real and imaginary parts we obtain

$$r = \frac{\omega^2 L_p^2 R}{R^2+(\omega L_p)^2} \tag{3.74}$$

and

$$\omega L_s = \frac{\omega L_p R^2}{R^2+(\omega L_p)^2} \tag{3.75}$$

Dividing (3.75) by (3.74) gives

$$\frac{\omega L_s}{r} = \frac{R}{\omega L_p} = Q \tag{3.76}$$

Equation (3.76) provides an alternative definition of the Q-factor: the ratio of the equivalent parallel resistance of the inductor to its equivalent parallel reactance.

Now (3.74) and (3.75) may be re-written in terms of Q, thus,

$$r = \frac{R}{\dfrac{R^2}{\omega^2 L_p^2}+1} = \frac{R}{Q^2+1} \simeq \frac{R}{Q^2} \quad (Q^2 \gg 1) \tag{3.77}$$

and

$$\omega L_s = \frac{\omega L_p}{1+\dfrac{\omega^2 L_p^2}{R^2}} = \frac{\omega L_p}{1+\dfrac{1}{Q^2}} \simeq \omega L_p \quad (Q^2 \gg 1) \tag{3.78}$$

We see from these relations that, in the transformation from series to parallel equivalent form, the inductance remains substantially unchanged but that the resistance is altered by a factor equal to Q^2. Hence we may write, to a very good approximation,

$$L_s = L_p = L \tag{3.79}$$

and

$$R = Q^2 r \tag{3.80}$$

(b) *Capacitors*

Power loss in a capacitor arises because of leakage and other effects in the dielectric between the plates and because of resistance in the plates themselves. To account for leakage, it is natural to model the capacitor by the parallel combination of fig. 3.34(*a*). However, as for the inductor, losses from all sources can be accounted for, at one frequency, by either a series or a parallel model. A similar procedure to that carried out above for the inductor leads to the following relations:

$$Q = \frac{\text{Reactance}}{\text{Series resistance}} = \frac{1/\omega C_s}{r} \tag{3.81}$$

$$= \frac{R}{1/\omega C_p} \left(\frac{\text{Parallel resistance}}{\text{Reactance}} \right)$$

$$C_s = C_p = C \tag{3.82}$$

and

$$R = Q^2 r \tag{3.83}$$

The quality of a capacitor can also be expressed in terms of the phase angle ϕ between current and applied voltage (fig. 3.35). For a pure capacitance the phase angle is 90° but any losses cause a reduction in the phase angle by a small amount (angle δ_d in the diagram). It is usual to take the cosine of the phase angle as a suitable measure of the losses since this is

Fig. 3.34. Lumped circuit models for a capacitor.

(*a*) Parallel model (*b*) Series model

the factor by which the product VI must be multiplied in order to calculate the power dissipated in the capacitor. We call this the *power factor* (see section 4.4).

Figs. 3.35(b) and (c) show the phasor relationships appropriate to the parallel circuit model of fig. 3.35(a). The two components of the total current I are $I_R(=V/R)$ flowing through R, and $I_C(=V/(1/\omega C))$ flowing through C. Hence,

$$\text{Power factor} = \cos\phi = \frac{V/R}{\sqrt{[(V/R)^2 + (\omega CV)^2]}} = \frac{1}{\sqrt{[1 + (\omega CR)^2]}} \quad (3.84)$$

But, for a good quality capacitor, $\omega CR \gg 1$, therefore,

$$\text{Power factor} = \cos\phi = \frac{1}{\omega CR} = \frac{1}{Q} \quad (3.85)$$

We see that there is a simple reciprocal relationship between power factor and Q-factor. Precisely the same relationship may be derived on the basis of the series circuit model of fig. 3.34(b).

In the type of capacitor commonly used in resonant circuits at radio frequencies, the losses occur mainly in the dielectric material of the capacitor; the greater these dielectric losses, the greater the angle δ_d in fig. 3.35(b). Consequently, the angle δ_d is sometimes referred to as the *loss-angle* of the dielectric. This loss-angle can be determined for a dielectric material by making measurements on a capacitor containing the dielectric using an a.c. bridge technique. By this means the parameters R and C are found at a particular operating frequency ω and the loss-angle may then be calculated from the expression

Fig. 3.35. Phasor relationships for the parallel circuit model of a capacitor. The power factor of the capacitor is $\cos\phi$, and the loss angle is δ_d.

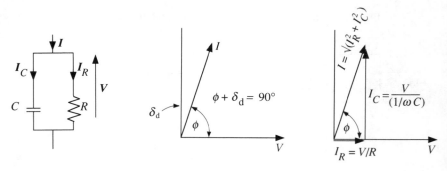

$$\tan\delta_d = \frac{1}{\omega CR} \tag{3.86}$$

This expression is readily derived by considering the geometry of the phasor diagrams in fig. 3.35.

3.13.2 The series resonant circuit

When the frequency at which a circuit is excited is such that the reactances of individual inductive and capacitive elements are comparable, the branch impedances of the circuit change rapidly with change of frequency and so do the currents and voltages. The phenomena that are exhibited when this takes place are classed under the general heading of resonance, and the frequencies at which these phenomena occur are called the resonant frequencies. Of the many forms of resonant circuit the simplest is the series resonant circuit shown in fig. 3.36(a). This comprises an inductor, a capacitor and a practical voltage source, all connected in series.

The losses of the inductor and capacitor are represented by r_L and r_C respectively, while r_S includes the resistance of the source together with any other resistances that may have been included in the circuit for any purpose. (The current in a series resonant circuit may be monitored, for example, by using an oscilloscope to detect the voltage developed across a small-value resistor connected into the circuit.) For the purposes of analysis, the separate resistances may be represented by a single equivalent resistance r as shown in fig. 3.36(b).

The impedance of the circuit is $r + j(\omega L - 1/\omega C)$ hence the current is given by

Fig. 3.36. The series resonant circuit.

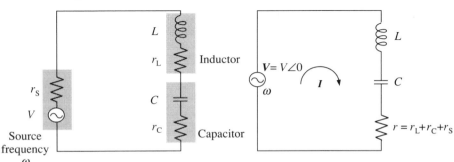

(a) Series connected circuit   (b) Circuit model

152 Alternating current circuits

$$I = \frac{V}{r + j\left(\omega L - \dfrac{1}{\omega C}\right)} \tag{3.87}$$

The magnitude of the current is therefore

$$I = \frac{V}{\sqrt{\left[r^2 + \left(\omega L - \dfrac{1}{\omega C}\right)^2\right]}} \tag{3.88}$$

and the phase angle ϕ of the current with respect to the voltage is given by

$$\phi = -\tan^{-1}\frac{(\omega L - 1/\omega C)}{r} \tag{3.89}$$

Let us now consider the situation where the voltage of the source is maintained at a constant value whilst the angular frequency is varied. Fig. 3.37 illustrates how the inductive and capacitive reactances, and their difference, vary with frequency. Clearly there is an angular frequency $\omega = \omega_0$ at which the total reactance of the circuit is zero. This occurs when

$$\omega_0 L = \frac{1}{\omega_0 C}$$

Fig. 3.37. The reactance of a series resonant circuit as a function of angular frequency ω; ω_0 is the frequency at which *current resonance* occurs.

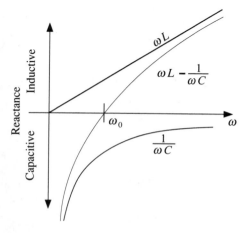

Resonant circuits

or

$$\omega_0 = \frac{1}{\sqrt{(LC)}} \tag{3.90}$$

At this frequency the circuit is purely resistive and the current has a maximum value $I_0 = V/r$. The circuit is said to exhibit *current resonance* and $\omega_0/2\pi$ is called the *resonant frequency*. (We shall see later that other definitions for the resonant frequency of a series circuit are possible.)

The magnitude and phase of the current, given respectively by (3.88) and (3.89), are shown plotted as functions of angular frequency (in the vicinity of ω_0) in fig. 3.38.

At frequencies below resonance the capacitive reactance is greater than the inductive reactance and the circuit is therefore predominantly capacitive. In consequence the current leads the source voltage, the angle of lead increasing asymptotically to the value $\pi/2$ with decreasing frequency. Above resonance the circuit is inductive and the current lags the source voltage.

The shape of the resonance curves shown in fig. 3.38 are governed by the relative values of the reactance and the resistance of the circuit and we now define a quality factor Q_0, applicable to the complete resonant circuit, which enables us to describe in a succinct fashion the shapes of the resonance curves and other properties characteristic of the resonant circuit. The quality factor Q_0 of the *complete* circuit is defined by:

$$Q_0 = \frac{\omega_0 L}{r} \equiv \frac{1}{\omega_0 C r} \tag{3.91}$$

Now $r = r_S + r_L + r_C$, therefore,

$$Q_0 = \frac{\omega_0 L}{r_S + r_L + r_C} = \frac{1/\omega_0 C}{r_S + r_L + r_C} \tag{3.92}$$

and

$$\frac{1}{Q_0} = \frac{r_S}{\omega_0 L} + \frac{r_L}{\omega_0 L} + \frac{r_C}{1/\omega_0 C}$$

that is

$$\frac{1}{Q_0} = \frac{1}{Q_S} + \frac{1}{Q_L} + \frac{1}{Q_C} \tag{3.93}$$

where $Q_S = \omega_0 L/r_S$, and Q_L and Q_C are the quality factors of the inductor and capacitor respectively at the resonant frequency. Thus, given the quality factors of each component separately, and a knowledge of the

154   Alternating current circuits

source resistance, the quality factor of the complete circuit may be found using (3.93). In most practical resonant circuits the quality factor of the capacitor is much greater than that of the inductor and Q_0 depends mainly on Q_S and Q_L. It must be remembered that although the quality factor for the complete circuit is defined strictly at the resonant frequency, it is substantially constant over the narrow band of frequencies with which we are concerned in the theory of resonant circuits.

We now take as a measure of the sharpness of the resonance peak (that is,

Fig. 3.38. Variation of magnitude and phase of the current in a series resonant circuit in the vicinity of resonance.

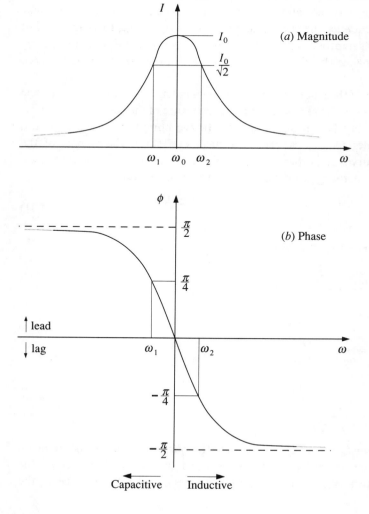

Resonant circuits

the sensitivity of the current amplitude to changes in frequency above and below ω_0), the increment of frequency between the two points at which the amplitude of the resonance curve (fig. 3.38(a)) falls to $1/\sqrt{2}$ of its peak value. The frequencies at these two points are designated ω_1 and ω_2.

At these frequencies the power dissipated in r is just half the power dissipated when $\omega = \omega_0$, and they are referred to as the *half-power frequencies*. The frequency increment $\omega_2 - \omega_1$ is called the *half-power bandwidth* or, simply, the *bandwidth*.

Rewriting (3.88) we obtain

$$I = \frac{V/r}{\left[1 + \left(\frac{\omega L - \frac{1}{\omega C}}{r}\right)^2\right]^{\frac{1}{2}}} = \frac{I_0}{\left[1 + \left(\frac{\omega L - \frac{1}{\omega C}}{r}\right)^2\right]^{\frac{1}{2}}} \quad (3.94)$$

But at the half-power frequencies $I/I_0 = 1/\sqrt{2}$ thus, the denominator in (3.94) is equal to $\sqrt{2}$ and it follows that

$$\omega_2 L - \frac{1}{\omega_2 C} = r$$

or

$$\omega_2^2 - \frac{r}{L}\omega_2 - \frac{1}{LC} = 0$$

Similarly

$$\frac{1}{\omega_1 C} - \omega_1 L = r$$

or

$$\omega_1^2 + \frac{r}{L}\omega_1 - \frac{1}{LC} = 0$$

Hence

$$\omega_1 = -\frac{r}{2L} + \sqrt{\left[\left(\frac{r}{2L}\right)^2 + \frac{1}{LC}\right]} \quad (3.95)$$

$$\omega_2 = \frac{r}{2L} + \sqrt{\left[\left(-\frac{r}{2L}\right)^2 + \frac{1}{LC}\right]} \quad (3.96)$$

Therefore, in terms of the angular frequency, the bandwidth is given by

156 Alternating current circuits

$$\omega_2 - \omega_1 = \frac{r}{L} = \omega_0 \frac{r}{\omega_0 L} = \frac{\omega_0}{Q_0} \qquad (3.97)$$

In terms of the resonant frequency $f_0 = \omega_0/2\pi$:

$$\text{Bandwidth} = \frac{f_0}{Q_0} \text{ Hz} \qquad (3.98)$$

A circuit having a large Q-factor (described as a high-Q circuit) has a small bandwidth; for a low-Q circuit the bandwidth is large. In fig. 3.38(a) the resonance curve is shown as being substantially symmetrical about ω_0, but detailed examination of (3.88) shows that the curve starts at the origin with zero amplitude and falls to zero again only at a theoretically infinite frequency. Considered over all frequencies the resonance curve is not therefore symmetrical, although near resonance the slight departure from symmetry is not normally noticeable.

Multiplying (3.95) by (3.96) we obtain, after some manipulation and reduction,

$$\omega_1 \omega_2 = \omega_0^2 \qquad (3.99)$$

Thus, we see that the resonant frequency is the geometric mean of the half-power frequencies, and it follows that $(\omega_0 - \omega_1) \neq (\omega_2 - \omega_0)$. However, as the bandwidth decreases with increasing Q, the geometric mean and the arithmetic mean approach the same value; for most practical purposes the half-power frequencies can be considered as being symmetrically disposed about ω_0.

As indicated in fig. 3.38(b), at the half-power frequencies the phase of the current with respect to voltage is just $\pm \pi/4$. This follows from (3.89) and the fact that at these frequencies the net reactance is equal to the resistance.

An important property of the resonant circuit arises from the fact that the voltages developed individually across the inductor and the capacitor, at or near resonance, can be many times larger than the voltage of the source itself. (The *sum* of the voltages across the capacitance and inductance in the circuit of fig. 3.36(b) is, of course, zero at resonance since these voltages are of equal magnitude and opposite in phase.) An expression for the voltage V_C across the capacitor at resonance may be obtained by taking the product of the magnitude of the current at resonance and the impedance of the capacitor, that is,

$$V_C = \frac{V}{r} \sqrt{\left[r_C^2 + \left(\frac{1}{\omega_0 C} \right)^2 \right]}$$

$$\approx \frac{V}{r \omega_0 C} \text{ for } \frac{1}{\omega_0 C} \gg r_C$$

or

$$V_C = VQ_0 \tag{3.100}$$

We see that the magnitude of the capacitor voltage is approximately Q_0 times the source voltage. For this reason Q_0 is also known as the *circuit magnification factor*.

When energy is taken from a mains power line, which represents a source of practically zero resistance, a resonant circuit can produce dangerously high voltages. Suppose, for example, $V = 240$ volts, $r = 40$ ohms, and $f = 50$ hertz ($\omega = 2\pi \times 50 = 314$ radians/s), $L = 1$ henry: a capacitor of about 10 microfarads will cause resonance. The current at resonance is $240/40 = 6$ amps, and the reactance of the capacitor is $1/(314 \times 10^{-5}) = 318$ ohms. The voltage across the capacitor is then approximately $6 \times 318 = 1908$ volts. The same voltage appears across the inductor.

The situation is different when the source of energy has an appreciable resistance. A common type of low-frequency laboratory oscillator has an internal resistance of 600 ohms and a maximum output of about 30 volts. Even with a short circuit across its terminals this device can supply only about 0.05 amps. So for the same circuit and frequency considered in the preceding paragraph, the capacitor and inductor voltages could not exceed about $318 \times 0.05 = 16$ volts.

The frequency selective and magnification properties of resonant circuits play an important part in all forms of telecommunications equipment, and we have seen that these properties can be neatly described by means of the quality factor. When, for instance, a resonant circuit is used in a broadcast receiver to select one station from among many others, it is important to use a high-Q circuit so that adjacent stations are not also received at the same time. This situation is illustrated schematically in fig. 3.39. Each broadcasting station transmits information over a narrow band of frequencies $\Delta\omega$ (typically 6 kHz on the medium wave a.m. broadcasting band of 0.5–1.5 MHz).

Let us suppose that the resonant circuit of a receiver is 'tuned' so that its resonant frequency ω_0 coincides with the centre frequency of the $\Delta\omega_2$ band transmitted by station 2. Using a high-Q circuit means that the response to stations 1 and 3 will be negligible because these frequency bands coincide with parts of the resonance curve where the response has fallen practically to zero. A low-Q circuit, on the other hand, will be unselective and allow stations 1 and 3 to be received also. A high quality factor is, therefore, a desirable property for this type of application, but it must also be remembered that if the resonant circuit is given too high a quality factor (dotted line in fig. 3.39), wanted frequencies in the selected signal will be attenuated and information will be lost. It should be stressed that the situation shown in fig. 3.39 is highly simplified; in practice not one but several resonant circuits would be used in a broadcast receiver, each tuned to a slightly different frequency so as to achieve an overall response tailored to meet the requirements of a particular broadcasting band.

So far we have considered the criterion for resonance in a series circuit as being the frequency for which the current rises to its maximum value. However, in many cases it is the voltage developed across the capacitor that is of interest. At first sight it might be expected that since the capacitor voltage is the product of current and impedance, this too would rise to a maximum at the same frequency ω_0 given by (3.90). We now show that this is very nearly but not quite the case.

To find the critical frequency at which the voltage V_C across the capacitor reaches its maximum value we express V_C as a function of ω, making the simplifying assumption that the losses in C are small and that r_C can therefore be neglected. We then have

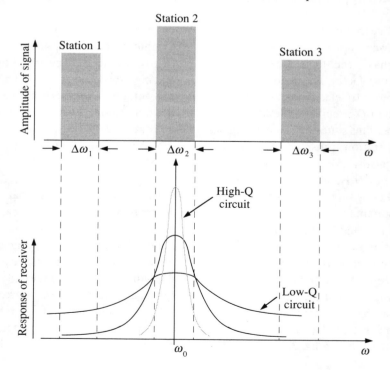

Fig. 3.39. Use of a resonant circuit in broadcast reception.

$$V_C = I\frac{1}{\omega C} = \frac{V}{\sqrt{\left[r^2 + \left(\omega L - \frac{1}{\omega C}\right)^2\right]}} \cdot \frac{1}{\omega C}$$

This is a maximum when

$$\omega^2 C^2 \left[r^2 + \left(\omega L - \frac{1}{\omega C}\right)^2\right] \qquad (3.101)$$

is a minimum. Expanding (3.101), differentiating with respect to ω, and equating to zero; we find that V_C is a maximum when $\omega = \omega_0'$ such that

$$\omega_0' = \frac{1}{\sqrt{LC}} \sqrt{\left(1 - \frac{r^2 C}{2L}\right)}$$

or

$$\omega_0' = \omega_0 \sqrt{\left(1 - \frac{1}{2Q_0^2}\right)} \qquad (3.102)$$

It is clear from this expression that, if Q is large ($Q \geqslant 10$), ω_0' and ω_0 are very nearly the same.

The reason for this slight difference between the values of ω corresponding to maxima of I and V_C will be apparent if we consider the shape of the resonance curve of fig. 3.38(a) in the vicinity of its maximum.

Fig. 3.40 shows the top of the resonance curve, on an enlarged scale, together with a curve depicting the variation of the reactance of the capacitance. From (3.94) we see that, over the range of frequencies for which

Fig. 3.40. Variation of current and capacitive reactance near resonance in a series resonant circuit.

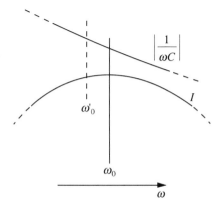

$\left(\omega L - \dfrac{1}{\omega C}\right) \ll r$, the current I is virtually equal to V/r and is independent of frequency; in other words, the curve is rather flat. Thus, although I is a maximum at ω_0 we shall get a higher value of V_C by going to a slightly lower value of ω, where the current has not decreased appreciably and the reactance of the capacitor is slightly larger.

In the whole of the above theory it has been assumed that the circuit parameters L and C are fixed and that the circuit is being brought to resonance by varying the angular frequency ω of the source. However, in broadcast reception, it is nearly always the capacitor that is varied to bring the circuit to resonance at a particular incoming signal frequency. By differentiating (3.101) with respect to C as the variable, we find that in this case the capacitor voltage is maximum at a frequency given by:

$$\omega_0 \sqrt{\left(1 - \dfrac{1}{Q_0^2}\right)}$$

Other criteria for resonance may be found, giving slightly different resonant frequencies, depending on which quantity is being measured or detected and which parameter is being varied. The term resonance must therefore be taken to refer to a class of phenomena and not simply to a unique condition. When, however, the Q-factor of the circuit is large ($Q \geq 10$), the different conditions for resonance are so nearly the same that for most purposes we need not bother to distinguish between them.

It may have become apparent from the discussion so far that fig. 3.38 represents in a general way the behaviour of any series RLC circuit. Rather than having curves that apply to circuits with specific component values, it is useful to construct universal curves in which the quantities plotted are dimensionless ratios. The first step toward obtaining such curves is to define a new dimensionless quantity δ, the *fractional mistuning*:

$$\delta = \dfrac{\omega - \omega_0}{\omega_0} \quad (\delta \ll 1) \tag{3.103}$$

This quantity allows us to effect a considerable simplification in the equations describing the behaviour of a circuit near resonance. Re-writing (3.87) we obtain an expression for the current in the series resonant circuit expressed as a dimensionless ratio:

$$\dfrac{I}{I_0} = \dfrac{1}{1 + \dfrac{j}{r}\left(\omega L - \dfrac{1}{\omega C}\right)} \tag{3.104}$$

Resonant circuits

Now remembering that $\omega_0^2 = 1/LC$ and that $Q_0 = \omega_0 L/r$, the term $(\omega L - 1/\omega C)$ in the denominator of (3.104) may be written

$$\omega L - \frac{1}{\omega C} = L\left(\omega - \frac{1}{\omega LC}\right)$$

$$= \omega_0 L\left(\frac{\omega}{\omega_0} - \frac{\omega_0}{\omega}\right) \quad (3.105)$$

But from (3.103) $\omega/\omega_0 = 1 + \delta$ hence

$$\omega L - \frac{1}{\omega C} = \omega_0 L[1 + \delta - (1+\delta)^{-1}]$$

$$\approx \omega_0 L[1 + \delta - (1-\delta)]$$

or to a very good approximation,

$$\omega L - \frac{1}{\omega C} = 2\omega_0 L\delta \quad (3.106)$$

Using this result in (3.104) we have, near resonance,

$$\frac{I}{I_0} = \frac{1}{1 + j2\left(\frac{\omega_0 L}{r}\right)\delta}$$

or in terms of the Q-factor

$$\frac{I}{I_0} = \frac{1}{1 + j2Q_0\delta} \quad (3.107)$$

The magnitude of I/I_0 is

$$\frac{I}{I_0} = \frac{1}{\sqrt{[1 + (2Q_0\delta)^2]}} \quad (3.108)$$

Then, if we plot I/I_0 versus $Q_0\delta$ we obtain the universal resonance curve. Similarly, a graph of $\tan^{-1}(-2Q_0\delta)$ yields a curve showing the phase of the current as a function of $Q_0\delta$. The half-power frequencies are those for which $Q_0\delta = 1/2$. Fig. 3.41 shows such universal curves derived from (3.107). They should be compared with figs. 3.38(*a*) and (*b*).

Alternative dimensionless representations of the series resonant circuit characteristics may be obtained by using the rationalized form of (3.104). Thus, the real and imaginary components of I/I_0 may be plotted against $Q_0\delta$ as shown in fig. 3.42.

162 *Alternating current circuits*

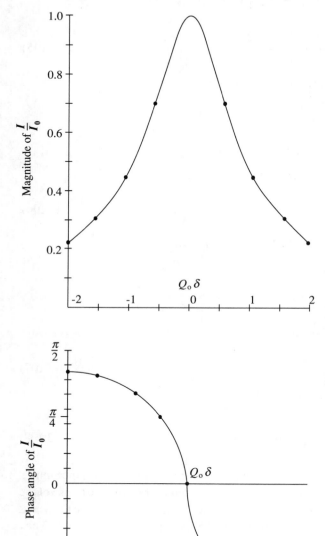

Fig. 3.41. Universal resonance curves: magnitude and phase.

Resonant circuits 163

It should be noted that the simplification achieved by the use of the fractional mistuning parameter δ in the equations describing resonance with frequency as the variable, can also be achieved when some other parameter in the circuit is varied. For example, if the circuit is being brought to resonance by varying C, at a fixed ω, we may write $\delta = (C - C_0)/C_0$, where C_0 is the capacitance required for resonance. In this case, instead of the approximation (3.106), we obtain $\left(\omega L - \dfrac{1}{\omega C}\right) = \delta/\omega C_0$.

Fig. 3.42. Universal resonance curves: real and imaginary components.

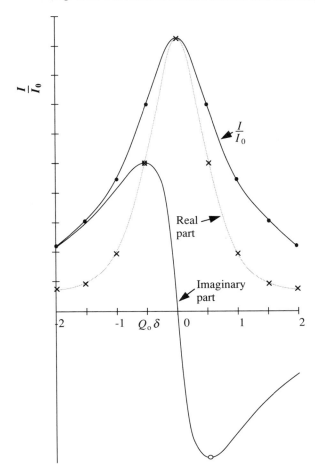

164 *Alternating current circuits*

3.13.3 The parallel resonant circuit

The basic form of the parallel resonant circuit is shown in fig. 3.43. The inductor and capacitor are in this case represented by their parallel equivalents and the circuit is driven by a practical current source. As shown in fig. 3.43(b), the three parallel resistances may be combined to form a single resistance R (or conductance $G = 1/R$). The voltage V across the circuit is given by

$$V = \frac{I_s}{G + j\left(\omega C - \dfrac{1}{\omega L}\right)} \tag{3.109}$$

Comparing this equation with the corresponding equation for the series resonant circuit (3.87), we recognize that the two circuits are duals; with the appropriate change of symbolism, therefore, the theory developed for the series resonant circuit applies in its entirety to the parallel resonant circuit. The frequency at which voltage resonance occurs in a parallel circuit is the same as that for current resonance in a series circuit (both containing the same L and C), namely, $\omega_0 = 1/\sqrt{(LC)}$, the curves of V and phase angle versus ω are similar to those in fig. 3.38, and the universal resonance curves

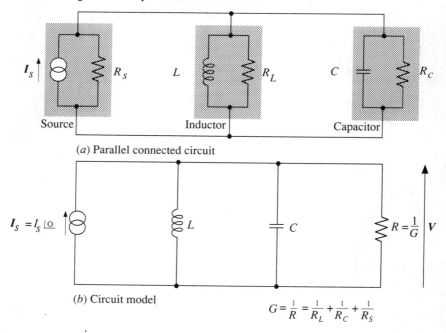

Fig. 3.43. The parallel resonant circuit.

(a) Parallel connected circuit

(b) Circuit model

$$G = \frac{1}{R} = \frac{1}{R_L} + \frac{1}{R_C} + \frac{1}{R_S}$$

Resonant circuits

of fig. 3.41 likewise apply if the dimensionless ratio V/V_0 is used instead of I/I_0. The parallel resonant circuit exhibits *current magnification*; the current in either L or C of fig. 3.43 being Q_0 times the current delivered by the source. This phenomenon is of importance in the type of resonant circuit used in the output stage of a broadcast transmitter. The very heavy currents circulating in such a circuit (the so-called 'tank' circuit) necessitate the use of massive water-cooled conductors.

The principle of duality renders it unnecessary to discuss the parallel circuit of fig. 3.43 in any greater detail, however, we conclude by considering a special case of practical interest for a parallel circuit in which the only significant losses arise in the inductor. In this case the inductor may be represented conveniently by means of its series model, as shown in fig. 3.44. The admittance across the source is

$$Y = j\omega C + \frac{1}{r + j\omega L}$$

$$= j\omega C + \frac{r - j\omega L}{r^2 + \omega^2 L^2}$$

$$= \frac{r - j(\omega L - \omega C(r^2 + \omega^2 L^2))}{r^2 + \omega^2 L^2} \qquad (3.110)$$

Then

$$V = \frac{I_s}{Y} = I_s \left[\frac{r^2 + \omega^2 L^2}{r - j(\omega L - \omega C(r^2 + \omega^2 L^2))} \right] \qquad (3.111)$$

For this circuit, zero phase angle (indicating that the circuit is a pure conductance) and the maximum value of V do not occur at exactly the same frequency. If we define resonance as the conditions for which V and I_s are in phase, then the imaginary part of the denominator in (3.111) must be zero.

Fig. 3.44. Parallel resonant circuit with lossless capacitor.

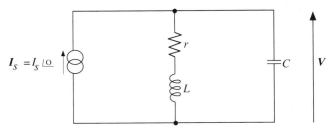

The resonant frequency is then found from

$$\omega_0' L = \omega_0' C (r^2 + {\omega_0'}^2 L^2)$$

which gives

$$\omega_0' = \frac{1}{\sqrt{LC}} \sqrt{\left(1 - \frac{r^2 C}{L}\right)} = \omega_0 \sqrt{\left(1 - \frac{1}{Q_0^2}\right)} \quad (3.112)$$

So the resonant frequency defined in this way differs from the previous definition by the factor $\sqrt{(1 - 1/Q_0^2)}$, but for $Q_0 \geq 10$, ω_0' and ω_0 differ by less than 1%.

Equation (3.112) shows that if $r^2 C > L$, ω_0' becomes imaginary. For this condition, corresponding to $Q < 1$, there will be no frequency for which V and I_s are in phase. Of course, as Q decreases below the numerical value of 1 the assumptions made in the preceding derivation become less valid. For example, if $Q_0 = 2$, then from (3.112), $\omega_0' = \omega_0 \sqrt{(1 - \frac{1}{4})} = 0.87 \omega_0$. It is not usually the practice to use coils at frequencies so high that the Q-factor of the coil has a value much below 1.

The impedance Z_0 of the circuit at resonance may be found from (3.110) by putting the imaginary term equal to zero with $\omega = \omega_0'$ thus

$$Z_0 = \frac{r^2 + (\omega_0' L)^2}{r} = r + \frac{(\omega_0' L)^2}{r}$$

But from (3.112),

$$(\omega_0' L)^2 = \frac{L}{C} - r^2$$

hence

$$Z_0 = r + \frac{1}{r}\left(\frac{L}{C} - r^2\right) = \frac{L}{Cr}$$

The quantity L/Cr, which is called the *dynamic resistance*, may be also expressed in terms of the Q-factor:

$$\text{Dynamic resistance} = \frac{L}{Cr} = Q_0 \omega_0 L = \frac{Q_0}{\omega_0 C} \quad (3.113)$$

3.13.4 Worked example

A coil and a variable capacitor are connected to a voltage generator to form a series resonant circuit. The coil has an inductance of 0.2 mH and a Q-factor of 150; the power factor of the capacitor is 4×10^{-4}. The frequency of the voltage generator is 1 MHz, its internal resistance is

Resonant circuits

2Ω, and its unloaded output voltage is 2 V. Find (a) the value of the capacitor required to tune the circuit to resonance, (b) the effective resistance and Q-factor of the complete circuit at resonance, (c) the complex voltage across the inductor both at the resonant frequency and at a frequency 10 kHz above resonance.

Solution

(a) At the resonant frequency of 1 MHz we have from (3.90)

$$\omega_0 = \frac{1}{\sqrt{(LC)}} = 2\pi \times 10^6$$

therefore,

$$C = \frac{1}{(2\pi \times 10^6)^2 \times 0.2 \times 10^{-3}} = 126.5 \,\text{pF}$$

(b) At 1 MHz the reactance of the inductor is $\omega_0 L = 2\pi \times 10^6 \times 0.2 \times 10^{-3} = 1256\,\Omega$. From the definition of Q-factor (equation (3.73)) we have

$$Q_L = \frac{\omega_0 L}{r_L} = \frac{1256}{r_L} = 150$$

hence

$$r_L = \frac{1256}{150} = 8.373\,\Omega$$

For the capacitor, using (3.85),

$$Q_C = \frac{1}{\text{power factor}} = \frac{1}{4 \times 10^{-4}} = 2500$$

but

$$Q_C = \frac{1/\omega_0 C}{r_C} = \frac{1256}{r_C} = 2500$$

hence

$$r_C = \frac{1256}{2500} = 0.502\,\Omega$$

The total resistance of the circuit is therefore

$$r = r_L + r_C + r_S = 8.373 + 0.502 + 2.0 = 10.875\,\Omega$$

From the definition (3.91), the Q-factor for the complete circuit is

168 Alternating current circuits

$$Q_0 = \frac{\omega_0 L}{r} = \frac{1256}{10.875} = 115.5$$

Alternatively the Q-factor of the complete circuit may be found using (3.93):

$$\frac{1}{Q_0} = \frac{1}{Q_S} + \frac{1}{Q_L} + \frac{1}{Q_C}$$

In this expression

$$Q_S = \frac{\omega_0 L}{r_S} = \frac{1256}{2} = 628,$$

hence

$$\frac{1}{Q_0} = \frac{1}{628} + \frac{1}{150} + \frac{1}{2500}$$

giving

$$Q_0 = 115.5 \text{ as before.}$$

(c) The complex voltage across the inductor is given by

$$V_L = I(r_L + j\omega L)$$

where

$$I = \frac{V}{r + j\left(\omega L - \frac{1}{\omega C}\right)} = \frac{V}{r + j2\omega_0 L\delta}$$

using the approximation (3.106) involving the fractional mistuning δ. We note that the Q-factor of the inductor is high hence $r_L \ll \omega L$. For frequencies near resonance the above expressions may therefore be written more simply as

$$V_L \simeq \frac{V}{r\left(1 + j2\frac{\omega_0 L}{r}\delta\right)} j\omega L \simeq \frac{V}{(1 + j2Q_0\delta)} \frac{j\omega_0 L}{r}$$

or

$$V_L \simeq \frac{jVQ_0}{1 + j2Q_0\delta}$$

At resonance the fractional mistuning $\delta = 0$, hence,

$$V_L = j2 \times 115.5 = j231 = 231\underline{/90}$$

Resonant circuits

That is, the magnitude of the voltage across the inductor is 231 V leading the generator e.m.f. by 90°.

At 10 kHz above resonance, $\delta = 10 \times 10^3/10^6 = 10^{-2}$, hence

$$V_L = \frac{j2 \times 115.5}{1+j(2 \times 115.5 \times 10^{-2})} = \frac{j231}{1+j2.31}$$

$$V_L = \frac{231\underline{/90}}{2.52\underline{/66.6}} = 91.66\underline{/90-66.6} = 91.66\underline{/23.4}$$

†3.13.5 Definition of Q-factor in terms of stored energy

The phenomenon of resonance in an electrical circuit involves the continuous interchange of energy between the inductances and capacitances in the circuit, and during this interchange some energy is lost in the resistances of the circuit. Taking as a specific example the series resonant circuit of fig. 3.36 the current at resonance is $I_0 = V/r$, therefore, the power in r is $(V/r)^2 r$. The energy dissipated per cycle is then given by

$$\left(\frac{V}{r}\right)^2 r \cdot \frac{2\pi}{\omega_0} = \frac{2\pi V^2}{r\omega_0}$$

Now the maximum energy stored in the inductor is, from (1.42),

$$\tfrac{1}{2}L\left(\sqrt{2}\frac{V}{r}\right)^2 = L\frac{V^2}{r^2}$$

therefore,

$$\frac{2\pi \text{ (Maximum energy stored)}}{\text{Energy dissipated per cycle}} = \frac{\omega_0 L}{r} = Q_0 \qquad (3.114)$$

This definition of the quality factor, although derived for a special case, is of general application. It can be applied to mechanical and acoustical vibrating systems as well as to more complicated electrical ones.

†3.13.6 Multiple resonance

Networks containing combinations of more than two independent reactive elements can resonate at more than a single frequency; a phenomenon known as *multiple resonance*. Three examples of such circuits are shown in fig. 3.45. For the purpose of this discussion it is easier to consider the components in these circuits as being lossless so that the resonant frequencies may be deduced from consideration of the reactances only. The results so obtained will be substantially correct provided the Q-factors of the components exceed 10 or so.

Considering first the circuit of fig. 3.45(a), this consists of two sections: first the inductance L_1 with reactance X_1, second the parallel resonant circuit, formed by L_2 and C_1, with reactance X_p. The variation of reactances X_1 and X_p is shown in fig. 3.46. The curve of reactance X_1 with frequency is, of course, a straight line through the origin (fig. 3.46(a)). The reactance X_p (fig. 3.46(b)) is zero at $\omega = 0$, because the inductance L_2 forms a short circuit, and is asymptotic to zero at $\omega \to \infty$ because the capacitor forms a short circuit. Resonance occurs at some intermediate frequency $\omega = \omega_p$ where the reactance rises to a theoretically infinite value. The form of the variation of the reactance $X = (X_1 + X_p)$ for the complete circuit is obtained by combining the individual curves, as shown in fig. 3.46(c). It is seen that a second resonant frequency ω_2 occurs where the reactance of X_1 (inductive)

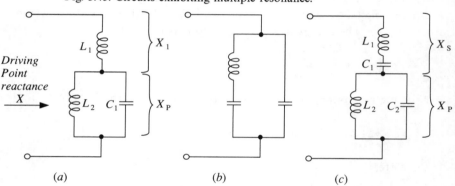

Fig. 3.45. Circuits exhibiting multiple resonance.

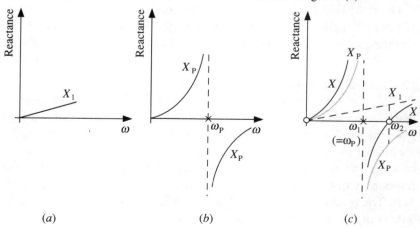

Fig. 3.46. Variation of reactance in the circuit of fig. 3.45(a).

Resonant circuits

and the reactance X_p (capacitive) cancel. The resonant frequency $\omega_1 = \omega_p$ is unchanged. Similar behaviour is exhibited by the circuit of fig. 3.45(b) since this is the dual of the circuit in fig. 3.45(a); for the dual circuit, however, all the curves shown in fig. 3.46 must be interpreted in terms of susceptance rather than reactance.

We now consider the circuit shown in fig. 3.45(c). This circuit again has two sections: L_2 and C_2 forming a parallel resonant circuit with resonant frequency ω_p, and L_1 and C_1 forming a series resonant circuit with resonant frequency ω_s. The variation of the total reactance $X = (X_s + X_p)$ of this circuit is indicated by the solid line in fig. 3.47, which is obtained by combining the curve for the series circuit (fig. 3.37) with that for the parallel circuit (fig. 3.46(b)). We see that the complete circuit resonates at three different frequencies: at ω_1 and ω_3 the reactance is zero, and at $\omega_2(=\omega_p)$ the reactance is infinite. The curves have been drawn on the assumption that $\omega_s < \omega_p$ but it is easy to see that the same conclusions will apply for the conditions $\omega_s \geqslant \omega_p$. Three resonant frequencies will always occur, the two frequencies for which the reactance is zero lying on either side of the frequency for which the reactance is infinite.

Expressions for the resonant frequencies $\omega_1, \omega_2, \omega_3$, are now derived in terms of the circuit parameters. The driving-point reactance function of the circuit in fig. 3.45(c) is

$$X = X_s + X_p$$

Fig. 3.47. Variation of reactance in the circuit of fig. 3.45(c).

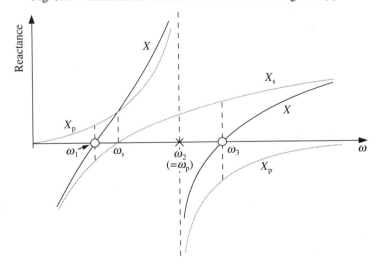

$$X = \left(\omega L_1 - \frac{1}{\omega C_1}\right) + \frac{\omega L_2\left(-\frac{1}{\omega C_2}\right)}{\left(\omega L_2 - \frac{1}{\omega C_2}\right)}$$

$$= \frac{L_1}{\omega}\left(\omega^2 - \frac{1}{L_1 C_1}\right) - \frac{\frac{L_2}{C_2}}{\frac{L_2}{\omega}\left(\omega^2 - \frac{1}{L_2 C_2}\right)}$$

$$= \frac{\left(\omega^2 - \frac{1}{L_1 C_1}\right)\left(\omega^2 - \frac{1}{L_2 C_2}\right) - \frac{\omega^2}{L_1 C_2}}{\frac{\omega}{L_1}\left(\omega^2 - \frac{1}{L_2 C_2}\right)}$$

$$= \frac{\omega^4 - \omega^2\left(\frac{1}{L_1 C_1} + \frac{1}{L_2 C_2} + \frac{1}{L_1 C_2}\right) + \frac{1}{L_1 C_1 L_2 C_2}}{\frac{\omega}{L_1}\left(\omega^2 - \frac{1}{L_2 C_2}\right)}$$

In this equation the numerator is a quadratic in ω^2. If $\omega_1{}^2$ and $\omega_3{}^2$ are solutions of this, we may write:

$$X = \frac{(\omega^2 - \omega_1{}^2)(\omega^2 - \omega_3{}^2)}{\frac{\omega}{L_1}(\omega^2 - \omega_2{}^2)} \qquad (3.115)$$

where $\omega_2{}^2 = 1/L_2 C_2$.

The expression (3.115) shows that the reactance function X is zero at $\omega = \omega_1$ and at $\omega = \omega_3$, and that it is infinite at $\omega = \omega_2$. We describe this by saying that *zeros* of X occur at ω_1 and ω_3, and that a *pole* occurs at ω_2. It will be observed that a pole also occurs at the origin ($\omega = 0$) for this circuit. Poles and zeros are indicated by the crosses and circles on the ω-axis in figs. 3.46(c) and 3.47.

Circuits of greater complexity give rise to a greater number of poles and zeros but the driving-point reactance (or susceptance) function of a circuit always conforms to the general pattern of behaviour described above and which is illustrated in fig. 3.48. The slopes of the curves of reactance or susceptance are everywhere positive, and poles and zeros occur alternately along the frequency axis. If there is a continuous path through the circuit from terminal to terminal formed by one or more inductive elements (as in fig. 3.45(a)), then a zero occurs at the origin, otherwise there is a pole. Likewise, if there is a continuous path formed by capacitive elements, there

Resonant circuits

is a zero at $\omega = \infty$, otherwise there is a pole. The alternative possibilities of either a pole or a zero occurring at the extremes of frequency is indicated in fig. 3.48 by the dotted lines. Note, however, that whatever happens at these extremes does not change the alternating pattern of poles and zeros. The poles and zeros other than those at the two extremes are called the *internal* poles and zeros, and it may be shown rigorously that the driving-point reactance function is uniquely specified, apart from a vertical scaling factor, by the location of its internal poles and zeros. If the value of the function at a single frequency is also specified thus determining the scaling factor, then the function is completely specified. This is known as *Foster's reactance theorem* (see reference 2). It is useful for the purposes of synthesizing networks having specified characteristics.

†3.13.7 Inductively coupled resonant circuits

In telecommunications equipment, radio receivers for example, high-frequency signals are amplified by means of active devices such as transistors. The signals usually occupy a relatively narrow frequency band disposed about some large central frequency, and to amplify this narrow band it is usual to employ tuned coupling networks between successive stages of amplification. Such networks often take the form of two inductively coupled coils each coil forming part of a resonant circuit. A typical arrangement is shown in fig. 3.49. In this circuit the output of the transistor is represented by a practical current source. Coupling to the next stage is achieved by means of L_1, L_2 and M, which together constitute a *radio-frequency* or *high-frequency transformer*.

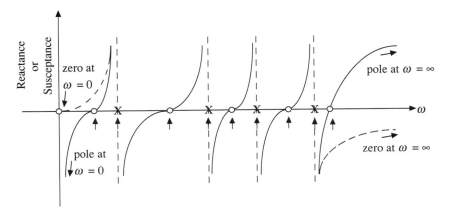

Fig. 3.48. Poles and zeros of a driving-point reactance (or susceptance) function. (Internal poles and zeros are indicated by arrows.)

The frequency response characteristic of the complete circuit is governed by the frequencies to which the primary and secondary circuits are tuned, their Q-factors, and the degree of coupling between them. Tuning is achieved by fine adjustment of the values of the capacitors or inductors; the mutual inductance is usually fixed once the designer has decided upon a suitable value.

If we apply a variable frequency drive voltage of constant amplitude to the input of the circuit of fig. 3.49, with the primary and secondary circuits tuned to the *same* resonant frequency, we observe an output voltage of the form shown in fig. 3.50. Here we see curves for three different values of the mutual inductance. When the mutual inductance is very small, the normal type of resonance curve with a single peak at the resonant frequency is

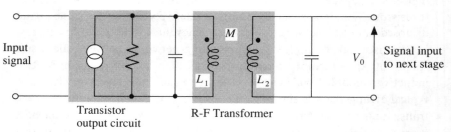

Fig. 3.49. Basic circuit for one stage of a tuned radio-frequency amplifier.

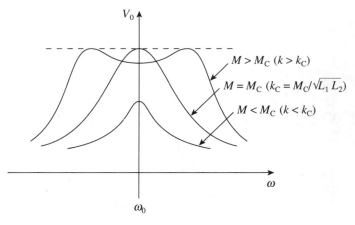

Fig. 3.50. Curves of output voltage as a function of frequency for the circuit of fig. 3.49 with mutual inductance (or coupling coefficient) as a parameter. M_c is the critical value of mutual inductance.

Resonant circuits

observed. As the mutual inductance is increased the peak of the curve rises until a critical value is reached when no further increase is obtained, instead the peak splits into two separate peaks which move further apart as the mutual inductance is increased beyond the critical value. This phenomenon is known as *double-humping*.

For reasons that will become apparent it is convenient to express the degree of coupling in terms of the coupling coefficient $k = M/\sqrt{(L_1 L_2)}$. The critical value for the coupling coefficient is approximately 0.01, when the Q-factors of the primary and secondary circuits are of the order 100.

We shall now proceed to investigate the behaviour of this circuit analytically but in order to do so we first simplify the circuit by applying the Thévenin–Norton transformation to the left-hand portion of the circuit consisting of the current generator and the capacitor. When this is done we obtain the circuit shown in fig. 3.51 in which R_1 accounts for both the primary coil loss-resistance and the output resistance of the transistor. R_2 accounts for all loss resistance in the secondary circuit. The circuit now consists of two separate series resonant circuits coupled by the mutual inductance. It will be sufficient for our purpose to solve for the current I_2 since for small frequency changes about resonance, the output voltage will be very nearly proportional to the modulus of this current. Using a procedure similar to that established in section 3.12 we obtain by mesh analysis:

for the primary circuit

$$\left(R_1 + j\omega L_1 + \frac{1}{j\omega C_1}\right)I_1 - j\omega M I_2 = V$$

and for the secondary circuit

$$-j\omega M I_1 + \left(R_2 + j\omega L_2 + \frac{1}{j\omega C_2}\right)I_2 = 0$$

Fig. 3.51. Circuit model for fig. 3.49.

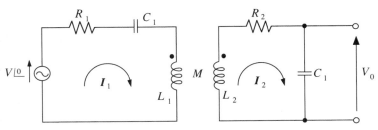

176 *Alternating current circuits*

Putting $X_1 = \left(\omega L_1 - \dfrac{1}{\omega C_1}\right)$, $X_2 = \left(\omega L_2 - \dfrac{1}{\omega C_2}\right)$ and solving for the modulus of I_2 we obtain

$$I_2 = \dfrac{V}{\left\{\left(\dfrac{R_1 X_2 + R_2 X_1}{\omega M}\right)^2 + \left(\omega M + \dfrac{R_1 R_2}{\omega M} - \dfrac{X_1 X_2}{\omega M}\right)^2\right\}^{\frac{1}{2}}} \qquad (3.116)$$

Now consider the denominator of this expression at the resonant frequency; both X_1 and X_2 will be zero since primary and secondary circuits are tuned alike to the same frequency. In this case (3.116) reduces to

$$I_2 = \dfrac{V}{\omega_0 M + \dfrac{R_1 R_2}{\omega_0 M}}$$

Regarding M as the variable parameter we see that there will be a minimum value of the denominator, and therefore a maximum in I_2, when M is equal to a critical value $M = M_c$ such that

$$\dfrac{d}{dM}\left(\omega_0 M + \dfrac{R_1 R_2}{\omega_0 M}\right) = 0$$

that is when $\omega_0^2 M_c^2 = R_1 R_2$

The above relationship may be expressed alternatively in terms of the coupling coefficient $k_c = M_c/\sqrt{(L_1 L_2)}$. We have in this case

$$\omega_0^2 M_c^2 = \omega_0^2 k_c^2 L_1 L_2 = R_1 R_2$$

or

$$k_c^2 = \dfrac{R_1 R_2}{\omega_0^2 L_1 L_2}$$

Recalling our definition of Q-factor (3.91) this may be expressed as

$$k_c = \dfrac{1}{\sqrt{(Q_1 Q_2)}} \qquad (3.117)$$

where Q_1 and Q_2 are the quality factors of the primary and secondary circuits. We see that the use of the coupling coefficient leads to a particularly simple expression.

It is possible to show that if the coupling is less than critical, then I_2 falls monotically either side of the resonance frequency. If the coupling is greater than critical ($k \geqslant k_c$), then the current rises to a maximum on either side of the resonant frequency; the location of these maxima may be found by differentiation of the complete denominator of (3.116). The procedure is simplified if the assumption is made that ωM remains substantially constant, with value $\omega_0 M$, over the narrow band of frequencies of interest around ω_0. It is convenient also to introduce the fractional mistuning δ, (3.103); then, by (3.106), $X_1 = 2\omega_0 L_1 \delta$ and $X_2 = 2\omega_0 L_2 \delta$. The denominator of (3.116) may then be written:

$$\frac{1}{\omega_0 M} \left\{ [R_1(2\omega_0 L_2 \delta) + R_2(2\omega_0 L_1 \delta)]^2 + [\omega_0^2 M^2 + R_1 R_2 - (2\omega_0 L_1 \delta)(2\omega_0 L_2 \delta)]^2 \right\}^{1/2}$$

Differentiating the square of this expression and equating the result to zero gives

$$[2R_1 \omega_0 L_2 + 2R_2 \omega_0 L_1]^2 - 8\omega_0^2 L_1 L_2 [\omega_0^2 M^2 + R_1 R_2 - 4\omega_0^2 L_1 L_2 \delta^2] = 0$$

We may express this equation in terms of the quality factors for the primary and secondary circuits: multiplying through by the factor $1/(\omega_0^2 L_1 L_2)^2$ and putting $M^2 + k^2 L_1 L_2$ gives

$$\left[\frac{2}{Q_1} + \frac{2}{Q_2} \right]^2 - 8 \left[k^2 + \frac{1}{Q_1 Q_2} - 4\delta^2 \right] = 0$$

Further simplification is obtained if it is assumed that the quality factors for the primary and secondary circuits are equal. Then $1/Q_1 = 1/Q_2 = k_c$, and the above expression becomes

$$16k_c^2 - 8(k^2 + k_c^2 - 4\delta^2) = 0$$

which gives, finally,

$$4\delta^2 = k^2 - k_c^2$$

or

$$\delta = \pm \tfrac{1}{2}\sqrt{(k^2 - k_c^2)} \qquad (k \geqslant k_c) \tag{3.118}$$

The over-coupled condition with primary and secondary circuits tuned to the same frequency produces a desirable band-pass characteristic but it is found in practice to be difficult to set up the circuit correctly because the primary and secondary circuits interact strongly in the over-coupled condition. It is more usual therefore to design cascaded radio-frequency amplifiers using under-coupled transformers with primary and secondary circuits each tuned to slightly different frequencies to produce the desired overall band-pass characteristic. This is known as *stagger tuning*.

The phenomenon of double-humping described in this section is not confined to tuned coupled circuits; the same phenomenon arises in many other physical systems which exhibit oscillatory characteristics. Such systems fall under the general heading of *coupled oscillators*.

3.14 Summary

A.C. networks consist of interconnected elements of resistance, inductance, mutual inductance and capacitance excited by ideal sources producing sinusoidal voltage and current waveforms of the same frequency. The theory developed for such networks refers to steady-state conditions in which the amplitudes and phases of all waveforms are time invariant.

Representation of sinusoidal voltages and currents by means of the complex exponential leads to a succinct notation ($\mathbf{V} = V\underline{/\theta}$ or $\mathbf{I} = I\underline{/\beta}$) for describing the amplitudes and phases of currents and voltages in every branch of a network. Each branch in a network is then characterized by a complex impedance ($\mathbf{Z} = Z\underline{/\phi}$), being the ratio of complex voltage to complex current at that branch. Complex voltages, currents and impedances may be combined and manipulated according to the normal rules of linear network analysis to provide a complete description of network behaviour at any terminal pair or port. The use of the complex exponential (or phasor) notation also provides a convenient means of illustrating graphically (using the Argand diagram as a basis) the relationships between voltages and currents in parts of a network.

For many important types of circuit, including filters and bridge circuits, it is of primary interest to determine the behaviour of the circuit as a function of frequency. Such circuits often have two ports in which case the relationship between input and output port parameters can be described by a transfer function $H(j\omega) = |H(j\omega)|\underline{/\phi(\omega)}$. The transfer function is conveniently illustrated by means of the polar or locus diagram (based on the Argand diagram), or by the Bode diagram in which amplitude and phases are plotted separately on graphs having linear–log scales.

Some types of circuit exhibit marked changes in their impedance properties at certain critical frequencies known as the resonant frequencies. In a circuit containing a single inductance L and a single capacitance C, there is one resonant frequency only given by $\omega_0 = 1/\sqrt{(LC)}$.

The properties of a resonant circuit are determined by the inductance capacitance and resistance associated with the circuit elements of which it is composed. The Q-factor is a mathematically convenient parameter for describing these properties. If, in a series connected circuit the effective resistance is R, the Q-factor is given by $Q_0 = \omega_0 L/R$ where ω_0 is the resonant frequency. Multiple resonances can occur in circuits containing several inductive and capacitive elements, or in circuits containing mutual inductance.

3.15 Problems

1. A d.c. generator of e.m.f. E and an a.c. generator of e.m.f. $E_m \sin \omega t$, each having negligible internal impedance, are connected in series to a load resistance R. Obtain an expression for the power dissipated in R, and hence an expression for the equivalent r.m.s. e.m.f. in the circuit.

2. Calculate the r.m.s. value and form factor for:
(a) a triangular wave of unit amplitude;
(b) a sine wave of unit amplitude

3. Find the equivalent complex impedance and admittance of the circuits shown in fig. 3.52, in both Cartesian and polar forms.

4. Represent the following voltages on a phasor diagram, hence find their sum, both graphically and by calculation.
 (i) $200 \sin(\omega t + \pi/6)$ V,   (ii) $150 \cos(\omega t + \pi/6)$ V,
 (iii) $200 \sin(\omega t + 5\pi/6)$ V,   (iv) $-150 \cos(\omega t + 5\pi/6)$ V.

5. The three elements shown in fig. 3.53 are connected in series across a 50 Hz supply, and the current drawn from the supply is found to be 7 A. The voltages are $V_R = 120$, $V_L = 240$ and $V_C = 160$. Find the supply voltage and

Fig. 3.52. Circuit for problem 3.

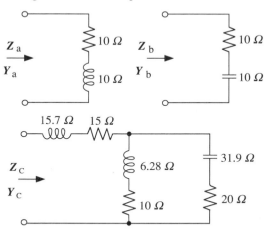

Fig. 3.53. Circuit for problem 5.

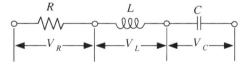

its phase relative to the current. Find also the values of the three elements.

6. What 50 Hz voltage V must be applied to the circuit of fig. 3.54 to produce a steady-state current of 5 A in the capacitor?

(Manchester University)

7. Figure 3.55 shows a circuit used in TV and radar amplifiers. Show that for $L=0$ the ratio V_o/V_i falls from 0.5 at very low frequencies to 0.35 at a frequency of 1.59 MHz, but that if $L=400\,\mu$H this drop is reduced, the ratio then being 0.47 at 1.59 MHz. What are the phase angles between V_o and V_i at this frequency when $L=0$ and when $L=400\,\mu$H?

(Hint: apply voltage divider principle (admittance formulation, table 3.1).)

8. In fig. 3.56 V is a variable frequency, constant-voltage source. Find an expression for the frequency at which the voltage across AB is in phase with V.

(Liverpool University)

9. Show for the circuit of fig. 3.57 that

$$\text{as } \omega \to 0, \quad \frac{V_o}{V_i} \to \frac{R_2}{R_1+R_2} = m_0$$

$$\text{as } \omega \to \infty, \quad \frac{V_o}{V_i} \to \frac{C_1}{C_1+C_2} = m_\infty$$

Fig. 3.54. Circuit for problem 6.

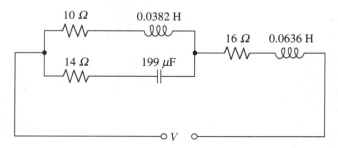

Fig. 3.55. Circuit for problem 7.

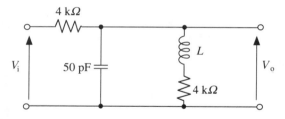

that $\left|\dfrac{V_o}{V_i}\right| = \sqrt{m_o m_\infty}$ when $\omega^2 = \dfrac{R_1 + R_2}{C_1(C_1 + C_2)R_1^2 R_2}$

and that the phase lead of V_o on V_i is

$$\theta = \tan^{-1}(\omega C_1 R_1) - \tan^{-1}\left(\dfrac{m_o}{m_\infty}\omega C_1 R_1\right)$$

Discuss the case of $R_1 C_1 = R_2 C_2$
(Manchester University)

10. Show for the circuit of fig. 3.58 that as the frequency is raised from zero, $\left|\dfrac{V_o}{V_i}\right|$ increases from $\dfrac{r}{2R+r}$ at $\omega = 0$ to 1 at $\omega \to \infty$, without a maximum or minimum value between these extremes.

Show that the output lags the input by $\theta = \tan^{-1}\omega CR + \tan^{-1}\dfrac{\omega CRr}{2R+r}$, or leads by $(\pi - \theta)$.
(Manchester University)

11. Sketch the straight line approximation for the frequency response of i_2/i_1 where $i_1 = I\sin\omega t$ in the circuit of fig. 3.59. You may assume that C is of

Fig. 3.56. Circuit for problem 8.

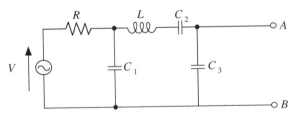

Fig. 3.57. Circuit for problem 9.

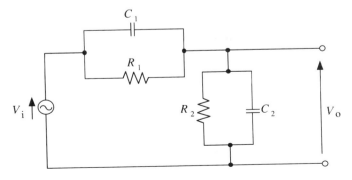

such a value that any frequency-dependent effects it introduces occur at frequencies which are very much higher than those at which frequency-dependent effects due to L are evident. The effects of C and L on the frequency response may thus be considered separately. Give expressions for any 'corner frequencies' shown in the response and values of the ratio $|i_2/i_1|$ on any horizontal parts of the response.
(Lancaster University)

12. The Schering bridge shown in fig. 3.60 is used for measuring the power loss in dielectrics. The specimens are in the form of discs 0.3 cm thick and have a dielectric constant of 2.3. The area of each electrode is 314 cm² and the bridge frequency is 50 Hz. The bridge balances when $R_3 = 1000\,\Omega$, $C_1 = 50\,\text{pF}$, and $C_4 = 1960\,\text{pF}$. Find the value of R_4, C_2, R_2 and the power factor for the dielectric.
(Liverpool University)

13. Show that there is zero output from the bridged-T section of fig. 3.61(a) if

$$Z = -(Z_1 + Z_2) - \frac{Z_1 Z_2}{Z_3}$$

Fig. 3.58. Circuit for problem 10.

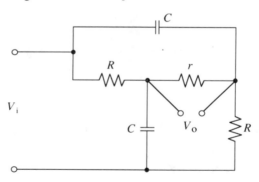

Fig. 3.59. Circuit for problem 11.

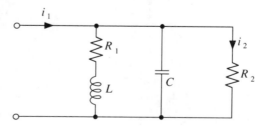

Determine the form of Z in the section of fig. 3.61(b) if there is to be zero output at angular frequency ω_0.
(Manchester University)

14. Show for the circuit of fig. 3.62 that

$$I_1 = \left[\frac{R_2 + j\omega(L_2 \pm M)}{R_1 R_2 + j\omega(L_1 R_2 + L_2 R_1) - \omega^2(L_1 L_2 - M^2)}\right]V$$

$$I_2 = \left[\frac{R_1 + j\omega(L_1 \pm M)}{R_1 R_2 + j\omega(L_1 R_2 + L_2 R_1) - \omega^2(L_1 L_2 - M^2)}\right]V$$

and that the p.d. between A and B is zero when

$$\frac{R_1}{R_2} = \frac{L_1 \pm M}{L_2 \pm M}$$

(Manchester University)

Fig. 3.60. Circuit for problem 12.

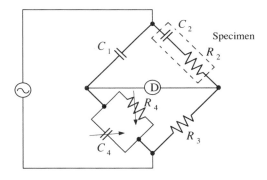

Fig. 3.61. Circuit for problem 13.

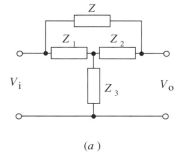 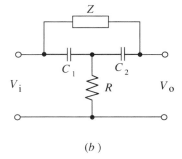

(a) (b)

15. Give expressions for the self and mutual impedances of the two meshes in the circuit of fig. 3.63.

16. A parallel circuit consists of a coil of resistance R and inductance L and a variable capacitance C. A fixed-frequency voltage source, of angular frequency ω, is connected to this circuit. If $Q = \omega L/R$ and $C_0 = 1/(\omega^2 L)$ calculate the value of C, in terms of Q and C_0, such that the current drawn from the supply is a minimum.
(Newcastle University)

17. Why is a parallel resonant circuit sometimes known as a rejector circuit?

Show that the resonant frequency f_0 and dynamic resistance R_d of a parallel resonant circuit consisting of a capacitor C in parallel with a coil of inductance L and resistance R are given by

$$f_0 = \frac{1}{2\pi} \sqrt{\left(\frac{1}{LC} - \frac{R^2}{L^2}\right)}; \qquad R_d = \frac{L}{CR}$$

Fig. 3.62. Circuit for problem 14.

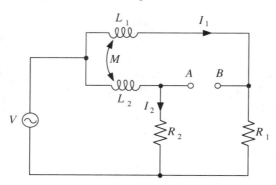

Fig. 3.63. Circuit for problem 15.

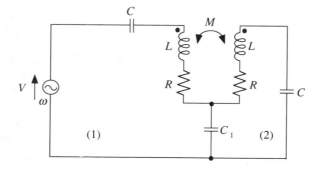

In the circuit of fig. 3.64, the capacitor C has been adjusted until the voltage E_L is at its minimum value. Calculate the source frequency and E_L under these conditions.
(Newcastle University)

18. In the circuit of fig. 3.65, $R = 20\,\Omega$, $L_1 = 2\,\text{H}$, $C_1 = 10^{-2}\,\text{F}$, $C_2 = 5 \times 10^{-3}\,\text{F}$.

(a) Calculate the complex impedance $Z(j\omega)$ seen by a voltage source $v(t) = 80\cos 10t$. Find $i(t)$.

(b) Construct the phasor diagram for I and V and the impedance diagram for $Z(j\omega)$.

(c) What are the resonant frequencies of this circuit, that is $X(j\omega_1) = 0$ and $X(j\omega_2) = \infty$? [$X(j\omega)$ is the reactance.]
(Sheffield University)

19. The response of two non-identical tuned circuits coupled together by

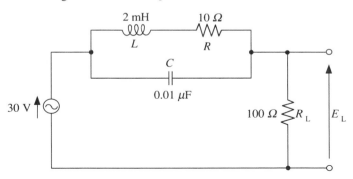

Fig. 3.64. Circuit for problem 17.

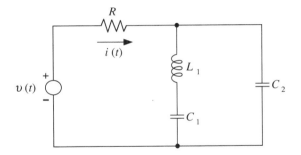

Fig. 3.65. Circuit for problem 18.

Fig. 3.66. Circuit for problem 19.

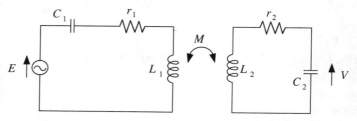

the mutual inductance $M = k\sqrt{(L_1 L_2)}$, as shown in fig. 3.66, is given by

$$\left|\frac{V}{E}\right| = \frac{k\sqrt{(C_1/C_2)}}{\sqrt{\left[\left(\frac{1}{Q_1 Q_2} + k^2 - x^2\right)^2 + x^2\left(\frac{1}{Q_1} + \frac{1}{Q_2}\right)^2\right]}}$$

where

$$x = \frac{\omega}{\omega_0} - \frac{\omega_0}{\omega}; \quad Q_1 = \frac{\omega_0 L_1}{r_1} \text{ and } Q_2 = \frac{\omega_0 L_2}{r_2}$$

Show that for critical coupling

$$k^2 = \tfrac{1}{2}\left(\frac{1}{Q_1} + \frac{1}{Q_2}\right)^2 - \frac{1}{Q_1 Q_2}$$

and the $-3\,\text{dB}$ points on the critically coupled response curve are given by

$$x = \pm \frac{1}{\sqrt{2}}\left(\frac{1}{Q_1} + \frac{1}{Q_2}\right)$$

Find the 3 dB bandwidth if $f_0 = 1\,\text{MHz}$, $Q_1 = 100$ and $Q_2 = 80$.
(Glasgow University: Third year)

4

Power and transformers in single-phase circuits

4.1 Introduction

For the sinusoidal steady state, one can calculate the total power supplied to a circuit consisting of linear elements by adding directly the power absorbed by each individual resistive element in the circuit. However, it is often more convenient to express power in terms of the voltage across and the current supplied to the input terminals of a circuit whose detailed configuration is unknown or is of no interest.

In all electric power distribution networks voltage and frequency are maintained substantially constant. A given load will draw a current whose amplitude and phase (relative to the power line voltage) depend upon the load impedance. On the other hand in electronic and telecommunication networks, signal power rather than voltage is fixed and we are concerned more with arranging source and load conditions to achieve maximum power transfer from one part of a circuit to another.

For the above reasons the treatment of power in electrical circuits depends to a marked extent on the type of circuit under consideration. In this chapter we develop general methods for determining the power and total energy supplied to, or dissipated within, a circuit. We also consider one of the most important components involved in the utilization and transmission of power; namely, the transformer.

4.2 Average power

Consider a network or load as shown in fig. 4.1, supplied at voltage V (r.m.s. magnitude V) and drawing current I (r.m.s. magnitude I). If the network contains reactive elements, voltage and current will differ in phase by some angle ϕ. It is convenient, particularly in power distribution calculations, to take the voltage as phase reference. The instantaneous current and voltage are then described by $v = V_m \sin\omega t$ and $i = I_m \sin(\omega t + \phi)$, and the instantaneous power to the load is

$$p = vi = V_m I_m \sin\omega t \sin(\omega t + \phi) \tag{4.1}$$

In fig. 4.2, p, v, and i are shown as functions of time. Instantaneous power is sinusoidal with a frequency double that of the supply voltage. By our previously established sign convention (section 1.4) p is positive when energy flows into the load.

The total energy W supplied during the interval t_0 to t is (see section 1.4)

$$W = \int_{t_0}^{t} p \, dt$$

therefore the net area under one complete cycle of the instantaneous power curve represents the energy supplied to the load during a time equal to this period (equal to half the period of the supply frequency). When $\phi = 0$ (corresponding to a pure resistive load), this area is everywhere positive. As ϕ increases in magnitude, the negative area increases until for $|\phi| = \pi/2$ (corresponding to a pure reactive load), positive and negative areas are equal and there is no net transfer of energy to the load. For passive linear elements ϕ is always within the range $-\pi/2 \leq \phi \leq \pi/2$, so that the net area will never become negative.

The *average power* P supplied to the load is equal to the net area under one cycle of p in fig. 4.2 divided by the period of p. Thus the average power is maximum when $\phi = 0$ and zero when $|\phi| = \pi/2$.

An expression for P in terms of conveniently measurable quantities may be derived by expanding (4.1) using the appropriate trigonometrical identities, viz.,

$$p = \tfrac{1}{2} V_m I_m (\cos\phi - \cos 2\omega t \cos\phi + \sin 2\omega t \sin\phi) \tag{4.2}$$

Fig. 4.1. Instantaneous and phasor voltages and currents at the terminals of a reactive network or 'load'. Current leads voltage by phase angle ϕ.

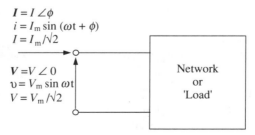

$I = I \angle \phi$
$i = I_m \sin(\omega t + \phi)$
$I = I_m/\sqrt{2}$

$V = V \angle 0$
$v = V_m \sin \omega t$
$V = V_m/\sqrt{2}$

Average power

If T is the period of p, the average power is

$$P = \frac{1}{T}\int_0^T p\,dt = \tfrac{1}{2}V_m I_m \cos\phi = VI\cos\phi \tag{4.3}$$

In this expression V and I are the r.m.s. values. It is an important convention that all power line voltages and all voltages and currents marked on a.c. appliances such as a motor, for example, are r.m.s. values. Thus the European standard voltage of 240 V had a maximum value of $240\sqrt{2} = 364$ V, similarly the standard in many parts of the U.S.A. is 120 V, which rises to $120\sqrt{2} = 169$ V. The same applies to the ratings of fuses and circuit breakers; a fuse marked 5 A must carry a peak current of just over 7 A. In all a.c. problems the data are assumed to refer to r.m.s. quantities unless specifically stated otherwise.

Because, as noted above, $-\pi/2 \leqslant \phi \leqslant \pi/2$, $\cos\phi$ in (4.3) is always positive or zero, so P is never negative. Note also that the expression (4.3) is independent of the choice of the quantity taken as phase reference.

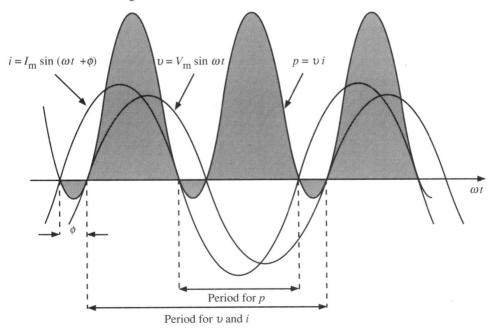

Fig. 4.2. Instantaneous values of voltage, current and power for the circuit of fig. 4.1.

4.3 Reactive power and apparent power

Further insight into the details of power flow comes from examination of (4.2). The three terms on the right and their sum are shown in fig. 4.3. The sum corresponds, of course, to the instantaneous power curve in fig. 4.2.

The first two terms on the right can be written

$$\tfrac{1}{2}V_m I_m(\cos\phi - \cos 2\omega t \cos\phi) = VI\cos\phi(1 - \cos 2\omega t)$$

The sum of just these two terms represents the *instantaneous real power*; it never becomes negative, oscillating in amplitude between $2VI\cos\phi$ and zero. The average value is $VI\cos\phi$, which we have identified as the average power P. This quantity is often referred to as the *real power* or *active power*.

The third term on the right, namely,

$$\tfrac{1}{2}V_m I_m \sin 2\omega t \sin\phi = VI\sin\phi\sin 2\omega t,$$

represents energy that oscillates between the power source and the purely reactive elements in the load. Its average value is zero and its magnitude is $VI\sin\phi$. This quantity is referred to as *reactive power*; however, since it does

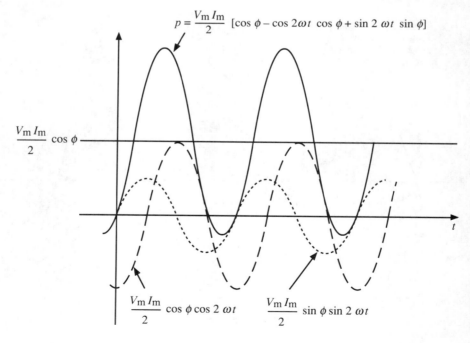

Fig. 4.3. The three components of instantaneous power (equation 4.2).

Reactive power and apparent power

not represent energy actually absorbed by the load, the designation *volt–amperes reactive* (abbreviated vars) is also commonly employed.

The sign of this term will depend on the sign of the phase angle ϕ, which in turn will depend on whether voltage or current is chosen as phase reference. In this instance, with voltage as phase reference, a capacitive load will draw a leading current, that is, ϕ will be positive and such a load is said to draw positive vars. An inductive load will draw negative vars. This convention has been adopted by the International Electrotechnical Convention, but the reader should be aware that the alternative convention (capacitive vars: negative; inductive vars: positive) is often used.

The symbol for reactive power or volt–amperes reactive is Q, thus

$$\text{Reactive power } Q = VI\sin\phi \tag{4.4}$$

When a load on a power system has a phase angle other than zero, the resulting reactive power represents a requirement in system current capacity in addition to that necessary to supply the average power, that is, the actual power used to produce work or heat. It is customary, therefore, to specify the *apparent power* required by a load. This is simply the product of the effective (r.m.s.) voltage and effective current at the terminals of the load; it is usually designated by the symbol S. Thus,

$$\text{Apparent power } S = VI \tag{4.5}$$

From the definitions of P, Q and S we see that the apparent power is given by quadrature addition of real and reactive power, that is,

$$P^2 + Q^2 = (VI\cos\phi)^2 + (VI\sin\phi)^2 = (VI)^2 = S^2$$
$$S = \sqrt{(P^2 + Q^2)} \tag{4.6}$$

Fig. 4.4. Diagrams illustrating the relationships between real, apparent and reactive powers.

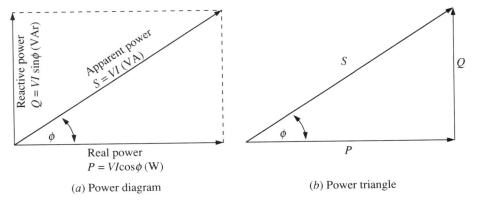

(a) Power diagram

(b) Power triangle

The relationship of these quantities may be represented diagrammatically by means of either the power diagram or power triangle as shown in fig. 4.4. Observe that only the power P is properly expressed in watts. Reactive power is expressed in vars (unit symbol VAr) and apparent power is expressed in volt–amperes (unit symbol VA). In the following sections and throughout the remainder of this book, when reference is made simply to *power* (without qualification) we shall mean the average, real, or active power.

Any two-terminal circuit containing resistive and reactive elements may, according to Thévenin's theorem, be reduced to a single equivalent complex impedance or admittance. This implies that a load consuming energy may be characterized by either of the elementary series or parallel circuits shown in fig. 4.5. in which the R are pure resistances and the X are pure reactances. For either circuit, values for the components can be found such that any given impedance $Z = V/I$ is produced at the terminals.

For power distribution calculations the parallel equivalent circuit of fig. 4.5(a) is often convenient. The phasor diagram for this circuit is shown in fig. 4.6(a). As before, the terminal voltage is chosen as reference, and ϕ is the angle by which the current leads the voltage. We see that the current I may be resolved into two components: (a) a current $I_p = I\cos\phi$, called the *in-phase* component, which flows through R_p; (b) a current $I_q = I\sin\phi$, called the *quadrature* component, which flows through X_p. The power diagram fig. 4.6(c) may be derived directly from this phasor diagram simply by multiplying the total current I and each of its components by the magnitude of the terminal voltage V.

Since $I_p = V/R_p$, $I_q = V/X_p$, and $I = V/Z$, the real, reactive and apparent powers may be expressed as:

$$\text{Power } P = VI\cos\phi = V\left(\frac{V}{R_p}\right) = \frac{V^2}{R_p}$$

Fig. 4.5. Equivalent circuits representing a complex load Z.

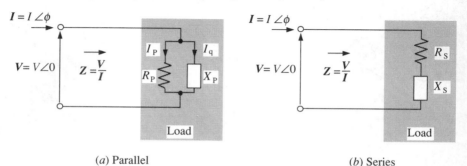

(a) Parallel (b) Series

Reactive power $Q = VI\sin\phi = V\left(\dfrac{V}{X_p}\right) = \dfrac{V^2}{X_p}$ (4.7)

Apparent power $S = VI = V\left(\dfrac{V}{Z}\right) = \dfrac{V^2}{Z}$

We now turn to the series equivalent circuit of fig. 4.5(b). Since this circuit presents the same complex impedance Z at its terminals as that of the parallel circuit of fig. 4.5(a), the same terminal voltage will cause the same current to flow; the power diagram of fig. 4.6(c) must therefore apply also to this circuit.

Using a similar procedure to that adopted in relation to the current we may resolve the voltage into two components (fig. 4.6(b)): that across R_s (in phase with I) is $IR_s = V\cos\phi$; and that across X_s (in quadrature with I) is $IX_s = V\sin\phi$. Expressions analogous to (4.7) are then

$$\begin{aligned} Power\ P &= VI\cos\phi = I(IR_s) = I^2 R_s \\ Reactive\ power\ Q &= VI\sin\phi = I(IX_s) = I^2 X_s \\ Apparent\ power\ S &= VI = I(IZ) = I^2 Z \end{aligned}$$ (4.8)

Although the expressions (4.7) and (4.8) have been derived above in relation to the parallel and series equivalent circuits of a general complex two-terminal network, it will be readily apparent that they apply also to the individual elements of which the network is composed if voltages and currents are interpreted appropriately. For example, if a current I_b flows through a series branch within the network containing a resistance R and a capacitance C, with reactance $1/\omega C$, then the power in the resistance is $I_b^2 R$ and the reactive power in the capacitance is $I_b^2(1/\omega C)$. If the currents in

Fig. 4.6. Phasor and power diagrams for the circuits of fig. 4.5. (a) and (b). Resolution of current and voltage into in-phase and quadrature components. (c) Power diagram.

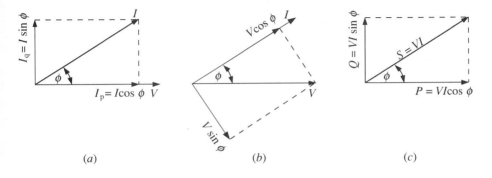

every branch of a network are known, then the real and reactive powers associated with every element of that network can be simply calculated using the appropriate expressions.

Now we have seen that the real power at the terminals of a network is a measure of the average rate of energy flow associated with the resistive elements of the network, this energy being dissipated in the form of heat. The reactive power is a measure of the average rate of energy flow associated with the reactive elements, this energy oscillating between source and network. It follows from the principle of conservation of energy that the power, in watts, absorbed by all the resistive elements in a network must equal the total power supplied at its terminals. Likewise, the reactive power, in vars, associated with all of the reactive elements must equal the reactive power supplied at the terminals. Symbolically we have for any network:

$$\text{Total power (watts)} = \Sigma(I^2 R)_{\text{branch}} \\ \text{Total reactive power (vars)} = \Sigma(I^2 X)_{\text{branch}} \quad (4.9)$$

The expressions (4.9), are sometimes loosely referred to as the *principle of conservation of watts* and *vars*.* (The use of this principle is illustrated, in relation to three-phase power systems, in the worked example of section 5.5.)

4.4 Power factor

The quantity $\cos\phi$ in (4.3) is the *power factor*. When energy is being drawn from a distribution system, the most desirable condition is $\cos\phi = 1$, because then the apparent power and the real power are identical and the current requirement for a given amount of delivered power is a minimum. The excess current requirement represented by $\cos\phi \neq 1$ means that the power distribution lines must be capable of supplying the additional current. It follows that there will be increased heating in the conductors and a corresponding decrease in the efficiency of the overall system. It follows further that an installation having a low power factor may reasonably be required to pay more for each unit of energy delivered than would an installation having a power factor close to unity.

Operators of establishments that use substantial amounts of power find it worthwhile to 'improve' the power factor, that is, to install equipment that will bring the power factor closer to unity. This improvement may be accomplished by installing in parallel with the load a device that draws a

---

* The expressions (4.9) may be derived also from a more general theorem known as Tellegen's Theorem. This states that if v_k and i_k are the voltage across and the current through the kth element of a network, then $\Sigma v_k i_k = 0$ where the sum is taken over all elements of the network including sources. (For further details see reference 4.)

Power factor

quadrature current of a sign opposite to that drawn by the load. Most industrial loads are predominantly inductive and draw a lagging current as shown in the phasor diagram of fig. 4.7(a). Such a load is said to be a *lagging load* and to possess a *lagging power factor*. The corrective measure usually consists in adding capacitance at the terminals of the load; the resulting arrangement being electrically similar to the parallel tuned circuit discussed in section 3.13.3. In fig. 4.7(b), I_c is the current drawn by the capacitance, the new (reduced) line current is I' and the new power factor is $\cos\phi'$. Looked at in another way, the usual load draws negative vars; therefore, one places in parallel a circuit element that draws positive vars. The power diagram shown in fig. 4.7(c), corresponding to the phasor diagram of fig. 4.7(b), illustrates power factor improvement from this viewpoint. We see that the total apparent power is reduced; the actual load power is, of course, unchanged. Usually one does not attempt to make the overall power factor unity (the circuit is not quite tuned to resonance) because, (1) the power factor of the load may vary as load conditions in a large installation change, so such exact correction would require continual adjustment, and (2) the cost of the added capacitance must be weighed against the saving that may be expected as a result of its installation.

Where large values of capacitance are required the capacitors may be rotating machines. An over-excited synchronous motor draws a leading current and so has the electrical characteristics of a capacitor.

The following worked example is intended to illustrate the principles

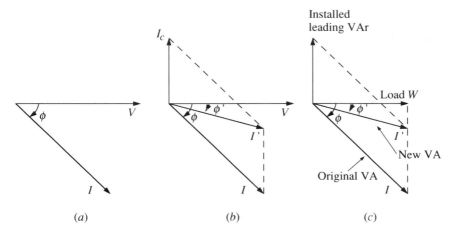

Fig. 4.7. Power factor correction. (a) Phasor diagram for inductive (lagging) load. (b) Modified phasor diagram showing the result of power factor correction. (c) Power diagram corresponding to (b).

discussed above, but it should be remembered that power factor improvement is usually applied to three-phase rather than single-phase systems. Power factor and its improvement in relation to three-phase system is discussed in sections 5.4 and 5.5.

4.5 Worked example

A factory draws 18 kW of power from a 240 V, 50 Hz distribution system. The power factor is 0.75 lagging. The feeder wires to the factory have a total resistance of 0.35 Ω. The factory operates 24 hours, every day.
(a) Calculate the current supplied to the factory. If energy costs 1p/kW hr, determine the annual cost of the energy lost in heating the feeder wires. (Note: It is still common practice in power systems analysis to measure energy in units of kilowatt hours rather than joules.)
(b) If the power factor is improved to 0.95, how much money will be saved annually by the reduction of feeder wire loss?
(c) Determine the size and nature of the unit required to correct the power factor to 0.95.

Solution
(a)

$$\text{Power } P = VI\cos\phi$$

therefore

$$\text{current } I = \frac{P}{V\cos\phi} = \frac{18 \times 10^3}{240 \times 0.75} = 100 \text{ A}$$

Power loss in feeder $= I^2 R_f = 100^2 \times 0.35 = 3.5$ kW

Annual cost $= 3.5$ kW $\times$ (24 hr/day) $\times$ (365 day/yr) $\times$ (1p/kW hr)
$= £306$

(b) With a new power factor of 0.95 the current becomes

$$I = \frac{18 \times 10^3}{240 \times 0.95} = 78.9 \text{ A}$$

Power loss in feeder is then $78.9^2 \times 0.35 = 2.18$ kW.
New annual cost:

$$306 \times \frac{2.18}{3.5} = £190.7$$

Saving:

$$£(306 - 191) = £115 \text{ per annum}$$

(c) The line voltage and initial current are shown in the phasor diagram of fig. 4.8(a). The in-phase current is 75 A and the quadrature current is 66.14 lagging. The initial phase angle is 41.4°. With the power factor improved to 0.95 the new phase angle is $\cos^{-1} 0.95 = 18.2°$. The new total current is 78.9 A and the new quadrature current is $78.9 \sin 18.2 = 24.62$ A. To achieve this improvement a capacitor must be connected across the load terminals which will draw a current of $66.14 - 24.62 = 41.3$ A. The reactance of the capacitor will be $X_c = 240/41.3 = 5.8\,\Omega$. But $X_c = 1/(2\pi f C)$ hence

$$C = 1/(2\pi \times 50 \times 5.8) = 550\,\mu F$$

An alternative approach to this problem is shown in fig. 4.8(b). Here the calculation is carried out in terms of real, apparent and reactive powers. We see that the capacitor is required to draw 9.96 kVAr, hence, from (4.7), $X_c = 240^2/9960 = 5.8\,\Omega$ as before.

Observe that addition of the capacitor does not reduce the reactive current *in the load*. The inductive load still draws vars equal to 15.87 kVAr. However, with the capacitor in place most of the quadrature current flows between the capacitor and the load. Power loss in the feeder wire conductors is thus greatly reduced.

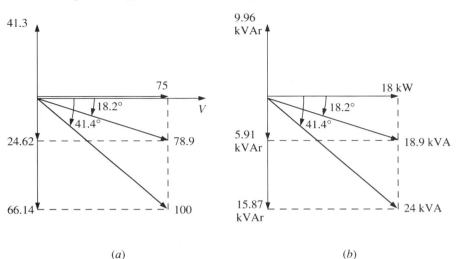

Fig. 4.8. Diagrams for worked example (section 4.5).

4.6 Complex power

For the purposes of calculation it is often convenient to express the relationship between power, reactive power and apparent power in the form of a complex number, as follows:

$$S = P + jQ = VI\cos\phi + jVI\sin\phi = VIe^{j\phi} \qquad (4.10)$$

The power diagram of fig. 4.4(a) may then be interpreted in terms of the Argand diagram as shown in fig. 4.9. Now suppose that the complex voltage and current are known at the terminals of a load; these quantities having been found by previous calculation. Let

$$V = V\underline{/\alpha} = Ve^{j\alpha} \quad \text{and} \quad I = I\underline{/\alpha + \phi} = Ie^{j(\alpha + \phi)}$$

If we form the product $VI(=VIe^{j(2\alpha+\phi)})$, we fail to obtain the complex power as defined in (4.10). However, if we form the product V^*I, where V^* indicates the complex conjugate of V, namely, $Ve^{-j\alpha}$, we obtain

$$V^*I = Ve^{-j\alpha}Ie^{j(\alpha+\phi)} = VIe^{j\phi} = S \qquad (4.11)$$

Therefore, if we have the complex expressions for the voltage V across a load and the current I drawn by the load and if we form the product V^*I, the real part of this product is the power P and the imaginary part is the reactive power Q.

Note that (4.11) gives the correct sign for reactive power using the convention adopted here, namely, that capacitive vars are positive, inductive vars negative. If the alternative convention is used, (capacitive vars negative, inductive vars positive), then the product VI^* must be formed to obtain the correct sign for the reactive power. The real part of either V^*I or VI^* will give the power P.

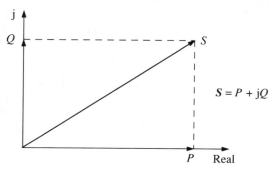

Fig. 4.9. Apparent power expressed as a complex quantity.

An alternative expression for P, which is sometimes useful, is obtained as follows:

$$V^*I = VIe^{j\phi}$$

and

$$VI^* = Ve^{j\phi}Ie^{-j(\alpha+\phi)} = VIe^{-j\phi}$$

Adding

$$V^*I + VI^* = VI(e^{j\phi} + e^{-j\phi}) = 2VI\cos\phi = 2P$$

Hence

$$P = \tfrac{1}{2}(V^*I + VI^*) \tag{4.12}$$

4.7 The ideal transformer

In section 1.10 we described inductively coupled circuits. A useful application of inductive coupling is the transformer, a device that transfers energy from one circuit to another without direct connection between the two. A *real* transformer consists of two or more coils of wire wound on a common core. The core usually is of ferromagnetic material in order to achieve as nearly as possible a coefficient of coupling of unity. Such a device has losses in the resistance of the windings and in the magnetic core material. Real transformers will be considered in detail in a later section.

Useful information about transformer characteristics results from a study of the *ideal* transformer which is assumed to have

  (a) negligible energy loss in the windings and in the core material
  (b) perfect coupling so that the coupling coefficient $k = 1$
  (c) very large self-inductance in each coil.

A transformer may have any number of independent windings. We shall consider the two-winding device shown schematically in fig. 4.10.

Let the supply voltage be connected at terminals AA' of winding number

Fig. 4.10. Schematic diagram for an ideal two-winding transformer.

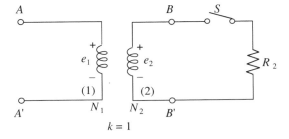

(1). It is customary to refer to the input windings as the *primary* winding (or simply the primary) and to the other windings as *secondary* windings (or secondaries). In general, energy may be supplied to any winding, so 'primary' may not always refer to the same winding of a particular transformer.

In fig. 4.10 with switch S *open*, an applied voltage e_1 requires, according to Faraday's Law, a flux ϕ_1 such that $e_1 = N_1 d\phi_1/dt$, $N_1\phi_1$ being the flux linkage in winding (1). But if the coupling coefficient is unity, the flux in winding (2) will also be ϕ_1, and the induced voltage $e_2 = N_2 d\phi_1/dt$. It follows then that:

$$\frac{e_1}{e_2} = \frac{N_1}{N_2} \tag{4.13}$$

and the voltage ratio is just equal to the ratio of numbers of turns. Because we have specified very large coil inductance, we may assume that with S open a negligibly small amount of current is required to establish the flux ϕ_1.

Now let the switch be *closed*. The secondary voltage e_2 will then give rise to a current $i_2 = e_2/R_2$ in winding (2), which will establish a flux ϕ_2. We recall, from Amperes circuital law of electromagnetic theory, that this flux ϕ_2 is proportional to the product $N_2 i_2$ so that we may write $\phi_2 = c N_2 i_2$ where c is a constant of proportionality. Further, from Lenz's law, the current i_2 will be in such a direction as to oppose the change of magnetic flux producing it, in other words, the direction of the flux ϕ_2 will be opposite to that of ϕ_1. Because e_1 is still applied to winding (1), the net flux in the core must still be ϕ_1, consequently there must now be a current i_1 in winding (1) of sufficient magnitude to produce a forward flux (that is, in the direction of the initial flux ϕ_1) just equal in magnitude to ϕ_2. Therefore, $\phi_2 = c N_1 i_1$ and we have the relation:

$$\phi_2 = c N_2 i_2 = c N_1 i_1$$

or

$$\frac{i_1}{i_2} = \frac{N_2}{N_1} \tag{4.14}$$

With the switch closed, the secondary current and voltage are related by $e_2/i_2 = R_2$, hence, from (4.13) and (4.14) we obtain for the primary:

$$\frac{e_1}{i_1} = \left(\frac{N_1}{N_2}\right)^2 \frac{e_2}{i_2} = \left(\frac{N_1}{N_2}\right)^2 R_2 \tag{4.15}$$

The ratio e_1/i_1 is the effective resistance at the primary, thus a resistance R_2

in the secondary is reflected into the primary as a resistance whose value depends upon R_2 and upon the square of the ratio of primary turns to secondary turns.

Equation (4.15) describes the resistance-transforming property of the transformer. Because the relation holds equally for reactive circuit elements, we may conclude that the transformer has impedance transforming properties. This property is useful in applications such as the following.

4.8 Worked example

Loudspeakers have resistance of the order of 10 ohms. Neither vacuum tube nor many simple transistor amplifier circuits operate satisfactorily with such small values of load resistance. Therefore a transformer is employed to raise the impedance level of the speaker to a value compatible with the requirements of the amplifier circuit. What turns ratio n is required to match a 16 Ω speaker to a transistor circuit that is designed to have a load resistance of 400 Ω?

Solution: When the 16 Ω resistance is connected to the secondary, it should be reflected into the primary as 400 Ω, therefore,

$$(N_1/N_2)^2 = R_1/R_2 \text{ or } (N_1/N_2)^2 = 400/16 = 25$$

and

$$n = (N_1/N_2)^{\frac{1}{2}} = 25 = 5$$

4.9 Single-phase power transformers

Although some transmission lines operate at high (hundreds of kilovolts) direct voltage, all power distribution systems use alternating voltage. Power transformers are essential parts of these systems. Such devices, which may operate at high voltage and carry large currents, cannot be represented by the simple model of fig. 4.10. In this section we develop a model for a power transformer. In succeeding sections we see how the transformer is used in a real circuit and describe methods of determining experimentally the characteristics of a power transformer.

In a power transformer, a core of high permeability ferromagnetic material provides a path for magnetic flux. The separate circuits are wound in such a fashion that practically all the flux produced by one winding links with all the other windings. Sufficient ferromagnetic material is provided so that it does not become saturated under normal operating conditions. Fig. 4.11 shows three frequently used transformer constructions. We shall confine our discussion to transformers having two windings, although often several independent secondary circuits may be supplied from a single primary winding.

The simple ideal transformer model of fig. 4.10 is inappropriate for the power transformer. We require a model that takes into account: (1) the *exciting current* which is defined as the primary current when the secondary current is zero; (2) the energy loss in the conductors that comprise the windings; and (3) the presence of *leakage* flux (the magnetic flux from one winding that does not link the other winding).

Let the transformer consist of a primary and secondary wound upon a high permeability core as shown schematically in fig. 4.12. Let the secondary be open circuit so that the secondary current is zero. If the flux in the primary is sinusoidal, of the form $\phi = \Phi_m \sin\omega t$, then the primary induced voltage is

$$e_1 = N_1 \frac{d\phi}{dt} = N_1 \frac{d}{dt}(\Phi_m \sin\omega t)$$

$$= \omega N_1 \Phi_m \cos\omega t \qquad (4.16)$$

Fig. 4.11. Transformer construction.

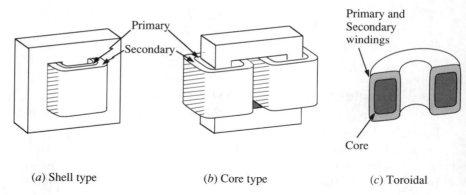

(a) Shell type   (b) Core type   (c) Toroidal

Fig. 4.12. Two-winding transformer with ferromagnetic core. With the secondary open circuit a small exciting current i_ϕ flows in the primary.

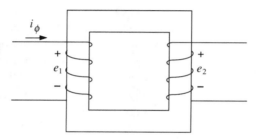

Single-phase power transformers

and e_1 is, therefore, also of sinusoidal form leading ϕ by 90° (fig. 4.13(a)). The r.m.s. magnitude of the induced voltage is

$$E_1 = \frac{2\pi f}{\sqrt{2}} N_1 \Phi_m = 4.44 f N_1 \Phi_m \qquad (4.17)$$

Thus, an applied sinusoidal voltage of this magnitude will cause a sinusoidal flux to be established whose maximum value (if the effects of conductor resistance and leakage flux are negligible) will be given by

$$\Phi_m = \frac{E_1}{4.44 f N_1}$$

In order for the flux to exist, there must be an exciting current i_ϕ. For an air-cored coil, flux is directly proportional to current, so with a sinusoidal applied voltage, i_ϕ will be sinusoidal. In a closed ferromagnetic core, *flux density B* and *magnetizing force H* are not linearly related and one must describe the core's magnetic properties by the familiar *B–H* curve. Now

$$B = \frac{\phi}{A} \quad \text{and} \quad H = \frac{Ni}{\ell} \qquad (4.18)$$

where N is the number of turns in the winding, and A and ℓ are, respectively, the cross-sectional area and the length of the flux path in the magnetic material. For a specific device, then, one may plot ϕ v. i_ϕ as shown in fig. 4.13(b). Then we may use a graphical method as shown in fig. 4.13(c) to derive the graph of i_ϕ v. time. This current is the exciting current of the transformer. Because of the non-linear magnetic characteristic of the transformer core, i_ϕ is non-sinusoidal, consisting of a fundamental and a set of odd harmonics (see section 7.8 for a discussion of harmonic (Fourier) analysis). Thus a sinusoidal applied voltage results in a non-sinusoidal exciting current.

In comparison with the rated load current of the transformer the exciting current is small. Unless we are interested specifically in its harmonic content, we may confine our attention to the fundamental, sinusoidal frequency component of i_ϕ. This fundamental may be resolved into two components. One component, called the *magnetizing current*, is in phase with ϕ, and so in quadrature with e_1. The other component, called the *core loss current*, leads ϕ by 90° and so is in phase with e_1. The core loss current thus represents power supplied to the transformer. It accounts for the work required periodically to magnetize the core first in one direction and then in the other. On the *B–H* plane, the area enclosed by the hysteresis loop is a measure of the work required to carry the magnetic material through one

Fig. 4.13. Primary voltage e_1, core flux ϕ, and primary exciting current i_ϕ for the transformer of fig. 4.12.

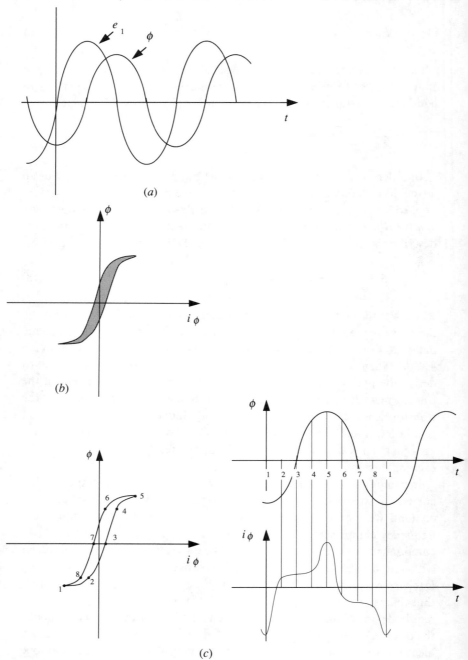

complete cycle of magnetization. The core loss current also accounts for the power loss due to eddy currents induced in the transformer core.

The phasor diagram of fig. 4.14(b) shows the applied voltage E_1 and the exciting current I_ϕ along with its two components, I_m and I_c. The exciting current may be accounted for in the transformer model by adding the conductance g_c and the susceptance b_m to the ideal transformer model as shown in fig. 4.14(a), where

$$g_c = I_c/E_1 \quad \text{and} \quad b_m = I_m/E_1 \tag{4.19}$$

When an impedance is connected to the secondary, currents will flow in secondary and primary circuits 4.15(a). The total current I_1 flowing in the primary circuit will be the sum of the load current I_1' and the exciting current I_ϕ as shown in the phasor diagram of fig. 4.15(b). In this diagram the length of the I_ϕ phasor has been exaggerated in relation to that of the I_1'

Fig. 4.14. Circuit model of the transformer including elements g_c and b_m to account for the exciting current I_ϕ.

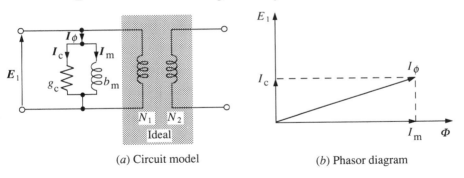

(a) Circuit model   (b) Phasor diagram

Fig. 4.15. Illustrating the effect of adding a load to the secondary winding in the circuit model of fig. 4.14(a).

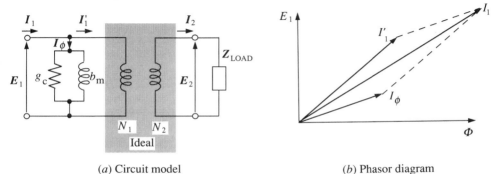

(a) Circuit model   (b) Phasor diagram

phasor; the full load current may be twenty times the exciting current in a power transformer.

To complete the circuit model of the transformer we must introduce elements which account for the resistance of the windings and the leakage flux. Consider first the primary leakage flux, that is, the flux which links with the primary winding but not the secondary winding; let this flux be ϕ_{l1}. The path which this flux takes lies partly in the core and partly in the air space surrounding the core; the primary winding will, therefore, as far as this flux is concerned, behave very nearly like an air-cored inductor, and the

Fig. 4.16. Transformer circuit model including elements X_1, X_2, R_1, R_2 to account for leakage reactances and winding resistances.

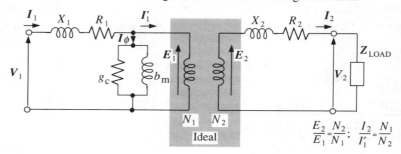

(a) Circuit model

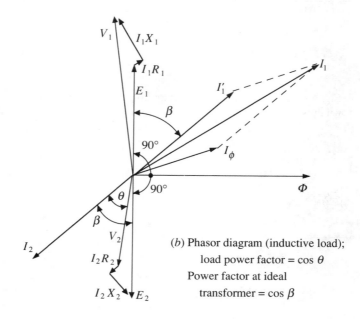

(b) Phasor diagram (inductive load);
load power factor = $\cos \theta$
Power factor at ideal
transformer = $\cos \beta$

flux linkage $N_1\phi_{l1}$ will be proportional to the primary current. (This is in contradistinction to the mutual flux which, as we have seen, is independent of the load currents flowing in the windings.) Hence, from the definition of inductance given in section 1.9, we may write

$$N_1\phi_{l1} = L_1 i_1 \qquad (4.20)$$

where the constant of proportionality L_1 is called the *primary leakage inductance*.

This inductance is represented in the circuit model of fig. 4.16(a) by the series reactance $X_1 = \omega L_1$. Similarly a reactance X_2 in the secondary circuit represents the secondary leakage inductance. Resistances R_1 and R_2 account for the resistances of the windings. The complete phasor diagram, drawn for an inductive load (lagging current) corresponding to this circuit model is shown in fig. 4.16(b). Note that the lengths of the phasors representing reactive and resistive voltage drops have been exaggerated, in relation to the main voltage phasors, for the sake of clarity. The relative phase of E_1 and E_2 can be either zero or 180°, depending on the relative winding directions of the transformer; the latter has been chosen – again for clarity in the diagram.

We may employ the impedance transforming property of the transformer to simplify the circuit of fig. 4.16(a). Resistance and reactance parameters are transferred across the ideal transformer by multiplying them by the square of the turns ratio. Fig. 4.17(a) shows the simplified circuit model

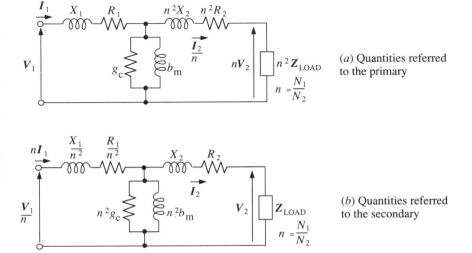

Fig. 4.17. Alternative transformer circuit models.

(a) Quantities referred to the primary

(b) Quantities referred to the secondary

resulting when secondary quantities are 'referred to the primary'. Fig. 4.17(b) is the circuit model with quantities 'referred to the secondary'. Of course, in any calculation, both models give the same final result; the choice of model will depend on the particular problem in hand.

Calculations using the circuit models are simplified if the shunt elements representing the exciting current are shifted to the left in fig. 4.17 so that they are directly across the input voltage as shown in fig. 4.18. A small but usually insignificant error is introduced by neglect of the voltage drops in R_1 and X_1 caused by the exciting current.

4.10 Worked example

A single-phase transformer with turns ratio $n = N_1/N_2 = 10$ has an output of 200 volts when supplying a load of 10 kVA at 0.8 power factor lagging. Resistance and leakage reactance are $4\,\Omega$ and $7\,\Omega$ respectively for the primary and $0.04\,\Omega$ and $0.08\,\Omega$ for the secondary. The exciting current is 0.5 A at 0.2 power factor lagging. Calculate the input voltage and the efficiency of the transformer with this load.

Solution: The appropriate circuit is found by calculating numerical values for the elements shown in fig. 4.18, and combining resistances and reactances in series.

$$R = R_1 + n^2 R_2 = 4 + (10)^2 (0.04) = 8\,\Omega$$
$$X = X_1 + n^2 X_2 = 7 + (10)^2 (0.05) = 15\,\Omega$$

The load current is

$$I_2 = 10 \times 10^3 / 200 = 50 \text{ A}$$

Then,

$$nV_2 = 2000 \text{ V} \quad \text{and} \quad I_2/n = 5 \text{ A}$$

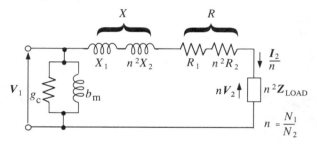

Fig. 4.18. Simplified transformer circuit model with quantities referred to primary.

Assume $I_2/n = 5 + j0$ (phase reference). Then, since the power factor of the load is 0.8,

$$nV_2 = 2000(0.8) + j2000(0.6) = 1600 + j1200 \text{ V}$$

Hence,

$$V_1 = (I_2/n)(R + jX) + nV_2 = 5(8 + j5) + (1600 + j1200) = 1640 + j1275$$

and

$$V_1 = (1640^2 + 1275^2)^{\frac{1}{2}} = 2080 \text{ V}$$

The efficiency of the transformer is found by dividing the output power by the input power. Furthermore, the input power may be written as (output power + losses). Now,

$$\text{Output power} = V_2 I_2 \cos\theta = 10^4(0.8) = 8 \times 10^3 \text{ watts}$$
$$\text{Copper losses} = (I_2/n)^2 R = 5^2 \times 8 = 200 \text{ W}$$
$$\text{Core loss} = V_1 I_e \cos\theta' = (2080)(0.5)(0.2) = 208 \text{ W}$$

Therefore,

$$\text{Efficiency} = (8000)/(8000 + 408) = 0.951 \text{ or } 95.1\%$$

4.11 Transformer tests

All the elements in the circuit model of fig. 4.18 may be determined experimentally by performing two simple tests using voltmeters, an ammeter and a wattmeter. Connections are shown in fig. 4.19.

Open circuit test: The appropriate circuit model for this test is shown in fig. 4.20(a). Rated voltage V_{1o} is applied at the primary terminals, and the output voltage V_{2o}, the input current I_{1o}, and the input power P_o are measured.

Fig. 4.19. Connections for transformer tests.

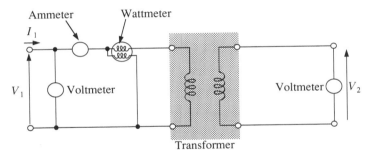

Since the secondary is open circuit, $I_2 = 0$ and so $I_1' = 0$. Therefore $E_1 = V_{1o}$ and $E_2 = V_{2o}$ and the turns ratio is, by (4.13),

$$\frac{N_1}{N_2} = \frac{V_{1o}}{V_{2o}} \tag{4.21}$$

Since $I_1 = 0$, the current I_{1o} is the exciting current. Therefore

$$g_c = \frac{P_o}{V_{1o}^2} \quad \text{and} \quad \frac{I_{1o}}{V_{1o}} = \sqrt{(g_c^2 + b_m^2)} \tag{4.22}$$

from which b_m may be calculated.

Short circuit test: The secondary winding is shorted (fig. 4.20(b)) and the primary voltage raised carefully from a low value to give rated current. The input voltage V_{1s}, the input current I_{1s} and the power P_s are measured. Because all impedances are referred to the primary in fig. 4.20, when $V_2 = 0$ then $E_2 = 0$ and $E_1 = 0$. So the voltage V_{1s} is across the impedance $R + jX$. For a typical short circuit test the input voltage may be only one-tenth the rated value. Then the loss in the magnetizing circuit will be only $(0.1)^2 = 0.01$ of the loss at rated voltage and may be neglected in these measurements. R and X may be determined from:

$$R = \frac{P_s}{I_{1s}^2} \quad \text{and} \quad \frac{V_{1s}}{I_{1s}} = \sqrt{(R^2 + X^2)} \tag{4.23}$$

For most transformer calculations it is not necessary to subdivide R and X to represent the separate contributions of the primary and the secondary windings.

Fig. 4.20. Circuit models for open and short circuit tests on the transformer. R and X are total resistance and reactance referred to the primary, as in fig. 4.18.

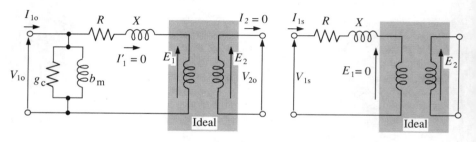

(a) Open circuit test

(b) Short circuit test

4.12 Voltage regulation

An important characteristic of a transformer is the *voltage regulation*, defined as

$$\text{Regulation} = \frac{V_{2o} - V_2}{V_{2o}} \qquad (4.24)$$

where V_{2o} is the no-load output voltage and V_2 is the output voltage under a specified load. We shall derive an expression for regulation in terms of the transformer properties and the phase angle of the load. We use the circuit model of fig. 4.21 where resistance and reactance are referred to the secondary. In fig. 4.21

$$R' = R_2 + (N_2/N_1)^2 R_1 \qquad X' = X_2 + (N_2/N_1)^2 X_1$$

If

$$I_2 = 0,$$

then

$$V_{2o} = (N_2/N_1) V_1$$

If there is a load current I_2 such that the phase angle for the secondary is ϕ, the appropriate phasor diagram is as shown in fig. 4.22. In fig. 4.22

$$a = I_2 R' \cos\phi \quad ; \quad b = I_2 R' \sin\phi$$
$$c = I_2 X' \cos\phi \quad ; \quad d = I_2 X' \sin\phi$$

Then

$$\left[\frac{N_2}{N_1} V_1 \right]^2 = (V_2 + a + d)^2 + (c - b)^2$$

Fig. 4.21. Circuit model for calculation of regulation.

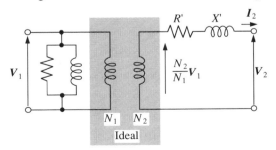

In a well-designed transformer the losses are small (the voltage drops I_2R' and I_2X' have been greatly exaggerated in relation to V_2 in fig. 4.22) and so we neglect the difference term $(c-b)^2$. Then

$$\frac{N_2}{N_1} V_1 = V_2 + I_2(R'\cos\phi + X'\sin\phi)$$

therefore

$$\frac{N_2}{N_1} V_1 - V_2 = I_2(R'\cos\phi + X'\sin\phi)$$

But

$(N_2/N_1)V_1 = V_{2o}$, hence from (4.24)

$$\text{Regulation} = \frac{I_2(R'\cos\phi + X'\sin\phi)}{V_{2o}} \quad (4.25)$$

It is often convenient to express regulation in terms of quantities referred to the primary. Multiplying numerator and denominator of (4.25) by $(N_1/N_2)^2$ gives

$$\text{Regulation} = \frac{I_2\left[R'\left(\frac{N_1}{N_2}\right)^2\cos\phi + X'\left(\frac{N_1}{N_2}\right)^2\sin\phi\right]\frac{N_2}{N_1}}{\frac{N_1}{N_2}V_{2o}}$$

Now from (4.21), $V_1 = (N_1/N_2)V_{2o}$ and, if the exciting current is small, we have $I_1 = (N_2/N_1)I_2$. Furthermore, $R_1'(N_1/N_2)^2 = R$ and $X'(N_1/N_2)^2 = X$. Hence,

Fig. 4.22. Phasor diagram, corresponding to fig. 4.21, for determining regulation.

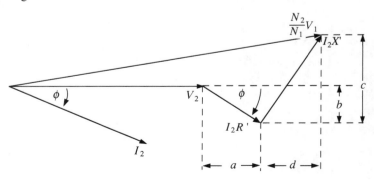

$$\text{Regulation} = \frac{I_1(R\cos\phi + X\sin\phi)}{V_1} \qquad (4.26)$$

For a lagging power factor, ϕ (as defined in fig. 4.22) is positive and so the regulation is positive meaning that the voltage falls with increasing load current. However, for a leading power factor ϕ is negative in (4.26). There is then the possibility that the regulation may be negative, corresponding to an increased output voltage under load.

4.13 Conditions for maximum efficiency

The efficiency of a transformer can be expressed as

$$\eta = \frac{\text{output power}}{\text{input power}} = \frac{\text{input power} - \text{losses}}{\text{input power}}$$

or

$$\eta = 1 - \frac{\text{losses}}{\text{input power}} \qquad (4.27)$$

Losses are of two kinds: core losses (hysteresis and eddy current) and copper losses. The core losses depend on the magnitude of the flux in the core, which in turn depends on the applied voltage, and on the frequency. For fixed voltage and frequency the core losses in a transformer will be substantially constant irrespective of load conditions. The copper losses depend on the current carried by the windings of the transformer and will vary with the load.

Let the fixed core losses be P_0, and the variable copper losses be $I_1^2 R$, where R is the total equivalent resistance referred to the primary and I_1 is the primary current.

Then if the primary voltage is V_1 and the power factor is $\cos\phi$, (4.27) becomes

$$\eta = 1 - \frac{P_0 + I_1^2 R}{V_1 I_1 \cos\phi} = 1 - \frac{P_0}{V_1 I_1 \cos\phi} - \frac{I_1 R}{V_1 \cos\phi} \qquad (4.28)$$

Clearly, when the transformer is unloaded, the input power will be P_0 (neglecting the small copper loss due to the exciting current) and the efficiency will be zero. As the load is increased from zero the efficiency improves but copper losses become an increasingly large proportion of the total losses. Eventually, the third term in (4.28) predominates and the efficiency falls. The load current at which maximum efficiency occurs is obtained by differentiating (4.28) and equating to zero:

$$\frac{d\eta}{dI_1} = \frac{P_0}{V_1 I_1^2 \cos\phi} - \frac{R}{V_1 \cos\phi} = 0$$

214 *Power and transformers in single-phase circuits*

whence

$$P_0 = I_1^2 R \tag{4.29}$$

Therefore, at a given power factor, maximum efficiency is obtained when the fixed core losses equal the variable copper losses. Under these conditions the primary load current is, from (4.29), $I_1 = \sqrt{(P_0/R)}$ and the efficiency is then

$$\eta_{max} = 1 - \frac{2\sqrt{(P_0 R)}}{V_1 \cos\phi} \tag{4.30}$$

The maximum possible efficiency is attained when the power factor, $\cos\phi$, becomes unity in the above expression.

In practice, transformer efficiencies range between 95% for small, single-phase units to better than 98% for large, three-phase units of the type employed in power distribution systems.

4.14 The autotransformer

Figure 4.23(a) shows a conventional two-winding transformer, with a primary/secondary turns ration of 2. We assume that the transformer is ideal. The primary is supplied from the 240 V a.c. line. The voltage across the secondary winding $B'C'$ is then 120 V and a 10 ohm load resistor will draw a secondary current of 12 A. The corresponding primary current is 6 A. The primary current is downward and the secondary current is upward as indicated by the arrows.

Now let point C and C' be connected together and let point B' be connected to point B, which is located midway between points A and B (so that the number of turns between A and B is equal to the number of turns

Fig. 4.23. Currents in a two-winding transformer, and an autotransformer with identical input and load conditions.

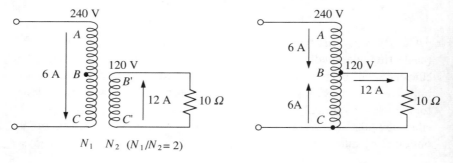

(a) Two-winding transformer   (b) Autotransformer

The autotransformer

between *B* and *C*). These connections will cause no change in the circuit because $V_{BC} = V_{B'C'}$ and the two voltages are in phase. Let winding $B'C'$ be removed. The new situation is shown in fig. 4.23(*b*) with currents as indicated; the net upward current in *BC* is now 6 A. By using the connections of fig. 4.23(*b*), we have replaced two windings carrying, respectively, 6 A and 12 A by a single winding carrying 6 A. We have eliminated the weight and cost of the secondary winding without affecting the transfer of energy from primary to secondary. (In addition, we have eliminated the loss of energy resulting from the resistance of the secondary winding, although this was assumed to be negligibly small in this case.) The transformer shown in fig. 4.23(*b*) is an *autotransformer*.

Because the currents are in opposite directions in the two halves of winding *AC*, it is the traditional approach to call section *AB* the primary and section *BC* the secondary. With these definitions it follows that for fig. 4.23(*b*)

$$\text{Power in primary} = (120 \text{ V})(6 \text{ A}) = 720 \text{ W}$$
$$\text{Power in secondary} = (120 \text{ V})(6 \text{ A}) = 720 \text{ W}$$
$$\text{Load power} = (120 \text{ V})(12 \text{ A}) = 1440 \text{ W}$$

Thus we may say that for this transformer ratio half the power is *transformed* and half the power is supplied *conductively* from the input directly to the load.

Let us now generalize our approach by considering the autotransformer of fig. 4.24, where the two segments of the winding have, respectively, N_1 and N_2 turns. Let $N_2/(N_1 + N_2) = m$. The two segments of the winding carry the same flux, hence by the arguments leading to (4.13) the voltage ratio will be

$$\frac{V_2}{V_1} = \frac{N_2}{N_1 + N_2} = m \tag{4.31}$$

Fig. 4.24. Circuit model for an autotransformer with variable output voltage.

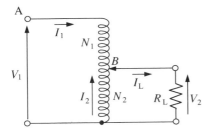

Also, from the arguments leading to (4.14) we have

$$I_1 N_1 = I_2 N_2 = (I_L - I_1) N_2$$

or

$$I_1 (N_1 + N_2) = I_L N_2$$

Hence the current ratio is:

$$\frac{I_L}{I_1} = \frac{N_1 + N_2}{N_2} = \frac{1}{m} \qquad (4.32)$$

The current I_2 in the N_2 turns may be written

$$I_2 = I_L - I_1 = I_L - m I_L = (1 - m) I_L \qquad (4.33)$$

As $m \to 1$ the current I_2 becomes smaller and is zero when $m = 1$, corresponding to a direct connection between terminals A and B. As m gets smaller, $I_2 \to I_L$, and the saving resulting from the use of the autotransformer becomes negligible.

If the point B in fig. 4.24 is moveable, then the autotransformer is a convenient source of variable alternating voltage. A common type of autotransformer designed for use in electronics laboratories has the winding configuration shown in fig. 4.25. By adjustment of the moveable tap the user may obtain output voltages from zero to a few volts above the line voltage.

Although the circuit diagram of the autotransformer resembles that of the resistance-type voltage divider, the principles of operation of the two circuits are quite different. In the voltage divider a substantial fraction of the input power appears as heat in the resistor. Except for small losses all the input power to the autotransformer appears in the load.

Care must be exercised in using the autotransformer because there is a direct connection between input and output. When they are used in the

Fig. 4.25. Autotransformer connections to provide output voltage greater than input voltage.

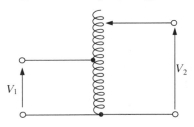

laboratory, autotransformers often are supplied from a unity turns ratio isolating transformer as a safety measure.

Autotransformers find applications in power systems where small voltage changes are required. For example, if it is desired to get 2000 volts from a 2400 volt line, one may use an autotransformer with $m = 2000/2400 = 5/6$. Such a transformer is considerably cheaper in first cost and in operation than a conventional two-winding device. The lack of isolation between primary and secondary is of no concern in this application.

4.15 Maximum power transfer

The transmission of an a.c. signal through an electrical network is accompanied inevitably by a loss of signal power, and it is often important to ensure that the loss is as small as possible. We now consider the factors affecting the transfer of power from one section of a network to another, and derive the conditions for which the power transfer is a maximum.

In fig. 4.26 a practical voltage source is shown connected to a load. The source may represent an actual signal source, for example, some form of inductive or capacitive transducer, or it may be the Thévenin equivalent of a section of a complex network. Likewise, the load Z may be the equivalent impedance at the terminals of a complex network to which the transducer or first network is connected. It is shown below that the efficiency of power transfer depends only on the relative values (magnitudes and angles) of the source and load impedances.

For the circuit of fig. 4.26

$$I = \frac{V}{Z_0 + Z} = \frac{V}{(R_0 + R) + j(X_0 + X)}$$

Fig. 4.26. Basic circuit for derivation of the maximum power transfer theorem.

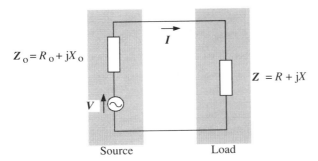

so,

$$I = \frac{V}{\sqrt{[(R_0+R)^2 + (X_0+X)^2]}}$$

Therefore, power to the load is

$$P = I^2 R = \frac{V^2 R}{(R_0+R)^2 + (X_0+X)^2} \tag{4.34}$$

We now assume that Z_0 is fixed while Z is variable. There are three cases to consider appertaining to the way in which the load Z is constrained to vary.

Case 1: R and X are independently variable.

In this case, since X appears only in the denominator of (4.34), the value of X that maximizes P is

$$X = -X_0 \tag{4.35}$$

This means simply that the circuit must be brought to a state of resonance to maximize power transfer.

The power is then

$$P = \frac{V^2 R}{(R_0+R)^2}$$

To find the optimum value of R, we differentiate this expression, set $dP/dR = 0$ and solve for R. This gives

$$R = R_0 \tag{4.36}$$

So, if both the resistance and reactance of the load are adjustable, maximum power is delivered to the load when the load impedance Z is the complex conjugate of the source impedance Z_0, that is,

$$Z = Z_0^* \tag{4.37}$$

The relationship (4.37) is referred to as the *maximum power theorem*, and the load is said to be *matched* to the source. For this condition, equal amounts of power are absorbed by the load and the internal resistance of the source, and the efficiency of the system is at best only 50%. It will be obvious from (4.35) that, if the source contains reactance, matching can be achieved only at one particular frequency.

Case 2: R is variable, X is fixed.

To find the optimum value of R we may differentiate the expression (4.34) directly, however, it is somewhat easier to rearrange this expression so that R appears in the denominator only; thus

Maximum power transfer

$$P = \frac{V^2}{\frac{1}{R}[(R_0+R)^2+(X_0+X)^2]} \quad (4.38)$$

The value of R that maximizes the power is then found from

$$\frac{d}{dR}\left[\frac{R_0^2}{R}+2R_0+R+\frac{1}{R}(X_0+X)^2\right]=0$$

or

$$R = \sqrt{[R_0^2+(X_0+X)^2]} \quad (4.39)$$

In this case the efficiency of power transfer is less than 50%.

Case 3: magnitude of Z is variable, angle of Z is fixed.
The angle of Z is $\tan^{-1}(X/R)=$ constant, therefore $X/R=$ constant $=a$, say, or $X=aR$. Substitution in (4.38) gives

$$P = \frac{V^2}{\frac{1}{R}[(R_0+R)^2+(X_0+aR)^2]}$$

Setting the differential of the denominator of the above expression to zero (as in Case 2 above) yields

$$R^2+X^2 = R_0^2+X_0^2$$

or, taking the square root of both sides of the expression,

$$Z = Z_0 \quad (4.40)$$

For this case we see that we must set the *magnitudes* of the source and load impedances equal to obtain maximum power transfer. Again the efficiency is less than 50%.

The impedance transforming property of the transformer (equation (4.15)) may be utilized to achieve the condition specified by (4.40). For example, if the generator impedance is $8+j6$ so that $Z=10\,\Omega$, and if the load is a pure resistance of $2100\,\Omega$, then maximum power will be delivered to the load if a transformer is employed such that:

$$\left(\frac{N_2}{N_1}\right)^2 = \frac{2100}{10}, \quad \text{or} \quad \frac{N_2}{N_1} = \sqrt{210} = 14.49$$

In practice the turns ratio must be a whole number so we choose a ratio of either 14 or 15. The curve of P versus Z has a broad maximum rather than a sharp peak, therefore, either ratio would be equally satisfactory. For the same reason the assumption of an ideal transformer (implicit in the use of (4.15)) will not in practice give rise to any significant error.

Matched conditions may also be achieved by inserting between source and load a two-port network the elements of which are arranged to provide the requisite impedance transformation. If the elements are purely reactive, no power will be absorbed by the network itself. An 'L-section' combination of inductance and capacitance may, for example, be used to match a resistive load to a resistive source, as indicated in fig. 4.27. Values of L and C may be found such that the impedance Z_{AB} looking into terminals AB will equal the source resistance R_0 when the two port is terminated by load resistance R. For this matched condition we have

$$Z_{AB} = R_0 = j\omega L + \frac{R(1/j\omega C)}{R + 1/j\omega C}$$

which gives

$$R_0 + j\omega C R R_0 = R - \omega^2 LCR + j\omega L$$

Equating real and imaginary parts:

$$R_0 = R - \omega^2 LCR \quad \text{or} \quad LC = \frac{R - R_0}{\omega^2 R}$$

and

$$\omega C R R_0 = \omega L \quad \text{or} \quad \frac{L}{C} = R R_0$$

Combining these expressions we obtain

$$C = \frac{1}{\omega R} \sqrt{\left(\frac{R - R_0}{R}\right)}$$
$$L = \frac{R_0}{\omega} \sqrt{\left(\frac{R - R_0}{R}\right)} \qquad (4.41)$$

Fig. 4.27. Matching source and load by means of a reactive two-port network.

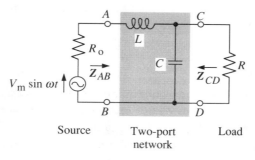

The transformer bridge

Note that these expressions are functions of ω; matching is achieved, therefore, at one particular frequency only. A result identical to (4.41) is obtained by putting $Z_{CD} = R$.

†4.16 The transformer bridge

Impedances of *like* kind may be compared by means of a conventional four-arm, a.c. bridge (section 3.10) using the circuit shown in fig. 4.28(a). Resistances R_1 and R_2 form a voltage divider across the a.c. source and the standard impedances Z_3 and the unknown impedance Z_4 are placed in opposite arms. If the impedances can each be represented by a series combination of resistance and reactance, the balance conditions are:

$$\frac{R_1}{R_2} = \frac{Z_3}{Z_4} = \frac{R_3 + jX_3}{R_4 + jX_4}$$

which gives

$$\frac{R_1}{R_2} = \frac{R_3}{R_4} = \frac{X_3}{X_4} \qquad (4.42)$$

Two difficulties arise in connection with the use of such a bridge in practice. Firstly, from (4.42) we see that in order to evaluate the unknown reactance X_4 and resistance R_4, the ratio R_1/R_2 must be known accurately. If, however, the bridge is to be used for comparing a wide range of values of impedance against a single standard impedance this ratio must be adjustable over a correspondingly large range. But it is technically difficult and expensive to manufacture precision resistance ratio arms that can be made variable over a wide range, and in practice the range of the bridge

Fig. 4.28. A.C. bridges using: (a) resistance ratio arms; (b) inductively coupled ratio arms.

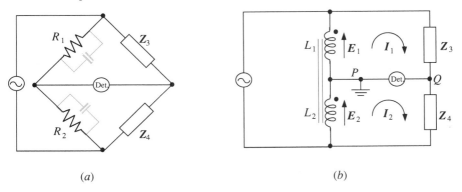

(a)   (b)

must be extended by providing a large number of different standard impedances.

The second difficulty that arises in connection with the circuit of fig. 4.28(a) is the presence of stray capacitances across the two resistance arms (indicated by the dotted components) which will have the effect of altering the balance conditions in an unpredictable way.

Both of these difficulties may, to a very large extent, be circumvented by the arrangement shown in fig. 4.28(b) in which R_1 and R_2 have been replaced by a pair of coupled coils. The coils are closely wound on a core of high permeability so that the coupling coefficient closely approaches unity. If for the moment we assume that the windings have negligible resistance, the two coils will behave like an ideal transformer and the voltages E_1 and E_2 across the coils will be in direct ratio to their turns, that is,

$$\frac{E_1}{E_2} = \frac{N_1}{N_2} \tag{4.43}$$

Now with corresponding ends of the coils arranged as shown, and taking E_1 and E_2 as phase reference, we may write

$$I_1 Z_3 + (I_1 - I_2) R_D = E_1$$
$$(I_2 - I_1) R_D + I_2 Z_4 = E_2 \tag{4.44}$$

Where R_D is the resistance of the detector.

At balance the current in the detector is zero, that is, I_1 and I_2 must be equal in magnitude and phase, hence from (4.44) and (4.43) we obtain

$$\frac{Z_3}{Z_4} = \frac{E_1}{E_2} = \frac{N_1}{N_2}$$

or

$$\frac{N_1}{N_2} = \frac{R_3}{R_4} = \frac{X_3}{X_4} \tag{4.45}$$

We see from this expression that the balance condition is dependent only upon the ratio of the turns on the two coils, which can be fixed in manufacture to very high precision and which, unlike resistance ratio arms, is not subject to the influence of temperature changes or ageing of the components. Moreover, by providing tapping on the two windings at suitably arranged intervals the bridge ratio may be changed in precisely defined steps over a very wide range.

An impedance connected across either of the windings will draw current, but because the ratio of the voltages across the two windings is fixed by the

The transformer bridge

turns ratio, the bridge balance will remain unchanged. Stray capacitances across the windings will, therefore, have no effect on the operation of the bridge. Also, at balance, both points P and Q will be at the same potential so if P is connected to earth, stray capacitances from these points to earth will also have no effect.

So far we have assumed that the resistances of the two coils are zero and that the coupling between them is perfect. In practice there will be resistance and leakage reactance associated with both windings so that current drawn by an additional impedance connected across one coil will produce a voltage drop. This drop however will, by transformer action, cause a drop in the other coil and by proper design the two can be made to compensate so that the ratio of the voltages across the coils remains virtually unchanged.

A practical form of the transformer ratio-arm bridge is shown in fig. 4.29. In this circuit two transformers are used each having an additional winding, to which source and detector are connected. This arrangement avoids earth loops which might affect bridge balance, and allows the various sections of the bridge to be more efficiently screened from one another.

At balance the two currents flowing through Z_3 and Z_4 create equal and opposite ampere turns in the transformer on the detector side of the bridge. Thus, by arranging a series of tappings on this transformer the bridge ratio may be further multiplied, and an overall ratio ranging from 1 to 10^6 may be readily achieved. A single adjustable standard usually suffices for this type of bridge. A further advantage of this arrangement is that by connecting the standard to the opposite side of the transformer on the detector side, as indicated by the dotted line, it is possible to compare unlike impedances; thus, an inductive impedance may be measured using a capacitance standard. The parameters of three-terminal networks, both active and passive, may also be conveniently measured with this type of bridge.

Fig. 4.29. Practical form of transformer bridge.

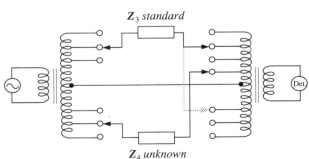

The transformer bridge allows impedances to be compared in the frequency range 1–10 kHz with a precision typically of 1 part in 10^4, and a precision of better than 1 in 10^6 is achievable. By the use of ferrite cores in the construction of the transformer, measurements may be made at frequencies of up to 250 MHz with a precision of a few per cent.

4.17 Summary

In a network containing reactive as well as resistive elements the voltage V and the current I at a pair of terminals will, in general, differ in phase by some angle ϕ. The current may be resolved into two components: a component $I\cos\phi$, called the in-phase current, and a component $I\sin\phi$, called the quadrature current. The product of voltage and in-phase current, gives the power at the terminals:

$$\text{Power} = VI\cos\phi \quad \text{watts} \tag{4.3}$$

This quantity is called the real (or active) power to distinguish it from two other related quantities:

$$\text{Reactive power} = VI\sin\phi \quad \text{vars} \tag{4.4}$$

and

$$\text{Apparent power} = VI \quad \text{volt–amperes} \tag{4.5}$$

The real power is simply the apparent power multiplied by $\cos\phi$ – the power factor. For a purely resistive network the power factor is unity while for a purely reactive network it is zero. The relationship between real, reactive and apparent powers may be shown diagramatically by means of the power diagram or power triangle (fig. 4.4). It is also sometimes convenient to express the relationship in complex form:

$$S = P + jQ \tag{4.10}$$

where S is the apparent power, P the real power, and Q the reactive power.

The principle of conservation of energy applies to both real and reactive powers, which implies that the total real power flowing into a network must equal the sum of the powers dissipated in the individual resistances within the network; likewise the total reactive power must equal the sum of the reactive power associated with the reactances within the network. (Principle of conservation of watts and vars, equation 4.9.) Reactive powers associated with inductance and capacitance carry opposite sign for the purposes of calculating total reactive power.

When energy is drawn from a distribution system, it is desirable to operate at unity power factor because the current requirement for a given power delivered is then a minimum. Industrial loads are predominantly inductive and, therefore, draw a lagging current. Connection of a capacitor, drawing a leading current from the supply, will reduce the total reactive power thereby improving the power factor. Such improvement is often economically justifiable in the case of large industrial power consumers.

The transformer is of fundamental importance as a component in power distribution systems; it is also used in electronic and communications equipment of all kinds. The analysis of the transformer as a circuit element is greatly simplified by reference to the loss-free *ideal* transformer in which the voltage ratio is in direct proportion to the turns ratio between primary and secondary windings and the current ratio is in indirect proportion to the turns ratio (equations (4.13) and (4.14)). Real transformers may then be characterized by circuit models based on the ideal transformer with components added to account for winding resistances, leakage inductances and an exciting current. By employing the impedance transforming properties of the transformer, a simplified circuit model may be derived in which all elements are referred to either the primary circuit or the secondary circuit. This results in a circuit model consisting of just two series elements and two shunt elements (fig. 4.18). These elements may be determined experimentally from simple short-circuit and open-circuit tests.

In electrical and electronic circuits intended for signal transmission it is important to ensure that the loss of signal power is as small as possible. Optimum power transmission between a source of impedance Z_0 and a load impedance Z is achieved by matching source to load according to the maximum power theorem:

$$Z = Z_0^* \qquad (4.37)$$

where Z_0^* is the complex conjugate of Z_0. For purely resistive circuits this condition implies that source and load resistances should be the same. The impedance transforming properties of the transformer or combinations of reactive elements can be utilized to achieve matched conditions.

4.18 Problems

1. For the circuit of fig. 4.30 determine.
(a) the power dissipated in each branch of the circuit.
(b) the watts and vars to the whole circuit.
(c) the power factor of the circuit.
2. Find in the circuit of fig. 4.31 the pure reactance or reactances X that will make the overall power factor 0.8.
3. A load which takes 3 MW at 0.6 power factor lagging is fed by a line whose inductive reactance is five times its resistance. In order to provide a

load voltage of 75 kV it is found that the input to the line must be 90 kV. Find the line current.

If a capacitor is connected in parallel with the load to bring the power factor to 0.9 lagging, what must then be the input voltage to the line?

4. Two impedances, Z_1 and Z_2, are connected in parallel. The resistive component of the first branch is 5 Ω. When the parallel combination is connected to a supply voltage of 240 V, the first branch takes a lagging current of 21.5 A and the second branch takes a leading current at power factor 0.6. The total power supplied is 3.69 kW.

Determine:
(a) the branch and total currents.
(b) the impedances of the two branches.
(c) the impedance of the parallel combination.
(London University)

5. If the alternating voltage across a certain load is represented by the complex number V and the alternating current through the load is given by I, demonstrate that the power is the real part of either VI^* or V^*I.

A voltage of $(100\sin\omega t + 20\sin(2\omega t + \pi/2))$ is applied to a circuit consisting of a resistor and a capacitor in series. The impedance of the circuit at angular frequency ω is $(10-j20)$ ohms. Calculate the r.m.s. current, the power dissipated and the reactive power.
(Oxford University)

Fig. 4.30. Circuit for problem 1.

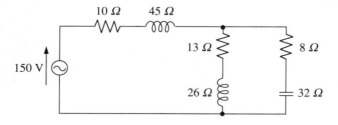

Fig. 4.31. Circuit for problem 2.

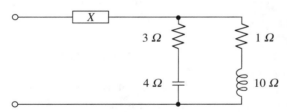

Problems

6. In the circuit shown in fig. 4.32 the transformer is to be assumed ideal with turns ratio 1:2.

   (a) Show that at angular frequency $\omega = 10^2$ rad/s the current I_2 is in phase with the voltage across the transformer secondary winding.

   (b) Given that $I_0 = 2$A at 100 rad/s determine the magnitudes of the currents I_1, I_2 and I_3 and their phase angles relative to the current I_0.
   (London University)

7. Describe briefly how an equivalent circuit for a power transformer can be derived from measurements made in open- and short-circuit tests.

   A 415/240 V, 50 Hz, single-phase transformer has winding resistance 0.15 Ω and leakage reactance j1.0 Ω, both referred to the high-voltage winding.

   Estimate the terminal voltage on the low-voltage winding when the transformer supplies the following load from a 415 V source:

   (a) a resistance of 6 Ω.

   (b) a resistance of 6 Ω in parallel with a 500 μF capacitor.
   (Cambridge University: First year)

8. The maximum efficiency for a single-phase, 50 Hz transformer rated at 1000 kVA, 2000/250 V is obtained when it is supplying, at the secondary side, 70% of full load at unity power factor and 250 V. The following data are available for the transformer:

   Turns ratio, 8:1.
   Primary winding resistance, 0.04 Ω.
   Secondary winding resistance, 0.001 Ω.
   Leakage reactance referred to the primary winding, 1.04 Ω.

   Estimate the maximum efficiency and calculate the magnitude of the in-phase component of the current when the secondary winding is on open circuit. Estimate also the readings on the measuring instruments used in a test on the transformer with full-load current in the primary and secondary short circuited.
   (Cambridge University: Second year)

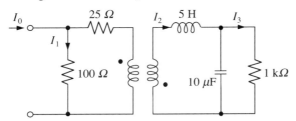

Fig. 4.32. Circuit for problem 6.

9. How are the Thévenin and Norton equivalents of a two-terminal network related to each other? Show that if a load is connected at the two terminals, the calculated load current is the same, whichever equivalent circuit is used.

In the circuit of fig. 4.33 the resistor R is adjusted until it dissipates maximum power. Find:
  (a) the ohmic value of R.
  (b) the current and power in R.
  (c) the total power drawn from the 8 V and 2 A sources.
(Newcastle University)

10. A vibration measuring instrument is equivalent to a 10 mV source in series with an impedance of $(900+j1200)\,\Omega$. What is the maximum power it can supply to an amplifier whose input impedance is an adjustable resistor?

What would be the maximum power if a suitable capacitor were connected in series with the input? What should be the value of this capacitor if the vibration frequency is 1 kHz?

11. A practical voltage source has an output voltage of E on open circuit and an internal impedance of $Z_1 = R_1 + jX_1$. It is connected to a load impedance $Z_2 = R_2 + jX_2$, whose magnitude may be changed without

Fig. 4.33. Circuit for problem 9.

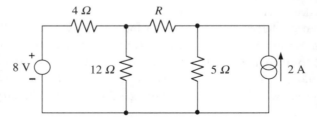

Fig. 4.34. Circuit for problem 12.

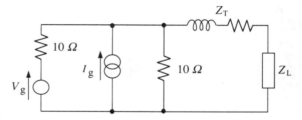

change of angle, i.e. $\theta = \tan^{-1}(X_2/R_2) = $ constant. Show that maximum power transfer from source to load occurs for the condition:

$$|Z_2| = |Z_1|.$$

An electromechanical vibration transducer has an impedance of $(500+j600)\,\Omega$ and an output of 0.2 mV on open circuit. It is to be transformer-coupled to an amplifier having an input resistance of 100 kΩ. Determine the transformer turns ratio required to establish the maximum voltage at the input of the amplifier. What is the magnitude of this voltage? It may be assumed that the transformer is ideal.
(Cambridge University: Second year)

12. In the circuit shown in fig. 4.34, a variable load impedance Z_L is supplied through a transmission line $Z_T = (5+j12)\,\Omega$ from a voltage generator $V_g = 10\underline{/60°}$ and a current generator $I_g = 5\underline{/0}$.

(a) Determine the value of impedance Z_L to absorb maximum power.

(b) Calculate the power absorbed by the load calculated in part (a).
(Sheffield University: First year)

13. A 1.0 MHz generator having an open-circuit voltage of 10 V and an output resistance of 50 Ω supplies a variable impedance $R+jX$, as shown in fig. 4.35(a). Derive an expression for the maximum power P_m which can be supplied to the load, and find the corresponding values of R and X.

Fig. 4.35. Circuit for problem 13.

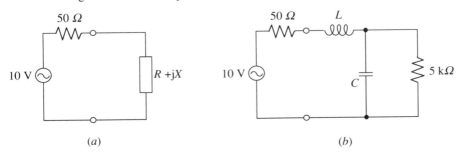

(a)   (b)

Fig. 4.36. Circuit for problem 14.

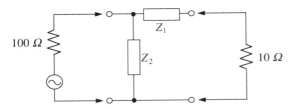

Show that the circuit shown in fig. 4.35(b) can be so designed that the power delivered to the 5.0 kΩ resistor has the same value P_m, and determine the appropriate values of L and C.
(London University)

14. In the circuit of fig. 4.36 the network of pure reactors Z_1 and Z_2 is to be used to transfer maximum power from the signal generator of output resistance 100 Ω on the load of resistance 10 Ω at an angular frequency of 10^5 rad/s. Find the components required for Z_1 and Z_2.

In what ratio is the power supplied to the load reduced if the Z_1, Z_2 network is omitted?
(Cambridge University: Second year)

15. In the circuit of fig. 4.37 impedances are given in ohms at the operating angular frequency ω of the signal source S. Find the reactance of the capacitor C which will give maximum power transfer to the 30 Ω load.
(Cambridge University: Second year)

Fig. 4.37. Circuit for problem 15.

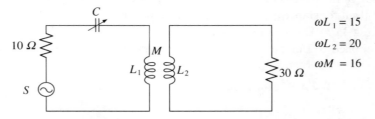

$\omega L_1 = 15$
$\omega L_2 = 20$
$\omega M = 16$

5
Three-phase alternating current circuits

5.1 Introduction

In chapters 3 and 4, circuits were considered in which the alternating energy sources possess the same frequency but, in general, different voltages and internal impedances, and arbitrary phase relationships. Each voltage source in such circuits may be thought of as being generated by the interaction between a stationary coil of wire and a properly shaped rotating magnetic field, as shown schematically in fig. 5.1(a).* Now, suppose that instead of a single coil (generally referred to as a winding) there are n windings symmetrically disposed on the stator of the machine. If the windings are identical, their impedances are equal and the amplitudes of their induced voltages are equal. The voltages will all be of the same frequency, determined by the angular speed of the rotating magnet, and the phase relations among them will be fixed. The phase difference between the voltages of two successive windings will be $2\pi/n$ radians or $360/n$ degrees. A machine constructed in this fashion is an n-phase generator.

The majority of power systems throughout the world utilize the three-phase generator, shown schematically in fig. 5.1(b). In fig. 5.2 are shown (a) a phasor diagram for the three-phase generator, and (b) graphs of voltage v. time for the three phases. The three voltages differ in phase by $360/3 = 120°$.

The generators in a power system are connected to a series of step-up and step-down transformers that provide voltage levels appropriate for the efficient transmission, distribution and consumption of the power generated. The three-phase transformers used possess primary and secondary circuits each consisting of three identical windings electrically (although

---

* A discussion of rotating a.c. machines is beyond the scope of this text; see reference 12 for a general introduction to the theory and practice of electrical machines and associated equipment.

232 Three-phase alternating current circuits

not mechanically) similar to those in the three-phase generator. Generator and transformer windings are connected phase-to-phase, the voltages impressed across the phase windings of any transformer in the system being of similar form to those shown in fig. 5.2.

An important feature of any polyphase generator is the fact that at any instant the sum of the individual phase voltages is zero. An examination of fig. 5.2 shows the validity of this statement for three phases. In fig. 5.2(a) the resultant of the three phasors is obviously zero. In fig. 5.2(b), at any instant, the sum of the three voltages is exactly zero.

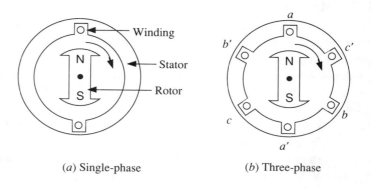

Fig. 5.1. Schematic representation of an a.c. generator or alternator.

(a) Single-phase   (b) Three-phase

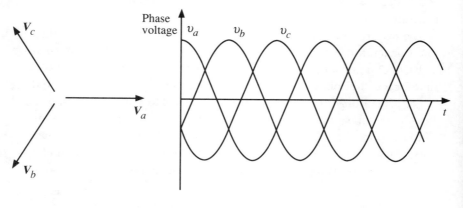

Fig. 5.2. The phase voltages of a three-phase generator.

(a) Phasor diagram   (b) Instantaneous values

5.2 Advantages of three-phase systems

Among important advantages of three-phase systems for power distribution may be numbered the following.

(1) Economy in energy conversion equipment.

The generation and utilization of electrical energy in a large-scale power system depends on the efficient operation of large numbers of rotating a.c. machines. Polyphase windings make better use of the space available in a machine of a given physical size and therefore capital costs are lower. The load capacity of a two-phase machine is some 40% greater than that of a single-phase machine while that of three-phase machines is 50% greater. Polyphase machines with four or more phases offer a negligible increase in load capacity over the three-phase case which, therefore, represents the optimum arrangement.

(2) Economy in transmission equipment.

The transmission of power in a single-phase system requires two conductors, two- and three-phase systems both require three conductors, and a four-phase system four conductors. The three-phase system has only two-thirds of the transmission loss of the single-phase system for the same power delivered to a load, consequently it offers the greatest economy among the possible polyphase systems.

(3) Provision of constant power and torque.

Polyphase systems provide constant instantaneous power to a load, which implies that both generators and motors exhibit a constant torque characteristic. By contrast, in a single-phase system power goes to zero twice every cycle so that the torque of a single-phase machine is pulsating.

This characteristic of polyphase systems, shared of course by the three-phase system, is essential for large-scale power generation since it allows the use of constant-torque prime movers such as the steam and water turbine. The provision of constant torque is also essential in many industrial applications of electric motors.

5.3 Three-phase circuits

5.3.1 Phase and line voltages

To begin our study of three-phase circuits consider the arrangement shown in fig. 5.3 in which a three-phase generator or transformer is represented by three identical, ideal voltage sources. Each source is joined to a load impedance through two wires each of impedance Z_l. It will be apparent that nothing is lost if we form a common point by connecting together the three ends x, y and z, of the individual sources. We may also join points x', y' and z'. Then we have reduced our circuit from a 6-wire to a 4-wire system (fig. 5.4), with the common points n and n' joined by a *neutral* wire of impedance Z_n.

Fig. 5.3. Source-to-load connection using 6 conductors.

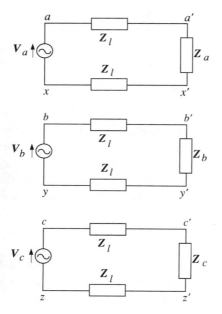

Fig. 5.4. Source-to-load connection using 4 conductors.

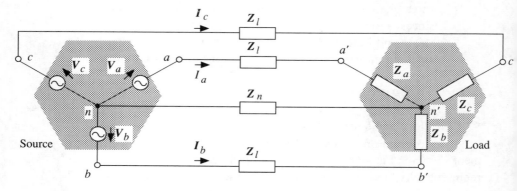

Three-phase circuits

With the interconnections shown in fig. 5.4, the phasor diagram representing the three source voltages is as shown in fig. 5.5(a). The three phasors each have magnitude V_p; thus, a voltmeter connected between the *neutral point n* and any of the points a, b or c, will read V_p volts. The voltages V_a, V_b and V_c are called the *phase* voltages. If V_a is taken as the reference phasor, we may write the phase voltages as

$$V_a = V_p \underline{/0}$$
$$V_b = V_p \underline{/-120} \qquad (5.1)$$
$$V_c = V_p \underline{/+120}$$

Other voltages of significance in fig. 5.4 are the *line*-to-*line* (or simply *line*) voltages V_{ab}, V_{bc} and V_{ca}. The relationships among line and phase voltages are readily determined with the aid of fig. 5.5(b). With V_a as the reference phasor we have

$$V_{ab} = V_a - V_b = V_a + (-V_b)$$

Fig. 5.5. Phasor diagrams for the source shown in fig. 5.4.

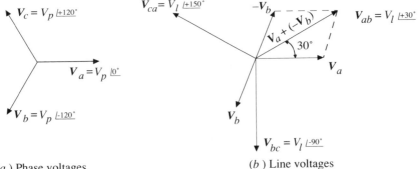

(a) Phase voltages

(b) Line voltages

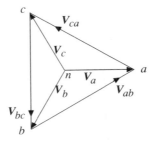

(c) Line and phase voltage relationships

or

$$V_{ab} = V_p\underline{/0} + V_p\underline{/60} = V_p\left(1 + \tfrac{1}{2} + j\frac{\sqrt{3}}{2}\right) = \sqrt{3}V_p\underline{/30}$$

Similarly it may be shown that $V_{bc} = \sqrt{3}V_p\underline{/-90}$ and $V_{ca} = \sqrt{3}V_p\underline{/150}$. Let V_l be the magnitude of the line voltage, then we may write (with $V_a = V_a\underline{/0}$ as phase reference)

$$\begin{aligned} V_{ab} &= V_l\underline{/30} \\ V_{bc} &= V_l\underline{/-90} \\ V_{ca} &= V_l\underline{/+150} \end{aligned} \quad (5.2)$$

where

$$V_l = \sqrt{3}V_p$$

A compact (and easily remembered) form of phasor diagram, showing the relationships among the complete sets of line and phase voltages, is shown in fig. 5.5(c).

5.3.2 Balanced load

We now assume that the three load impedances in fig. 5.4 are identical so that

$$Z_a = Z_b = Z_c = Z$$

Such an arrangement is referred to as a *balanced* load.* In fig. 5.4, the line currents are designated I_a, I_b and I_c. By applying Kirchhoff's voltage law around the appropriate loops we obtain

$$\begin{aligned} V_a &= I_a Z_l + I_a Z + (I_a + I_b + I_c)Z_n \\ V_b &= I_b Z_l + I_b Z + (I_a + I_b + I_c)Z_n \\ V_c &= I_c Z_l + I_c Z + (I_a + I_b + I_c)Z_n \end{aligned} \quad (5.3)$$

We have already shown that $V_a + V_b + V_c = 0$. It follows then that the sum of the terms on the right-hand side of equations (5.3) is zero. The result of this summation is

$$(I_a + I_b + I_c)(Z_l + Z + 3Z_n) = 0$$

Therefore,

$$I_a + I_b + I_c = 0 \quad (5.4)$$

---

* Much of the electrical equipment used in industry (three-phase motors for example) constitutes a balanced load, and it is possible in a large power distribution system to maintain conditions where loads on generators and transformers are essentially balanced.

So, for the balanced load, there is no current in the neutral wire and, consequently, no voltage drop across the impedance Z_n.

From (5.3) and (5.4) the line currents under balanced conditions are given by

$$I_a = \frac{V_p \underline{/0}}{Z_l + Z}; \quad I_b = \frac{V_p \underline{/-120}}{Z_l + Z}; \quad I_c = \frac{V_p \underline{/+120}}{Z_l + Z} \quad (5.5)$$

In practice it is not possible to maintain perfect balance among loads in a power distribution system, particularly so at the local distribution level where loads drawn by small commercial and domestic consumers, distributed between the three separate phases of the system, are continually varying. A neutral wire is always provided in this case to carry the small out-of-balance currents that arise. This neutral wire is connected to earth at some point (usually at the sub-station transformer) and provides a common reference for the whole of the local distribution network.

The provision of a neutral wire has the further advantage of providing consumers with two alternative voltages: line voltage and phase voltage differing, as shown above, by a factor of $\sqrt{3}$. In the UK these voltages are 415 V and 240 V respectively. The higher of these is more suitable for commercial and industrial consumers operating, for example, machine tools powered by three-phase induction motors. For a given installed load the higher voltage results in a lower current, which reduces the capital costs associated with wiring and switchgear. These considerations are of less importance for the domestic consumer where a single-phase supply (one phase of the three-phase supply) is adequate for low-power lighting and heating purposes. The lower voltage is also safer in situations where portable electrical apparatus is used extensively.

As mentioned previously, the average load on a large-scale power system is always arranged to be as nearly as possible in balance. The generators in a power station operate under essentially balanced conditions, as do the main distribution transformers and overhead lines. In the case of long-distance lines the neutral wire is dispensed with, although an additional wire is necessary to connect the transmission towers together and to earth. (This normally carries no current except in the event of a lightning strike.)

5.3.3 Worked example

For the three-phase system shown in fig. 5.4 the line voltage of the source is 415 V, and the load consists of three equal impedances $Z = 20 + j10 (= 22.4\underline{/26.5})\Omega$. The line impedance $Z_l = 2 + j4$ $(= 4.47\underline{/53.4})\Omega$. (Note that in this example the line impedance Z_l is chosen to be unrealistically high compared with Z so that its effect will not be negligible in the calculations.) Find the line current, the voltage across the load, the power delivered to the load, and the power lost in the transmission line.

Solution. Because this is a balanced system, we may make calculations for one phase alone and find all the required information from the results of these computations.

The magnitude of the phase voltage is, from (5.2), $V_p = 415/\sqrt{3} = 240$ V. We may choose any phasor to be our reference, i.e. to have phase angle zero, so letting $V_a = 240\underline{/0°}$:

$$I_a = \frac{V_a}{Z_l + Z} = \frac{240\underline{/0}}{22 + j14} = \frac{240\underline{/0}}{26.1\underline{/32.5}} = 9.2\underline{/-32.5}$$

The load voltage is

$$V_{a'} = I_a Z = (9.2\underline{/-32.5})(22.4\underline{/26.5}) = 206\underline{/-6}$$

Because the system is balanced, the three line currents all have magnitude 9.2 A and the three load voltages have magnitudes 206 V. Furthermore, the angle between adjacent voltages and between adjacent currents is 120°. We may therefore write

$I_b = 9.2\underline{/-152.5}$ $V_{b'} = 206\underline{/-126}$
$I_c = 9.2\underline{/87.5}$ $V_{c'} = 206\underline{/114}$

The relationships among currents and voltages are shown in the phasor diagram of fig. 5.6.

Fig. 5.6. Phasor diagram for worked example (section 5.4).

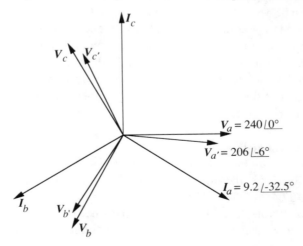

Three-phase circuits

The power factor for one element of the load is the cosine of the angle between I_a and $V_{a'}$. This is $\cos(-6° - (-32.5°)) = \cos 26.5° = 0.895$. So, the average power supplied to one element of the load is

$$P_a = V_{a'} I_a \cos 26.5° = 206 \times 9.2 \times 0.895 = 1696 \text{ W}$$

The total power to the load is $P = 3P_a = 5088$ W.
The power loss in one line is $I_a^2 R_{\text{line}} = 9.2^2 \times 2 = 169$ W.
Then the total line loss is $3 \times 169 = 507$ W.

As a check, let us calculate the power delivered by the source. For one phase of the source, the phase of the angle is $(-32.5°)$; so the power delivered by phase a is

$$P_a = V_a I_a \cos(-32.5°)$$
$$= 240 \times 9.2 \times 0.843 = 1861 \text{ W}$$

The total power supplied by the source is $3P_a = 3 \times 1861 = 5583$ W. The total power absorbed is the sum of the line loss and the power delivered to the load, that is, $5088 + 507 = 5595$ W. Within the limits of error of our calculations, this equals the power supplied by the source.

5.3.4 Star and delta connections

In general, for an n-phase system the configuration in which one end of each element (either source or load) is tied to a common point is called a *star* connection. For a three-phase system this configuration is also called a Y connection (sometimes spelled out as *wye*). In fig. 5.4 both the source and the load are Y connected.

An alternative way of connecting the elements of the source or load is one in which they are joined to form a closed path. In the case of the voltage sources this connection is possible because, as we have seen, the sum of the phase voltages is zero, consequently, such an interconnection will not result in a circulating current through the sources. For a three-phase system this interconnection is called a delta, often written Δ. It is represented in fig. 5.7.

It is usual for the alternators in a large power station to be connected in Y, but transformers are connected in Y or Δ depending on their particular function in the system. When sources are connected in Δ there is, of course, no neutral point.

A three-phase load may be connected in either Y or Δ regardless of how the source is connected. Indeed, in connecting a three-phase load one may have no information concerning the source. The usual situation is that one has three lines (plus, perhaps, a neutral wire) and information about the *phase sequence*, that is information concerning the order in which the line voltages (or phase voltages, if there is a neutral) reach their maximum

240 *Three-phase alternating current circuits*

Fig. 5.7. Delta connection of three-phase voltage sources.

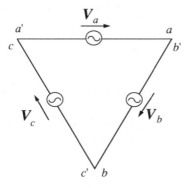

Fig. 5.8. Relationship among line and phase currents for a balanced delta-connected load.

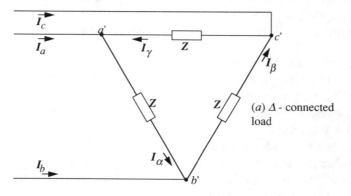

(a) Δ - connected load

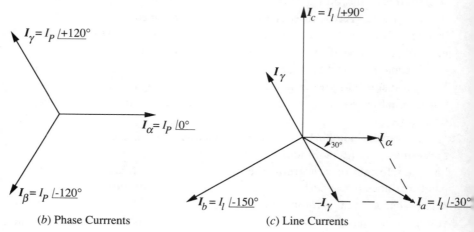

(b) Phase Currents

(c) Line Currents

Three-phase circuits

values. One also knows, or can measure the magnitude of the line voltage (or of the phase voltage, if appropriate). The consumer who connects a load to the three-phase line usually is not concerned about losses in the transmission line or voltages and currents at the source; it is rather the authority responsible for supplying the power that is interested in these quantities.

Fig. 5.8(a) shows a Δ-connected balanced load in which the load impedances carry currents I_α, I_β and I_γ. Because the load is balanced these phase currents are of equal magnitude and mutually spaced on the phasor diagram at 120°, as shown in fig. 5.8(b). Let the magnitudes of the phase and line currents be I_p and I_l respectively, and let I_α be the phase reference. Then at node a'

$$I_a = I_\alpha + (-I_\gamma) = I_p\underline{/0} + I_p\underline{/-60} = I_p\left(1 + \tfrac{1}{2} - j\frac{\sqrt{3}}{2}\right) = \sqrt{3}I_p\underline{/-30}$$

The three line currents must also be at mutual phase angles of 120°, hence, we may write

$$I_a = I_l\underline{/-30°}; \qquad I_b = I_l\underline{/-150°}; \qquad I_c = I_l\underline{/90°} \tag{5.6}$$

where

$$I_l = \sqrt{3}I_p$$

The relationship among line and phase currents for the balanced Δ-connected load is shown in fig. 5.8(c). This relationship is similar to that which exists between line and phase voltages in a balanced Y-connected system (see fig. 5.5).*

5.3.5 Worked example

For the balanced three-phase system shown in fig. 5.9 the line voltage at the source is 415 V, the line impedance $Z_l = (2 + j4)\,\Omega$, and the load impedance $Z = (20 + j10)\,\Omega$. (These parameters are identical to those in the previous worked example.) Find the line currents, the voltage across the load, the power delivered to the load and the power lost in the transmission line.

Solution. In fig. 5.9 we have defined three mesh currents I_1, I_2 and I_3. The currents that we wish to find; namely, the line currents I_a, I_b and I_c and the three load currents, I_α, I_β and I_γ are readily expressed in terms of these mesh currents. Before substituting numerical values, expressions for I_1, I_2 and I_3 are found.

---

* The Y- and Δ-connected systems are duals.

Using mesh analysis we obtain:

$$I_1(2Z_l+Z) \quad -I_2Z_l \quad -I_3Z = V_a-V_b$$
$$-I_1Z_l \quad +I_2(2Z_l+Z) \quad -I_3Z = V_c-V_a$$
$$-I_1Z \quad -I_2Z \quad +I_3(3Z)=0$$

Simultaneous solution of these equations yields

$$I_1 = \frac{2(V_a-V_b)+(V_c-V_a)}{Z_0}$$

$$I_2 = \frac{(V_a-V_b)+2(V_c-V_a)}{Z_0}$$

$$I_3 = (I_1+I_2)/3 = \frac{(V_a-V_b)+(V_c-V_a)}{Z_0}$$

where

$$Z_0 = 3Z_l + Z$$

But

$$(V_a-V_b) = V_{ab} \text{ and } (V_c-V_a) = V_{ca}$$

hence

$$I_1 = (2V_{ab}+V_{ca})/Z_0$$
$$I_2 = (V_{ab}+2V_{ca})/Z_0 \quad\quad (5.7)$$
$$I_3 = (V_{ab}+V_{ca})/Z_0$$

If V_a is chosen to have zero phase angle then, by (5.1),

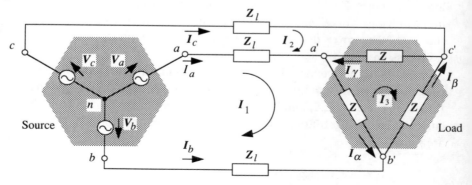

Fig. 5.9. Wye-connected generator with balanced delta-connected load (worked example, section 5.3.5).

Three-phase circuits

$$V_{ab} = 415\underline{/30} = 415(0.866 + j0.5)$$
$$V_{ca} = 415\underline{/150} = 415(-0.866 + j0.5)$$

Also,

$$Z_0 = 3(2+j4) + 20 + j10 = 26 + j22 = 34.1\underline{/40.3}\,\Omega$$

Substituting these numerical values in (5.7) gives

$$I_1 = 21.1\underline{/19.7}\,\text{A}; \qquad I_2 = 21.1\underline{/79.7}\,\text{A}; \qquad I_3 = 12.2\underline{/49.7}\,\text{A}$$

From fig. 5.9:

$$\begin{aligned}I_a &= I_1 - I_2 & I_\alpha &= I_1 - I_3 \\ I_b &= -I_1 & I_\beta &= -I_3 \\ I_c &= I_2 & I_\gamma &= I_2 - I_3\end{aligned}$$

Substituting values of I_1, I_2 and I_3 in these equations gives

$$\begin{aligned}I_a &= 21.1\underline{/-40.3}\,\text{A} & I_\alpha &= 12.2\underline{/-10.3}\,\text{A}\\ I_b &= 21.1\underline{/-160.3}\,\text{A} & I_\beta &= 12.2\underline{/-130.3}\,\text{A}\\ I_c &= 21.1\underline{/79.7}\,\text{A} & I_\gamma &= 12.2\underline{/109.7}\,\text{A}\end{aligned}$$

A considerable saving in the amount of calculation required to find the line and phase currents may be achieved by utilizing (5.6) and the relationship shown in fig. 5.8(c). Specifically, a knowledge of either I_1 or I_2 makes it possible to write down line and load (phase) currents directly. Thus, having solved (5.7) for current I_1, then

$$I_b = -I_1 = -21.1\underline{/19.7} = 21.1\underline{/-160.3}$$

and we may write

$$I_a = 21.1\underline{/-160.3 + 120} = 21.1\underline{/-40.3}$$
$$I_c = 21.1\underline{/-160.3 - 120} = 21.1\underline{/79.7}$$

Furthermore, the phase current I_α is given by

$$I_\alpha = (I_a/\sqrt{3})\underline{/+30} = (21.1/\sqrt{3})\underline{/-40.3 + 30} = 12.2\underline{/-10.3}$$

from which it follows that

$$I_\beta = 12.2\underline{/-10.3 - 120} = 12.2\underline{/-130.3}$$
$$I_\gamma = 12.2\underline{/-10.3 + 120} = 12.2\underline{/109.7}$$

Now that the load currents have been found, the load voltages are readily obtained

$$V_{a'b'} = I_\alpha Z = (12.2\underline{/-10.3})(22.4\underline{/26.5}) = 273\underline{/16.2}\,\text{V}$$

244 Three-phase alternating current circuits

Similarly,

$$V_{b'c'} = 273\underline{/-103.8} \text{ V, and } V_{c'a'} = 273\underline{/136.2} \text{ V}$$

Because the load is balanced, the total power delivered to the load is three times the power in one load impedance. For example, the power delivered by current I_a is $I_a^2 R_{\text{load}}$, where R_{load} is the resistive part of the load impedance, in this case 20 ohms. The total power delivered to the load is then

$$P_{\text{load}} = 3 \times 12.2^2 \times 20 = 8930 \text{ W}$$

The power loss is three times the loss in one line. Therefore

$$P_{\text{line}} = 3 \times 21.1^2 \times 2 = 2671 \text{ W}$$

Finally, we calculate the power supplied by the source. Again the fact that we have a balanced load makes it possible to calculate for one phase and multiply the result by three. Thus the total power supplied by the source is

$$3 \times V_a I_a \cos(-40.3) = 3 \times 240 \times 21.2 \times 0.763 = 11\,591 \text{ W}$$

This checks with the sum of the load power and line loss: $8930 + 2671 = 11\,601$ W (within the limits of error of the calculations).

5.3.6 Use of Y–Δ transformation

An alternative approach to the solution of the above example is to use a Y–Δ transformation (section 2.9.2) of the load impedance. The result of this transformation is shown in fig. 5.10. Because the system is balanced, points n and n' are at the same potential and may, for purposes of calculation, be connected by a conductor of zero impedance. Then from fig. 5.9,

Fig. 5.10. Load of fig. 5.9 after delta to wye transformation.

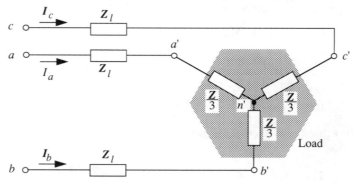

Three-phase circuits

$$I_a = \frac{V_a}{Z_l + Z/3} = \frac{240\angle 0}{(2+j4)+(20+j10)/3} = \frac{3 \times 240\angle 0}{6+j12+20+j10}$$
$$= 21.1\angle -40.3$$

The remaining line currents and the three load currents are found, as before, by invoking (5.6) and the relationship shown in fig. 5.8(c).

The Y–Δ transformation is especially useful when there are mixed Y- or Δ-connected loads that must be combined to find the total line current. It is convenient then to use the transformation so that the loads will be all Δ-connected or all Y-connected. After line currents and line voltages at the load have been calculated, the transformation may be used in the opposite direction to permit calculation of individual load currents.

5.3.7 Unbalanced load

If the individual load impedances are not identical (that is, if the load is *unbalanced*) then in general neither the line currents nor the line voltages at the load will have equal magnitudes. Moreover, the phase sequence will affect both the magnitudes and phase angles of currents and voltages in the circuit.

The worked example that follows outlines a procedure for analyzing an unbalanced Y-connected load. A similar procedure may be used for a Δ-connected load if the load is first converted to a Y-configuration by means of the Y–Δ transformation.

5.3.8 Worked example

Fig. 5.11 shows a Y-connected source, supplying a Y-connected unbalanced load. The source has a phase voltage of 240 V, and the load and line impedances have the following values:

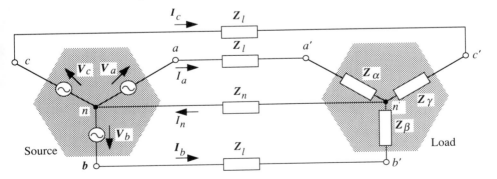

Fig. 5.11. An unbalanced wye-connected load (worked example, section 5.3.8).

$$Z_\alpha = 20 + j10 = 22.4\underline{/26.5} \qquad Z_l = 2 + j4 = 4.47\underline{/63.4}$$
$$Z_\beta = 8 + j2 = 8.25\underline{/14.1} \qquad Z_n = 0 + j1 = 1\underline{/90}$$
$$Z_\gamma = 15 - j5 = 15.8\underline{/-18.5}$$

Calculate the line currents (a) if the source has phase sequence abc; (b) if the source has phase sequence acb.

Solution. Since the load is unbalanced there will be a current in the neutral line and, consequently, the impedance Z_n now appears in the calculations. We may find the line currents directly using the method of mesh analysis, but this involves the solution of three simultaneous equations with complex coefficients. The use of nodal analysis to find the voltage at the node n' provides a somewhat easier approach.

Let $V_{n'}$ be the voltage of node n' with respect to the neutral point n, then the nodal equation at n' is

$$\frac{V_{n'} - V_a}{Z_\alpha + Z_l} + \frac{V_{n'} - V_b}{Z_\beta + Z_l} + \frac{V_{n'} - V_c}{Z_\gamma + Z_l} + \frac{V_{n'}}{Z_n} = 0$$

or

$$V_{n'}\left[\frac{1}{Z_\alpha + Z_l} + \frac{1}{Z_\beta + Z_l} + \frac{1}{Z_\gamma + Z_l} + \frac{1}{Z_n}\right]$$
$$= \frac{V_a}{Z_\alpha + Z_l} + \frac{V_b}{Z_\beta + Z_l} + \frac{V_c}{Z_\gamma + Z_l}$$

Inserting numerical values for the impedances:

$$V_{n'}\left[\frac{1}{22+j14} + \frac{1}{10+j6} + \frac{1}{17-j1} + \frac{1}{0+j1}\right]$$
$$= \frac{V_a}{22+j14} + \frac{V_b}{10+j6} + \frac{V_c}{17-j1} \tag{5.8}$$

(a) With phase sequence abc* the source phase voltages are

$$V_a = 240\underline{/0} = 240 + j0 \text{ (reference)}$$
$$V_b = 240\underline{/-120} = -120 - j208$$
$$V_c = 240\underline{/120} = -120 + j208$$

---

* Note that the phase sequence specified refers to the cyclic order in which the voltage phasors must appear in the *stationary* phasor diagram when moving round the diagram in a *clockwise* direction. The rotating phasor diagram (see section 3.5) will then cause phasors to sweep past a given reference direction in the correct order.

Three-phase circuits

Inserting these values in (5.8) gives

$$V_{n'} = 17\underline{/-88.8} \simeq 0 - j17$$

The line currents are then given by

$$I_a = \frac{V_a - V_{n'}}{Z_\alpha + Z_l} = \frac{(240 + j0) - (0 - j17)}{22 + j14} = 9.2\underline{/-28.4}\,\text{A}$$

$$I_b = \frac{V_b - V_{n'}}{Z_\beta + Z_l} = \frac{(-120 - j208) - (0 - j17)}{10 + j6} = 19.3\underline{/-153}\,\text{A}$$

$$I_c = \frac{V_c - V_{n'}}{Z_\gamma + Z_l} + \frac{(-120 + j108) - (0 - j17)}{17 + j1} = 14.9\underline{/121}\,\text{A}$$

The current in the neutral wire is simply

$$I_n = \frac{V_{n'}}{Z_n} = \frac{17\underline{/-88.8}}{1\underline{/90}} = 17\underline{/-179}$$

(b) With phase sequence *acb* the source phase voltages are

$$V_a = 240\underline{/0} = 240 + j0 \text{ (reference)}$$
$$V_b = 240\underline{/120} = -120 + j208$$
$$V_c = 240\underline{/-120} = -120 - j208$$

Inserting these values in (5.8) gives

$$V_{n'} = 3.29\underline{/140.6} = -2.54 + j2.09$$

The line currents and the neutral current are found as in part (a):

$$I_a = 9.3\underline{/-32.9}\,\text{A}; \quad I_b = 20.3\underline{/88.7}\,\text{A};$$
$$I_c = 14.1\underline{/-116}\,\text{A}; \quad I_n = 3.29\underline{/50.6}\,\text{A}$$

We note that the different phase sequences produce different magnitudes and phase angles for the line currents. Phasor diagrams for the two phase sequences are shown in fig. 5.12. Fig. 5.12(*b*) and (*d*) demonstrate that in each case the sum of the line currents is equal to the current in the neutral wire. Such diagrams provide an approximate graphical check on the calculations.

It may be remarked, finally, that calculations for three-phase circuits, whether balanced or unbalanced, are considerably simplified if line impedances, or the impedance in the neutral wire, can be assumed to be zero. For instance, in the above example involving an unbalanced load, if Z_n in fig. 5.11 is zero, than the source phase voltages appear directly across the series combinations of line and load impedances and expressions for the line currents can be written down directly, without recourse to nodal analysis (see problem 5.2).

Similarly, in fig. 5.9, if the line impedances Z_l are zero, then the source line voltages are applied directly to the loads and, again, expressions for the load currents can be written down directly without recourse to mesh analysis. In this case, since the load is balanced, only one current need be evaluated; all other currents can be deduced from the known relationships between line and phase currents (see problem 5.1).

5.4 Power, reactive power and apparent power in balanced loads

It has been shown in section 5.3 that for a balanced Y-connected load the voltage across one phase is equal to $V_l/\sqrt{3}$, where V_l is the line voltage. The apparent power in one phase is therefore $(V_l/\sqrt{3})I_l$, where I_l is the line current (equal to the phase current in a Y connection). If ϕ is the phase angle between the phase voltage and current, the average power in one phase will be $(V_l/\sqrt{3})I_l\cos\phi$, and the reactive power will be $(V_l/\sqrt{3})I_l\sin\phi$. Multiplying by a factor of 3 gives totals for all three phases, hence using the notation of section (4.3) we may write:

$$\begin{aligned}\text{Power } P &= \sqrt{3}V_lI_l\cos\phi \\ \text{Reactive power } Q &= \sqrt{3}V_lI_l\sin\phi \\ \text{Apparent power } S &= \sqrt{3}V_lI_l\end{aligned} \quad (5.9)$$

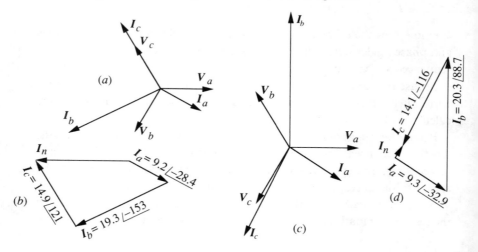

Fig. 5.12. Phasor diagrams for the circuit of fig. 5.11.

These expressions apply also to a balanced Δ-connected load. In this case the full line voltage is applied to one phase of the load, but the phase current is $I_l/\sqrt{3}$ consequently the product of voltage and current in one phase is the same as for the Y-connected load.

It will be observed that the expressions (5.9) are, except for the factor of $\sqrt{3}$, the same as the expressions (4.3), (4.4) and (4.5) appertaining to the single-phase case. The power diagram of fig. 4.4 is equally applicable to balanced 3-phase systems provided the factor of $\sqrt{3}$ is taken into account. Problems involving power factor correction may be solved using the expressions (5.9) together with the power diagram.

5.5 Worked example: power factor correction

An industrial plant draws a balanced electrical load of 9.8 MW at a lagging power factor of 0.8. The plant is supplied over a 3-phase, 50 Hz line having a maximum rating (load carrying capacity) of 660 A at 11 kV (line voltage).

(a) Calculate the apparent power and reactive power drawn by the load.
(b) Additional equipment drawing a load of 1.5 MW and 0.7 MVAr lagging is to be installed in the plant. Find the minimum rating in MVAr of the power-factor correction capacitor that must be installed if the rating of the line is not to be exceeded. What then is the system power factor?
(c) If the capacitor is to consist of three sections connected in delta across the line, calculate the capacitance required in each section.

Solution
(a) The power diagram for the original load condition is shown in Fig. 5.13(a). The phase angle ϕ_1 is $\cos^{-1} 0.8 = 36.87°$, therefore from (5.9)

Fig. 5.13. Power diagrams for worked example (section 5.5).

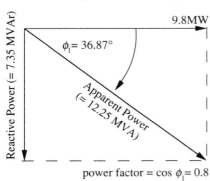

(a) Original load conditions

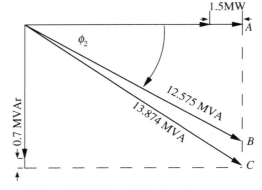

(b) Load including additional equipment

$$\text{Apparent power} = \frac{\text{power}}{\cos\phi_1} = \frac{9.8}{0.8} = 12.25 \text{ MVA}$$

$$\text{Reactive power} = (\text{power}) \times (\tan\phi_1) = 9.8 \times 0.75 = 7.35 \text{ MVAr}$$

(b) With the additional load installed

$$\text{Total power} = 9.8 + 1.5 = 11.3 \text{ MW}$$
$$\text{Total reactive power} = 7.35 + 0.7 = 8.05 \text{ MVAr}$$
$$\text{Total apparent power} = \sqrt{(11.3^2 + 8.05^2)} = 13.874 \text{ MVA}$$

The maximum load that the line can carry is

$$\sqrt{3} V_l I_l = \sqrt{3} \times 11 \times 10^3 \times 660 = 12.575 \text{ MVA}$$

The power diagram for the plant plus additional load is shown in Fig. 5.13(b). To bring the total apparent power within the load carrying capacity of the line a capacitor must be installed drawing leading vars equivalent to length BC on the diagram.

Now

$$BC = AC - AB = 8.05 - \sqrt{(12.575^2 - 11.3^2)}$$

hence, rating of capacitor = 2.53 MVAr.

$$\text{System power factor} = \cos\phi_2 = \frac{11.3}{12.575} = 0.9.$$

(c) Since the capacitor is in three sections, each section will account for $2.53/3 = 0.843$ MVAr. The capacitor sections are connected in delta and so will carry the full line voltage of 11 kV. If X_c is the reactance of each section, then we have, using (4.7),

$$\frac{V_l^2}{X_c} = \frac{(11 \times 10^3)^2}{X_c} = 0.843 \times 10^6$$

which gives $X_c = 143.5 \, \Omega$

The capacitance C is given by

$$X_c = \frac{1}{2\pi \times 50 \times C} = 143.5$$

whence

$$C = 22.2 \, \mu F$$

5.6 Three-phase power measurement

5.6.1 Alternating current meters

A moving-coil instrument that is used frequently at standard power frequencies (50–60 hertz) is the electrodynamometer. In this instrument the magnetic field is supplied by the current that flows in a fixed coil of a few turns of wire of large cross-sectional area. There is no magnetic material. The moving coil, which carries a pointer and is held in the zero position by a spring, is supported in the field of the fixed coil by jewelled bearings so that it is free to rotate in response to any torque that it experiences. Fig. 5.14 is a schematic representation of an electrodynamometer type movement.

A meter movement of this type may be used as a voltmeter. The two coils are connected in series and an external resistor is provided to limit the current. Since the same current is in both coils the instantaneous torque will be proportional to the square of the applied voltage. The mechanical constants of the moving coil are such that it cannot follow variations of torque for power frequencies. Instead, the coil assumes a deflection proportional to the average torque and thus proportional to the average of the square of the voltage. If therefore a square root scale is used, the pointer will read effective voltage.

The electrodynamometer cannot be used directly as a high current ammeter because the moving coil can carry only a small current. It is possible by using shunts to accommodate large currents. Care must be exercised to make sure that no phase shift is introduced by the use of shunts.

Fig. 5.14. Schematic diagram of electrodynamometer meter movement.

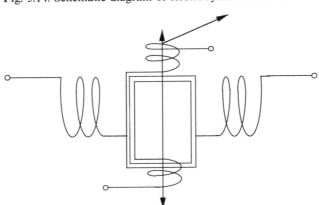

When the electrodynamometer movement is adapted to power measurement, the resulting instrument is called a wattmeter. The wattmeter has two pairs of terminals, one for each coil. Fig. 5.15 shows the wattmeter connected to measure power supplied to a single-phase load, Z_L. The fixed (current) coil (terminals labelled I, I') carries the load current, i, while the moving (potential) coil (terminals V, V') carries a current proportional to the potential difference, v, across the load. Thus the torque on the moving coil is proportional at any instant to the product vi, which is equal to the instantaneous power.

If v and i are exactly in phase, the torque is never negative. If there is a phase difference between v and i, the torque will be negative for part of each cycle. Because the coil cannot follow variations of torque at power frequencies, the coil assumes a deflection proportional to the average torque and so proportional to the average power in the load. It is possible, therefore, to calibrate the meter directly in watts.

One terminal of each terminal pair is marked with the symbol $\pm$. These two terminals must be connected to the same side of the power line as shown in fig. 5.15 so that the two coils will be at essentially the same potential. If, in fig. 5.15, the connection of the potential coil is reversed then the full line potential difference exists between the coils. Electrostatic forces may then introduce errors in the meter reading. In addition, the high potential difference may damage the coil insulation.

Ideally, the current coil has zero resistance and the potential coil has infinite resistance. Then it makes no difference whether the voltage coil is connected to the line or the load side of the current coil. In practice the voltage coil is usually connected to the load side as shown in fig. 5.15. Then, even with no load, the meter will indicate the power drawn by the potential coil. This power is small and can usually be neglected. For the highest

Fig. 5.15. Wattmeter connected to single-phase load.

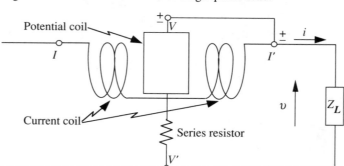

accuracy it is possible to obtain a *compensated* meter so constructed that its reading is corrected for meter loss.

5.6.2 Methods of power measurement

In a three-phase system where the load is Y-connected and there is a neutral wire, three wattmeters are required to measure the load power. In the special case of a balanced load the three load currents deliver identical amounts of power and so it is necessary to measure power in only one phase and multiply by 3 to get the total power delivered.

In the case of a three-phase system with no neutral, only two wattmeters are required to measure the load power, regardless of whether or not the load is balanced. We now justify this statement.

Wye-connected load

Fig. 5.16 shows a wye-connected three-phase load. Two wattmeters are shown: wattmeter 1 carries line current i_c in its current coil and has voltage $v_{c'a'}$ across its potential terminals; wattmeter 2 has line current i_b in its current coil and has potential difference $v_{b'a'}$ across its potential terminals. The total instantaneous power to the load is

$$p = v_{a'n'} i_a + v_{b'n'} i_b + v_{c'n'} i_c$$

At node n', by Kirchhoff's current law,

$$i_a + i_b + i_c = 0$$

Fig. 5.16. Two-wattmeter method for measurement of power in a wye-connected load.

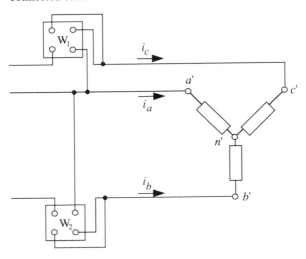

so

$$i_a = -(i_b + i_c)$$

then

$$p = -v_{a'n'}i_b - v_{a'n'}i_c + v_{b'n'}i_b + v_{c'n'}i_c$$
$$= i_c(v_{c'n'} - v_{a'n'}) + i_b(v_{b'n'} - v_{a'n'})$$

Now,

$$v_{b'n'} - v_{a'n'} = v_{b'n'} + v_{n'a'} = v_{b'a'}$$

and

$$v_{c'n'} - v_{a'n'} = v_{c'n'} + v_{n'a'} = v_{c'a'}$$

therefore

$$p = v_{c'a'}i_c + v_{b'a'}i_b$$

But wattmeter 1 responds to $v_{c'a'}i_c$ and wattmeter 2 responds to $v_{b'a'}i_b$, hence the sum of the responses is equal to the instantaneous power.

If voltages and currents are sinusoidal then, by (4.3), the *average* power measured by each wattmeter will be:

$$W_1 = V_{c'a'}I_c \cos\theta_1$$

$$W_2 = V_{b'a'}I_b \cos\theta_2 \qquad (5.10)$$

where θ_1 and θ_2 are the phase angles between voltages and currents. The average power supplied to the load (balanced or unbalanced) is

$$P = W_1 \pm W_2$$

We now explain the reason for the $\pm$ sign in the above equation.

In our study of the Y-connected source (section 5.3.1, figs. 5.4 and 5.5) it was found that line and phase voltages differ in phase by 30°. It follows from this that if the load is purely resistive, line currents will be out of phase with line voltages at the load by 30°. This is shown in the phasor diagram of fig. 5.17. It is seen that current I_c lags $V_{c'a'}$ by 30° and current I_b leads $V_{b'a'}$ by 30°. Equations (5.10) are then

$$W_1 = V_{c'a'}I_c \cos(-30°)$$
$$W_2 = V_{b'a'}I_b \cos(+30°)$$

Suppose now that the loads are complex and balanced. Then equations (5.10) become

Three-phase power measurement

$$W_1 = V_{c'a'} I_c \cos(-30 + \theta)$$
$$W_2 = V_{b'a'} I_b \cos(30 + \theta)$$

where θ is the angle of the load impedance. No matter whether the load is capacitive or inductive, if the magnitude of θ exceeds 60° (corresponding to a power factor of 0.5) one wattmeter will read negative. Then, if both wattmeters are connected to read up-scale, the net power to the load will be the arithmetic difference of the two wattmeter readings.

Delta-connected load

Fig. 5.18 shows a delta-connected load with two wattmeters connected in a similar fashion to those shown in fig. 5.16. The total instantaneous power in the load is

$$p = i_\alpha v_{a'b'} + i_\beta v_{b'c'} + i_\gamma v_{c'a'}$$

By Kirchhoff's voltage law

$$v_{a'b'} + v_{b'c'} + v_{c'a'} = 0$$

or

$$v_{b'c'} = -(v_{a'b'} + v_{c'a'})$$

Fig. 5.17. Phasor diagram for a balanced wye-connected resistive load.

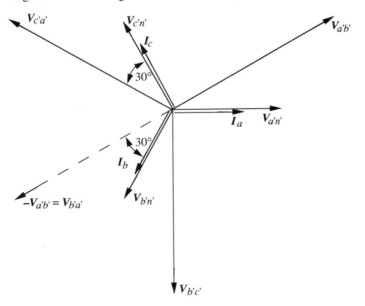

Then

$$p = v_{c'a'}(i_\gamma - i_\beta) + v_{a'b'}(i_\alpha - i_\beta)$$

But

$$i_\gamma - i_\beta = i_c \text{ and } i_\beta - i_\alpha = i_b$$

so

$$p = v_{c'a'} i_c + v_{b'a'} i_b = w_1 + w_2$$

The average power supplied to the load is, therefore

$$P = W_1 \pm W_2$$

The choice of sign is made just as in the case of the Y-connected load.

The construction of a wattmeter is such that it deflects up-scale or down-scale depending upon the direction of power flow. Since the meter reads in only one direction, it is necessary to determine which direction of power flow causes up-scale deflection. Such a determination may be made by connecting the meter to a resistive load and recording the connections that give an up-scale deflection. Now, if the two wattmeters are connected so that they indicate, by up-scale deflections, power flow to the load and if either the current connection or the voltage connection must be reversed for one of them in order to get an up-scale reading, the wattmeter readings have opposite signs.

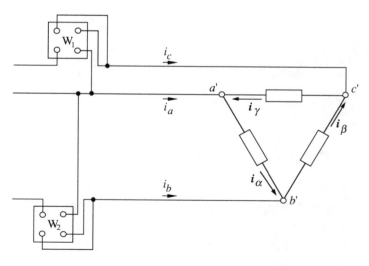

Fig. 5.18. Two-wattmeter method for measurement of power in a delta-connected load.

Three-phase power measurement

With regard to power measurement, two points should be emphasized. First, in a three-phase, three-wire system two wattmeters will measure the power to the load whether or not the load is balanced; however, as has already been stated, care must be exercised to determine whether the individual meter readings indicate power supplied to or drawn from the load. The second significant fact is that in any polyphase system using n conductors, the load power may be measured by the use of $(n-1)$ wattmeters so arranged that all the potential coils are connected together to the one conductor into which no wattmeter coil is introduced.

5.6.3 Worked example

In the circuit of fig. 5.9 two wattmeters are used to measure the power delivered to the load. The wattmeter current coils are arranged to carry the currents I_b and I_c while the potential coils share a common connection at the node a'. Determine the readings of the two wattmeters, and verify that the sum of these readings equals the total power in the load. The voltages of the source and the load impedances are as given in the worked example of section 5.3.5.

Solution
From (5.10) the wattmeters will read
in phase b

$$W_1 = V_{b'a'} I_b \cos\theta_1$$

and in phase c

$$W_2 = V_{c'a'} I_c \cos\theta_2$$

where θ_1 and θ_2 are in each case the phase angles between voltage and current.

From the results of the previous worked example we have

$$I_b = 21.1 \underline{/-160.3} \text{ and } I_c = 21.1 \underline{/79.7}$$

and for the line voltages at the load

$$V_{a'b'} = 273 \underline{/16.2} \text{ and } V_{c'a'} = 273 \underline{/136.2} \text{ and}$$

therefore

$$V_{b'a'} = -V_{a'b'} = -273 \underline{/16.2} = 273 \underline{/-163.8}$$

and the angles θ_1 between $V_{b'a'}$ and I_b is $(163.8 - 160.3) = 3.5°$. The angle θ_2 between $V_{c'a'}$ and I_c is $(136.2 - 79.7) = 56.5°$. (Note that it is the absolute difference between phase angles that is required for evaluation of the power factor in (5.10).)

The readings of the wattmeters are then

$$W_1 = 273 \times 21.1 \times \cos 3.5 = 5750;$$
$$W_2 = 273 \times 21.1 \times \cos 56.5 = 3179$$

and

$$(W_1 + W_2) = 8929 \text{ W}.$$

The current in each load is, from previous results, 12.2 A and the resistive component of each load impedance is 20 Ω. The total power in all three phases of the load is, therefore $(3 \times 12.2^2 \times 20) = 8930$ W, which agrees with the power measured by the two wattmeters.

5.7 Transformers for three-phase systems

5.7.1 Applications

Transformers are important components of all electric power distribution systems. They make it possible to raise or lower the voltages at various points in a system efficiently, thereby permitting economical transmission over long distances and safe distribution to industrial and residential users. The fact that transformers allow such control of transmission and distribution voltages was responsible for the decision, near the turn of the century, to use alternating rather than direct voltage for power systems.

When electric power is to be transmitted over great distances it is advantageous to use high voltage and low current. Heating loss in the line is proportional to the square of the current. Thus, the conductor size that is required to keep this loss to a desired fraction of total power transmitted decreases with the square of the current. Small conductors require less material. In addition, because they are lighter in weight, the cost of the supporting structures for overhead lines is correspondingly reduced.

The output voltage of a generator in a large power system is typically 22 kV; transformers then raise this voltage to several hundreds of kilovolts (fig. 5.19). In the U.K., power is transmitted over long distances using overhead lines operating at 400 kV; in North America, where greater transmission distances are encountered, overhead lines operate at up to 735 kV. Such high voltages are not practicable for the distribution of power in an urban area. Transformers are, therefore, again used to reduce the voltage to values suitable for power distribution to various categories of consumer. The voltage reduction is accomplished in several stages. Feeder

Transformers for three-phase systems

circuits operate at tens of kilovolts, the final step being a reduction, in the U.K., to 415/240 V. In most parts of North America the final distribution voltage is 208/120 V. Industrial consumers are usually supplied at 33 or 22 kV, further stages of reduction being accomplished by transformers situated within the industrial complex itself.

In a three-phase system one may use either three single-phase power transformers (as described in chapter 4) or a single transformer having three separate sets of primary and secondary windings. A single three-phase unit has some advantages in size and initial cost. In addition, the windings of the three phases may be connected internally, thus reducing the number of high voltage leads that must be brought out. A disadvantage of the single unit is that if one phase fails (and this is the way in which three-phase transformers often do fail) then service is interrupted while the whole unit is replaced. If single-phase units are used, then failure of one unit does not necessarily shut down the service completely. Replacement of one single-phase transformer is cheaper than replacement of a three-phase unit. Also, especially in small systems, less capital is invested in the inventory of spares if single-phase transformers are employed. The three single-phase units should be identical, otherwise, the transformers may introduce unbalance even in a system with balanced loads.

Regardless of the choice of transformers, there are four possible interconnections of the three pairs of phase windings leading to various combinations of primary and secondary voltages as shown in table 5.1.

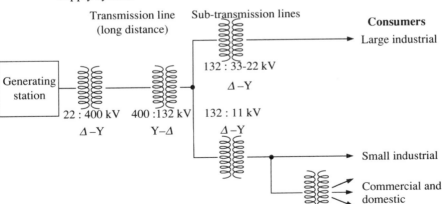

Fig. 5.19. Schematic showing arrangement of transformers in a power supply system.

Table 5.1

Connection		Primary voltage		Secondary voltage	
Primary	Secondary	Line	Phase	Line	Phase
Δ	Δ	1	1	N_2/N_1	N_2/N_1
Δ	Y	1	1	$(N_2/N_1)\sqrt{3}$	N_2/N_1
Y	Y	1	$1/\sqrt{3}$	N_2/N_1	$(N_2/N_1)/\sqrt{3}$
Y	Δ	1	$1/\sqrt{3}$	$(N_2/N_1)/\sqrt{3}$	

(N_2/N_1) = secondary/primary turns ratio.

Three phase transformers may be of the *shell* type, with the windings surrounded by the iron, or of the *core* type, where the windings surround the magnetic material. These two designs correspond to the two single-phase designs illustrated in fig. 4.11. The shell design has the slight advantage of providing more nearly identical flux paths for the three phases. Furthermore, in the shell configuration the magnetic paths of the three phases are more nearly independent than is the case with the core design.

On occasion, one phase of a three-phase transformer connection may develop a short circuit or an open circuit. If the connection consists of three separate units, or if it is a single unit of the shell type, it is possible to operate with only two phases and still supply 58% of the rated load. This arrangement is called an open-delta or V-connection. (A core-type three-phase unit may operate in open-delta if the fault is an open circuit. However, if the fault is a short circuit the other two phases cannot operate because the return paths for their fluxes are through the core of the damaged winding. The short-circuit currents in the damaged winding effectively block the fluxes of the other two phases.) Occasionally in the United States when the initial demand for power from a new installation is small, service will be provided by use of only two single-phase transformers connected in open-delta. When demand increases, a third transformer is installed.

When a Y–Y connection is used, certain constraints exist that do not arise in single-phase operation. One constraint is that with the Y–Y connection it is not possible, unless the primary neutral is connected to the system ground, to obtain any significant amount of power if only one phase of the secondary is loaded. With the Y–Y connection current for one phase of the secondary is supplied by currents in all three phases of the primary. A load on only one secondary phase leaves the other two phases with open secondaries and therefore with high primary impedances. Since only the

Transformers for three-phase systems

magnetizing current flows in the primary of these phases, no appreciable primary current is available for the single phase from which it is desired to draw power.

A second constraint upon operation with the Y–Y connection is related to the requirement for third harmonic current to maintain a sinusoidal flux in the core; this is discussed later in section 5.7.4.

5.7.2 Equivalent circuit parameters

Calculations on three-phase transformers are usually concerned with balanced loads, and circuits are analyzed on a per-phase basis using a transformer equivalent circuit essentially the same as that derived for the single-phase case in section 4.9. The parameters of this equivalent circuit are found similarly by open- and short-circuit tests. Values of series resistance and leakage reactance per phase are derived from the short-circuit test, and values of shunt resistance (or conductance) and reactance (or susceptance) per phase, accounting for core losses and magnetizing current, are derived from the open-circuit test. For the short-circuit test it is usual to short circuit the low-voltage heavy-current windings of the transformer, measurements being made on the high-voltage side of the transformer. Conversely, for the open-circuit test it is usual to open circuit the high-voltage windings and make measurements on the low-voltage side. These procedures minimize the required current and voltage ratings of the test supplies and measuring instruments used.

Parameters derived from these tests are referred to either side of the transformer, as convenient, by multiplying by the square of the turns ratio. Care must be exercised, however, when interpreting test data, particularly where phases are connected differently on the two sides of the transformer (for example Δ–Y) and where the open- and short-circuit tests are performed on different sides. In such cases the $\sqrt{3}$ conversion factor between line and phase values operates differently on the two sides of the transformer.

5.7.3 Worked example

A 6600/22 000 V,* three-phase transformer, rated at 2500 kVA, has its low-voltage windings connected in delta and its high-voltage windings connected in wye. Open- and short-circuited tests conducted on the transformer produced the following data.

---

* Unless stated otherwise, voltages specified for three-phase systems are always *line* voltages.

Three-phase alternating current circuits

Short-circuit test	Open-circuit test
(LV side short circuit)	(HV side open circuit)
Line voltage = 712 V	Line voltage = 6600 V
Line current = 64 A	Line current = 6.52 A
Total power = 38.34 kW	Total power = 35.1 kW

(a) Deduce the per-phase values of equivalent series resistance and reactance referred to the HV and LV sides.

(b) Deduce the per-phase values of the equivalent shunt conductance and susceptance referred to the LV and HV sides.

(c) If the transformer supplies a load of 1500 kW at 22 000 V and 0.8 p.f. lagging, calculate the line voltage and line current on the LV side.

(d) Calculate the efficiency and regulation at the above load.

(e) Estimate the unity power factor load for which the transformer efficiency will be a maximum; compare the efficiency at this load with the efficiency at full rated output.

Solution

(a) In the short-circuit test, the LV side is short circuited and measurements are made on the HV side. Let subscripts 1 and 2 denote the LV side and HV sides respectively. Since the HV side is connected in wye the phase current I_{p2} and line current I_{l2} are the same, but the phase voltage V_{p2} will be $V_{l2}/\sqrt{3}$. Therefore, from the short-circuit test data, we have

$$I_{p2s} = I_{l2s} = 64 \text{ A}; \quad V_{p2s} = \frac{V_{l2s}}{\sqrt{3}} = \frac{712}{\sqrt{3}} = 411.1 \text{ V}$$

Power per phase $P_{p2s} = \dfrac{38.34}{3} = 12.78$ kW.

Using (4.23), with appropriate subscripts, the equivalent series resistance and reactance referred to the HV side are given by

$$R_{(HV)} = \frac{P_{p2s}}{(I_{p2s})^2} = \frac{12.78 \times 10^3}{64^2} = 3.12 \, \Omega$$

and

$$\sqrt{(R_{(HV)}^2 + X_{(HV)}^2)} = \frac{V_{p2s}}{I_{p2s}} = \frac{411.1}{64} = 6.42 \, \Omega$$

whence

$$X_{(HV)} = 5.61 \, \Omega$$

Values of resistance and reactance referred to the LV side are found by multiplying the above values by the square of the phase-to-phase (LV/HV) turns ratio which is given by

$$\frac{N_{p1}}{N_{p2}} = \frac{V_{p1}}{V_{p2}} = \frac{6.6 \times 10^3}{(22 \times 10^3)/\sqrt{3}} = 0.5196$$

Therefore $R_{(LV)} = 3.12 \times 0.5196^2 = 0.84\,\Omega$; $X_{(LV)} = 5.61 \times 0.5196^2 = 1.51\,\Omega$.

(b) Values for the equivalent conductance and susceptance are found from the open circuit test. In this test the HV side is open circuit and measurements are made on the Δ-connected LV side, for which the phase and line voltages are the same but phase and line currents differ by a factor of $\sqrt{3}$. The test data give

$$V_{plo} = V_{llo} = 6600\,\text{V}; \quad I_{plo} = \frac{I_{llo}}{\sqrt{3}} = \frac{6.52}{\sqrt{3}} = 3.764\,\text{A}$$

Power per phase $P_{plo} = \dfrac{35.1}{3} = 11.7\,\text{kW}$.

Using (4.22), with the appropriate subscripts, the equivalent conductance and susceptance referred to the LV side are given by

$$g_{(LV)} = \frac{P_{plo}}{(V_{plo})^2} = \frac{11.7 \times 10^3}{(6.6 \times 10^3)^2} = 0.269\,\text{mS}$$

and

$$\sqrt{(g_{(LV)}^2 + b_{(LV)}^2)} = \frac{I_{plo}}{V_{plo}} = \frac{3.764}{6.6 \times 10^3} = 0.5704\,\text{mS}$$

whence

$$b_{(LV)} = 0.503\,\text{mS}$$

Values of conductance and susceptance referred to the HV side are found by *dividing* by the square of the (HV/LV) phase-to-phase turns ratio, which is the same as multiplying by the square of the ratio (N_{p1}/N_{p2}) found above (see section 4.9, fig. 4.17). Performing this operation gives

$$g_{(HV)} = 0.0726\,\text{mS} \quad \text{and} \quad b_{(HV)} = 0.136\,\text{mS}$$

(c) The equivalent circuit for one phase of the system is shown in fig. 5.20(a) (subscripts referring to phase values, and HV and LV sides are unnecessary here and have been omitted for simplicity). The load is supplied at a line voltage of 22 kV, hence, $V_2 = (22 \times 10^3)/\sqrt{3} = 12.7\,\text{kV}$. For one phase the load power is $1500/3 = 500\,\text{kW}$ at 0.8 p.f. lagging, therefore,

$$V_2 I_2 \cos\phi = 12.7 \times 10^3 \times I_2 \times 0.8 = 500 \text{ kW}$$

from which

$$I_2 = 49.2 \text{ A}$$

The phasor diagram appropriate to the HV side is shown in fig. 5.20(b). It is convenient in this type of problem to select I_2 as the reference phasor. Then the voltage E_2 is given by

$$E_2 = V_2 \cos\phi + jV_2 \sin\phi + RI_2 + jXI_2$$

where ϕ is the angle between V_2 and I_2.

Fig. 5.20. Diagrams for worked example (section 5.7.3).

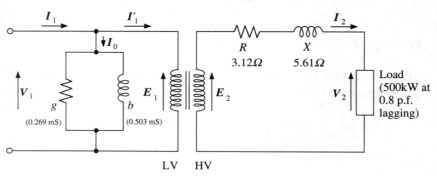

(a) Equivalent circuit for one phase; series resistance and reactance referred to the HV side.

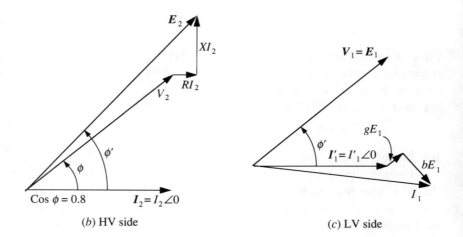

(b) HV side

(c) LV side

Transformers for three-phase systems

Inserting numerical values obtained from part (a) for the equivalent resistance and reactance referred to the HV side, the above expression becomes

$$E_2 = (12.7 \times 10^3 \times 0.8) + j(12.7 \times 10^3 \times 0.6) + 49.2(3.12 + j5.61)$$
$$= (10.16 \times 10^3) + j(7.62 \times 10^3) + 153.5 + j276$$
$$E_2 = (10.313 + j7.896) \times 10^3 = 12.99\underline{/37.44}\,\text{kV}$$

Now, the calculated LV/HV turns ratio is 0.5196, therefore, the line voltage on the LV side is

$$V_1 = E_1 = 12.99 \times 0.5196\underline{/37.44} = 6.75\underline{/37.44}\,\text{kV}$$

The magnitude of the line voltage is therefore 6.75 kV since the LV side is Δ-connected and line and phase voltages are identical.

The phasor diagram for the LV side is shown in fig. 5.20(c). The current I_1' is simply I_2 divided by the (LV/HV) turns ratio: (see section 4.7 for properties of the ideal transformer)

$$I_1' = \frac{49.2\underline{/0}}{0.5196} = 94.69\underline{/0}$$

and the main phase current is

$$I_1 = I_1' + I_0 = I_1' + E_1(g - jb)$$

Inserting numerical values for conductance and susceptance referred to the LV side, obtained from part (b), this expression becomes

$$I_1 = 94.69 + (6.75 \times 10^3\underline{/37.44})(0.269 - j0.503) \times 10^{-3}$$

whence

$$I_1 = 98.2 - j1.59 = 98.2\underline{/-0.9}$$

The magnitude of the line current is $98.2\sqrt{3} = 170$ A.

In the above calculation we have used values of g and b derived in an earlier part of the problem, however, it is possible to obtain the current I_0, and consequently I_1, directly from the test data without finding g and b specifically.

From the test data we have:

Magnitude of line (phase) voltage = 6600 V

Magnitude of phase current $= \dfrac{6.52}{\sqrt{3}} = 3.764$ A

Power per phase $= \dfrac{35.1}{3} = 11.7$ kW

If θ is the angle between line voltage and phase current, then

$$6.6 \times 10^3 \times 3.764 \times \cos\theta = 11.7 \times 10^3$$

and

$$\theta = 61.9°$$

The current will lag the voltage by 61.9° since the system is inductive.

Now, the open-circuit test was carried out at an applied voltage of 6600 V whereas our calculations have shown that, under the given load conditions, the actual voltage on the LV side is 6750 V. The magnitude of the current I_0 will, therefore, be $3.764 \times (6750/6600) = 3.849$ A. Its angle with respect to the line voltage (i.e. with respect to E_1) is $-61.9°$ so, with E_1 as the reference phasor, we may write $I_0 = 3.849\underline{/-61.9}$. Also the current I_1', which lags E_1 by angle 37.44°, may be expressed as $I_1' = 94.69\underline{/-37.44}$. Hence,

$$I_1 = I_1' + I_0 = 94.69\underline{/-37.44} + 3.849\underline{/-61.9}$$

or

$$I_1 = 98.2\underline{/-38.3}$$

which agrees with our previous calculation.

(Note that the phase angle of the current given here is with respect to E_1 not I_1' as before.)

(d) From the open-circuit test data the core loss is 35.1 kW at an applied line voltage of 6600 V. The actual line voltage at the given load condition is, from section (c), 6750 V. Since the power is proportional to the square of the applied voltage we have

$$\text{Total core loss} = 35.1 \times \left(\frac{6750}{6600}\right)^2 = 36.7 \text{ kW}$$

Likewise, the copper loss, which is proportional to the square of the load current, is

$$\text{Total copper loss} = 38.34 \times \left(\frac{49.2}{64}\right)^2 = 22.66 \text{ kW}$$

Therefore,

$$\text{Efficiency} = \frac{\text{output}}{\text{output} + \text{losses}} = \frac{1500}{1500 + 36.7 + 22.66} = 96.2\%$$

Note that the small change in the core loss between the no-load (test) condition and the loaded condition has an insignificant effect on the

Transformers for three-phase systems

calculation of efficiency; for most practical calculations the core loss can be assumed constant.

From (4.25) the regulation is expressed by

$$\text{Regulation} = I_2 \frac{(R\cos\phi + X\sin\phi)}{V_{20}}$$

where (in this problem) R and X are the per-phase values of the equivalent series resistance and reactance of the transformer referred to the HV side. At the specified load condition, $I_2 = 49.2$ A and, (with 6.75 kV on the LV side) $V_{20} = 12.99$ kV. The values of R and X have been calculated in part (a) above. Hence,

$$\text{Regulation} = \frac{49.2(3.12 \times 0.8 + 5.61 \times 0.6)}{12.99 \times 10^3} = 2.22\%$$

(e) The transformer efficiency will be a maximum (according to (4.29)) when the core losses and copper losses are equal. Assuming that the core losses are constant at the value given by the open circuit test, namely, 35.1 kW, the load current I_2 for which the efficiency will be maximum is given by

$$3I_2^2 R = 3I_2^2 \times 3.12 = 35.1 \times 10^3$$

whence,

$$I_2 = 61.24 \text{ A}.$$

If the load voltage is 22 kV, then the power delivered is

$$\sqrt{3} \times 22 \times 10^3 \times 61.24 = 2333 \text{ kW}$$

$$\text{Efficiency} = \frac{2333}{2333 + (2 \times 35.1)} = 97.1\%$$

At the full rated output of 2500 kVA at unity power factor, the load current is given by

$$\sqrt{3} \times 22 \times 10^3 \times I_2 = 2500 \times 10^3$$

whence

$$I_2 = 65.6 \text{ A}$$

The total copper loss is then $3 \times 65.6^2 \times 3.12 = 40.29$ kW and

$$\text{Efficiency} = \frac{2500}{2500 + 40.29 + 35.1} = 97.1\%$$

It is evident from these results that although the maximum efficiency occurs at a load rather less than full rated load, the difference in efficiency is insignificant. The efficiency of a transformer remains very nearly constant over a wide range of load conditions.

The principle of conservation of watts and vars (see section 4.3) may be used with advantage in this type of calculation. The method is particularly rapid if it can be assumed that the core loss current and magnetizing currents are constant under change of load conditions. In the following calculations this assumption introduces a negligible error in the value for the line current at the input terminals of the transformer.

The real power in the load is 1500 kW at 0.8 p.f. lagging, from which it may be deduced that the reactive power is 1125 kVAr. The load is supplied at 22 kV, therefore the line current I_{l2} is, by (5.9)

$$\sqrt{3} \times 22 \times 10^3 \times I_{l2} \times 0.8 = 1500 \times 10^3$$

whence

$$I_{l2} = 49.2 \text{ A}$$

The real and reactive powers in the referred series resistance and reactance per phase are $I_{l2}^2 R$ and $I_{l2}^2 X$ respectively. The total real power P_t and the total reactive power Q_t are, therefore,

$$P_t = 1500 + 3(49.2^2 \times 3.12)/10^3 = 1523 \text{ kW}$$

and

$$Q_t = 1125 + 3(49.2^2 \times 5.61)/10^3 = 1166 \text{ kW}$$

By (4.6), the apparent power is $\sqrt{(P_t^2 + Q_t^2)} + 1918$ kVA and since there is no loss in the ideal transformer, this must be the apparent power on the LV side of the transformer. But the *line* current on the LV side is $49.2 \times (22/6.6) = 164$ A, therefore the line voltage on the LV side is given (by (5.9)) as

$$\sqrt{3} \times V_{l1} \times 164 = 1918 + 10^3$$

whence

$$V_{l1} = 6.75 \text{ kV}$$

Also, from the open-circuit test data, the real and reactive powers associated with the transformer core are 35.1 kW and 65.75 kVAr respectively; adding these components to the values of P_t and Q_t obtained above we obtain

Total apparent power $=\sqrt{[(1523+35.1)^2 + (1166+65.75)^2]} = 1986$ kVA.

Therefore, $\sqrt{3} \times 6.75 \times 10^3 \times I_{l1} = 1986 \times 10^3$ from which $I_{l1} = 170$ A. It will be noted that this method avoids the necessity of converting between line and phase valves during the calculations.

†5.7.4 Harmonic currents

We recall from section 4.9, that when a sinusoidal voltage is applied to an iron-cored transformer the resulting magnetic flux will be sinusoidal, but, because of the non-linear character of the magnetic core material, the magnetizing current is non-sinusoidal. It includes odd harmonics of which the third has much the greatest amplitude (see fig. 4.13(c)).

Now consider the separate phase voltages of a three-phase generator. It is quite likely that these output voltages will contain odd harmonics of the fundamental frequency, but because of the symmetrical construction of the alternator, even harmonics are not likely to appear. The three voltages of fundamental frequency are mutually 120° apart in phase, as explained in section 5.1. Except for the third harmonic and multiples thereof, all the odd harmonics will likewise be 120° out of phase with one another, although the phase sequence of some will be opposite to that of the fundamentals. It will be readily appreciated that, because of the 120° phase differences among the fundamentals, the third harmonic voltages (and the 9th, 15th and so on) will be in phase.

Let the three lines from the generator be joined to Y-connected transformer windings with no neutral connection, as shown in fig. 5.21. Now suppose third harmonic currents exist in the three lines as shown in the figure. These currents must add to zero at point N. This, however, is impossible for three currents that are in phase. It follows, therefore, that third harmonic currents cannot exist in the transformer windings.

Fig. 5.21. Third harmonic currents.

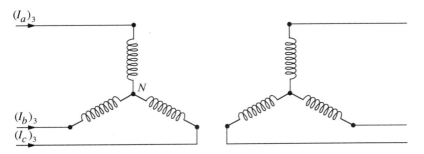

Because no third harmonic current can exist in the connection of fig. 5.21, the magnetizing currents in the three windings will be sinusoidal (if we ignore the higher harmonics that are of very small amplitude). Then because of non-linearity in the iron core, the flux will contain appreciable third harmonic. This non-sinusoidal flux induces third harmonic voltages in the transformer windings that are significantly greater in magnitude than the fundamental voltages and these may damage the transformer insulation.

It can be shown that if the transformer is a three-phase core-type then the flux will be essentially sinusoidal even in the absence of third harmonic currents. For other transformer configurations there are several ways to provide the required third harmonic currents. One way is to supply the Y-connected primary with a connection to the system neutral. Then there will be return paths for the third harmonic currents in fig. 5.21. However, even if the system neutral were available at the transformer, the third harmonic currents that result from a neutral connection could be great enough to interfere with nearby telephone circuits.

A second possibility is to connect the three secondary windings in delta, thus providing a closed path for circulation of the third harmonic current. As far as the iron is concerned, it is immaterial what path the third harmonic current takes; it simply is necessary that a path for it exist. Of course, this solution to the problem is not available if the transformer is arranged in a Y–Y connection.

A common method of supplying a path for the third harmonic current is that of including a third, delta-connected set of transformer windings. The third harmonic currents induced in these windings will be in phase, and there will be a short-circuit path for them in the delta connection. If each component of the 'tertiary' winding has a number of turns equal to that of its primary then the third harmonic currents in these windings will be just equal to the current that would exist in the primary were it not suppressed by the Y connection of the primary winding.

†5.8 Phase transformation

Transformers may be used to convert a set of polyphase voltages to another set with a different number of phases. A practical application of this property of transformers is in producing six-phase voltages for large mercury or semiconductor rectifiers. Successive voltages of the six-phase set differ in phase by 60°. Rectifiers supplied from such voltages show significantly less ripple than exists when three-phase power is used. The required connection is shown in fig. 5.22(a). The primaries of the three transformers are connected in delta. Each secondary winding is centre

Phase transformation

tapped to provide a common node as shown. The output voltage phasors are shown in fig. 5.22(*b*). By using transformers with more windings one may get 12 output voltages equal in magnitude and having 30° phase difference between successive voltages.

A second example of phase conversion is the Scott connection between three-phase and two-phase systems. (Two-phase voltages are equal in magnitude and differ in phase by 90°. Unlike other polyphase voltages, they do not form a symmetrical pattern when represented as phasors. Sometimes the designation 'quarter phase' is used for these voltages.) Two transformers are required for the Scott connection, as shown in fig. 5.23. Transformer *A* has primary and secondary windings *AA'* and *aa'*

Fig. 5.22. Transformer connection for three-phase to six-phase conversion.

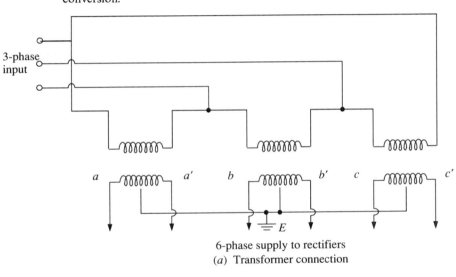

6-phase supply to rectifiers
(*a*) Transformer connection

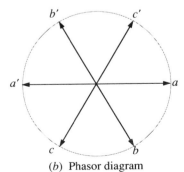

(*b*) Phasor diagram

respectively. Transformer B has primary winding BB', which is centre tapped at C, and secondary winding bb'. The tap C is connected to A', and the ends of the two secondary windings a' and b are also connected together. The turns ratios of the two transformers are different and provided these are correctly arranged a balanced three-phase set of voltages applied to $AB'B$ will result in a balanced two-phase voltage set at abb' or vice versa. We now determine the two transformer ratios required. For the purposes of the following argument it will be convenient to consider the secondary side energized with balanced two-phase voltages, the phasor diagram for which is shown in fig. 5.24(a). Phasors $V_{aa'}$ and $V_{bb'}$ have equal magnitudes.

On the primary side the three voltages appearing at terminals $AB'B$ are

$$\left. \begin{array}{l} V_{AB'} = V_{AC} + V_{CB'} = V_{AC} + V_{BC} \\ V_{B'B} = V_{B'C} + V_{CB} = 2V_{CB} \\ V_{BA} = V_{BC} + V_{CA} \end{array} \right\} \tag{5.11}$$

These are shown in the phasor diagram of fig. 5.24(b); if they are to form a balanced three-phase set, their magnitudes must be equal and angles ϕ must be 30°. It follows that

$$\frac{V_{AC}}{V_{AB'}} = \frac{V_{AC}}{V_{B'B}} = \cos 30 = \frac{\sqrt{3}}{2} \tag{5.12}$$

We see from this expression that the voltages across the two primary windings, V_{AC} and $V_{B'B}$, must be in the ratio of $\sqrt{3}/2 = 0.866$. This implies that if transformer B has a primary-to-secondary turns ratio of N_1/N_2, transformer A must have a turns ratio of $0.866 N_1/N_2$ in order to produce

Fig. 5.23. Scott connection for three-phase to two-phase conversion.

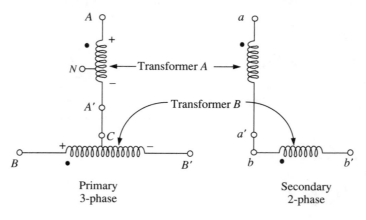

balanced voltages on the three-phase side. It may be shown that for such an arrangement, if the load is balanced on one side of the system, then it will be balanced on the other.

A point properly chosen on the primary of transformer A provides a neutral point N for the three-phase side. To locate N we observe that the magnitude of the voltage V_{AN} must be equal to the phase voltage of the three-phase balanced set, that is $V_{AN}=V_{AB'}/\sqrt{3}$. Therefore, from (5.12),

$$V_{AN}=\frac{1}{\sqrt{3}}\cdot\frac{2}{\sqrt{3}}V_{AC}=\tfrac{2}{3}V_{AC}$$

So, the tap N must be located two-thirds of the way from A to A'.

The Scott connected transformer arrangement has found application in the foundry industry where it is used in connection with certain types of induction channel furnace.

†5.9 Instantaneous power to balanced load

In section 5.2 it was stated that an advantage of a three-phase system is that power to a load is constant and therefore the torque produced by, for example, a three-phase motor is uniform. It is for this reason, as well as ones of efficiency, that almost all industrial machine tools are powered by three-phase motors.

To show that the power in a balanced three-phase system is constant we

Fig. 5.24. Phasor diagrams relating to the Scott connection shown in fig. 5.23.

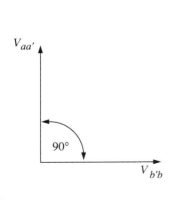

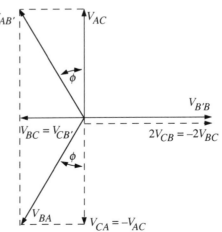

(a) Secondary Phasors   (b) Primary phasors

274   Three-phase alternating current circuits

form the products of the instantaneous voltages and currents in the three separate phases of the system.

Let the amplitudes of the phase voltages and currents be V_m and I_m respectively. Then the r.m.s. magnitudes will be $V_p = V_m/\sqrt{2}$ and $I_p = V_m/\sqrt{2}$. Let the phase voltage in phase a be the phase reference. Then the instantaneous power in phase a is given by:

$$p_a = V_m \sin \omega t \times I_m \sin(\omega t + \phi)$$

where ϕ is the phase angle.

Similarly in phases b and c

$$p_b = V_m \sin(\omega t - 120) \times I_m \sin(\omega t + \phi - 120)$$
$$p_c = V_m \sin(\omega t + 120) \times I_m \sin(\omega t + \phi + 120)$$

Now, using the identity $\sin A \sin B = \frac{1}{2}[\cos(A - B) - \cos(A + B)]$, we may obtain:

$$p_a = \frac{V_m I_m}{2} [\cos(\omega t - \omega t - \phi) - \cos(\omega t + \omega t + \phi)]$$

$$= V_p I_p [\cos(-\phi) - \cos(2\omega t + \phi)]$$

Similarly,

$$p_b = V_p I_p [\cos(-\phi) - \cos(2\omega t + \phi - 240)]$$
$$p_c = V_p I_p [\cos(-\phi) - \cos(2\omega t + \phi + 240)]$$

Let $\theta = (2\omega t + \phi)$ then, using the identity $\cos(A - B) = \cos A \cos B + \sin A \sin B$, we have

$$\cos(2\omega t + \phi - 240) = -\tfrac{1}{2}\cos\theta + \frac{\sqrt{3}}{2} \sin\theta;$$

and

$$\cos(2\omega t + \phi + 240) = -\tfrac{1}{2}\cos\theta - \frac{\sqrt{3}}{2} \sin\theta$$

The total power p_t may then be expressed as

$$p_t = p_a + p_b + p_c = V_p I_p \left[3\cos(-\phi) - \cos\theta \right.$$
$$\left. - \left(-\tfrac{1}{2}\cos\theta + \frac{\sqrt{3}}{2}\sin\theta\right) - \left(-\tfrac{1}{2}\cos\theta - \frac{\sqrt{3}}{2}\sin\theta\right) \right]$$

Therefore, $p_t = 3 V_p I_p \cos\phi = \text{constant}$ \hfill (5.13)

5.10 Summary

Three-phase systems possess important economic and technical advantages over single-phase and other polyphase systems, and they are employed universally for the generation, transmission and distribution of electrical energy. The generators in a three-phase system produce a three-component set of voltages, equal in magnitude and differing in phase by 120°. Two basic circuit configurations are encountered in three-phase systems:
(a) the wye (or Y) connection in which generators (or loads) are connected together at a common or neutral point;
(b) the delta (or Δ) connection in which generators (or loads) are connected to form a closed path.

Generators, or transformer secondaries, connected in the wye configuration provide a four-wire system (three lines plus neutral) which is widely used for local power distribution because it offers the choice to consumers of two voltages: line-to-line or line-to-neutral. These voltages differ by a factor of $\sqrt{3}$.

If the loads on the three phases of a system are equal, then the system is said to be balanced. Under balanced conditions no current flows in the neutral wire of a Y-connected system. Large-scale power distribution systems are operated as closely as possible to balanced conditions, and long-distance power transmission lines utilize only three-wires, a neutral wire being unnecessary. The analysis of three-phase circuits is often considerably simplified if loads are balanced since then it is necessary to analyze only one phase of the complete circuit.

The power in a balanced three-phase system is governed by the relations:

$$\left. \begin{array}{r} \text{Power } P = \sqrt{3} V_l I_l \cos\phi \\ \text{Reactive power } Q = \sqrt{3} V_l I_l \sin\phi \\ \text{Apparent power } S = \sqrt{3} V_l I_l \end{array} \right\} \quad (5.9)$$

where V_l and I_l are the magnitudes of the line voltages and line currents respectively, and $\cos\phi$ is the power factor. The relations (5.9) are particularly useful for calculations concerning power factor correction.

The accurate measurement of power in a three-phase system is of considerable importance. Wattmeters of the electrodynamometer type, which read the time average of the product of the instantaneous voltage and current applied to their terminals, are commonly employed for this purpose. For a balanced Y-connected system a single wattmeter in one phase of the system will suffice to measure total power, it being necessary

276   Three-phase alternating current circuits

only to multiply the wattmeter reading by a factor of 3. For an unbalanced system with a neutral wire, three wattmeters, one in each phase, are necessary to measure total power. In a system with no neutral, two wattmeters only are required, regardless of whether the load is balanced or unbalanced.

Transformers constitute essential elements in a three-phase power system. A variety of technical and economic factors govern the choice of a transformer for a given application in the system. An important consideration concerns the requirement for a third-harmonic circulating current in the windings of transformers used in the system. This is necessary in order to maintain purity of the sinusoidal voltage waveforms throughout the system.

Transformers are also employed in a variety of configurations to provide three-phase to polyphase conversions, a technique used in power control and in a.c.-to-d.c. converter systems.

5.11   Problems

1. In the circuit of fig. 5.9 the line impedance Z_l is zero, and the load impedance Z is $(20+j10)\Omega$. The phase voltages are $V_a = 240 \underline{/0}$; $V_b = 240 \underline{/-120}$; $V_c = 240 \underline{/120}$.

Determine:
(a) the load currents I_α, I_β and I_γ;
(b) the line currents I_a, I_b and I_c;
(c) the load power factor;
(d) the power in the load;
(e) the impedance per phase of an equivalent Y-connected load that draws the same power at the same power factor.

2. In the circuit of fig. 5.11 the impedance Z_n is zero. Find the magnitude of the current in the neutral wire. The phase voltage of the source is 240 V and the phase sequence is *abc*. The line and load impedances are as given in the example of section 5.3.8.

3. Three identical impedances form a Y-connected balanced load on a 415 V, 3-phase power line. By ammeter and wattmeter measurements it is determined that the line current is 8 A and the total power taken by the load is 3 kW. Find the resistive and reactive components of the phase impedances.

4. Three equal impedances, $(2-j1)\Omega$, are connected in delta across a 240 V, 3-phase circuit. At the same point, three other equal impedances, $(1.5+j1)\Omega$, are connected in Y across the circuit.

Calculate:
(a) the line current;

(b) the power factor of the two loads together;
(c) the total power supplied.

5. Three impedances are connected in delta to lines *ABC* of a 415 V, 3-phase system. From *A* to *B* the impedance (Z_{AB}) is 24 Ω with p.f. 0.5 lagging; from *B* to *C* the impedance (Z_{BC}) is represented by $(8\sqrt{3}+j8)\,\Omega$; and between *C* and *A* there is a capacitor of 24 Ω reactance (Z_{CA}). The phase sequence is *ABC*. Wattmeters are connected with their current coils in lines *A* and *B*, their potential coils being connected from *A* to *C* and from *B* to *C* respectively. Find the readings of the wattmeters.
(Liverpool University)

6. In the system shown in fig. 5.25 the line voltage is 440 V at 50 Hz.
(a) Determine the line and phase currents.
(b) Sketch a complete phasor diagram of currents and voltages.
(c) Determine the voltage across the inductor and its phase relationship to V_{ab}.

7. A wattmeter measuring power into a load containing some reactance is connected as shown in fig. 5.26. It indicates power equal to *W*. The meter is a type that can be switched also to be an ammeter or a voltmeter to measure

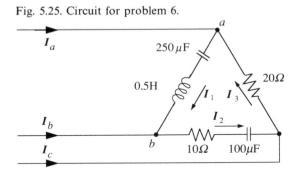

Fig. 5.25. Circuit for problem 6.

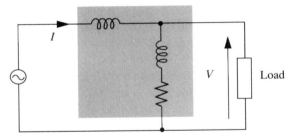

Fig. 5.26. Circuit for problem 7.

the current and voltage respectively as I and V. How can the true power taken by the load be determined if the internal resistance and reactance of the wattmeter between the voltage measuring terminals is known?
(Cambridge University)

8. A 3-phase transmission line has impedance $(3+j12)\,\Omega$. It supplies a load of 80 MVA at 0.8 p.f. lagging at 132 kV. Calculate:
(a) the power loss in the line;
(b) the power input;
(c) the voltage required at the sending end.
(Liverpool University: Second year)

9. (a) Show that the total power flow from 3 power lines to a 3-phase load may always be measured by using only two wattmeters.
(b) Explain why low power factors are objectionable in the operation of high power plants.

A factory is supplied with 3-phase 50 Hz power at 11 kV. The factory loading can be represented by the following balanced loads:
 (i) 3.0 MW at 0.9 lagging p.f.
 (ii) 0.7 MW at unity p.f.
 (iii) 4.0 MVA at 0.9 lagging p.f.
 (iv) 1.0 MVA at 0.8 leading p.f.

A star-connected capacitor bank is required to correct the overall p.f. to 0.98 lagging at full load. Find the capacitance per phase of the bank.
(Cambridge University: Second year)

10. Explain, briefly, the nature of the losses in a power transformer and describe the steps taken in the design of transformers to minimize these losses.

A 2 MVA 3-phase transformer, ratio 33/6.6 kV, delta/star, has a primary resistance of 8.3 Ω per phase and a secondary resistance of 0.08 Ω per phase. The regulation at full-load current, unity power factor, is 1.2%. If the primary is supplied at 33 kV, estimate the secondary voltage at full-load current, 0.75 power factor lagging. If the iron losses at rated voltage amount to 18 kW, estimate the efficiency at full load, 0.75 power factor lagging.
(Cambridge University: Second year)

11. A transformer supplies a variable current to a load at constant voltage and power factor. If the iron losses in the transformer remain substantially constant show that the transformer is most efficient when the copper loss is equal to the iron losses.

A 3-phase, 11 000/415 V, 50 Hz, star/delta connected transformer gave the following results for tests carried out on the high voltage winding.

Problems **279**

	Line voltage	Line current	3-phase power
Open circuit test	11 000 V	1.5 A	5000 W
Short circuit test	650 V	26.3 A	8000 W

At what value of line current on the low voltage side will the efficiency be a maximum? Calculate the secondary terminal voltage when supplying this current at 0.8 p.f. lagging, the primary voltage being maintained at 11 000 V. (Cambridge University: Second year)

6

Transient and steady-state analysis

6.1 Introduction

In chapters 2 and 3 we found the steady-state response of linear circuits when they are driven by direct (d.c.) or sinusoidal (a.c.) voltages or currents. In this chapter we shall look at the conditions arising in a circuit during the time required for it to reach the steady state. What occurs is called the *transient* behaviour of the circuit.

Consider the simple series circuits of fig. 6.1, and assume that the switches have been closed for a long time so that the circuits have reached steady-state conditions. For the d.c. circuit (fig. 6.1(a)) the voltage across the inductance is zero; therefore, the steady state current is $i_{ss} = V/R$. For the a.c. circuit (fig. 6.1(b)) the inductive reactance is ωL and the steady-state current is

$$i_{ss} = \frac{V_m}{\sqrt{[R^2 + (\omega L)^2]}} \sin(\omega t - \theta) \text{ where } \theta = \tan^{-1}\frac{\omega L}{R} \qquad (6.1)$$

Fig. 6.1. Inductive (RL) circuits with direct and alternating voltage driving sources.

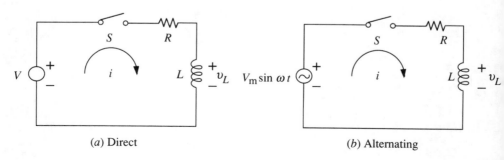

(a) Direct (b) Alternating

Qualitative analysis of the RL circuit

Now consider again the circuits of fig. 6.1, but this time assume that the switches are initially open so there is no current in either circuit. At time $t=0$ let the switches be closed. What will be the current in each circuit the instant after the switch is closed? The answer is zero for both circuits. We have seen in section 1.9 that the energy stored in an inductance cannot change instantaneously. Since the initial stored energy is zero in both circuits, and since the stored energy depends upon the current in the inductance, it follows that for both circuits the current immediately after the switch is closed must be zero. It will be convenient to use $t=0^+$ to designate the time immediately after a switching operation has been completed and before there has been any change in energy storage in any circuit element.

6.2 Qualitative analysis of the RL circuit

Let us examine in detail how the current in the circuit of fig. 6.1(a) rises from zero to its final steady-state value. At $t>0$,

$$v_R+v_L=V \quad \text{or} \quad iR+L\frac{di}{dt}=V$$

But at $t=0^+$, $i=0$, and so $v_R=iR=0$, hence

$$L\frac{di}{dt}=V \quad \text{or} \quad \frac{di}{dt}=\frac{V}{L}$$

and i is increasing. As i increases, v_R is no longer zero, so for $t>0$

$$\frac{di}{dt}=\frac{V}{L}-i\frac{R}{L}$$

We see then that the rate of change of current depends upon the current already in the circuit. When $i=V/R$, $di/dt=0$ and the current is no longer changing, having reached its steady-state value. Thus there is an interval of time during which the current rises at a decreasing rate toward its final value. Because di/dt depend upon i, the current cannot reach the final value in a finite length of time; therefore, i approaches the value V/R asymptotically.

The information that we now have enables us to sketch qualitatively the curve i v. t. This is shown in fig. 6.2(a). The voltage across the resistance has exactly the same time dependence as i. At every instant $v_L=V-iR$. Therefore, the voltage across the inductance starts at V and approaches zero asymptotically as shown in fig. 6.2(b).

The currents in and the voltages across the circuit elements during the interval while the current rises from zero to V/R are referred to collectively as the *transient response* of the circuit. It represents the smooth transition from the initial state ($i=0$) to the final state ($i=V/R$). In some circuits, for example the lights in a building, or motors operating household appliances, the transient response is probably of no interest; in other circuits the transient response is the only feature of interest. Transient voltages and currents may, for example, generate useful waveforms or they may be used to provide precisely known time delays in circuits.

There is also a transient response if, after the steady state is reached, the switch in fig. 6.1(a) is opened. When the conducting path is completely broken, i must be zero. However, at $t=0^+$, $i=V/R$ because current in the inductance cannot change instantaneously. When L is large, the rapid decrease in current as the switch contacts part results in a large induced voltage in the coil, a voltage that may be high enough to make the air between the contacts become conducting. Thus, the current path is not broken in zero time, but in a time determined by the rate at which energy initially stored in the coil is dissipated in the resistance and in the conducting arc established between the switch contacts. Again, the transient response provides a smooth transition from the initial state of $i=V/R$ to the final state $i=0$.

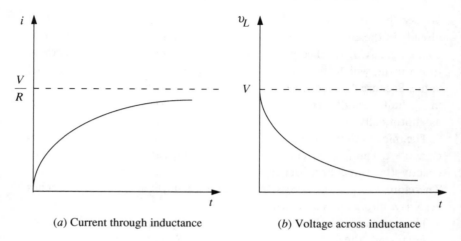

Fig. 6.2. Qualitative analysis of the circuit of fig. 6.1(a).

(a) Current through inductance

(b) Voltage across inductance

6.3 Mathematical analysis of the *RL* circuit

Now let us find an explicit expression for the current in the circuit of fig. 6.1(*a*). Assume that the switch is initially open and is closed when $t=0$. Then for $t>0$,

$$L\frac{di}{dt}+iR=V \tag{6.2}$$

Obviously, the steady state current, $i_{ss}=V/R$ is a solution of (6.2). The complete solution, however, contains another term that goes to zero as t increases. The complete solution describes the transient and reduces to the steady state solution as $t\to\infty$.

It is easy to write the solution of (6.2) by separating the variables, integrating, and using the initial condition $i=0$ at $t=0^+$. The result is

$$i=\frac{V}{R}-\frac{V}{R}e^{-Rt/L} \tag{6.3}$$

The current then is the sum of two terms. The first term is the steady state current that is independent of time. The second term represents an exponentially decaying current. The two terms and their sum are shown in fig. 6.3. We see that the total current has the type of time dependence that was predicted in the qualitative analysis of the circuit.

When there is a sinusoidal driving voltage as shown in fig. 6.1(*b*), the differential equation is

$$L\frac{di}{dt}+iR=V_m\sin\omega t \tag{6.4}$$

Before finding a solution for (6.4) we add a phase angle to the driving voltage. This is convenient because the solution will depend upon the value of the driving voltage at the instant the switch is closed. With the phase angle included in the voltage, switching can always occur at $t=0$. The phase angle λ may then be used to specify the value of the applied voltage at $t=0$. With this addition, and putting $R/L=\alpha$, (6.4) becomes

$$\frac{di}{dt}+\alpha i=\frac{V_m}{L}\sin(\omega t+\lambda) \tag{6.5}$$

To solve this equation we multiply by the integrating factor

$$e^{\int\alpha dt}=e^{\alpha t}$$

Then,

$$e^{\alpha t}\frac{di}{dt}+e^{\alpha t}\alpha i=e^{\alpha t}\frac{V_m}{L}\sin(\omega t+\lambda)$$

Integration gives

$$ie^{\alpha t}=\frac{V_m}{L}\int e^{\alpha t}\sin(\omega t+\lambda)\,dt$$

The right-hand side of the equation may be integrated by parts or by consulting tables.

$$ie^{\alpha t}=\frac{V_m e^{\alpha t}}{L(\alpha^2+\omega^2)}[\alpha\sin(\omega t+\lambda)-\omega\cos(\omega t+\lambda)]+K$$

Now multiply through by $e^{-\alpha t}$ and simplify the trigonometric expression in brackets by using the identity

$$A\sin\beta-B\cos\beta=\sqrt{(A^2+B^2)}\sin(\beta-\theta) \text{ where } \theta=\tan^{-1}\frac{B}{A}$$

Then

$$i=\frac{V_m}{\sqrt{[R^2+(\omega L)^2]}}\sin(\omega t+\lambda-\theta)+Ke^{-\alpha t}$$

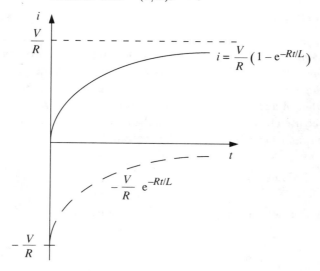

Fig. 6.3. Quantitative analysis of the circuit of fig. 6.1(a): equation (6.3). The current is the sum of a steady-state term V/R and a transient term $-(V/R)e^{-Rt/L}$.

Mathematical analysis of the RL circuit

To evaluate K, use the condition $i=0$ at $t=0^+$. Then

$$K = -\frac{V_m}{\sqrt{[R^2+(\omega L)^2]}}\sin(\lambda-\theta)$$

and

$$i = \frac{V_m}{\sqrt{[R^2+(\omega L)^2]}}\sin(\omega t+\lambda-\theta) - \frac{V_m e^{-\alpha t}}{\sqrt{[R^2+(\omega L)^2]}}\sin(\lambda-\theta) \quad (6.6)$$

As in the case with d.c. driving voltage, i is the sum of a steady-state term and a transient term that decreases exponentially and eventually becomes zero. The two terms on the right of (6.6) and their sum are shown in fig. 6.4. Except for the arbitrary phase angle λ the first of these terms is identical to (6.1).

As the circuits under consideration become more complex it is useful to take advantage of the fact that the complete solution of the differential equation for a linear circuit is the sum of two responses. That is,

$$i = i_{ss} + i_n \quad (6.7)$$

where i_{ss} is the *steady-state response* or *forced response*, which we know how to find for d.c. and sinusoidal driving sources, and i_n is the *transient response* or *natural response* of the circuit. (The steady-state response and the natural response correspond respectively to the particular integral and the complementary function in the mathematical solution of the circuit

Fig. 6.4. Response of an *RL* circuit to a sinusoidal driving source: plot of equation 6.6 with $\lambda=90°$, $\theta=45°$.

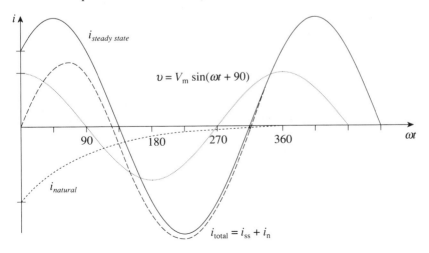

286 Transient and steady-state analysis

differential equation.) The natural response describes the behaviour of the circuit as energy initially stored in one or more elements is dissipated in the resistive elements of the circuit. To find the natural response one simply finds the solution to the differential equation for the circuit when the driving force is set equal to zero. For linear circuit elements the forced response has the same time dependence as the forcing function, and its amplitude (and in the case of a.c., its phase) is completely determined by the circuit parameters. In contrast, the time dependence of the natural response is independent of the forcing function; all currents and voltages have the same time dependence, which is always of the form $r_n = Ae^{st}$. In this expression, s is determined by the circuit configuration and the values of the circuit elements; the constant A depends upon the conditions obtaining at the instant the change occurs (e.g. the throwing of a switch) that initiates the transient behaviour.

Our procedure will be first to find the natural response of some simple circuits. The natural response will then be added to the forced response to obtain the total response. Initial conditions are applied to the total response in order to evaluate the constants that appear in the natural response. For any circuit, then, the natural response provides the smooth transition between the initial state of the circuit and the steady state response to a time dependent driving function.

6.4 Time constant

Consider time dependence of the form

$$y = Ae^{-\alpha t} \tag{6.8}$$

Here, y has value A at $t=0$ and decreases exponentially approaching zero asymptotically as $t \to \infty$. The constant α is a measure of how rapidly y decreases from its initial value. When $t = 1/\alpha$, $y = Ae^{-1} = 0.368\,A$. $1/\alpha$ is called the *time constant* and represents the time required for y to fall to 36.8% of its initial value, A. The time constant usually is designated by τ and is commonly expressed in seconds. (There are however some systems for which the time constant is more appropriately expressed in minutes or hours.) If for (6.8) we plot the ratio y/A against time, expressed as multiples of τ, we have a relation between two dimensionless quantities that is applicable to any equation of the form of (6.8). This is shown in fig. 6.5(a).

The concept of time constant is applicable also to

$$u = B(1 - e^{-t/\tau}) \tag{6.9}$$

when $t = \tau$, $u = B(1 - 0.368) = 0.632B$. So for this time dependence, the time constant represents the time required for u to reach 63.2% of its final value.

Time constant 287

Fig. 6.5. Illustrating the time constant τ: dimensionless plots of exponential waveforms.

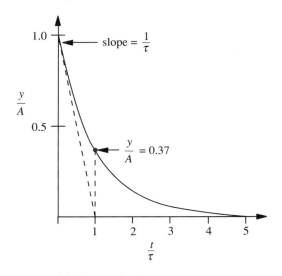

(a) Exponential decay

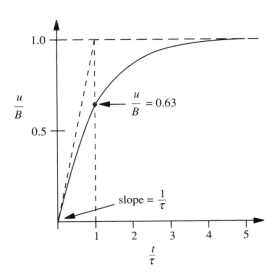

(b) Exponential rise

The curve of u/B v. t/τ is shown in fig. 6.5(b).

The initial slope of the curve represented by (6.8) is

$$\frac{\mathrm{d}(y/A)}{\mathrm{d}t} = -\alpha \mathrm{e}^{-\alpha t}\bigg|_{t=0} = -\frac{1}{\tau}$$

If y/A continued to decrease linearly at the initial rate it would reach zero in time equal to τ. This is shown in fig. 6.5(a) where the tangent drawn at $t=0$ is extended to intersect the t-axis. Similarly, in fig. 6.5(b) the initial slope is $+1/\tau$. Then the tangent drawn at $t=0$ intersects the line $u/B=1$ at time $t=\tau$.

For both (6.8) and (6.9), when $t=5\tau$, the dependent variable is within less than 1% of its final value. Therefore for practical purposes one may assume the final value has been achieved when $t > 5\tau$.

6.5 Natural response of some basic series circuits

6.5.1 RL circuit

In fig. 6.6 switch S_1 has been closed for a long time and S_2 is open. The current through R and L in series is, therefore, $I_0 = V/(R_1 + R)$.

At $t=0$, S_2 is closed and, simultaneously, S_1 is opened. Current now flows in the part of the circuit completed by S_2, and the energy initially stored in the inductance $(=\frac{1}{2}LI_0^2)$ is dissipated over a period of time in the resistance. Since the driving source voltage V is disconnected, the forced response in the part of the circuit which is active for $t>0$ must be zero.

For $t>0$, Kirchhoff's voltage law gives

$$v_L + v_R = 0$$

or

$$L\frac{\mathrm{d}i}{\mathrm{d}t} + Ri = 0 \tag{6.10}$$

The solution must be such that i and $\mathrm{d}i/\mathrm{d}t$ have the same time dependence; the only appropriate function is the exponential.

Let $i = A\mathrm{e}^{st}$; substitution in (6.10) then gives:

$$sLA\mathrm{e}^{st} + RA\mathrm{e}^{st} = 0$$

or

$$sL + R = 0 \tag{6.11}$$

(In the mathematical theory of differential equations this equation is referred to as the *auxiliary* equation.)

Natural response of some basic series circuits

From (6.11) we obtain

$$s = -R/L, \text{ hence}$$
$$i = Ae^{-Rt/L}$$

The current therefore decays exponentially with time constant L/R.
The constant A is evaluated from the initial condition

$$i = -I_0 = -\frac{V}{R_1 + R} \text{ at } t = 0^+$$

Note that the negative sign appearing, in this expression, arises because of the assignment of current in a clockwise direction in fig. 6.6.
The solution is then

$$i = -\frac{V}{R_1 + R} e^{-Rt/L} \qquad (6.12)$$

This equation is represented by the dimensionless plot of fig. 6.5 with $A = |-V/(R_1 + R)|$ and $\tau = L/R$.

Equation (6.12) gives the natural *current* response of the circuit. The natural *voltage* response across either R or L may be written immediately using (6.12). For the voltage across R we have

$$v_R = iR = -\frac{VR}{R_1 + R} e^{-Rt/L} \qquad (6.13)$$

and for the voltage across L we have

$$v_L = L\frac{di}{dt} = L\left[\frac{R}{L}\frac{V}{(R_1 + R)} e^{-Rt/L}\right] = \frac{VR}{R_1 + R} e^{-Rt/L} \qquad (6.14)$$

This expression also follows from the fact that $v_L = -v_R$.

Fig. 6.6. Circuit for calculating natural RL response. I_0 is the magnitude of the initial current through L (S_1 closed, S_2 open).

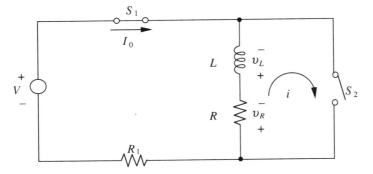

6.5.2 RC circuit

Referring to fig. 6.7, the voltage source V_0 is connected to the capacitance C and S_2 is open. At $t=0$, S_2 is closed and S_1 is opened. At this instant the voltage across the capacitance remains unchanged at V_0 since the stored energy cannot change instantaneously. If we assign current i in a clockwise direction, then v_C will have the polarity indicated and initially (at $t=0^+$) $v_C = -V_0$.

From Kirchhoff's voltage law,

$$v_C + v_R = 0$$

or

$$\frac{1}{C}\int i\, dt + Ri = 0 \tag{6.15}$$

Note that, after S_1 is opened, there is no driving source in the circuit so the right-hand side of this equation is zero. (The initial voltage V_0 on the capacitor is not to be confused with a driving source voltage.)

Differentiating (6.15) we obtain

$$\frac{di}{dt} + \frac{1}{CR} i = 0$$

Following a procedure similar to that in the previous section for the RL circuit, the solution of this equation is found to be:

$$i = A e^{-t/RC} \tag{6.16}$$

Now, referring to the directions of current and voltage shown in fig. 6.7, it is seen that

$$i = \frac{v_R}{R} = -\frac{v_C}{R} = \frac{V_0}{R} \text{ at } t=0^+$$

giving $A = V_0/R$. The natural current response is therefore

Fig. 6.7. Circuit for calculating natural RC response. V_0 is the magnitude of the initial voltage on C (S_1 closed, S_2 open).

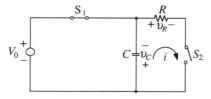

$$i = \frac{V_0}{R} e^{-t/RC} \tag{6.17}$$

This equation is also represented by the dimensionless plot of fig. 6.5 with the time constant $\tau = RC$.

The natural voltage response is given by

$$v_R = -v_C = iR = V_0 e^{-t/RC} \tag{6.18}$$

The *RL* circuit and the *RC* circuit each contain one energy storage element and, for this reason, are called *single-energy* circuits. They are also referred to as *first-order* circuits because their behaviour can be described by a first-order differential equation.

Commencing at $t = 0^+$ there is in each circuit a unidirectional current which continues, decreasing exponentially in amplitude, until all the energy that was initially stored is transformed into heat in the resistance. In the *RC* circuit, for example, the energy dissipated in the resistance is

$$W_R = \int_0^\infty i^2 R \, dt = \frac{V_0^2}{R} \int_0^\infty e^{-2t/RC} \, dt = \frac{V_0^2}{R}(-RC/2)[e^{-2t/RC}]_0^\infty$$
$$= \tfrac{1}{2} C V_0^2$$

which is just equal to the energy initially stored in the capacitance.

6.5.3 RLC circuit

The circuit shown in fig. 6.8 has two energy storage elements and is referred to as a *double-energy* or *second-order* circuit; a second-order differential equation is required to describe its behaviour.

Initially the switch is open and we assume that the capacitance is charged (by means of a circuit similar to that shown in fig. 6.7) to some voltage V_0. For the polarity of V_0 indicated, $v_C = -V_0$ initially.

At $t = 0$ the switch is closed, then for $t > 0$

$$v_L + v_C + v_R = 0$$

Fig. 6.8. *RLC* circuit; V_0 is the magnitude of the initial voltage on *C*.

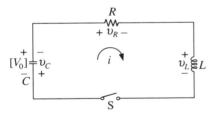

Transient and steady-state analysis

or

$$L\frac{di}{dt} + \frac{1}{C}\int i\,dt + Ri = 0 \tag{6.19}$$

Differentiating removes the integral sign and produces a second-order differential equation of *homogeneous form* (RHS of equation identically zero):

$$\frac{d^2i}{dt^2} + \frac{R}{L}\frac{di}{dt} + \frac{1}{CL}i = 0$$

Assume, as before, a solution of the form $i = Ae^{st}$; then the auxiliary equation becomes

$$s^2 + \frac{R}{L}s + \frac{1}{CL} = 0 \tag{6.20}$$

Solving this quadratic equation gives

$$s = -\frac{R}{2L} \pm \sqrt{\left[\left(\frac{R}{2L}\right)^2 - \frac{1}{LC}\right]} \tag{6.21}$$

In general, there will be two distinct values of s so

$$i = A_1 e^{s_1 t} + A_2 e^{s_2 t} \tag{6.22}$$

giving the two arbitrary constants required by the original second-order differential equation. (The special case where $s_1 = s_2$ will be considered later.)

Two initial conditions are required for the evaluation of these constants. Appropriate conditions are:

(1) at $t = 0^+$, $i = 0$.

This follows from the fact that current through the inductance cannot change instantaneously.

(2) at $t = 0^+$, $di/dt = V_0/L$.

This follows from the fact that when $i = 0$, then $v_R = 0$ and $v_L = -v_C$. But $v_L = L\,di/dt$ and $v_C = -V_0$, so $di/dt = V_0/L$.

Use of these conditions in (6.22) enables us to evaluate A_1 and A_2. Then the solution of (6.19) is

$$i = \frac{V_0}{L}\frac{1}{(s_1 - s_2)}(e^{s_1 t} - e^{s_2 t}) \tag{6.23}$$

Now, referring to (6.21), the quantity

$$\left(\frac{R}{2L}\right)^2 - \frac{1}{LC} \tag{6.24}$$

Natural response of some basic series circuits

called the *discriminant*, determines which of three particular forms the solution (6.23) takes.

If the discriminant is positive, that is, if $(R/2L)^2 > 1/LC$, s_1 and s_2 are real, negative and unequal. (Remember that R, L and C are intrinsically positive.)

Let $s_1 = -m$, $s_2 = -n$, and let $|n| > |m|$. Then (6.23) becomes

$$i = \frac{V_0}{L} \frac{1}{|n-m|} (e^{-mt} - e^{-nt}) \tag{6.25}$$

The solution is then the sum of two decaying exponentials, as shown in fig. 6.9(a).

If the discriminant is negative $((R/2L)^2 < 1/LC)$, s_1 and s_2 are complex conjugate numbers.

Let

$$\frac{R}{2L} = \alpha \text{ and } \frac{1}{LC} - \left(\frac{R}{2L}\right)^2 = \omega_n^2 \tag{6.26}$$

then,

$$s_1 = -\alpha + j\omega_n; \qquad s_2 = -\alpha - j\omega_n \tag{6.27}$$

and, using Euler's identity, (6.23) takes the form:

$$i = \frac{V_0 e^{-\alpha t}}{2j\omega_n L} (e^{j\omega_n t} - e^{-j\omega_n t}) = \frac{V_0 e^{-\alpha t}}{\omega_n L} \sin\omega_n t \tag{6.28}$$

which is an exponentially damped sine wave as shown in fig. 6.9(b).

Fig. 6.9. Natural current response for the *RLC* circuit of fig. 6.8.

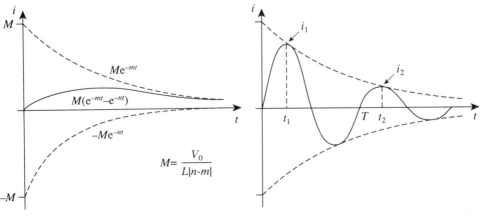

(a) Overdamped response: equation (6.25) (b) Underdamped response: equation (6.28)

294   *Transient and steady-state analysis*

When s_1 and s_2 are real, the current starts at zero, increases in magnitude, and then decreases to zero, but it is always in the same direction.

With complex values of s the current is oscillatory at a frequency

$$\omega_n = \sqrt{\left[\frac{1}{LC} - \left(\frac{R}{2L}\right)^2\right]} = \sqrt{(\omega_0^2 - \alpha^2)} \tag{6.29}$$

where $\omega_0^2 = 1/LC$.

ω_n is called the *damped natural frequency* and it represents the frequency of the periodic transfer of energy between the two storage elements. ω_0 is the resonant frequency of the circuit (discussed in section 3.13.2). The *damping constant* α governs the rate at which the amplitude of the oscillations approaches zero as the initial stored energy is dissipated as heat in the resistance. For small values of α, $\omega_n \simeq \omega_0$ and the oscillations last for many cycles. A circuit that has an oscillatory natural response is said to be *underdamped*.

When s_1 and s_2 are real, the circuit is said to be *overdamped*. The condition $s_1 = s_2$ (($R/2L)^2 = 1/LC$) represents the transition between the overdamped and the underdamped states, and it is called the condition of *critical damping*. To get a complete solution of (6.19) for this case one must use

$$i = (A_1 + tA_2)e^{st} \tag{6.30}$$

in order to have the required two constants of integration. Although it represents the condition for which the current reaches zero in minimum time, critical damping is of no special practical significance in electrical circuits. It usually is not worthwhile to select circuit components carefully enough to achieve exact critical damping. Moreover, there is the distinct possibility that ageing of carefully chosen circuit components may cause their values to change in such a way as to make the circuit oscillatory.

6.5.4   Q-factor and logarithmic decrement

For the oscillatory RLC circuit whose current is given by (6.28) it is useful to have a number that relates the damping to the natural frequency ω_n of the circuit. We denote this number by Q_n (the Q-factor) and we define it as*

---

* This quantity is different from Q_0 the Q-factor defined in the discussion of resonance in section 3.13.2. There we defined $Q_0 = \omega_0 L/R$ where $\omega_0 = 1/(LC)^{\frac{1}{2}}$. Since large values of Q_n are associated with small values of R, it follows from (6.29) that when Q_n is large α is small, $\omega_n \simeq \omega_0$, and $Q_n \simeq Q_0$.

Total response

$$Q_n = \frac{\omega_n}{2\alpha} = \frac{\omega_n L}{R} \tag{6.31}$$

$$\alpha = \frac{R}{2L} = \frac{\omega_n}{2Q_n} = \frac{\pi}{Q_n T} \tag{6.32}$$

where T is the period of the natural response as indicated on fig. 6.9. Then, putting $I_0 = V_0/\omega_n L$, (6.28) becomes

$$i = I_0 e^{-(\pi/Q_n T)t} \sin \omega_n t \tag{6.33}$$

We see from (6.33) that when t increases by one period, the amplitude decreases by the factor $e^{-\pi/Q_n}$. So Q_n is a measure of the damping per cycle. Furthermore, the time required for i to decrease by a factor $1/e$ is equal to $Q_n T/\pi$. Since Q_n is inversely proportional to R we expect that a large value of Q_n is characteristic of a circuit that requires a long time for the oscillations to die out.

We can determine the Q of the circuit by examining the oscillatory decay and measuring the amplitude of two successive peaks. In fig. 6.9

$$i_1 = I_0 e^{-(\pi/Q_n T)t_1}, \qquad i_2 = I_0 e^{-(\pi/Q_n T)t_2}$$

and

$$\frac{i_1}{i_2} = e^{(\pi/Q_n T)(t_2 - t_1)} \tag{6.34}$$

If $(t_2 - t_1) = T$, then $i_1/i_2 = e^{\pi/Q_n}$

and

$$\ln(i_1/i_2) = \pi/Q_n.$$

The quantity π/Q_n is the *logarithmic decrement*.

6.6 Total response

The natural responses of the circuits that we have considered so far are examples of transient behaviour for the special situation where the steady-state response is zero because there are no driving currents or voltages present. When such energy sources are part of the circuit the constants that appear in the natural response must be evaluated by applying the initial conditions to the *complete* solution. Depending upon the driving function and upon conditions that exist immediately after the switching operation (that is, at $t = 0^+$) the constants assume the values necessary to provide a smooth transition from the initial to the final state of the circuit.

6.6.1 RL circuit with sinusoidal driving voltage

The usefulness of writing

$$i = i_{ss} + i_n$$

is well illustrated by a reconsideration of the circuit of fig. 6.1(b). The differential equation is (6.5)

$$\frac{di}{dt} + \frac{R}{L} i = \frac{V_m}{L} \sin(\omega t + \lambda)$$

The steady state solution is

$$i_{ss} = \frac{V_m}{\sqrt{[R^2 + (\omega L)^2]}} \sin(\omega t + \lambda - \theta); \qquad \theta = \tan^{-1} \frac{\omega L}{R} \qquad (6.35)$$

The natural response is

$$i_n = A e^{-Rt/L} \qquad (6.36)$$

The total response then is the sum of (6.35) and (6.36)

$$i = i_{ss} + i_n = \frac{V_m}{\sqrt{[R^2 + (\omega L)^2]}} \sin(\omega t + \lambda - \theta) + A e^{-Rt/L} \qquad (6.37)$$

Applying the initial condition ($i = 0$ at $t = 0^+$),

$$A = -\frac{V_m}{\sqrt{[R^2 + (\omega L)^2]}} \sin(\lambda - \theta)$$

and

$$i = \frac{V_m}{\sqrt{[R^2 + (\omega L)^2]}} \sin(\omega t + \lambda - \theta) - \frac{V_m}{\sqrt{[R^2 + (\omega L)^2]}} e^{-Rt/L} \sin(\lambda - \theta) \qquad (6.38)$$

This is identical to (6.6). It is apparent that the approach just employed is more direct than that followed in deriving (6.6).

Fig. 6.10. *RC* circuit with constant voltage driving source.

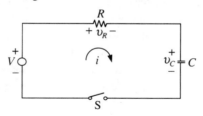

6.6.2 RC circuit with constant voltage source

In the circuit of fig. 6.10 we are interested in the voltage v_C on the capacitance after the switch is closed at $t=0$.

The natural response will, from the theory given in section 6.5.2, be of the form:

$$v_{Cn} = A e^{-t/RC} \tag{6.39}$$

For the steady state,

$$v_{Css} = V \tag{6.40}$$

The total response is the sum of (6.39) and (6.40).

$$v_C = A e^{-t/RC} + V$$

Assuming no initial charge on the capacitance, the constant A is evaluated from the initial condition:

$$v_C = 0 \text{ at } t = 0^+,$$

hence,

$$A = -V$$

and

$$v_C = V(1 - e^{-t/RC}) \tag{6.41}$$

This expression is represented by the dimensionless plot of fig. 6.5(b) with $B = V$ and $\tau = RC$.

The current is given by

$$i = C \frac{dv_C}{dt} = \frac{V}{R} e^{-t/RC} \tag{6.42}$$

6.6.3 Worked example

A circuit designed to fire a laser flash tube consists of the following (see fig. 6.11): a 12 V battery of internal resistance 10 Ω is connected via a switch s_1 to a resistance of 80 Ω in series with a relay coil of resistance 10 Ω and inductance 2 H. The relay operates when the current in the circuit reaches 50 mA. The operation of the relay closes a switch s_2 in another circuit so that a capacitor bank of 100 μF is charged up via a resistor of 1 kΩ in series with a 2 kV supply.

If the laser fires when the capacitor bank is charged up to 1 kV, find the time taken from the closing of s_1 to firing of the laser. Neglect the time required for the relay to operate.

298   Transient and steady-state analysis

Solution.
The diagrams for the two parts of the circuit are shown in fig. 6.11.

Relay circuit (fig. 6.11(a)): let R be the total resistance in the circuit and L be the inductance of the relay coil. The natural current response of the circuit is then, from section 6.5.1,

$$i_n = A e^{-Rt/L}$$

The steady state, or forced response, is obviously

$$i_{ss} = \frac{V}{R}$$

Hence, the total response is

$$i = A e^{-Rt/L} + \frac{V}{R}$$

At $t=0^+$, $i=0$ therefore $A = -V/R$ so the current in the relay circuit is given by

$$i = \frac{V}{R}(1 - e^{-Rt/L})$$

Inserting numerical values ($R = 100\,\Omega$; $L = 2\,\text{H}$) we find the time t_1 for the current to reach 50 mA

$$50 \times 10^{-3} = \frac{12}{100}(1 - e^{-50 t_1})$$

giving $e^{-50 t_1} = 0.583$ or $t_1 = 10.8$ ms.

Fig. 6.11. Circuits for worked example (section 6.6.3).

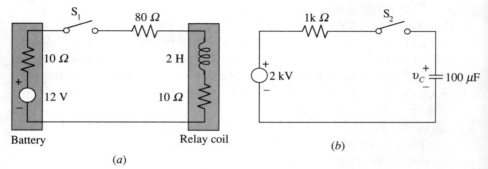

Capacitor circuit (fig. 6.11(b)): after the switch s_2 closes, the voltage on the capacitor is, from (6.41),

$$v_C = V(1 - e^{-t/RC})$$

If t_2 is the time for the voltage to reach 1 kV, then

$$1 \times 10^3 = (2 \times 10^3)(1 - e^{-10t_2})$$

giving $e^{-10t_2} = 0.5$ or $t_2 = 69.3$ ms.

So the total time from closing the first switch to the firing of the laser is $10.8 + 69.3 = 80.1$ ms.

6.6.4 RLC circuit with constant voltage source

In the circuit of fig. 6.12(a) the capacitor is initially uncharged. At $t=0$, S is closed. We require an expression for the current for $t>0$. The steady-state current i_{ss} is zero and the natural response is, from (6.22),

$$i_n = A_1 e^{s_1 t} + A_2 e^{s_2 t}$$

The initial conditions are:

$$i = 0 \text{ and } v_C = v_R = 0, \text{ so } v_L = V, \text{ or } di/dt = V/L$$

When these conditions are used to evaluate A_1 and A_2, we obtain

$$i = \frac{V}{L} \frac{1}{s_1 - s_2} (e^{s_1 t} - e^{s_2 t}) \tag{6.43}$$

This is identical with the expression contained in section 6.5.3 for the natural response of the circuit; a result which is to be expected since the forced response is zero in the present case.

As discussed previously in section 6.5.3, the current will be either the sum of two decreasing exponentials or oscillatory with exponentially decreasing amplitude, depending upon the relative magnitudes of $(R/2L)^2$ and $1/LC$.

Fig. 6.12. *RLC circuit with constant and sinusoidal driving voltages.*

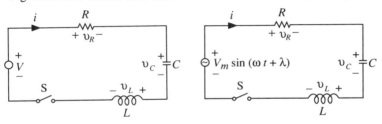

(a) Constant (d.c.) driving voltage   (b) Sinusoidal driving voltage

6.6.5 RLC circuit with sinusoidal driving voltage

When the *RLC* circuit is driven by a sinusoidal voltage, as in fig. 6.12(b), the forced response is (from our a.c. theory)

$$i_{ss} = \frac{V_m}{Z}\sin(\omega t + \lambda - \theta) \tag{6.44}$$

where

$$Z = \sqrt{[R^2 + (X_L - X_C)^2]} \text{ and } \tan\theta = \frac{X_L - X_C}{R}$$

Because the natural response can have different forms depending upon circuit constants, the transient response may exhibit wide variations. In every case, however, the *form* of the transient is determined by the circuit, and the *amplitude* is whatever is required to satisfy the initial conditions. These conditions depend upon the initial energy stored (if any) and upon the instant in the cycle of the applied voltage at which the switch is closed.

If the *RLC* circuit is overdamped, the natural response is given by (6.22)

$$i_n = A_1 e^{-mt} + A_2 e^{-nt} \tag{6.45}$$

where $-m$ and $-n$ are the two appropriate values of s in (6.22). The complete solution is then

$$i = i_{ss} + i_n = \frac{V_m}{Z}\sin(\omega t + \lambda - \theta) + A_1 e^{-mt} + A_2 e^{-nt} \tag{6.46}$$

Two initial conditions are required for evaluation of the constants A_1 and A_2.

For the underdamped case the natural response is, from (6.22), (6.26) and (6.27),

$$i_n = A_1 e^{s_1 t} + A_2 e^{s_2 t} = e^{-\alpha t}[A_1 e^{j\omega_n t} + A_2 e^{-j\omega_n t}] \tag{6.47}$$

Using Euler's identity this can be written

$$i_n = e^{-\alpha t}[B_1 \sin\omega_n t + B_2 \cos\omega_n t]$$

or

$$i_n = e^{-\alpha t} M \sin(\omega_n t + \phi) \tag{6.48}$$

where

$$M = \sqrt{(B_1^2 + B_2^2)}; \quad \phi = \tan^{-1}\frac{B_2}{B_1}$$

Total response

The complete solution is then

$$i = \frac{V_m}{Z}\sin(\omega t + \lambda - \theta) + e^{-\alpha t} M \sin(\omega_n t + \phi) \quad (6.49)$$

Note that the evaluation of the constants B_1 and B_2 is often easier than the direct evaluation of constants M and ϕ.

6.6.6. RLC circuit with sinusoidal drive and $\omega_0 \simeq \omega_n$

Referring to fig. 6.12(b), if $1/LC \gg (R/2L)^2$, the circuit is lightly damped, and the damped natural frequency ω_n and the resonant frequency ω_0 are very nearly equal. If there is no initial stored energy in the circuit, the transient response depends upon the phase angle λ of the applied voltage and also upon how the applied frequency ω compares with ω_n.

Case 1. Let $\lambda = 0$. The initial conditions are then:
(1) $i = 0$ (current in inductor cannot change instantaneously);
(2) $di/dt = 0$ ($v = v_R + v_C + v_L$, but $v = 0$ and also $v_R = 0$ and $v_C = 0$, therefore $v_L = L\,di/dt = 0$).

Case 1a: $\omega = \omega_n$. The circuit is resonant, therefore in (6.44) $X_L = X_C$, $Z = R$, and $\theta = 0$.
So, (6.44) becomes

$$i_{ss} = \frac{V_m}{R}\sin\omega t$$

and (6.49) becomes

$$i = \frac{V_m}{R}\sin\omega t + e^{-\alpha t} M \sin(\omega t + \phi) \quad (6.50)$$

Differentiating (6.50) gives

$$\frac{di}{dt} = \frac{\omega V_m}{R}\cos\omega t + e^{-\alpha t} M \omega \cos(\omega t + \phi) - \alpha e^{-\alpha t} M \sin(\omega t + \phi) \quad (6.51)$$

The first initial condition gives, by (6.50),

$$M \sin\phi = 0$$

If $M = 0$, there is no transient, so we take this condition to mean $\sin\phi = 0$. Hence, $\sin\phi = 0$ and so $\phi = 0$.

The second initial conditions gives, by (6.51),

$$0 = \omega\frac{V_m}{R} + M\omega, \text{ hence } M = -\frac{V_m}{R}$$

Then

$$i = \frac{V_m}{R}(1 - e^{-\alpha t})\sin\omega t \qquad (6.52)$$

The current is sinusoidal inside an envelope that starts at zero and approaches asymptotically the values $\pm V_m/R$ (see fig. 6.13).

Case 1b: $\omega \gg \omega_n$. The circuit is predominantly inductive and so to a close approximation we may write $Z = \omega L$ and $\theta = \pi/2$ rad. So (6.44) becomes,

$$i_{ss} = -\frac{V_m}{\omega L}\cos\omega t$$

Then

$$i = -\frac{V_m}{\omega L}\cos\omega t + e^{-\alpha t}M\sin(\omega_n t + \phi)$$

and

$$\frac{di}{dt} = \frac{V_m}{L}\sin\omega t - \alpha e^{-\alpha t}M\sin(\omega_n t + \phi) + e^{-\alpha t}\omega_n M\cos(\omega_n t + \phi)$$

Substitution of the initial conditions yields

$$\tan\phi = \frac{\omega_n}{\alpha}, \text{ and since } \omega_n \gg \alpha, \phi \simeq \pi/2$$

$$M = \frac{V_m}{\omega L}$$

Fig. 6.13. Response of *RLC* circuit driven at its natural frequency.

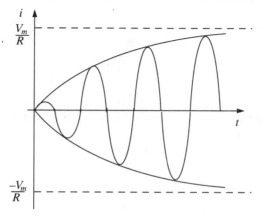

Therefore,

$$i = -\frac{V_m}{\omega L}(\cos\omega t - e^{-\alpha t}\cos\omega_n t) \tag{6.53}$$

Case 1c: $\omega \ll \omega_n$. In the steady state this circuit is capacitive with current leading the voltage by 90°. So (6.44) becomes:

$$i_{ss} = \omega C V_m \cos\omega t$$

Following the same procedure as that used for Case 1b, we obtain

$$i = \omega C V_m(\cos\omega t - e^{-\alpha t}\cos\omega_n t) \tag{6.54}$$

In neither Case 1b nor Case 1c does the current reach exceptionally high values; under no circumstances will it be greater than twice the steady-state value.

Case 2. Let $\lambda = \pi/2$ radians. Now the driving voltage has maximum value at $t=0^+$. Initial conditions are:

(1) $i = 0$
(2) $di/dt = V_m/L$

These lead to the following expressions for the current

Case 2a: $\omega = \omega_n$.

$$i = \frac{V_m}{R}(1 - e^{-\alpha t})\cos\omega t + \frac{V_m}{2\omega L}e^{-\alpha t}\sin\omega t \tag{6.55}$$

Case 2b: $\omega \gg \omega_n$.

$$i = \frac{V_m}{\omega L}\left(\sin\omega t - \frac{\omega_n}{\omega}e^{-\alpha t}\sin\omega_n t\right) \tag{6.56}$$

The transient term is negligibly small because of the multiplying factor (ω_n/ω).

Case 2c: $\omega \ll \omega_n$.

$$i = -\omega C V_m\left(\sin\omega t - \frac{\omega_n}{\omega}e^{-\alpha t}\sin\omega_n t\right) \tag{6.57}$$

Now the ratio (ω_n/ω) is large and so the amplitude of the transient component may be many times the steady-state amplitude.

6.7 The D-operator

In our study of transient analysis so far we have considered circuits mainly of a simple series form containing not more than two storage elements, and driven by constant (d.c.) or sinusoidal (a.c.) driving sources.

For more complicated circuits with arbitrary driving sources, the circuit integro-differential equations can be complex, and their solution correspondingly difficult. For this reason 'operational' methods have been devised which greatly simplify the process of formulating the circuit equations and which, for some important practical cases, provide elegant methods of solution. One such operational method is presented in this section. Our procedure will be to describe the method and illustrate its use with examples. A more complete description and mathematical justification of the method will be found in ref. 15.

6.7.1 The operators D and D^{-1}

We define the 'differential operator' D by

$$D = \frac{d}{dt} \tag{6.58}$$

so that we may write

$$Dx = \frac{dx}{dt} \tag{6.59}$$

and we interpret $D^n x$ to mean $d^n x/dt^n$, that is the symbol D^n operating on x signifies the process of differentiating n times.

Extending the notation further we interpret $1/D = D^{-1}$ as signifying the process of integration, that is,

$$\frac{1}{D}x = D^{-1}x = \int x \, dt \tag{6.60}$$

so that

$$D\frac{1}{D}x = \frac{d}{dt}\int x \, dt = x$$

Using this notation the voltage–current relationships for inductance and capacitance,

$$v_L = L\frac{di}{dt} \qquad v_C = \frac{1}{C}\int i \, dt$$

become,

$$v_L = L\,Di \qquad v_C = \frac{1}{CD}i \tag{6.61}$$

Similarly, the circuit equation appertaining to the general branch with a sinusoidal driving voltage (fig. 6.12(b)), namely,

$$L\frac{di}{dt} + \frac{1}{C}\int i\,dt + Ri = V_m \sin(\omega t + \lambda)$$

becomes

$$LDi + \frac{1}{CD}i + Ri = V_m \sin(\omega t + \lambda) \qquad (6.62)$$

The use of D and D^{-1} to indicate respectively differentiation and integration, would appear to be but a modest extension of the process of symbolic representation, however, it can be shown that for linear differential equations with constant coefficients the D-operator can be treated like a coefficient in an algebraic equation. Specifically, the operator obeys the distributive, commutative and associative laws of algebra. This implies, for example, that

$$D(x+y) = Dx + Dy$$

and

$$(D - m_1)(D - m_2) = (D - m_2)(D - m_1) = D^2 - (m_1 + m_2)D + m_1 m_2$$

Functions of D also obey the laws of algebra; for example,

$$F_1(D) \times F_2(D) = F_2(D) \times F_1(D) \text{ and } F(D)(x+y) = F(D)x + F(D)y$$

These algebraic properties allow us to multiply both sides of (6.62) by D (corresponding to differentiation term by term) to obtain

$$LD^2 i + \frac{1}{C}i + DRi = DV_m \sin(\omega t + \lambda)$$

or

$$\left(D^2 + \frac{R}{L}D + \frac{1}{LC}\right)i = \frac{D}{L}V_m \sin(\omega t + \lambda) \qquad (6.63)$$

6.7.2 Solution of differential equations by D-operator

In general, the differential equations that characterize linear circuits are of the form:

$$F(D)y = X \qquad (6.64)$$

where the function $F(D)$ is a polynomial, y represents voltage or current, and X represents a time-dependent driving source (voltage or current). We have already seen that such an equation may be solved in two stages: first, the complementary function (natural response of the circuit) is found from the homogeneous equation

$$F(D)y = 0 \tag{6.65}$$

Substitution of $y = Ae^{st}$ leads directly to the auxiliary equation $F(s) = 0$. For example, putting $i = Ae^{st}$ in (6.63) and using (6.65) we obtain

$$\left(D^2 + \frac{R}{L}D + \frac{1}{LC}\right)Ae^{st} = 0 \tag{6.66}$$

$$As^2 e^{st} + \frac{RA}{L} se^{st} + \frac{A}{LC} e^{st} = 0$$

or

$$s^2 + \frac{R}{L}s + \frac{1}{LC} = 0 \tag{6.67}$$

Comparing (6.66) and (6.67) we see that the auxiliary equation may be written directly merely by substitution of s for D.

The second stage in the solution of (6.64) is to find the particular integral (forced or steady-state response of the circuit). Now it can be shown that the D-operator method enables one to obtain the particular integral from

$$y = \frac{1}{F(D)} X \tag{6.68}$$

A variety of methods exist for solving this equation, depending upon the particular form of X; we consider three important cases.

Case 1. $X = x^n$ (n = positive integer)

In this case

$$y = [F(D)]^{-1} x^n \tag{6.69}$$

and $[F(D)]^{-1}$ is expressed as a polynomial in rising powers of D as far as D^n. (Any higher powers of D will yield zero.)

Example: $(D^2 - 4D + 4)y = x^2$

The P.I. is

$$y = \frac{1}{D^2 - 4D + 4} x^2 = \frac{1}{4(1 - D + D^2/4)} x^2$$

$$= \tfrac{1}{4}\left(1 - D + \frac{D^2}{4}\right)^{-1} x^2$$

Expanding by the binomial theorem:

The D-operator

$$y = \tfrac{1}{4}\left[1 + \left(D - \frac{D^2}{4}\right) + \left(D - \frac{D^2}{4}\right)^2 \cdots \right]x^2$$

$$= \tfrac{1}{4}\left(1 + D - \frac{D^2}{4} + D^2 \cdots \right)x^2$$

$$y = \tfrac{1}{4}(x^2 + 2x + \tfrac{3}{2})$$

Case 2. $X = e^{ax}$

We have $De^{ax} = ae^{ax}$; $D^2 e^{ax} = a^2 e^{ax}$; $D^n e^{ax} = a^n e^{ax}$ hence, if $F(D)$ is a polynomial,

$$F(D)e^{ax} = F(a)e^{ax} \qquad (6.70)$$

Now

$$\frac{1}{F(D)} F(D)e^{ax} = \frac{1}{F(D)} F(a)e^{ax} = F(a)\frac{1}{F(D)} e^{ax}$$

but

$$\frac{1}{F(D)} F(D)e^{ax} = e^{ax}$$

therefore

$$y = \frac{1}{F(D)} e^{ax} = \frac{1}{F(a)} e^{ax} \qquad (F(a) \neq 0) \qquad (6.71)$$

Example: $(3D^2 - 2D + 4)y = 36e^{-x}$

The P.I. is

$$y = \frac{36e^{-x}}{3D^2 - 2D + 4} = \frac{36e^{-x}}{3(-1)^2 - 2(-1) + 4} = 4e^{-x}$$

Case 3. $X = e^{ax}V(x)$ where $V(x)$ is a function of x only.

For this case it can be shown that

$$y = \frac{1}{F(D)} e^{ax} V(x) = e^{ax} \frac{1}{F(D+a)} V(x) \qquad (6.72)$$

Example: $(D^2 + D - 2)y = xe^x$

The P.I. is

$$y = \frac{1}{D^2 + D - 2} xe^x$$

Using theorem (6.72), the exponential is shifted to the left of the operator and D becomes $D+1$:

$$y = e^x \frac{1}{(D+1)^2 + (D+1) - 2} x = e^x \frac{1}{D(D+3)} x$$

$$= \frac{e^x}{3D}\left(1 + \frac{D}{3}\right)^{-1} x = \frac{e^x}{3D}\left(1 - \frac{D}{3} \ldots\right) x$$

$$y = \frac{e^x}{3D}(x - \tfrac{1}{3})$$

The D in the denominator means that the expression to its right is integrated once, thus

$$y = \frac{e^x}{3}\left(\frac{x^2}{2} - \frac{x}{3}\right)$$

Theorem (6.72) allows us to deal with situations for which theorem (6.71) of Case 2 breaks down, for example

$$y = \frac{1}{D^2 + 4D + 4} e^{-2x}$$

Simply replacing D by -2 as required by (6.71), gives zero in the denominator of this expression. However, if we take $V(x) = 1$ in (6.72) we obtain:

$$y = e^{-2x} \frac{1}{(D-2)^2 + 4(D-2) + 4}(1) = e^{-2x} \frac{1}{D^2}(1) = e^{-2x} \frac{x^2}{2}$$

Returning now to (6.62), appertaining to the general branch with sinusoidal excitation, we may use theorem (6.71) of Case 2 to derive the particular integral. Representing the RHS of (6.62) by the imaginary part of the complex exponential we may write

$$\left(LD + \frac{1}{CD} + R\right)i = \text{Im} V_m e^{j(\omega t + \lambda)}$$

Now, according to (6.68), the particular integral is given by

$$i = \frac{V_m}{LD + 1/CD + R} \text{Im} e^{j(\omega t + \lambda)} \tag{6.73}$$

and, by (6.71), D may be replaced by $j\omega$ to give

$$i = \text{Im} \frac{V_m}{j\omega L + 1/j\omega C + R} e^{j(\omega t + \lambda)} \tag{6.74}$$

The denominator of this expression is recognized as the complex impedance, which may be written $Ze^{j\theta}$ where $Z = [R^2 + (\omega L - 1/\omega C)^2]^{\frac{1}{2}}$ and

The D-operator

$\theta = \tan^{-1}(\omega L - 1/\omega C)/R$, hence

$$i = \text{Im} \frac{V_m}{Ze^{j\theta}} e^{j(\omega t + \lambda)} = \text{Im} \frac{V_m}{Z} e^{j(\omega t + \lambda - \theta)}$$

or

$$i = \frac{V_m}{Z} \sin(\omega t + \lambda - \theta)$$

The derivation of this result should be compared with the methods used in section 3.4.

6.7.3 D-impedance

The concepts of complex exponential and complex impedance were introduced in sections 3.3 and 3.4 in connection with steady-state a.c. circuit analysis. A related concept which is useful in the context of transient analysis will now be considered.

We observe that the 'jω' in the denominator of (6.74), which is the complex impedance, arise directly as a result of the 'D' in (6.73). This result follows whatever the combination of circuit elements under consideration, and by analogy we call the function of D appearing in the denominator of equations such as (6.73) the D-impedance. This concept offers a convenient approach to the setting up of circuit differential equations, which is exactly analogous to that used for setting up the steady-state a.c. circuit equations.

First, write down the a.c. impedance (or reactance) of each circuit element but with D in place of jω. Then combine the impedances and derive the circuit equations in the usual way. The ratio of voltage $v(t)$ to current $i(t)$ at any terminal pair of a network will be the D-impedance at that terminal pair. For pure inductive and capacitive elements the relations analogous to jωL and 1/jωC are LD and 1/CD. Notice that we place the operator D after the constant since in the full equations in which they occur, for instance (6.62), the D or 1/D will be operating upon a variable, either i or v, situated to the right-hand side of the operator. This convention need not be strictly adhered to in the course of algebraic manipulation but it makes for clarity in the interpretation of the end formulation of the circuit differential equations. This point will become apparent in the following worked example.

6.7.4 Worked example

For the circuit of fig. 6.14 derive differential (D-operator) equations for the current $i(t)$ and the voltage $v(t)$. If $R_1 = R_2 = 2\,\text{M}\Omega$ and $C = 0.5\,\mu\text{F}$, find explicit expressions for the steady state components of $i(t)$

and $v(t)$ given that the driving function $v_1(t)$ is: (a) V_0(constant); (b) t; (c) e^{-t}; (d) te^{-t}.

Solution:
Let $Z(D)$ be the impedance of $R_2//(1/CD)$ then

$$Z(D) = \frac{R_2(1/CD)}{R_2 + 1/CD} = \frac{R_2}{1 + R_2 CD}$$

R_1 and $Z(D)$ form a voltage divider, hence,

$$v = \frac{Z(D)}{R_1 + Z(D)} v_1 = \frac{R_2}{(1 + R_2 CD)R_1 + R_2} v_1$$

or

$$v = \frac{R_2}{R_1 R_2 CD + R_1 + R_2} v_1$$

A differential equation for v is then

$$(R_1 R_2 CD + R_1 + R_2)v = R_2 v_1$$

or

$$\left[D + \frac{R_1 + R_2}{CR_1 R_2} \right] v = \frac{1}{CR_1} v_1$$

Current i and voltage v are related by $v = (1/CD)i$, hence substituting in the above expression gives a differential equation for i:

$$\left[D + \frac{R_1 + R_2}{CR_1 R_2} \right] i = \frac{D}{R_1} v_1$$

If $R_1 = R_2 = 2\,\mathrm{M}\Omega$ and $C = 0.5\,\mu\mathrm{F}$, the above equations for i and v reduce to:

$$(D+2)v = v_1 \quad \text{and} \quad (D+2)i = \frac{10^{-6}}{2} D v_1$$

Fig. 6.14. Circuit for worked example (section 6.7.4).

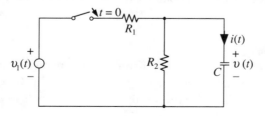

The D-operator

The steady state responses are then given by

$$v_{ss} = \frac{1}{D+2} v_1 \quad \text{and} \quad i_{ss} = \frac{10^{-6}}{2(D+2)} Dv_1$$

(a) $v_1 = V_0$

$$v_{ss} = \frac{1}{2(1+D/2)} V_0 = \tfrac{1}{2}\left(1+\frac{D}{2}\right)^{-1} V_0 = \tfrac{1}{2}\left(1-\frac{D}{2}\ldots\right) V_0 = \frac{V_0}{2}$$

$$i_{ss} = \frac{10^{-6}}{2(D+2)} DV_0 = 0 \quad [\text{Alternatively } i_{ss} = CDv_{ss} = 0]$$

(b) $v_1 = t$

$$v_{ss} = \tfrac{1}{2}\left(1-\frac{D}{2}\ldots\right)t = \tfrac{1}{2}\left(t-\frac{1}{2}\right) = \left(\frac{t}{2}-\frac{1}{4}\right)$$

$$i_{ss} = \frac{10^{-6}}{2(D+2)} Dt = \frac{10^{-6}}{4\left(1+\dfrac{D}{2}\right)} = \frac{10^{-6}}{4}\left(1+\frac{D}{2}\right)^{-1}$$

$$= \frac{10^{-6}}{4}\left(1-\frac{D}{2}\ldots\right) = \frac{10^{-6}}{4} \quad [\text{Alternatively } i_{ss} = CDv_{ss} = \frac{10^{-6}}{4}.]$$

(c) $v_1 = e^{-t}$

$$v_{ss} = \frac{1}{D+2} e^{-t} = \frac{1}{-1+2} e^{-t} = e^{-t}$$

$$i_{ss} = \frac{10^{-6}}{2(D+2)} De^{-t} = \frac{10^{-6}}{2(D+2)} (-e^{-t}) = -\frac{10^{-6}}{2} e^{-t}$$

(d) $v_1 = te^{-t}$

$$v_{ss} = \frac{1}{D+2} te^{-t} = e^{-t} \frac{1}{D-1+2} t = e^{-t}(1+D)^{-1} t = e^{-t}(1-D\ldots)t$$
$$= e^{-t}(t-1)$$

$$i_{ss} = \frac{10^{-6}}{2(D+2)} Dte^{-t} = \frac{10^{-6}}{2(D+2)} (e^{-t} - te^{-t})$$

$$= \frac{10^{-6}}{2} \frac{e^{-t}}{-1+2} - e^{-t} \frac{1}{D-1+2} t = \frac{10^{-6}}{2} e^{-t}(2-t)$$

Case (b) in this example may be used to illustrate an important point in connection with the response of circuits containing storage elements. For the component values given, the time constant of the circuit is $\frac{1}{2}$ second, and the complete solution for the voltage is

$$v = A\mathrm{e}^{-2t} + \frac{t}{2} - \frac{1}{4}$$

The initial condition is $v = 0$ at $t = 0^+$, therefore $A = \frac{1}{4}$ giving

$$v = \frac{1}{4}\mathrm{e}^{-2t} + \frac{t}{2} - \frac{1}{4} \tag{6.75}$$

The curve of v versus t, shown in fig. 6.15, is asymptotic to the line $\frac{t}{2} - \frac{1}{4}$ which is, of course, the steady state solution.

Without the capacitance the response of the circuit would simply be $\frac{t}{2}$ since $R_1 = R_2$. We see then that the addition of the capacitance has the effect of shifting the steady-state response bodily to the right by one time constant. This result is true in general; a circuit containing a single storage element will introduce a time delay between excitation and response equal

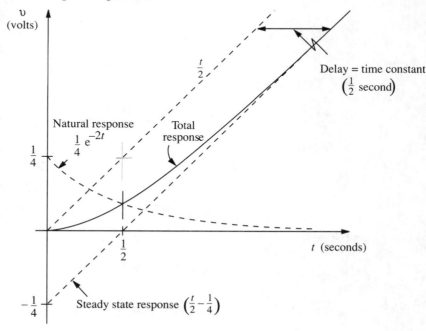

Fig. 6.15. Illustrating delay in the response of a circuit containing a single storage element.

6.7.5 Thévenin's theorem in transient analysis

We have seen that in the solution of the circuit differential equations the superposition theorem plays a central role; it allows us to find the natural and forced response separately which can be then added together to give the complete solution. The other linear circuit theorems discussed in chapter 2 are occasionally useful in the transient analysis of circuits, particularly Thévenin's theorem. The circuit of fig. 6.16(a) illustrates how this theorem can be used to simplify a circuit problem. In this circuit the capacitance is charged fully to the source voltage V_0 (constant); the switch is then closed at $t=0$. We wish to find the current $i(t)$ through R_2.

To apply Thévenin's theorem the circuit is broken at AA' in fig. 6.16(a) and the equivalent circuit to the left of AA' is found. The Thévenin equivalent e.m.f. (equal to the open-circuit voltage across AA') will be $V_0 R_2/(R_1+R_2)$, and the resistance looking into AA' (with V_0 reduced to a short circuit) is $R_1 R_2/(R_1+R_2)$. Hence, for the component values shown, the circuit of fig. 6.16(b) is obtained. The circuit is now reduced to a simple series form and it will be obvious that the steady-state value of v (voltage across AA') is $V_0/2$. With a circuit time constant of 2 seconds the voltage v is given by

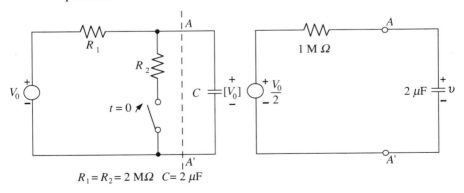

Fig. 6.16. The application of Thévenin's theorem to a transient problem.

(a) Original circuit   (b) Thévenin circuit

$$v = Ae^{-t/2} + \frac{V_0}{2}$$

We have not changed the circuit to the right of terminals AA' so the initial condition is still $v = V_0$ at $t = 0^+$, which gives $A = V_0/2$, hence

$$v = \frac{V_0}{2}(e^{-t/2} + 1)$$

Now the voltage v, which is that across C, is unchanged between circuits (*a*) and (*b*) in fig. 6.16 so that the current through R_2 will be v/R_2 giving finally,

$$i = \frac{V_0}{2 \times 10^6}(e^{-t/2} + 1)$$

The circuits of fig. 6.14 and fig. 6.16 are similar; the reader should compare the above approach with that adopted in section 6.14.

Thévenin's theorem is also useful if we wish to determine the effect of some modification to a circuit upon which an analysis has already been carried out. For example, suppose an additional resistance R_3 is switched into the circuit of fig. 6.14 at some instant $t = t_1$, as shown in fig. 6.17(*a*). We wish to find the voltage v across the circuit for $t \geqslant t_1$. With both switches closed, the circuit becomes as shown in fig. 6.17(*b*). To apply Thévenin's theorem the circuit is broken at AA' and the equivalent circuit to the left of AA' is found. The Thévenin equivalent e.m.f. e_T, that is, the open circuit voltage, is given by the solution (6.75) previously obtained for the unmodified circuit:

$$e_T = \frac{1}{4}e^{-2t} + \frac{t}{2} - \frac{1}{4}$$

The equivalent impedance, in terms of the D-operator notation, is

$$Z_T = \frac{R(1/CD)}{R + 1/CD} = \frac{R}{RCD + 1}$$

where $R = R_1 // R_2$.

The circuit is thus reduced to the form shown in fig. 6.17(*c*) and the voltage is given by

$$v = \frac{R_3}{R_3 + Z_T} e_T = \frac{RCD + 1}{RCD + 1 + R/R_3} e_T$$

For the component values shown the differential equation for v is

$$(D + 3)v = (D + 2)e_T = (D + 2)\left(\frac{1}{4}e^{-2t} + \frac{t}{2} - \frac{1}{4}\right) \tag{6.76}$$

The D-operator

Performing the operation on the RHS we obtain

$$(D+3)v = t$$

The steady-state solution is given by

$$v = \frac{1}{D+3}t = \frac{1}{3}\left(1+\frac{D}{3}\right)^{-1}t = \frac{1}{3}\left(1-\frac{D}{3}\cdots\right)t$$

or

$$v = \frac{t}{3} - \frac{1}{9}$$

We are here interested in the transient conditions after closing the switch S_2, that is, for $t > t_1$. Clearly, from the LHS of (6.76), the effective time constant is $\frac{1}{3}$ second so the complete solution may be written

Fig. 6.17. Application of Thévenin's theorem to a double switching problem.

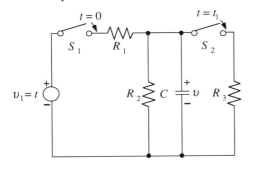

(a)

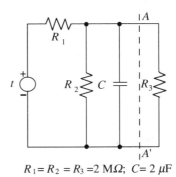

$R_1 = R_2 = R_3 = 2\ M\Omega;\ C = 2\ \mu F$

(b)

(c)

$$v = Ae^{-3(t-t_1)} + \frac{t}{3} - \frac{1}{9} \quad (t \geq t_1)$$

The constant A is found from the condition of v at the instant t_1

$$v|_{t=t_1} = \frac{1}{4}e^{-2t_1} + \frac{t_1}{2} - \frac{1}{4} = A + \frac{t_1}{3} - \frac{1}{9}$$

hence

$$A = \frac{1}{4}e^{-2t_1} + \frac{t_1}{6} - \frac{5}{36}$$

and

$$v = \left(\frac{1}{4}e^{-2t_1} + \frac{t_1}{6} - \frac{5}{36}\right)e^{-3(t-t_1)} + \frac{t}{3} - \frac{1}{9} \quad (t \geq t_1)$$

For this particular example the use of Thévenin's theorem does not effect a great saving in the amount of algebraic manipulation involved. This is because it would be particularly simple in this instance to incorporate the additional resistance R_3 within the formulation of the original circuit equation. However, for more complicated situations the Thévenin approach can offer significant advantages.

It may be remarked, finally, that for the simple circuits discussed in this and previous sections containing a single source, mesh analysis is *not* an efficient approach to the formulation of the circuit differential equations. The reader may care to consider, for example, the use of mesh analysis (rather than the 'voltage divider' approach) for the worked example of section 6.7.4 (fig. 6.14).

6.7.6 Differentiating and integrating circuits

The RC circuits of fig. 6.18 are frequently used to perform simple signal differentiation and integration. Consider the circuit of fig. 6.18(a); with voltage v_1 applied at its input. The output v_2 is

$$v_2 = \frac{R}{R + 1/CD} v_1 = \frac{RCD}{RCD + 1} v_1$$

or

$$(RCD + 1)v_2 = RCDv_1$$

For RC sufficiently small ($RCDv_2 \ll v_2$) we may write:

$$v_2 \simeq RCDv_1 \simeq RC\frac{d}{dt}(v_1)$$

Thus, the output is approximately equal to the derivative of the input. Note that the accuracy of differentiation will depend both on the magnitude of RC and upon v_1 and its rate of change, which implies that the accuracy is signal dependent.

For the circuit of fig. 6.18(b) we may show that, if RC is sufficiently large,

$$v_2 \simeq \frac{1}{CR}\int v_1 dt$$

Because of the requirement in the integrating circuit that CR should be large, the output is generally small, and for this reason the circuit is usually used in conjunction with an active device which amplifies the signal and improves the accuracy of integration by ensuring that the capacitance receives, effectively, a constant charging current through the resistance (see page 131 of reference 5).

6.8 The unit step and related driving functions

In this section we introduce the concept of the *step* function and its relatives, the *impulse* function and the *ramp* function. These are members of a class of functions, called *singularity* functions, that are of fundamental importance to the development of more advanced aspects of circuit theory. The singularity functions allow us to describe the behaviour of circuits subject to driving waveforms of arbitrary shape and of a discontinuous nature. Examples of the latter have already been encountered in which the action of a switch impresses a driving voltage on a circuit at the instant $t=0$. It is convenient to introduce the step function as a mathematical description of this discontinuous process although, as will be seen later, the concepts embodied in the step and its related functions extend far beyond this simple application.

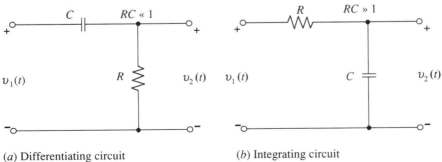

Fig. 6.18. RC circuits used for signal differentiation and integration.

(a) Differentiating circuit

(b) Integrating circuit

6.8.1 Step function

The unit step function (fig. 6.19) is defined by

$$u(t) = 1 \quad t \geq 0 \atop = 0 \quad t < 0 \} \tag{6.77}$$

It should be noted that according to this definition the function is zero at $t = 0^-$ and unity at both $t = 0$ and $t = 0^+$.

Fig. 6.20 illustrates the way in which the step function may be used to describe the action of a switch. The voltage source V and switch S of fig. 6.20(a) are replaced by an ideal voltage source, permanently connected to the circuit as shown in fig. 6.20(b), producing a driving voltage:

$$v(t) = Vu(t) \tag{6.78}$$

This expression signifies that for $t < 0$ the voltage impressed on the circuit is zero, for $t \geq 0$ the voltage is V. It will be appreciated that the representation shown in fig. 6.20 refers to an ideal switch; that is, one which has infinite resistance before closure, zero resistance after closure, and for which the transition between these states is of infinitesimally short duration. An absence of inductive and capacitive effects is also implied.

Fig. 6.19. The unit step function.

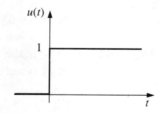

Fig. 6.20. Representation of switching action by means of the step function. (a) Original circuit with voltage source and switch. (b) Representation of circuit (a) by step-function source.

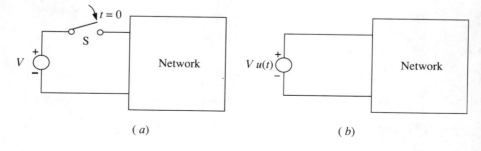

The representation of fig. 6.20 also assumes that the circuit is in the zero energy state at $t=0$; otherwise the voltage across the terminals to which V is applied may not be zero at $t=0^-$, thus invalidating the definition (6.77).

A current source switched into a circuit at $t=0$ can be represented in like manner by the unit step function. If I is the amplitude of the constant current source, then the switching action may be represented by:

$$i(t) = Iu(t) \qquad (6.79)$$

Voltages or currents which arise in a circuit subject to a unit step driving function are called the *step response*.

6.8.2 Impulse function

The unit impulse function (also known as the *Dirac* or *delta* function) is the derivative of the step function, and is defined by

$$\delta(t) = \frac{d}{dt}[u(t)] = u'(t) \qquad (6.80)$$

Now the slope of the step function is infinite at $t=0$, in other words the function is not, in the usual mathematical sense, differentiable at this point and is therefore singular. However, the meaning of (6.80) will become clear if we consider the approximate step function $g(t)$ shown in fig. 6.21(a). This is zero at $t=0$ and rises linearly to unit amplitude at $t=\Delta$. The derivative of $g(t)$ is the rectangular pulse shown in fig. 6.21(b). We see that as Δ is made smaller and allowed to approach zero, $g(t) \to u(t)$ (fig. 6.21(c)) while the

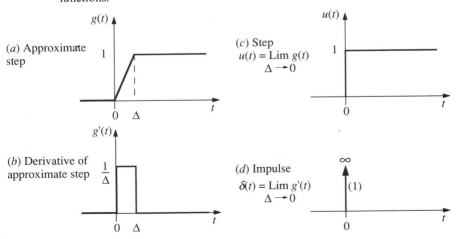

Fig. 6.21. Illustrating relationship between unit step and unit impulse functions.

(a) Approximate step

(b) Derivative of approximate step

(c) Step
$u(t) = \underset{\Delta \to 0}{\text{Lim}}\, g(t)$

(d) Impulse
$\delta(t) = \underset{\Delta \to 0}{\text{Lim}}\, g'(t)$

amplitude of its derivative $1/\Delta \to \infty$ (fig. 6.21(d)). However, although $g'(t)$ becomes infinite for $\Delta \to 0$, the area under the derivative curve remains finite and independent of Δ since

$$\frac{1}{\Delta} \times \Delta = 1$$

So, the unit impulse function is infinite at $t=0$ and zero elsewhere, while its area is unity. The symbolism of fig. 6.21(d) is used to indicate these properties of the unit impulse.

Since the impulse is the derivative of the step, the step must be the integral of the impulse. That this is so will be appreciated if we consider the area under the impulse function shown in fig. 6.21(d). To the left of the origin the function is zero so there is no contribution to the area. As we pass from $t=0^-$ through to $t=0^+$ we include unit area and the integral jumps to unity. There is no further contribution to the right of the origin so the value of the integral remains at unity. From this point of view the unit impulse may be defined by:

$$\int_{-\infty}^{\infty} \delta(t)\,\mathrm{d}t = 1$$

$$\delta(t) = 0, t \neq 0 \qquad (6.81)$$

It follows from the definitions of the unit step and unit impulse functions, that the derivative of a step of amplitude E is an impulse of area E; that is

$$\frac{\mathrm{d}}{\mathrm{d}t}[Eu(t)] = E\delta(t) \qquad (6.82)$$

The impulse function of voltage has dimensions of volt seconds; the impulse function of current has dimensions of amp seconds (coulombs). Voltages or currents which arise in a circuit subject to a unit impulse driving function are called the *impulse response*.

The importance of the impulse function lies in the fact that it can be used to represent functions and transforms of widely differing form. This will emerge fully when we deal with the theory of the convolution integral in the final section of this chapter. For the present we establish the circumstances under which the impulse function may be used to represent a single, short pulse of arbitrary shape. We begin by considering the response of a simple RC circuit (fig. 6.22) to: (a) a rectangular pulse, of duration Δ and unit area; (b) a unit impulse.

To find the response to the pulse we make use of the results obtained in section 6.6.2. There we found that the response of an RC circuit to an impressed voltage V was

The unit step and related driving functions

$$v_2(t) = V(1 - e^{-t/RC}) \tag{6.41}$$

For the case considered here the pulse has amplitude $1/\Delta$ so, putting $V = 1/\Delta$ in (6.41) we obtain

$$v_2(t)_{\text{pulse}} = \frac{1}{\Delta}(1 - e^{-t/\tau}) \qquad 0 < t < \Delta \tag{6.83}$$

where $\tau = RC$.

We may make use of (6.41) also to find the response to the unit impulse (see also section 6.9.6). This equation, with $V = u(t)$, gives the response to unit step; by differentiating this we obtain the response to unit impulse since the impulse is the derivative of the step. We have

$$v_2(t)_{\text{step}} = u(t)(1 - e^{-t/\tau}) \tag{6.84}$$

hence,

$$v_2(t)_{\text{impulse}} = \frac{d}{dt}[u(t)(1 - e^{-t/\tau})]$$

$$= \frac{d}{dt}u(t) - \frac{d}{dt}[u(t)e^{-t/\tau}]$$

$$= \delta(t) - u(t)\left(\frac{-e^{-t/\tau}}{\tau}\right) - e^{-t/\tau}\delta(t)$$

$$= (1 - e^{-t/\tau})\delta(t) + \frac{e^{-t/\tau}}{\tau}u(t)$$

Fig. 6.22. Response of an *RC* circuit to: (*a*) pulse input; (*b*) impulse input.

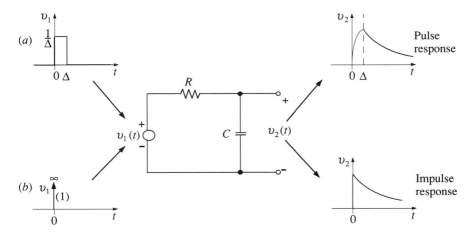

But $\delta(t)$ is zero for all $t \neq 0$, and at $t=0$ the coefficient of $\delta(t)$ is zero, so the first term in the expression vanishes to give

$$v_2(t)_{\text{impulse}} = \frac{1}{\tau} e^{-t/\tau} u(t) \tag{6.85}$$

In this expression $u(t)$ is to be regarded as a multiplying factor indicating that the response is zero for $t<0$ and $(1/\tau)e^{-t/\tau}$ for $t \geq 0$.

Now let us compare the response of the circuit to the two inputs for times equal to or greater than the duration of the pulse, that is for $t \geq \Delta$. For the pulse input the value of the output voltage at $t = \Delta$ will, according to (6.83), be

$$v_2(\Delta)_{\text{pulse}} = \frac{1}{\Delta}(1 - e^{-\Delta/\tau})$$

If it is assumed that the duration of the pulse is small in relation to the time constant of the circuit, this expression may be approximated by

$$v_2(\Delta)_{\text{pulse}} \simeq \frac{1}{\Delta}\left(1 - 1 + \frac{\Delta}{\tau} - \frac{1}{2}\left(\frac{\Delta}{\tau}\right)^2 + \ldots\right)$$

$$\simeq \frac{1}{\Delta}\left(\frac{\Delta}{\tau} - \frac{1}{2}\left(\frac{\Delta}{\tau}\right)^2 + \ldots\right)$$

$$v_2(\Delta)_{\text{pulse}} \simeq \frac{1}{\tau}\left(1 - \frac{\Delta}{2\tau} + \ldots\right) \tag{6.86}$$

For the impulse input the output at $t = \Delta$ is, by (6.85),

$$v_2(\Delta)_{\text{impulse}} = \frac{1}{\tau} e^{-\Delta/\tau}$$

which may be approximated by

$$v_2(\Delta)_{\text{impulse}} \simeq \frac{1}{\tau}\left(1 - \frac{\Delta}{\tau} - \frac{1}{2}\left(\frac{\Delta}{\tau}\right)^2 - \ldots\right)$$

$$\simeq \frac{1}{\tau}\left(1 - \frac{\Delta}{\tau} + \ldots\right) \tag{6.87}$$

Comparing (6.86) and (6.87), and neglecting second and higher order terms, we see that the difference ε between the impulse response and the pulse response (at $t = \Delta$) is

$$\varepsilon = \frac{1}{2}\frac{\Delta}{\tau} \tag{6.88}$$

The unit step and related driving functions

If, for example, the time constant of the circuit is a factor of ten greater than the pulse duration, then $\varepsilon = 0.05$. For $t > \Delta$ the difference between pulse and impulse responses will be less than that given by (6.88) since both response curves decay exponentially, with the same time constant, to zero and will therefore converge.

We may conclude that it is possible to predict the response of a simple RC circuit to a rectangular pulse of unit area (for times greater than the pulse duration) by determining the response to the unit impulse function. The shorter the pulse duration in relation to the circuit time constant the better the accuracy of prediction. It also follows that the response to a pulse of area A may be found from the response to an impulse of magnitude A. Provided the condition $\tau > \Delta$ is fulfilled, the ratio of pulse height to pulse width is immaterial. Indeed, the pulse may be of any shape since the magnitude of the equivalent impulse function depends only upon the area of the pulse. Consequently, we are able to determine the response of a circuit to a pulse of arbitrary shape (including pulses which cannot be expressed analytically) simply by finding the area enclosed by the pulse, and then determining the response of the circuit to the equivalent impulse function.

The foregoing argument has been developed on the basis of the simple RC circuit, but the same general conclusions are found to be true for any first order circuit. The conclusions are also valid for higher order circuits, but in such cases it is the shortest effective time constant of the particular circuit which must be used as the criterion.

6.8.3 Worked example

A photomultiplier tube, used in a scintillation counting system, produces at its output pulses of current of the form shown in fig. 6.23(*a*). The photomultiplier is connected to an amplifier whose input circuit can be modelled by a resistance of 1 MΩ in parallel with a capacitance of 30 pF. Estimate the form of the voltage response at the input of the amplifier subsequent to the arrival of a single pulse.

Solution: The photomultiplier can be regarded as having an infinite output resistance so that it behaves essentially as an ideal current source. The circuit model is therefore as shown in fig. 6.23(*b*) where $i(t)$ is of the form shown in fig. 6.23(*a*).

The time constant of the circuit is $RC = 10^6 \times 30 \times 10^{-12} = 30\,\mu\text{s}$ whereas the pulse duration is approximately one microsecond; consequently the pulse may be replaced by an impulse function at the origin.

The differential equation relating $v(t)$ and $i(t)$ is

$$v(t) = \frac{R(1/CD)}{R+(1/CD)} i(t) \quad \text{or} \quad \left(D + \frac{1}{RC}\right) v(t) = \frac{1}{C} i(t)$$

Putting $i(t)$ equal to the unit step function of current $u(t)$,

$$\left(D + \frac{1}{RC}\right) v_{step} = \frac{1}{C} u(t)$$

where v_{step} is the step response.

The steady state response to unit constant current will clearly be $v_{ss} = R$, and the natural response will be $v_n = Ae^{-t/RC}$, hence

$$v_{step} = v_n + v_{ss} = Ae^{-t/RC} + R$$

The initial condition is $v=0$ at $t=0^+$ giving $A = -R$, so;

$$v_{step} = R(1 - e^{-t/RC})$$

The impulse response is obtained by differentiating the step response. Using a procedure similar to that leading to (6.85) we find

$$v_{impulse} = \frac{1}{C} e^{-t/RC} u(t)$$

This is the response to *unit* impulse; the response to an impulse of magnitude Q will be $(Q/C)e^{-t/RC}$. With $Q = 2.3 \times 10^{-11}$ coulomb, the form of the voltage response at the input of the amplifier is

$$v = 0.76 e^{-t/30} \times 10^{-6}$$

Physically, we may interpret the result in the following way. During the pulse, lasting for about 1 μs, a charge of 2.3×10^{-11} C is delivered to the capacitance causing the voltage to rise to 0.76 V. The capacitance then discharges and the voltage decays with a time constant of 30 μs.

Fig. 6.23. Diagrams for worked example (section 6.8.3).

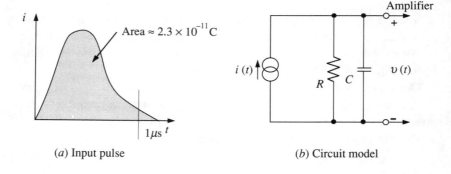

(a) Input pulse (b) Circuit model

6.8.4 Ramp and other singularity functions

The *ramp function* is the integral of the step function and is defined by:

$$\rho(t) = \int_{-\infty}^{t} u(t)\,dt = \int_{0}^{t} dt \qquad (6.89)$$

or

$$\left.\begin{array}{ll}\rho(t) = t & t \geq 0 \\ = 0 & t < 0\end{array}\right\} \qquad (6.90)$$

The integral of a step function of amplitude E corresponds to a ramp function of slope E, that is,

$$\int_{-\infty}^{\tau} Eu(t)\,dt = E\rho(t) \qquad (6.91)$$

Other singularity functions, useful in more advanced network analysis, are obtained by further differentiation or integration of the basic impulse, step, and ramp functions. For example, differentiation of the unit impulse function produces the *unit doublet* consisting of positive and negative going unit impulses at the origin. Integration of the ramp function produces the *unit parabola*.

The relationships among all of the functions mentioned in this section are shown in fig. 6.24. In this book only the impulse, step and ramp functions will be considered further.

Fig. 6.24. Relationships among the unit singularity functions.

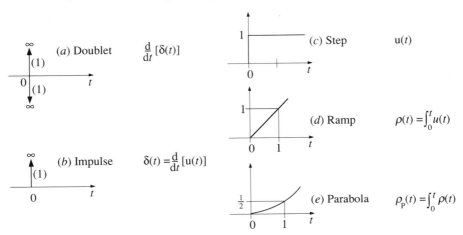

6.8.5 Delayed functions

The singularity functions considered so far have a point of discontinuity at time zero; however, for some purposes it is useful to extend the concept to embrace functions having a discontinuity at some other, positive, value of time. This may be accomplished simply by changing the argument of the function. Thus, if the unit step function is defined by

$$u(t-a) = 0 \quad (t-a) < 0 \\ = 1 \quad (t-a) \geq 0 \quad \quad (6.92)$$

Fig. 6.25. Delayed singularity functions.

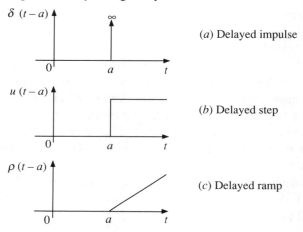

(a) Delayed impulse

(b) Delayed step

(c) Delayed ramp

Fig. 6.26. Use of delayed step function to provide time sectioning and time delay of a signal waveform.

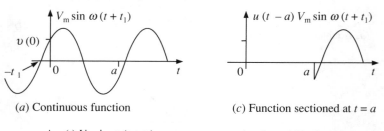

(a) Continuous function

(c) Function sectioned at $t = a$

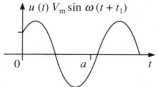

(b) Function sectioned at $t = 0$

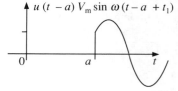

(d) Function delayed by $t = a$

the step will be delayed until $t=a$. The impulse and ramp functions may be treated similarly, as shown in fig. 6.25.

The delayed unit step function allows one to specify analytically the time at which a function commences; a process which is sometimes called *sectioning* of the function. For example, the sinusoid $v = V_m \sin\omega(t+t_1)$ shown in fig. 6.26(*a*) is continuous for all negative and positive time. Multiplication by $u(t)$ sections the function at the origin (fig. 6.26(*b*)) while multiplication by $u(t-a)$ sections the function at $t=a$ (fig. 6.86(*c*)).

Changing the argument of a function $f(t)$ to $f(t-a)$ and multiplying by $u(t-a)$ shifts the original function bodily along the time axis so that it commences at $t=a$ with the same value v_0 that it had at the origin. The result of this operation on the sinusoid is illustrated in fig. 6.26(*d*).

6.9 The Laplace transform

The method of analysis now to be described employs the Laplace transform by means of which functions of time are transformed into functions of a new variable s in such a way that what was initially a differential equation becomes an algebraic equation. For the functions of time normally encountered in linear circuit applications, these transformations are unique so that to each function of t there corresponds a function of s and, conversely, to each function of s there corresponds a function of t. Therefore, from the algebraic equation(s) we obtain a function of s that may, by the inverse Laplace transform, be converted to a function of t. This new function is the solution of the original differential equation.

The Laplace transform has applications to situations such as we have considered in this chapter where a known driving function is applied at $t=0$ to a circuit and it is desired to find the response of the circuit for all $t>0$, having given the conditions in the circuit at $t=0$ (the initial conditions). The method has several features in common with the D-operator approach to circuit analysis, in particular the method by which the circuit equations are set up and manipulated is essentially the same, however, the greater mathematical generality of the Laplace transform allows it to be used for a wider variety and range of problems. An important advantage of the method is that initial conditions are included automatically in the transformed circuit equations; a disadvantage is that it can involve a formidable amount of algebraic manipulation, and it sometimes tends to obscure the underlying physics of the problem under consideration.

The theory of the Laplace transform may be developed in relation to the Fourier series and Fourier integral which are used, respectively, in the representation of non-sinusoidal periodic functions and of pulses and other functions of finite duration. In our approach we shall simply define the Laplace transform; calculate the transforms for some functions of time that are commonly encountered, and show how the results may be applied to the solution of some specific problems. We shall not go into the question of the conditions that a function must satisfy in order that a transform should exist; we assume (with justification) that all the functions we use meet these criteria.

6.9.1 Definition of the Laplace transform

The Laplace transform $F(s)$ of a function $f(t)$ is defined by*

$$F(s) = \mathscr{L}\{f(t)\} = \int_0^\infty f(t) e^{-st}\, dt \qquad (6.93)$$

The symbol $\mathscr{L}\{f(t)\}$ is to be read: 'the Laplace transform of $f(t)$'. The variable s, which has dimensions of angular frequency, may be real or complex and is usually expressed by

$$s = \sigma + j\omega \qquad (6.94)$$

σ must be positive and sufficiently large to ensure that the integral converges. For the functions of t that we shall be concerned with, this condition is satisfied.

The inverse Laplace transform is defined by

$$f(t) = \mathscr{L}^{-1}\{F(s)\} = \frac{1}{2\pi j} \int_{\sigma-j\infty}^{\sigma+j\infty} F(s) e^{st}\, ds \qquad (6.95)$$

where the symbol $\mathscr{L}^{-1}$ indicates the process of finding the inverse of the function $F(s)$.

In the application of the Laplace transform to practical circuit problems, the integrals (6.93) and (6.95) are rarely used directly. The integral (6.93) has been evaluated for a large number of functions and one refers to tables of Laplace transform pairs to effect the appropriate transforms, both forward and inverse. We shall not, therefore, concern ourselves further with (6.95). A short table of transform pairs is given in Appendix D; we indicate below how some of the more useful of the entries in this table are derived.

6.9.2 Laplace transforms for some functions of time

Note that pair numbers given below refer to the table of Laplace transform pairs in Appendix D.

(1) $f(t) = t^n$

---

* The lower limit in the integral (6.93) is zero, and the definition used here is called the *one-sided* Laplace transform, which is applicable to functions that are zero for $t < 0$. The *two-sided* Laplace transform, which we do not deal with in this book, is defined by a similar integral but with lower limit $-\infty$.

The Laplace transform

Consider first $f(t) = t$ then using (6.93)

$$F(s) = \int_0^\infty t e^{-st} \, dt$$

Integrate by parts: let $t = u$ and $e^{-st} \, dt = dv$, so $du = dt$ and $v = -\frac{1}{s} e^{-st}$, then

$$\int u \, dv = uv - \int v \, du = \left[-\frac{t}{s} e^{-st} \right]_0^\infty + \int_0^\infty \frac{1}{s} e^{-st} \, dt = 0 + \frac{1}{s^2}$$

So,

$$\mathscr{L}\{t\} = \frac{1}{s^2}$$

Repeated application of integration by parts gives the general result:

$$\mathscr{L}\{t^n\} = \frac{n!}{s^{(n+1)}} \qquad (6.96)$$
[Pair No. 1]

(2) $f(t) = e^{at}$

$$F(s) = \int_0^\infty e^{at} e^{-st} \, dt = \int_0^\infty e^{-(s-a)t} \, dt = \frac{1}{s-a}$$

For the integral to converge, $s > a$. In the applications considered here, a is negative, so this inequality obtains. Then

$$\mathscr{L}(e^{at}) = \frac{1}{s-a}$$

and

$$\mathscr{L}(e^{-at}) = \frac{1}{s+a} \qquad (6.97)$$
[Pair No. 2]

(3) $f(t) = \sin \omega t$.

This is calculated conveniently by using the identity

$$\sin \omega t = \frac{1}{2j} (e^{j\omega t} - e^{-j\omega t})$$

together with the result (6.97)

$$F(s) = \frac{1}{2j} \int_0^\infty [e^{-(s-j\omega)t} \, dt - e^{-(s+j\omega)t} dt]$$

$$= \frac{1}{2j} \left[\frac{1}{s-j\omega} - \frac{1}{s+j\omega} \right]$$

So,
$$\mathcal{L}\{\sin\omega t\} = \frac{\omega}{s^2+\omega^2} \qquad (6.98)$$
[Pair No. 6]

(4) $f(t) = \cos\omega t$

Again express the function as the sum of exponentials and use the result (6.97)

$$\mathcal{L}\{\cos\omega t\} = \frac{s}{s^2+\omega^2} \qquad (6.99)$$
[Pair No. 8]

(5) $f(t) = c$ (constant)

$$F(s) = \int_0^\infty c e^{-st}\,dt = c\left[\frac{-e^{-st}}{s}\right]_0^\infty = \frac{c}{s}$$

So,
$$\mathcal{L}\{c\} = \frac{c}{s} \qquad (6.100)$$
[Pair No. 15]

(6) $f(t) = u(t)$ (unit step)

Since $u(t) = 1$ for $t \geq 0$ we may use the result (6.100)

$$\mathcal{L}\{u(t)\} = \frac{1}{s} \qquad (6.101)$$
[Pair No. 16]

(7) Transforms of $\dfrac{d}{dt}f(t);\quad \dfrac{d^2}{dt^2}f(t);\quad \dfrac{d^n}{dt^n}f(t)$

So far we have simply shown how, by a mathematical manipulation, it is possible to obtain a function of s that corresponds to some given function of t. The use of the transform in circuit problems requires that we also obtain transforms of derivatives of functions of t. To find these we must know the values of $f(t)$ and its derivatives at $t = 0^+$. Let

$$f(0^+) = f_0;\quad \frac{d}{dt}f(0^+) = f_1;\quad \frac{d^2}{dt^2}f(0^+) = f_2 \ldots \text{etc.}$$

Again using (6.93)

$$\mathcal{L}\left\{\frac{d}{dt}f(t)\right\} = \int_0^\infty \frac{df}{dt}e^{-st}\,dt$$

Integrate by parts: $e^{-st} = u$ and $\dfrac{df}{dt}dt = dv$; hence,

$$du = -se^{-st}\,dt \text{ and } v = f.$$

$$\mathcal{L}\left\{\frac{d}{dt}f(t)\right\} = [fe^{-st}]_0^\infty + s\int_0^\infty fe^{-st}\,dt$$

The Laplace transform

The first term on the right is $(-f_0)$ and the second term is just s times the Laplace transform of $f(t)$, therefore,

$$\mathscr{L}\left\{\frac{d}{dt}f(t)\right\} = sF(s) - f_0 \qquad (6.102)$$
[Pair No. 24]

So, to write the transform of the derivative of a function we write s times the transform of the function and subtract the value of the function at $t=0^+$.

The voltage–current relationship for inductance provides an important example of the use of this operational transform pair. We have

$$v = L\frac{di}{dt}$$

which, upon transforming both sides, becomes

$$V(s) = L(sI(s) - i_0) \qquad (6.103)$$

where $i_0 (\equiv i(0^+))$ is the initial value of the current in the inductance.

The transform of the second derivative of a function is also found by integrating by parts. With $e^{-st} = u$ and $\frac{d^2f}{dt^2}dt = dv$, then

$$\mathscr{L}\left\{\frac{d^2}{dt^2}f(t)\right\} = \left[\frac{df}{dt}e^{-st}\right]_0^\infty + s\int_0^\infty \frac{df}{dt}e^{-st}\,dt$$

When limits are substituted, the first term becomes $\frac{-df(0^+)}{dt} = -f_1$. The second term is s times the transform of $\frac{df}{dt}$, which we have just found. So,

$$\mathscr{L}\left\{\frac{d^2}{dt^2}f(t)\right\} = s^2 F(s) - sf_0 - f_1 \qquad (6.104)$$
[Pair No. 25]

Generalizing the above results we obtain:

$$\mathscr{L}\left\{\frac{d^n}{dt^n}f(t)\right\} = s^n F(s) - s^{n-1}f_0 - s^{n-2}f_1 - \ldots - f_{n-1} \qquad (6.105)$$
[Pair No. 26]

(8) Transforms of $\int f(t)\,dt$ and $\int_0^t f(t)\,dt$

$$\mathscr{L}\left\{\int f(t)\,dt\right\} = \int_0^\infty \left[\int f(t)\,dt\right] e^{-st}\,dt$$

Integrating by parts with $\int f\,dt = u$ and $e^{-st}\,dt = dv$ we obtain

$$\int u\,dv = uv - \int v\,du = \left[\left\{\int f\,dt\right\}\frac{e^{-st}}{-s}\right]_0^\infty - \int_0^\infty f\frac{e^{-st}}{-s}\,dt$$

The second term is simply $(1/s)F(s)$, and when limits are substituted in the first term we have

$$-\frac{1}{s}\left\{\left[\int f\,dt\right]e^{-\infty} - \left[\int f\,dt\right]_{t=0}\right\} = \frac{1}{s}\left[\int f\,dt\right]_{t=0}$$

But this is just $(1/s)$ times the value of the integral at the moment of the switching operation and therefore represents the initial condition. The transform is often written

$$\mathscr{L}\left\{\int f(t)\,dt\right\} = \frac{1}{s}F(s) + \frac{1}{s}f^{-1}(0^+) \qquad (6.106)$$
[Pair No. 27]

where $f^{-1}(0^+)$ represents the value of the integral at $t=0^+$.

The above expression gives the transform of the indefinite integral; the transform of the definite integral is obtained as follows:

$$\int_0^t f\,dt = \int f\,dt - \left[\int f\,dt\right]_{t=0} = \int f\,dt - f^{-1}(0^+)$$

Transforming term by term we have

$$\mathscr{L}\left\{\int_0^t f\,dt\right\} = \mathscr{L}\left\{\int f\,dt\right\} - \mathscr{L}\{f^{-1}(0^+)\}$$

The first term is the transform of the indefinite integral, which is given by (6.106). The second term represents the transform of a constant which is obtained from (6.100). Hence,

$$\left\{\int_0^t f\,dt\right\} = \frac{1}{s}F(s) + \frac{1}{s}f^{-1}(0^+) - \frac{1}{s}f^{-1}(0^+)$$

So the transform of the definite integral is given by:

$$\mathscr{L}\left\{\int_0^t f(t)\,dt\right\} = \frac{1}{s}F(s) \qquad (6.107)$$
[Pair No. 28]

Using this expression we may find the transform of the voltage-current relationship for capacitance with an initial voltage v_0 (Table 1.1 equation (1.31)),

$$v = \frac{1}{C}\int_0^t i\,dt + v_0$$

Transforming term by term we obtain

$$V(s) = \frac{1}{C}\frac{I(s)}{s} + \frac{v_0}{s} \qquad (6.108)$$

The same result may be obtained directly from (6.106) by noting that the initial value of the integral, $f^{-1}(0^+)$, is just the charge on the capacitance ($q_0 = Cv_0$).

As an example of the use of the transform relationships derived above, let us again consider the RL circuit of fig. 6.1. With the switch closed at $t=0$ the circuit equation for fig. 6.1(a) is

$$L\frac{di}{dt} + Ri = Vu(t)$$

where $u(t)$ indicates that the constant voltage V is switched into the circuit at $t=0$. Using (6.102) we find that the first term transforms to $L(sI(s) - i_0) = LsI(s)$ since i_0, the initial current, is zero. The second term transforms to $RI(s)$; and the term on the right, using (6.101), transforms to V/s. The complete transform equation is then

$$LsI(s) + RI(s) = \frac{V}{s}$$

or

$$I(s) = \frac{V}{s}\frac{1}{Ls+R} = \frac{V}{L}\left[\frac{1}{s(s+R/L)}\right] \quad (6.109)$$

Now referring to our table of transform pairs (Appendix D), we see that transform pair number 3 allows us to find the inverse of the expression in brackets directly, that is,

$$i(t) = \frac{V}{L} \cdot \frac{L}{R}(1 - e^{-Rt/L}) = \frac{V}{R}(1 - e^{-Rt/L})$$

This result is identical to (6.3).

In the case of the sinusoidal driving voltage (fig. 6.1(b)), the circuit equation is

$$L\frac{di}{dt} + Ri = V_m \sin\omega t \, u(t)$$

which upon transformation becomes

$$LsI(s) + RI(s) = \frac{V_m \omega}{s^2 + \omega^2}$$

Here we have used (6.98) to transform the sinusoidal driving voltage. Proceeding as before:

$$I(s) = \frac{V_m \omega}{L}\left[\frac{1}{(s+R/L)(s^2+\omega^2)}\right] \quad (6.110)$$

Now we refer to our table of transform pairs but we discover that in this case the appropriate form of the function in brackets does not appear. More extensive tables are available (see for example reference 9), and such tables would include the function we require. However, it will be instructive to consider how a table containing a relatively small number of transform pairs may be extended to allow the inversion of functions such as (6.110); this is the subject of the following section.

6.9.3 Partial fractions

Because we are dealing with linear circuit elements we may use the following property of the Laplace transform in our calculations.

$$\mathscr{L}\{C_1 f_1(t) + C_2 f_2(t)\} = C_1 \mathscr{L}\{f_1(t)\} + C_2 \mathscr{L}\{f_2(t)\} \qquad (6.111)$$

This equation means that if we have a function $F(s)$ for which the corresponding $f(t)$ is required, we may express $F(s)$ as the sum of several terms, find the inverse transform of each separately, and add the resulting functions of time to get the complete solution of the original differential equation. In order to express $F(s)$ as the sum of several terms we can often make use of the method of partial fractions.

The functions of s that we are concerned with in this text are in general *rational* functions, that is, they can be expressed as the ratio of two polynomials:

$$F(s) = \frac{N(s)}{D(s)} \qquad (6.112)$$

We assume that $F(s)$ is a proper fraction (numerator of a lower order in s than denominator).*

Now the fundamental theorem of algebra states that any polynomial in s with real coefficients may be expressed as the product of factors of one or both of the following types:
  (a) linear factors of the form, $as + b$
  (b) irreducible quadratic factors of the form $cx^2 + dx + e$, which does not have real, linear factors.

The coefficients a, b, c, d, e are real. If, therefore, $F(s)$ is the ratio of two polynomials, one may factorise the denominator. By the method of partial

---

* An improper fraction may be reduced by division to a form consisting of a polynomial plus a proper fraction. For example,

$$\frac{X^4 + 3X^2 + 2}{X^2 - 2X} = X^2 + 2X + 7 + \frac{14X + 2}{X^2 - 2X}$$

However, for all the problems that will concern us, $F(s)$ is a proper fraction.

The Laplace transform

fractions the original expression for $F(s)$ may then be replaced by the sum of a series of fractions whose denominators are found from the factors of the denominator of $F(s)$. The appropriate method of determining the numerators of the partial fractions depends upon the factors that appear in the denominator of $F(s)$. We consider three different cases.

Case A. If the denominator of $F(s)$ contains only linear factors none of which is repeated, then,

$$F(s) = \frac{N(s)}{(s-a)(s-b)(s-c)\ldots(s-m)} = \frac{A}{(s-a)} + \frac{B}{(s-b)} + \ldots + \frac{M}{(s-m)}$$

(6.113)

To find the coefficient A, multiply through by $(s-a)$ and let $s=a$. Then

$$(s-a)F(s) = \frac{N(s)(s-a)}{(s-a)(s-b)\ldots(s-m)} = A + \frac{B(s-a)}{(s-b)} + \ldots + \frac{M(s-a)}{(s-m)}$$

When $s=a$, all terms on the right are zero except A. So,

$$\begin{aligned} A &= [(s-a)F(s)]_{s=a} \\ B &= [(s-b)F(s)]_{s=b} \end{aligned}$$

(6.114)

and so on.

Example. Let

$$F(s) = \frac{7s-2}{s^3-s^2-2s} = \frac{7s-2}{s(s+1)(s-2)} = \frac{A}{s} + \frac{B}{(s+1)} + \frac{C}{(s-2)}$$

$$A = [(s)F(s)]_{s=0} = \frac{7s-2}{(s+1)(s-2)} = 1$$

$$B = [(s+1)F(s)]_{s=-1} = -3 \qquad C = [(s-2)F(s)]_{s=2} = 2$$

and

$$F(s) = \frac{1}{s} - \frac{3}{(s+1)} + \frac{2}{(s-2)}$$

The reader may care to derive the corresponding function of t, that is the inverse transform, using the table of transform pairs in Appendix D. (Answer: $1 - 3e^{-t} + 2e^{2t}$)

Case B. If a linear factor $(s-b)$ is repeated p times in the denominator, then the partial fraction expansion must include p terms of the form:

$$\frac{A_1}{(s-b)} + \frac{A_2}{(s-b)^2} + \ldots + \frac{A_p}{(s-b)^p}$$

(6.115)

The coefficient A_p is then found from

$$A_p = [(s-b)^p F(s)]_{s=b} \tag{6.116}$$

The other coefficients are found by repeated differentiation

$$A_{p-1} = \frac{d}{ds}[(s-b)^p F(s)]_{s=b}$$

$$A_{p-2} = \frac{1}{2!}\frac{d^2}{ds^2}[(s-b)^p F(s)]_{s=b} \tag{6.117}$$

$$A_{p-3} = \frac{1}{3!}\frac{d^3}{ds^3}[(s-b)^p F(s)]_{s=b}$$

and so on.

Example.

$$F(s) = \frac{s^2+4s-15}{s^3-3s^2+4} = \frac{s^2+4s-15}{(s+1)(s-2)^2} = \frac{A}{(s+1)} + \frac{B}{(s-2)} + \frac{C}{(s-2)^2}$$

$$A = [(s+1)F(s)]_{s=-1} = \left[\frac{s^2+4s-15}{(s-2)^2}\right]_{s=-1} = -2$$

$$C = [(s-2)^2 F(s)]_{s=2} = \left[\frac{s^2+4s-15}{(s+1)}\right]_{s=2} = -1$$

$$B = \frac{d}{ds}[(s-2)^2 F(s)]_{s=2} = \frac{d}{ds}\left[\frac{s^2+4s-15}{(s+1)}\right]_{s=2} = 3$$

Therefore,

$$F(s) = \frac{-2}{(s+1)} + \frac{3}{(s-2)} + \frac{-1}{(s-2)^2}$$

Again the reader may care to find the inverse of this function (Answer: $-2e^{-t} + 3e^{2t} - te^{2t}$)

Case C. Suppose there is an irreducible quadratic term of the form $s^2 + as + b^2$ in the denominator of $F(s)$. Because a and b are real numbers, this term gives two complex values of s that are complex conjugates of one another. That is

$$s^2 + as + b^2 = [s + (\alpha + j\omega)][s + (\alpha - j\omega)] \tag{6.118}$$

where α and ω are real and $\alpha = a/2$, $\omega^2 = b^2 - (a/2)^2$.

The partial fraction expansion then contains the following two terms:

The Laplace transform

$$\frac{A_1}{s+(\alpha+j\omega)} + \frac{A_2}{s+(\alpha-j\omega)} \tag{6.119}$$

where A_1 and A_2 are, in general, complex. Then

$$\begin{aligned} A_1 &= [\{s+(\alpha+j\omega)\}F(s)]_{s=-(\alpha+j\omega)} \\ A_2 &= [\{s+(\alpha-j\omega)\}F(s)]_{s=-(\alpha-j\omega)} \end{aligned} \tag{6.120}$$

Because $F(s)$ is a rational function in s with real coefficients, A_1 and A_2 must be complex conjugates, that is, $A_2 = A_1^*$. In polar form:

$$A_1 = A_1 e^{j\phi_1}$$

and

$$A_2 = A_2 e^{j\phi_2} = A_1 e^{-j\phi_1} \tag{6.121}$$

Now the inverse transform of the sum of the two terms in (6.119) is

$$\begin{aligned} f(t) &= A_1 e^{-(\alpha+j\omega)t} + A_2 e^{-(\alpha-j\omega)t} \\ &= A_1 e^{j\phi_1} e^{-(\alpha+j\omega)t} + A_1 e^{-j\phi_1} e^{-(\alpha-j\omega)t} \\ &= A_1 e^{-\alpha t}[e^{j(\omega t - \phi_1)} + e^{-j(\omega t - \phi_1)}] \end{aligned}$$

or

$$f(t) = 2A_1 e^{-\alpha t}\cos(\omega t - \phi_1) \tag{6.122}$$

Example.

$$F(s) = \frac{s^2+3s+7}{(s+1)(s^2+2s+5)} = \frac{A}{s+1} + \frac{B_1}{s+(1+j2)} + \frac{B_2}{s+(1-j2)}$$

Then

$$A = [(s+1)F(s)]_{s=-1} = \tfrac{5}{4} = 1.25$$
$$B_1 = [\{s+(1+j2)\}F(s)]_{s=-(1+j2)}$$
$$= \frac{(-1-j2)^2 + 3(-1-j2) + 7}{(-1-j2+1)(-1-j2+1-j2)} = -\frac{1}{8} + j\frac{1}{4}$$

In polar form, $B_1 = 0.28\underline{/116°}$.

The inverse transform of the last two terms is from (6.122)

$$2B_1 e^{-\alpha t}\cos(\omega t - \phi_1) = 2 \times 0.28 e^{-t}\cos(2t - 116°)$$

and the complete function of t is

$$f(t) = 1.25 e^{-t} + 0.56 e^{-t}\cos(2t - 116°)$$

We may use the foregoing theory to derive two useful transform pairs.

(1) Let $F(s)=\dfrac{1}{s^2+as+b^2}=\dfrac{1}{(s+\alpha)^2+\omega^2}$

where $\alpha = a/2$ and $\omega^2 = b^2 - (a/2)^2$.

By (6.118) and (6.119) we may write $F(s)$ as

$$F(s) = \frac{A_1}{s+(\alpha+j\omega)} + \frac{A_2}{s+(\alpha-j\omega)}$$

then,

$$A_1 = [\{s+(\alpha+j\omega)\}F(s)]_{s=-(\alpha+j\omega)}$$

$$= \frac{1}{-j2\omega} = \frac{1}{2\omega}\underline{/90°}$$

and, using (6.122), we obtain

$$f(t) = 2A_1 e^{-\alpha t}\cos(\omega t - \phi_1) = \frac{1}{\omega}e^{-\alpha t}\cos(\omega t - 90°)$$

or

$$f(t) = \frac{1}{\omega}e^{-\alpha t}\sin\omega t$$

We then have

$$\mathscr{L}\{e^{-\alpha t}\sin\omega t\} = \frac{\omega}{(s+\alpha)^2+\omega^2} \qquad (6.123)$$
[Pair No. 10]

(2) Let $F(s)=\dfrac{s}{s^2+as+b^2}=\dfrac{s}{(s+\alpha)^2+\omega^2}$

By an argument similar to that given above

$$A_1 = [\{s+(\alpha+j\omega)\}F(s)]_{s=-(\alpha+j\omega)}$$

$$= \frac{\alpha+j\omega}{j2\omega} = \frac{\sqrt{(\alpha^2+\omega^2)}}{2\omega}\underline{/\theta-90°}$$

where $\theta = \tan^{-1}(\omega/\alpha)$.

So, by (6.122),

$$f(t) = \frac{\sqrt{(\alpha^2+\omega^2)}}{\omega}e^{-\alpha t}\cos(\omega t - \theta + 90°)$$

$$= \frac{\sqrt{(\alpha^2+\omega^2)}}{\omega}e^{-\alpha t}\sin(\theta - \omega t)$$

$$= \frac{\sqrt{(\alpha^2+\omega^2)}}{\omega}e^{-\alpha t}(\sin\theta\cos\omega t - \sin\omega t\cos\theta)$$

The Laplace transform

But $\sin\theta = \omega/\sqrt{(\alpha^2+\omega^2)}$ and $\cos\theta = \alpha/\sqrt{(\alpha^2+\omega^2)}$ hence

$$f(t) = e^{-\alpha t}\left(\cos\omega t - \frac{\alpha}{\omega}\sin\omega t\right)$$

Hence

$$\mathscr{L}\left\{e^{-\alpha t}\left(\cos\omega t - \frac{\alpha}{\omega}\sin\omega t\right)\right\} = \frac{s}{(s+\alpha)^2+\omega^2} \qquad (6.124)$$
[Pair No. 12]

Let us now return to the problem of finding the inverse of (6.110), namely,

$$I(s) = \frac{V_m\omega}{L}\left[\frac{1}{(s+R/L)(s^2+\omega^2)}\right] \qquad (6.110)$$

The term in brackets may be expanded using the procedures detailed under Case C above. With $\alpha = 0$ in (6.119), and letting $R/L = a$, we have

$$F(s) = \frac{1}{(s+a)(s^2+\omega^2)} = \frac{A}{s+a} + \frac{B_1}{s+j\omega} + \frac{B_2}{s-j\omega}$$

By (6.114)

$$A = [(s+a)F(s)]_{s=-a} = \frac{1}{a^2+\omega^2}$$

and the inverse of the first term is $e^{-at}/(a^2+\omega^2)$.

By (6.120)

$$B_1 = [(s+j\omega)F(s)]_{s=-j\omega} = \frac{1}{(-j\omega+a)(-2j\omega)} = \frac{1}{-2\omega(\omega+ja)}$$

$$= -\frac{1}{2\omega\sqrt{(\omega^2+a^2)}}\underline{/-\phi_1} \quad \text{where } \phi_1 = \tan^{-1}\frac{a}{\omega}$$

Therefore, the inverse of the last two terms is, from (6.122),

$$-\frac{1}{\omega\sqrt{(\omega^2+a^2)}}\cos(\omega t + \phi_1)$$

and the complete expression for the inverse of (6.110) becomes

$$i(t) = \frac{V_m\omega}{L}\left[\frac{e^{-at}}{\omega^2+a^2} - \frac{1}{\omega\sqrt{(\omega^2+a^2)}}\cos(\omega t+\phi_1)\right]$$

Putting $a = R/L$

$$i(t) = \frac{V_m\omega L}{(\omega L)^2+R^2}e^{-Rt/L} - \frac{V_m}{\sqrt{[(\omega L)^2+R^2]}}\cos(\omega t+\phi_1) \qquad (6.125)$$

Now referring to (6.1), the phase angle for the circuit of fig. 6.1(b) is given by $\tan\theta = \omega L/R$, hence, $\sin\theta = \omega L/\sqrt{[(\omega L)^2+R^2]}$. Also, $\tan\phi_1 = R/\omega L$, hence,

$\phi_1 = 90 - \theta$. So, putting $\omega L/\sqrt{[(\omega L)^2 + R^2]} = \sin\theta$ in the first term of (6.125), and $\phi_1 = 90 - \theta$ in the second term of (6.125) gives

$$i(t) = \frac{V_m e^{-Rt/L}}{\sqrt{[(\omega L)^2 + R^2]}}\sin\theta + \frac{V_m}{\sqrt{[(\omega L)^2 + R^2]}}\sin(\omega t - \theta)$$

which is identical to (6.38) with $\lambda = 0$.

If one compares the above method of solving the circuit of fig. 6.1(b) with that given in section 6.6.1, it will be seen that it is algebraically more complicated. In general, the Laplace transform method is not to be recommended for solving first order circuits with constant voltage or sinusoidal driving sources. However, as will become apparent in the following sections, the method has advantages when dealing with circuits of second or higher order and having finite initial energy states. The Laplace transform method can also be advantageous for first order circuits containing driving sources other than constant voltage or sinusoidal.

6.9.4 Network analysis by Laplace transform

In applying the Laplace transform method to circuit problems two approaches are possible. The first is to set up the complete circuit integro-differential equations and then transform these into algebraic equations in s. With this approach difficulties can arise when taking into account initial energy states of the circuit, particularly if the differential equations contain second or higher order derivatives. The second approach, which we adopt here, is to transform the voltage–current relationships for each circuit element before setting up the circuit equations. Using this approach it is often helpful to reconstruct the original time-domain circuit in the s-domain. This new circuit will contain complete information concerning the initial energy states of the original circuit.

We have already found in section 6.9.2 (equation (6.103)) the transform corresponding to the voltage–current relationships for an inductance carrying initial current i_0:

$$v(t) = \frac{L\,di(t)}{dt} \Rightarrow V(s) = sLI(s) - Li_0 \tag{6.126}$$

The corresponding current–voltage relationship is

$$i(t) = \frac{1}{L}\int_0^t v(t)\,dt + i_0 \Rightarrow I(s) = \frac{V(s)}{sL} + \frac{i_0}{s} \tag{6.127}$$

The circuit interpretation of these relationships is shown in fig. 6.27. In (6.126) we see that the voltage $V(s)$ is the sum of two terms: (1) a voltage drop, $sL \times I(s)$, where sL is interpreted as a reactance (dimensions of ohms);

(2) a constant voltage Li_0. The s-domain circuit consists, therefore, of an inductance L in series with a constant voltage source, as shown in fig. 6.27(b). Likewise in (6.127) the current $I(s)$ is the sum of two terms: $V(s)/sL$ and i_0/s. The s-domain circuit consists, therefore, of an inductance L in parallel with a constant current source (fig. 6.27(c)).

It will be appreciated that the relationships (6.126) and (6.127) in the s-domain are mathematically identical; one can be derived from the other by simple algebraic manipulation. From the circuit point of view this manipulation corresponds to the Thévenin–Norton transformation (discussed in section 2.9.1). Fig. 6.27(b) is a Thévenin circuit (inductance in series with an ideal voltage source), while fig. 6.27(c) is a Norton circuit consisting of the same inductance in parallel with a current source the magnitude of which is given by $Li_0/sL = i_0/s$. It is of interest to note that the source Li_0 in the s-domain circuit of fig. 6.27(b) corresponds to an impulse in the time-domain circuit since the inverse transform of a constant is an impulse (transform pair No. 18). Likewise, the source i_0/s in fig. 6.27(c) corresponds to a step function in the time-domain circuit (transform pair No. 16).

The transform corresponding to the voltage–current relationship for capacitance, charged to an initial voltage v_0, was also derived in section 6.9.2 (equation 6.108):

Fig. 6.27. Time- and s-domain circuits for inductance and capacitance. i_0 and v_0 are initial values of current and voltage.

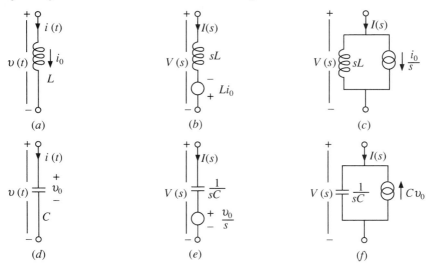

Time-domain   s-domain

$$v(t) = \frac{1}{C}\int i(t)\,dt + v_0 \Rightarrow V(s) = \frac{I(s)}{sC} + \frac{v_0}{s} \qquad (6.128)$$

The corresponding current–voltage relationship is:

$$i(t) = C\frac{dv(t)}{dt} \Rightarrow I(s) = sCV(s) - Cv_0 \qquad (6.129)$$

Figures 6.27(d), (e) and (f) show the circuit interpretation of these relationships; again it will be evident that the circuits of figs. 6.27(e) and (f) are related through the Thévenin–Norton transformation.

The relationships shown in fig. 6.27 enable one to transform any circuit into the s-domain. Once the s-domain circuit is established any of the formal procedures and techniques of steady-state circuit analysis may be applied. For the inexperienced reader it is recommended that inductive and capacitive elements with initial current and voltages be transformed using the series forms given in figs. 6.27(b) and (e). The Thévenin–Norton transformation can always be applied subsequently according to the demands of a particular problem.

As an example of how the relationships shown in fig. 6.27 are applied to derive the transform of a circuit, consider the time-domain circuit of fig. 6.28(a). The switch has been closed for an appreciable time so that C is charged to the source voltage V_0, and a constant current $i_0(=V_0/R_1)$ flows through L. At $t=0$ the switch is opened. We wish to find an expression for the current $i(t)$ in the circuit for $t>0$.

To derive the s-domain circuit of fig. 6.28(b) we consider each of the storage elements in turn. The inductance with its initial current i_0 is, according to fig. 6.27(b), transformed to an inductance connected in series

Fig. 6.28. Example of circuit transformation using the relationships shown in figs. 6.2.7 (a), (b) and (d), (e).

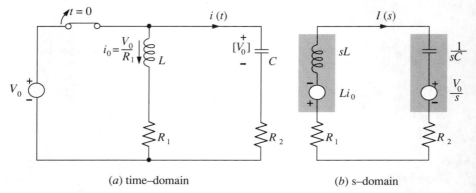

(a) time–domain   (b) s–domain

The Laplace transform

with an ideal voltage source. The capacitance with its initial voltage V_0 also transforms to a series combination of capacitance and ideal voltage source. We choose the series (Thévenin) circuit transformations in each case because this leads to the simplest possible single-mesh circuit in the s-domain. Note that care must be exercised to ensure that the correct polarities are assigned to the voltage sources in the s-domain. For the inductance, the polarity of its associated source must be such as to drive current in the *same* direction as that of the initial current in the time-domain circuit. For the capacitance the polarity of the s-domain source must be identical to that of the initial voltage in the time-domain. These observations apply irrespective of the directions of the assigned currents $i(t)$ or $I(s)$.

Applying Kirchhoff's voltage law to the s-domain circuit we have

$$\left(sL + \frac{1}{sC} + R_1 + R_2\right)I(s) = -\left(\frac{V_0}{s} + Li_0\right)$$

or

$$I(s) = \frac{-(V_0/s + Li_0)}{sL + 1/sC + R_1 + R_2}$$

Inversion of this expression yields the required function of current in the time domain.

A further example of the application of the Laplace transform method is shown in fig. 6.29. In fig. 6.29(a) C is charged to an initial voltage $v_0 (= I_0 R_1)$. The switch is closed at $t = 0$; we wish to find the voltage $v(t)$ for $t > 0$.

In this case it is slightly more convenient to use the parallel transformation of fig. 6.27(f) since this leads directly to an s-domain circuit with one

Fig. 6.29. Example of circuit transformation using (for the capacitance) the relationship shown in figs. 6.27 (d), (f).

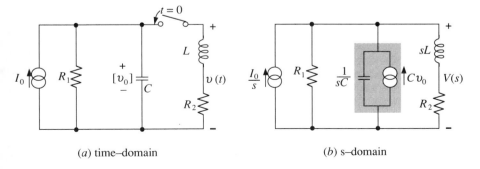

(a) time–domain

(b) s–domain

independent node for which a nodal analysis, to find $V(s)$, is clearly appropriate. (The choice of the series transformation of fig. 6.27(e) would have led, albeit indirectly, to precisely the same nodal analysis.) Note that the ideal current generator of magnitude I_0 in the t-domain circuit transforms to I_0/s in the s-domain (transform pair no. 15).

Applying nodal analysis to the s-domain circuit we obtain:

$$\left[\frac{1}{R_1} + sC + \frac{1}{sL + R_2}\right]V(s) - \frac{I_0}{s} - Cv_0 = 0$$

or

$$V(s) = \frac{I_0/s + Cv_0}{1/R_1 + sC + 1/(sL + R_2)}$$

Inversion of this function yields $v(t)$.

The transform relationships for inductance expressed by (6.126), and illustrated in figs. 6.27(a) and (b), may be readily extended to the case of mutual inductance. We have seen that an initial current i_0 in an inductance gives rise to a constant voltage source Li_0 in the s-domain. Referring to the circuit of fig. 6.30(a), in which there are two coils of inductances L_1 and L_2 coupled by mutual inductance M, if the first coil carries initial current i_{01}, then in the s-domain circuit this will give rise to a voltage source of magnitude $L_1 i_{01}$ in series with L_1. In addition, a source of magnitude $M i_{01}$ will arise in series with L_2. This follows from the theory of mutual inductance presented in section 1.10. The polarity of this source will depend upon the way in which the two coils are wound with respect to one another. This information is provided by the dot convention (see sections 1.10 and 3.12). Notice that in fig. 6.30(b), the polarities of the two sources $L_1 i_{01}$ and $M i_{01}$ bear precisely the same relationship with the corresponding ends of

Fig. 6.30. Time- and s-domain circuits for mutual inductance.

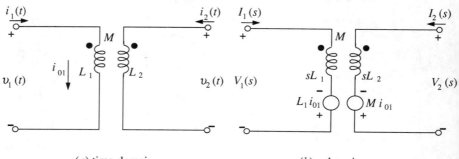

(a) time-domain   (b) s-domain

the two coils; that is, the negative side of the Li_{01} source is joined to the non-dotted end of L_1, and the negative side of the Mi_{01} is joined to the non-dotted end of L_2. (Of course, the same end result would be obtained if the positive sides of either or both sources were connected to the dotted ends of the coils.)

Finally, it follows that if there is an initial current i_{02} in L_2, this will produce in the s-domain circuit a source $L_2 i_{02}$ in series with L_2 and, additionally, a source Mi_{02} in series with L_1.

6.9.5 Worked example

The circuit of fig. 6.31(a) is in equilibrium with the switch open. At $t=0$ the switch is closed. Find the current $i_2(t)$ for $t>0$.

Solution: Since the circuit is in equilibrium for $t<0$, the current through L_1 at the instant of closing the switch must be $i_0 = V_0/R_1 = 5$ A for the given circuit values. The direction of i_0 is from left to right in the circuit diagram. In the s-domain circuit (fig. 6.31(b)) this current gives rise to the voltage source $L_1 i_0$ with its polarity such that it drives current from left to right (into the dotted end of L_1). In addition, i_0 gives rise to the source Mi_0, with polarity such that it also drives current into the dotted end of L_2.

The procedure for solving circuits containing mutual inductance is outlined in section 3.12. Currents $I_1(s)$ and $I_2(s)$ are assigned to the two meshes in the s-domain circuit. (In practical problems it is sufficient to denote currents by I_1, I_2 etc.)

Applying KVL to mesh (1) we obtain

$$(sL_1 + R_1)I_1 - R_1 I_2 + sMI_2 = \frac{V_0}{s} + L_1 i_0$$

and for mesh (2)

$$(R_1 + R_2 + sL_2)I_2 - R_1 I_1 + sMI_1 = Mi_0$$

Note that the terms due to mutual inductance, sMI_2 and sMI_1, are both positive because both assigned currents enter corresponding (dotted) ends of coils.

Substituting numerical values and rearranging the above two equations we obtain

$$(s+1)I_1 + (s-1)I_2 = \frac{5}{s} + 5$$

and

$$(s-1)I_1 + (s+2)I_2 = 5$$

346 Transient and steady-state analysis

Solution of these equations yields, after some algebraic manipulation,

$$I_2 = \frac{s+1}{s(s+0.2)}$$

By (6.113) we may write

$$I_2 = \frac{A}{s} + \frac{B}{s+0.2}$$

and by (6.114)

$$A = [sF(s)]_{s=0} = \frac{1}{0.2} = 5$$

$$B = [(s+0.2)F(s)]_{s=-0.2} = \frac{-0.2+1}{-0.2} = -4$$

Fig. 6.31. Circuits for worked example (section 6.9.5).

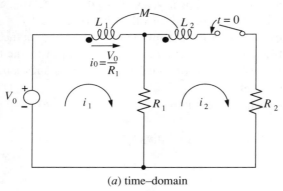

(a) time–domain

$(L_1 = L_2 = M = 1\text{H}; R_1 = R_2 = 1\Omega; V_0 = 5\text{ V})$

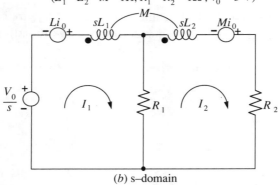

(b) s–domain

Therefore,

$$I_2(s) = \frac{5}{s} - \frac{4}{s+0.2}$$

The first term of this expression is inverted using transform pair No. 15, and the second term using transform pair No. 2. So,

$$i_2(t) = 5 - 4e^{-0.2t}$$

If the simultaneous equations above are solved for I_1, a similar procedure shows that the current in the first mesh is $i_1(t) = 10 - 6e^{-0.2t}$.

It may be remarked that the current $i_2(t)$ at the instant $t = 0^+$ is, according to the expression above, equal to 1 A, while the current $i_1(t)$ is 4 A. Before closing the switch, that is, at $t = 0^-$, the currents were 0 and 5 A respectively; evidently the currents in the inductances are changing instantaneously, which might be taken to imply that the energies are doing likewise. The reason for this apparent anomaly is that, if energy is to be conserved, the total flux linkage (inductance × current) must be conserved at the instant of switching. Both currents contribute to the total flux linkage. Thus, at $t=0^-$: $L_1 i_0 = (1)(5) = 5$. At $t=0^+$: $L_1 i_1 + M i_2 = (1)(4) + (1)(1) = 5$. The product $(L \times i)$ is often referred to as the *electrokinetic momentum* (see reference 2).

6.9.6 Generalized impedance, network function and impulse response

Consider the general branch driven by a voltage source $v(t)$ as shown in fig. 6.32(a). The circuit is initially dead so that the circuit transforms into that shown in fig. 6.32(b). In the s-domain the circuit equation is

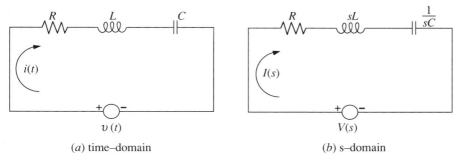

Fig. 6.32. Time- and s-domain circuits for the general *RLC* branch driven by an ideal voltage source.

(a) time–domain

(b) s–domain

$$\left(R+sL+\frac{1}{sC}\right)I(s)=V(s)$$

or

$$\frac{V(s)}{I(s)}=R+sL+\frac{1}{sC}$$

The ratio $V(s)/I(s)$ may be interpreted as an impedance and we can write

$$Z(s)=\frac{V(s)}{I(s)}=R+sL+\frac{1}{sC} \qquad (6.130)$$

Now in the detailed theory of the Laplace transform it is shown that s is complex with dimensions of angular frequency. (It has already been indicated in section 6.5.3 how the concept of complex frequency can arise in circuit theory.) The complex frequency is usually written

$$s=\sigma+j\omega \qquad (6.131)$$

If we let $\sigma=0$, then

$$Z(s)=Z(j\omega)=R+j\omega L+\frac{1}{j\omega C} \qquad (6.132)$$

which is recognized as the steady-state a.c. circuit impedance. We conclude that (6.132) is merely a special case of (6.130), and indeed, from this viewpoint the a.c. theory developed in chapter 3 can be regarded as a special case of the more general approach afforded by the Laplace transform.

In the s-domain the ratio of voltage to current at any terminal pair or port of a network is denoted by $Z(s)$ and is referred to as the *generalized impedance* (or sometimes the generalized driving point impedance). Likewise the ratio of current to voltage is called the *generalized admittance*.* (The word 'generalized' is often omitted when the context is clear.)

For example, the generalized admittance of a combination of R, L and C connected in parallel may be written

$$Y(s)=\frac{1}{R}+\frac{1}{sL}+sC \qquad (6.133)$$

It will be apparent, therefore, that the analytical techniques and terminology developed in chapter 3 for a.c. quantities may be translated directly in terms of these generalized concepts. In particular, the important

---

* The term *immittance* is often used when referring in a non-specific way to either impedance or admittance.

The Laplace transform

ideas concerning the transfer function of a network, introduced in section 3.9, may be extended to the s-domain, as indicated in fig. 6.33.

If $e(t)$ is a driving function or excitation in the time-domain network and $r(t)$ is the response to this excitation, and if $E(s)$ and $R(s)$ are the corresponding Laplace transforms, then we may define a *network function* $H(s)$ by

$$H(s) = \frac{R(s)}{E(s)}$$

or

$$R(s) = H(s)E(s) \tag{6.134}$$

$E(s)$ and $R(s)$ are called the *excitation function* and *response function* respectively.

If $R(s)$ and $E(s)$, which may be voltage or current functions in s, refer to the same port, then $H(s)$ is a driving point immittance (impedance or admittance). If they refer to different ports, then $H(s)$ is termed a transfer function.

As an example serving to illustrate some of the points discussed above, let us find the transfer function for the circuit shown in fig. 6.34, and its response to a ramp function input. The circuit forms a two-arm divider with parallel elements in each arm; the admittance divider formulation is therefore the most appropriate way of finding the transfer function. The admittances of the two arms are:

$$Y_1(s) = \frac{1}{R_1} + sC_1 = \frac{1 + sC_1R_1}{R_1}$$

Fig. 6.33. Definition of the network function.

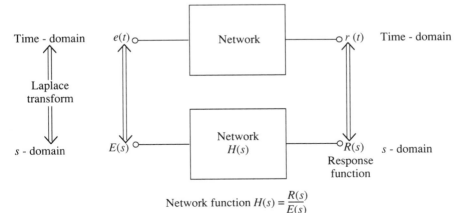

350 Transient and steady-state analysis

and

$$Y_2(s) = \frac{1}{R_2} + sC_2 = \frac{1 + sC_2R_2}{R_2}$$

Therefore, the transfer function is

$$H(s) = \frac{V_2(s)}{V_1(s)} = \frac{Y_1(s)}{Y_1(s) + Y_2(s)}$$

$$= \frac{(1 + sC_1R_1)/R_1}{(1 + sC_1R_1)/R_1 + (1 + sC_2R_2)/R_2} \tag{6.135}*$$

If the input $e(t)$ is a ramp function, then the excitation function $E(s)$ is (from transform pair No. 20) $1/s^2$. Thus, the response in the s-domain will be

$$R(s) = H(s)E(s) = \frac{(1 + sC_1R_1)/R_1}{(1 + sC_1R_1)/R_1 + (1 + sC_2R_2)/R_2} \cdot \frac{1}{s^2}$$

This example illustrates the way in which one can utilize the techniques of a.c. circuit analysis in the s-domain to obtain the response of a network not only to sinusoidal input waveforms but to any waveform whose Laplace transform can be found.

An important special case arises when the excitation $e(t)$ is the unit

Fig. 6.34. s-domain circuit for a voltage divider.

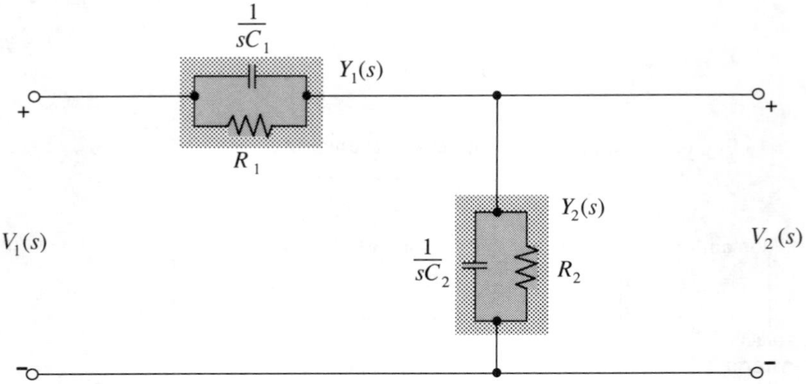

* If the time constants in the two arms are equal, that is, if $C_1R_1 = C_2R_2$, then (6.135) reduces to $H(s) = R_2/(R_1 + R_2)$, which is independent of frequency. Waveforms of arbitrary shape will, therefore, be transferred without distortion, (except for a scaling factor). For this reason the circuit with equal time constants is known as a frequency compensated divider. (This forms the basis of the well-known 'oscilloscope probe'.)

The Laplace transform

impulse function. In this case the response $r(t)$ is, by definition, equal to the impulse response $h(t)$ (see section 6.8.2). Now the transform of the unit impulse function is equal to unity, consequently the response in the s-domain is given by

$$R(s) = H(s)E(s) = H(s) \times 1 = H(s)$$

This means that the response function in the s-domain for unit impulse excitation in the time domain, is simply the network function itself. It follows from this that the impulse response $h(t)$ must be the inverse transform of the network function $H(s)$. The time and s-domain relationships for this particular case are depicted in fig. 6.35.

This relationship between impulse response and network function provides a relatively simple way of determining the impulse response of a network: first find the network function using the 'a.c. steady-state' approach described above; then find the inverse transform of the network function using tables of transform pairs. For example, let us find the impulse response of the simple RC circuit of fig. 6.36. For this circuit, the network (transfer) function is given by

$$H(s) = \frac{V_2(s)}{V_1(s)} = \frac{1/sC}{1/sC + R} = \frac{1}{\tau(s + 1/\tau)}$$

where $\tau = CR$. Hence, the impulse response is

$$h(t) = \frac{1}{\tau} e^{-t/\tau}$$

Fig. 6.35. Illustrating the relationships between the network function and impulse response.

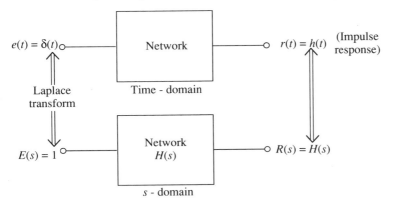

352 Transient and steady-state analysis

The reader should compare this procedure with that used earlier to obtain (6.85).

6.9.7 Third and higher order networks

We have observed in preceding sections that a circuit containing a single energy storage element leads to a first order differential equation, while a circuit containing two independent storage elements leads to a second order differential equation. For example, the equation for the general series branch, containing two storage elements is

$$L\frac{di}{dt} + Ri + \frac{1}{C}\int i\,dt = v$$

which upon differentiation gives the second order equation

$$L\frac{d^2i}{dt^2} + R\frac{di}{dt} + \frac{i}{C} = \frac{dv}{dt}$$

We now consider in a more general way the relationship between the number and types of storage element in a network, and the form of the network equation and its solution. Generalizing the above result for the RLC series branch, the equation relating the response $r(t)$ at any port in a network to the excitation $e(t)$ at the same or a different port of that network may be written in differential form as:

$$a_n\frac{d^n r}{dt^n} + a_{n-1}\frac{d^{n-1} r}{dt^{n-1}} + \ldots + a_1\frac{dr}{dt} + a_0$$
$$= b_m\frac{d^m e}{dt^m} + b_{m-1}\frac{d^{m-1} e}{dt^{m-1}} + \ldots + b_1\frac{de}{dt} + b_0 \quad (6.136)$$

where the a_i and b_i are functions of the network elements only, and n is equal to the number of independent storage elements.

Fig. 6.36. Simple RC circuit in the s-domain.

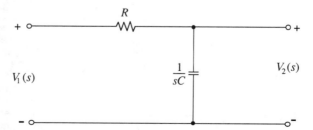

The Laplace transform

Setting the RHS of (6.136) to zero results in an equation that characterizes the natural behaviour of the network. Its solution gives the natural response which contains n terms of the form

$$r(t) = A_1 e^{s_1 t} + A_2 e^{s_2 t} + \ldots + A_n e^{s_n t} \tag{6.137}$$

in which $A_1 \ldots A_n$ are arbitrary constants (governed by the initial energy states of the network and by the form of the excitation function), and $s_1 \ldots s_n$ are the roots of the auxiliary equation:

$$s^n + \frac{a_{n-1}}{a_n} s^{n-1} + \ldots + \frac{a_1}{a_n} s + \frac{a_0}{a_n} = 0 \tag{6.138}$$

a polynomial of degree n.

The corresponding network equation may be formulated in the s-domain by transforming both sides of (6.136)

$$(a_n s^n + a_{n-1} s^{n-1} + \ldots + a_1 s + a_0) R(s)$$
$$= (b_m s^m + b_{m-1} s^{m-1} + \ldots + b_1 s_1 + b_0) E(s)$$

or

$$R(s) = \frac{b_m s^m + b_{m-1} s^{m-1} + \ldots + b_1 s + b_0}{a_n \left(s^n + \frac{a_{n-1}}{a_n} s^{n-1} + \ldots + \frac{a_1}{a_n} s + \frac{a_0}{a_n} \right)} E(s) \tag{6.139}$$

To find the response $r(t)$ from this expression requires the inverse transform to be found, and this in turn necessitates finding the roots of the denominator polynomial, which is of degree n and identical to (6.138).

Thus, whether the network equation is formulated in the time-domain or the s-domain, its solution entails finding the roots of a polynomial of degree n where n is the number of independent storage elements in the network. (Exceptions to this general rule are provided by certain network configurations containing elements of identical value, in which case the network equation may be of lower order than n.)

Before attempting to solve a network problem it is good practice to count the number of independent storage elements and to note their type. The following rules can then offer a general (although not infallible) guide to the form of the network equation and its solution. The polynomial referred to below is that given by (6.138) or the denominator of (6.139).

Rule 1 If n is the number of independent storage elements in a network, then the following parameters are numerically equal to n:
   (a) order of circuit differential equation
   (b) degrees of polynomial
   (c) number of roots

(d) number of exponential terms in the solution
(e) number of arbitrary constants.

Rule 2 If storage elements are all of one type, then the roots of the polynomial will be real and there will be no oscillatory terms in the natural response. If they are not of one type, the roots may be complex leading to oscillatory terms in the natural response.

Rule 3 Coefficients of the polynomial must be real and positive in all cases.

It must be emphasized that the number of independent storage elements in a network is not necessarily the same as the number of separate, identifiable elements. Two inductances in the same branch, for instance, combine to form a single independent storage element. Both circuits in fig. 6.37 contain four separate storage elements yet lead to a third order equation; in each case elements can be combined by series or parallel addition. In fig. 6.37(b) a Thévenin–Norton transformation is required to allow the two leftmost inductances to be combined in parallel.

Finding the roots of third and higher degree polynomials can constitute a major proportion of the work involved in solving complex networks. An algebraic formula is available for finding the roots of a cubic, but it is distinctly more difficult to apply than the quadratic formula. For quartic and polynomials of higher degree, resort has to be made to algebraic methods aimed at reduction of the polynomial in question into cubic, quadratic or linear components (Lin's method). Iterative methods also exist for finding the real roots of a polynomial to any desired degree of accuracy (for example, Horner's method) but the amount of repetitive manipulation involved is considerable.

For these reasons it has become commonplace to use numerical methods to find the roots of third and higher order polynomials. Program C4 in Appendix C, allows one to compute the real and complex roots of higher order polynomials.

Fig. 6.37. Circuits containing four storage elements: (a) reduces to third order by combining C_1 and C_2 in series, (b) reduces to third order by combining L_1 and L_2 in parallel.

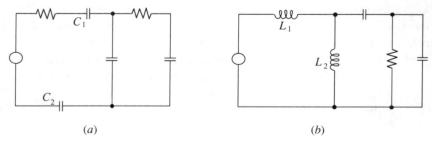

(a)   (b)

The Laplace transform

Although the amount of work involved in evaluating the network polynomial is the same irrespective of whether the network equation is formulated in the time domain or the s-domain, the latter has advantages when dealing with third and higher order networks. This stems from the fact that initial energy states can be incorporated quite simply into the s-domain network equation. The time-domain solution, on the other hand, requires the evaluation of arbitrary constants after the complete solution of the network differential equation has been found. This can be a difficult operation since it involves the evaluation of higher derivatives in terms of the initial energy states of the network.

6.9.8 Worked example

(a) Find the transfer function $H(s) = V_2(s)/V_1(s)$ for the circuit shown in fig. 6.38.

(b) Show that the expression for the natural response of this circuit contains three exponential terms, and determine the time constants associated with these terms for the given values of R and C.

(c) If $v_1(t)$ is a steady sinusoidal function, show that the phase shift of $v_2(t)$ with respect to $v_1(t)$ is zero for a particular frequency, and determine this frequency for the given values of R and C.

Solution

(a) Referring to the s-domain circuit (fig. 6.38(b)), the circuit contains three independent nodes O, P, Q. The voltage at O is the assigned voltage V_2; let the voltages at P and Q be V_P and V_Q respectively. Applying nodal analysis we have: at node O

$$\frac{V_2 - V_1}{R} + \frac{V_2 - V_P}{1/sC} = 0$$

$$(sCR + 1)V_2 - sCRV_P = V_1$$

Fig. 6.38. Circuits for worked example (section 6.9.8).

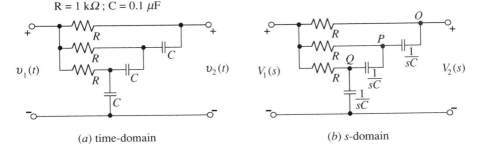

(a) time-domain   (b) s-domain

Let $sCR = a$ (The algebraic manipulation is thus greatly simplified.) Then

$$(a+1)V_2 - aV_p = V_1$$

Similarly, at nodes P and Q we obtain

$$-aV_2 + (2a+1)V_P - aV_Q = V_1$$

and

$$-aV_P + (2a+1)V_Q = V_1$$

Elimination of V_P and V_Q from the above three simultaneous equations yields:

$$V_2(a^3 + 6a^2 + 5a + 1) = V_1(6a^2 + 5a + 1)$$

The transfer function is therefore

$$H(s) = \frac{V_2(s)}{V_1(s)} = \frac{6s^2(RC)^2 + 5sRC + 1}{s^3(RC)^3 + 6s^2(RC)^2 + 5sRC + 1}$$

(b) The natural response is obtained from the roots of the denominator polynomial in the transfer function. Normalizing, by letting the product $RC = 1$, the polynomial becomes

$$s^3 + 6s^2 + 5s + 1$$

Using program C4 in Appendix C, we find that this has roots at $s = -5.05$; $s = -0.643$; $s = -0.308$.

In the time-domain the natural response is therefore of the form:

$$A_1 e^{-5.05t} + A_2 e^{-0.643t} + A_3 e^{-0.308t}$$

or

$$A_1 e^{-t/0.198} + A_2 e^{-t/1.56} + A_3 e^{-t/3.25}$$

For the given component values, $RC = 10^{-4}$, hence, the required time constants are 19.8 μs; 156 μs; 325 μs.

(c) For a steady sinusoidal input, the transfer function is

$$H(j\omega) = \frac{6(j\omega RC)^2 + 5j\omega RC + 1}{(j\omega RC)^3 + 6(j\omega RC)^2 + 5j\omega RC + 1}$$

$$= \frac{1 - 6(\omega RC)^2 + j5\omega RC}{1 - 6(\omega RC)^2 + j[5\omega RC - (\omega RC)^3]}$$

Let $1 - 6(\omega RC)^2 = p$; $5\omega RC = q$ and $5\omega RC - (\omega RC)^3 = r$, then,

$$H(j\omega) = \frac{p + jq}{p + jr} = \frac{(p + jq)(p - jr)}{p^2 + r^2}$$

The Laplace transform 357

$$= \frac{p^2+qr}{p^2+r^2} + j\frac{p(q-p)}{p^2+r^2}$$

If the phase shift is to be zero, then the imaginary term in this expression must vanish, that is $p(q-p)=0$. Putting $(q-p)=0$ leads to the trivial result $\omega=0$, therefore,

$$p = 1 - 6(\omega RC)^2 = 0$$

which gives

$$\omega = \frac{1}{\sqrt{6}RC}$$

For the given values of R and C, $\omega = 4082$ rad/s.

This circuit is one of a small number of RC circuits capable of producing a voltage step-up. For the zero phase-shift condition, the transfer function becomes

$$H(j\omega) = \frac{q}{r} = \frac{5\omega CR}{5\omega CR - (\omega RC)^3} = \frac{5/\sqrt{6}}{5/\sqrt{6} - (1/\sqrt{6})^3} = \frac{30}{29}$$

This property of the circuit has been utilized in one type of oscillator.

6.9.9 Further Laplace transform theorems

In this section we consider some useful theorems that allow us to extend the table of transform pairs and increase the facility with which transforms and inverse transforms may be found. The theorems are introduced without proof; interested readers will find proofs in references 1, 2 and 9.

Theorem 1: shift in the s-domain
If $F(s) = \mathscr{L}[f(t)]$,

then

$$F(s+\alpha) = \mathscr{L}[e^{-\alpha t}f(t)] \qquad (6.140)$$

Example: find $\mathscr{L}[e^{-\alpha t}\cos\omega t]$.
Since

$$\mathscr{L}[\cos\omega t] = \frac{s}{s^2+\omega^2} \qquad \text{[Pair No. 8]}$$

then

$$\mathscr{L}[e^{-\alpha t}\cos\omega t] = \frac{s+\alpha}{(s+\alpha)^2+\omega^2} \qquad \text{[Pair No. 11]}$$

Example: given $F(s) = \dfrac{1}{(s+2)^2}$, find $f(t)$.

Since

$$\mathscr{L}^{-1}\left[\dfrac{1}{s^2}\right] = t \qquad \text{[Pair No. 1]}$$

then

$$\mathscr{L}^{-1}\left[\dfrac{1}{(s+2)^2}\right] = f(t) = te^{-2t}$$

Theorem 2: shift in the time-domain (also known as 'real translation')
If $\mathscr{L}[f(t)] = F(s)$

then

$$\mathscr{L}[f(t-a)] = e^{-at}F(s) \qquad (6.141)$$

Example: find the Laplace transform of the delayed impulse function $\delta(t-a)$.

Since

$$\mathscr{L}[\delta(t)] = 1 \qquad \text{[Pair No. 18]}$$

then

$$\mathscr{L}[\delta(t-a)] = e^{-as} \qquad \text{[Pair No. 19]}$$

Theorem 3: differentiation with respect to s
If $\mathscr{L}[f(t)] = F(s)$

then

$$\mathscr{L}[tf(t)] = -\dfrac{\mathrm{d}}{\mathrm{d}s}(F(s)) \qquad (6.142)$$

Example: find the Laplace transform of $t\sin\omega t$. Since

$$\mathscr{L}[\sin\omega t] = \dfrac{\omega}{s^2+\omega^2} \qquad \text{[Pair No. 6]}$$

$$\mathscr{L}[t\sin\omega t] = -\dfrac{\mathrm{d}}{\mathrm{d}s}\left[\dfrac{\omega}{s^2+\omega^2}\right] = \dfrac{2s\omega}{(s^2+\omega^2)^2}$$

Theorems 4 and 5: initial and final value theorems
If $\mathscr{L}[f(t)] = F(s)$

then

Pole-zero methods

$$\lim_{t \to 0} f(t) = \lim_{s \to \infty} sF(s) \quad \text{initial value theorem} \quad (6.143)$$

and

$$\lim_{t \to \infty} f(t) = \lim_{s \to 0} sF(s) \quad \text{final value theorem} \quad (6.144)$$

These theorems are applicable only if $f(t)$ and its first derivative are Laplace transformable. In addition, for the final value theorem, complex factors in the denominator polynomial of $sF(s)$ must have positive real parts. The theorems are often useful for checking whether a function of current or voltage derived in the s-domain gives physically sensible results.

Example. Consider a step function of magnitude V applied to the RC circuit of Fig. 6.36. The transfer function is $1/(sRC+1)$ so that

$$V_2(s) = \frac{1}{(sRC+1)} \cdot \frac{V}{s} \quad \text{and} \quad sV_2(s) = \frac{V}{sRC+1}$$

Then, by the initial value theorem,

$$\lim_{t \to 0} v_2(t) = \lim_{s \to \infty} \left[\frac{V}{sRC+1} \right] = 0$$

and by the final value theorem

$$\lim_{t \to \infty} v_2(t) = \lim_{s \to 0} \left[\frac{V}{sRC+1} \right] = V$$

These results are what might be expected from physical considerations: since the voltage on the capacitance cannot change instantaneously, its voltage must be zero at $t=0^+$ (assuming zero initial voltage). It is also apparent that the capacitance will charge up to a final value V.

6.10 Pole-zero methods

Let us again consider the expression (6.139) relating the response and excitation functions of a network. Since the ratio $R(s)/E(s)$ is equal to the network function $H(s)$, (6.139) may be rewritten as

$$\frac{R(s)}{E(s)} = H(s) = \frac{b_m s^m + b_{m-1} s^{m-1} + \ldots + b_1 s + b_0}{a_n s^n + a_{n-1} s^{n-1} + \ldots + a_1 s + a_0}$$

$$= \frac{b_m}{a_n} \cdot \frac{\left(s^m + \frac{b_{m-1}}{b_m} s^{m-1} + \ldots + \frac{b_1}{b_m} s + \frac{b_0}{b_m} \right)}{\left(s^n + \frac{a_{n-1}}{a_n} s^{n-1} + \ldots + \frac{a_1}{a_n} s + \frac{a_0}{a_n} \right)} \quad (6.145)$$

Recalling that the a_i and b_i in this expression are functions only of the network Ls, Cs and Rs, we see that $H(s)$ is the ratio of two polynomials with real coefficients, that is, it is a rational function. Therefore, according to the fundamental theorem of algebra, the numerator and denominator polynomials may be factorized to give:

$$H(s) = H_0 \frac{(s-z_1)(s-z_2)\ldots(s-z_m)}{(s-p_1)(s-p_2)\ldots(s-p_n)} \qquad (6.146)$$

where $H_0 = \dfrac{b_m}{a_n}$ is a scaling factor, and the z_i and p_i are roots respectively of the numerator and denominator polynomials.

Now if we examine the behaviour of $H(s)$ as the complex frequency s is varied, we see that at the particular values of $s = p_1$, $s = p_2$ etc., each of the factors in the denominator polynomial of $H(s)$ in turn becomes zero and the function becomes infinite. We say that *poles* of the function exist at $s = p_1, p_2 \ldots$ or that $p_1, p_2 \ldots$ are poles of $H(s)$. Likewise, at the particular values of $s = z_1$, $s = z_2$ etc., $H(s)$ becomes zero and we refer to $z_1, z_2 \ldots$ as *zeros* of the function $H(s)$. Put another way: the poles are the roots of the denominator polynomial, the zeros are the roots of the numerator polynomial. If we trace the derivation of (6.146) from (6.139), we see that the poles of $H(s)$ determine the form of the natural response. For example, consider the function

$$H(s) = \frac{2s^3 - 4s^2 + 4s}{s^4 + 6s^3 + 13s^2 + 24s + 36}$$

which when factorized becomes

$$H(s) = \frac{2s(s-1+j1)(s-1-j1)}{(s+3)^2(s+j2)(s-j2)}$$

The first factor in the numerator, s, may be written $(s-0)$ from which it will be apparent that a zero exists at $s = 0$. Other zeros occur when*

$$s - 1 + j1 = 0 \qquad \text{i.e., when } s = 1 - j1$$

and

$$s - 1 - j1 = 0 \qquad \text{i.e., when } s = 1 + j1$$

Poles of the function occur when

---

* Strictly, a zero also occurs at $s = \infty$. This arises because as $s \to \infty$ the function $H(s) \to 1/s \to 0$. Poles and zeros at infinity are of importance only in the more advanced theory of the method.

Pole-zero methods

$(s+3)^2 = 0$ i.e. when $s = -3$ (twice)
$s+j2 = 0$ i.e. when $s = -j2$
$s-j2 = 0$ i.e. when $s = +j2$

The locations of the poles and zeros may be mapped in the complex frequency plane, or *s*-plane, as shown in fig. 6.39. We shall see that such a map, which is known as a *pole-zero diagram*, provides an extremely informative way of displaying the essential features of a network function, and of illustrating the characteristics of the network from which the function is derived. Note that the scaling factor ($\times 2$ in this case) does not affect the positions of the poles and zeros, and is often omitted from the pole-zero diagram.

In order to illustrate the use of the pole-zero diagram, we again consider the general series branch shown in fig. 6.32. The circuit equation is

$$\left(R + sL + \frac{1}{sC}\right) I(s) = V(s)$$

which may be written

$$I(s) = \left[\frac{1}{R + sL + \frac{1}{sC}}\right] V(s)$$

In the above expression $V(s)$ is the excitation function, $I(s)$ is the response function, and the term in brackets is the network function – in this case an admittance function

Fig. 6.39. Pole-zero diagram for the function $H(s)$; poles are indicated by crosses, zeros by circles.

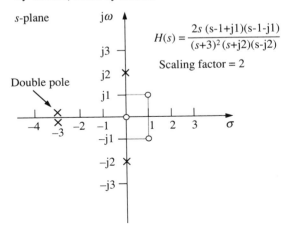

$$H(s) = \frac{2s\,(s-1+j1)(s-1-j1)}{(s+3)^2\,(s+j2)(s-j2)}$$

Scaling factor = 2

$$Y(s) = \frac{1}{R + sL + \dfrac{1}{sC}} = \frac{1}{L} \cdot \frac{s}{\left(s^2 + \dfrac{sR}{L} + \dfrac{1}{LC}\right)} \qquad (6.147)$$

The scaling factor for this function is $1/L$, and there is a single zero at $s=0$, that is, at the origin in the pole-zero diagram. To find the poles, it is necessary to factorize the denominator of (6.147). Let s_1, s_2 be the roots of $(s^2 + sR/L + 1/LC)$ then

$$Y(s) = \frac{s}{L(s-s_1)(s-s_2)}$$

where

$$s_1, s_2 = -\frac{R}{2L} \pm \sqrt{\left[\left(\frac{R}{2L}\right)^2 - \frac{1}{LC}\right]}$$

It will be convenient at this point to use the notation introduced in section 6.5.3. Let $R/2L = \alpha$, $1/LC = \omega_0^2$ then

$$s_1, s_2 = -\alpha \pm \sqrt{(\alpha^2 - \omega_0^2)} \qquad (6.148)$$

Now our study in section 6.5.3 of the behaviour of the *RLC* circuit revealed that three different types of natural response could occur: overdamped, underdamped and critically damped. Let us consider the admittance function $Y(s)$ in terms of these responses.

If $\alpha^2 > \omega_0^2$ in (6.148) then we have the overdamped case and the roots s_1, s_2 are negative, real and unequal. Again using the notation of section 6.5.3, let $s_1 = -m$ and $s_2 = -n$, then for the overdamped case the admittance function becomes:

$$Y(s) = \frac{1}{L} \frac{s}{(s+m)(s+n)}$$

Poles occur at $-m$ and $-n$ as shown in fig. 6.40(*a*). It follows from (6.148) that these poles are symmetrically disposed about the point $-\alpha$ on the σ-axis.

Next, consider critical damping; this occurs when $\alpha^2 = \omega_0^2$ in (6.148). Then $s_1 = s_2 = -\alpha$ and the admittance function becomes:

$$Y(s) = \frac{1}{L} \frac{s}{(s+\alpha)^2}$$

In this case a *repeated*, *double* or *second-order* pole is said to occur at $-\alpha$ (fig. 6.40(*b*)).

Pole-zero methods

Finally, consider underdamped or oscillatory response. In this case $\omega_0^2 > \alpha^2$ and

$$s_1, s_2 = -\alpha \pm j\omega_n$$

where ω_n is the damped natural frequency given by

$$\omega_n = \sqrt{(\omega_0^2 - \alpha^2)} \qquad (6.149)$$

Then

$$Y(s) = \frac{1}{L} \frac{s}{(s+\alpha+j\omega_n)(s+\alpha-j\omega_n)} \qquad (6.150)$$

and the poles are located as shown in fig. 6.40(c).

It should be noted that complex poles always occur as conjugate pairs, consequently, the pole-zero diagram is mirror symmetric about the σ-axis.

If the damping in the branch is varied by changing the value of R only (keeping L and C constant), then $\omega_0 [= 1/\sqrt{(LC)}]$ is constant and, from (6.149),

$$\omega_n^2 + \alpha^2 = \omega_0^2 = \text{const.}$$

Thus, the locus of the poles in the s-plane, as R is varied, is a semicircle of radius $1/\sqrt{(LC)}$.

The sequence of events as R is varied, so that the circuit changes from the overdamped case through critical damping to the underdamped case, is illustrated in the pole-zero diagram of fig. 6.41. Starting with a high value of R the poles are located at $(1, 1')$. As R is reduced, the poles move along the σ-axis converging towards point (2) located at $\sigma = -1/\sqrt{(LC)}$. On further reduction of R, the two poles diverge, moving along the semicircle of radius $1/\sqrt{(LC)}$ to points $(3, 3')$. As R becomes vanishingly small the poles lie on the $j\omega$-axis at points $(4, 4')$. This is not, of course, a practicable possibility

Fig. 6.40. Pole-zero diagrams for the admittance function of the general series branch.

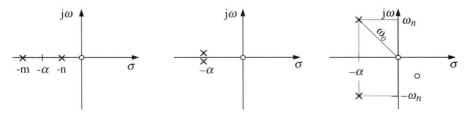

(a) Overdamped   (b) Critically damped   (c) Underdamped

364   *Transient and steady-state analysis*

for a passive circuit (except of the superconducting variety) and the poles will always lie somewhere in the left-hand half plane of the pole-zero diagram. Poles in the right-hand half plane imply exponentially increasing functions in the time-domain, which are possible only with circuits containing active devices.* (From this point of view the pole-zero diagram plays an important role in the theory of control systems and the conditions that must obtain for their stability.)

It is sometimes helpful in the interpretation of the pole-zero diagram to think of the modulus of the function under consideration as a surface above the s-plane. In the present instance the surface of $|Y(s)|$, corresponding to the overdamped and underdamped cases, would look somewhat as depicted in fig. 6.42. While this visualization can be informative, it must be appreciated that it is the *map* of the poles and zeros in the s-plane, showing the way in which they move with change of particular network parameters, that provides the all-important information.

An alternative way of representing a network function in the s-plane is obtained by expressing the factors of the function in polar form. In the expression (6.146) for $H(s)$, a factor such as $(s - z_1)$ in the numerator, which is the difference between two complex quantities, can be written as

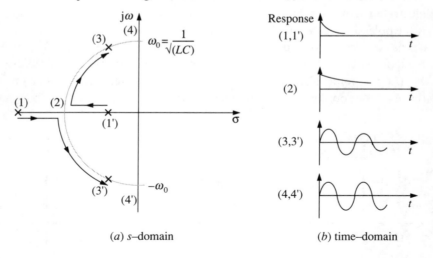

Fig. 6.41. Illustrating the effect of variation of R on the natural response of the general RLC series branch (L and C constant).

(a) s-domain   (b) time-domain

---

* The detailed theory for passive circuits shows that zeros may exist in the right-hand half plane, but only in the case of transfer functions. Neither poles nor zeros may exist in this region in the case of driving point functions.

Pole-zero methods

$$(s - z_1) = N_1 \underline{/\psi_1}$$

where $N_1 = |s - z_1|$ and $\psi_1 = \arg(s - z_1)$.

In the pole-zero diagram (fig. 6.43), $(s - z_1)$ is a line segment of length N_1 directed from the point z_1 to the point s, making an angle ψ_1 with the horizontal. (This follows from the normal laws of vector addition.)

A factor $(s - p_1)$ in the denominator of (6.146) may likewise be expressed by

$$(s - p_1) = D_1 \underline{/\theta_1}$$

with a similar interpretation in the pole-zero diagram.

Expression (6.146) may then be written

$$H(s) = H_0 \frac{(s - z_1)(s - z_2) \ldots}{(s - p_1)(s - p_2) \ldots}$$

$$= H_0 \frac{(N_1 N_2 \ldots)}{(D_1 D_2 \ldots)} \underline{/(\psi_1 + \psi_2 \ldots) - (\theta_1 + \theta_2 \ldots)} \quad (6.151)$$

As an example consider the transfer function

$$H(s) = \frac{(s - 1 + j1)(s - 1 - j1)}{(s + 3)(s + 2 + j4)(s + 2 - j4)}$$

which has the pole-zero diagram shown in fig. 6.44(a). Suppose s takes the particular value $(0 + j3)$, that is, the point s lies on the $j\omega$ axis as shown in fig. 6.44(b). In polar form the function is

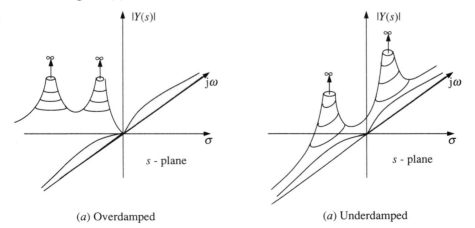

Fig. 6.42. Representation of the modulus of the network function $Y(s)$ in the s-plane. (a) Corresponding to fig. 6.40(a); (b) corresponding to fig. 6.40(c).

(a) Overdamped

(a) Underdamped

$$H(s) = \frac{N_1 N_2}{D_1 D_2 D_3} \underline{/(\psi_1 + \psi_2) - (\theta_1 + \theta_2 + \theta_3)}$$

Converting each of the factors of $H(s)$ to polar form we obtain:

$$(s - 1 + j1) = -1 + j4 = N_1 \underline{/\psi_1} = 4.12 \underline{/104°}$$
$$(s - 1 - j1) = -1 + j2 = N_2 \underline{/\psi_2} = 2.24 \underline{/111.6°}$$
$$(s + 3) = 3 + j3 = D_1 \underline{/\theta_1} = 4.24 \underline{/45°}$$
$$(s + 2 + j4) = 2 + j7 = D_2 \underline{/\theta_2} = 7.28 \underline{/74°}$$
$$(s + 2 - j4) = 2 - j1 = D_3 \underline{/\theta_3} = 2.24 \underline{/-26.7°}$$

(Note that θ_3 is negative.)

Hence,

$$H(s) = \frac{4.12 \times 2.24}{4.24 \times 7.28 \times 2.24} \underline{/(104 + 116.6) - (45 + 74 - 26.7)}$$

$$= 2.36 \underline{/-128.3°}$$

The above complex arithmetic may also be accomplished graphically using the construction shown in fig. 6.44(b).

This approach is particularly useful if one wishes to examine the steady-state behaviour of a network function as a function of frequency ω. In this case s lies on the $j\omega$ axis, as in the above example, and the vectors from the various poles and zeros change in length and angle as the frequency is varied.

From (6.146), with $s = j\omega$, the steady-state behaviour of the network function is given by

Fig. 6.43. Polar representation of factors in the network function.

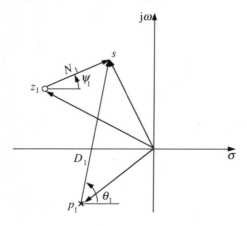

Pole-zero methods

$$H(j\omega) = H_0 \frac{(j\omega - z_1)(j\omega - z_2)\ldots}{(j\omega - p_1)(j\omega - p_2)\ldots}$$
$$= M(\omega)\underline{/\phi(\omega)} \qquad (6.152)$$

where

$$M(\omega) = |H(j\omega)| = \frac{(N_1 N_2 \ldots)}{(D_1 D_2 \ldots)}$$

and

$$\phi(\omega) = \arg H(j\omega) = (\psi_1 + \psi_2 \ldots) - (\theta_1 + \theta_2 \ldots)$$

$M(\omega)$ is called the *amplitude response* function and $\phi(\omega)$ the *phase response* function. Evaluation of $M(\omega)$ and $\phi(\omega)$ at a few spot frequencies can quickly give one an idea of how the network behaves as a function of frequency. By shifting, adding or deleting poles and zeros in the diagram, the effect of changes to the circuit can be assessed; the circuit designer is thus able to produce the circuit characteristics required.

To illustrate these ideas we examine the admittance function of the general series branch relating to the underdamped case (equation (6.150) and fig. 6.40(c)).

$$Y(s) = \frac{1}{L} \frac{s}{(s + \alpha + j\omega_n)(s + \alpha - j\omega_n)}$$

In polar form with $s = j\omega$ this becomes

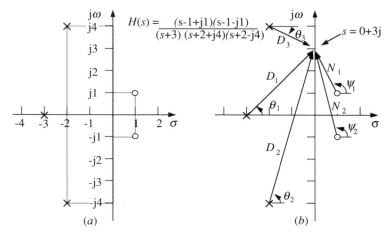

Fig. 6.44. (a) Pole-zero diagram for the function $H(s)$. (b) Vectors drawn to the point $s = 0 + j3$.

$$Y(j\omega) = \frac{1}{L} \frac{N_1}{D_1 D_2} \underline{/\psi_1 - (\theta_1 + \theta_2)}$$

So,

$$M(\omega) = |Y(j\omega)| = \frac{1}{L} \frac{N_1}{D_1 D_2}$$

and

$$\phi(\omega) = \arg Y(j\omega) = \psi_1 - (\theta_1 + \theta_2)$$

The vectors for the admittance function are shown for three different frequencies in fig. 6.45.

Consider first variation of the amplitude response function $M(\omega)$. For $\omega = 0$: $N_1 = 0$ (because the zero is at the origin), therefore $M(\omega) = 0$.

For $\omega = \omega_n$: $N_1 = \omega_n$, $D_1 = [\alpha^2 + (2\omega_n)^2]^{\frac{1}{2}}$, $D_2 = \alpha$ therefore $M(\omega) = \omega_n / L\alpha(\alpha^2 + 4\omega_n^2)^{\frac{1}{2}}$.

For $\omega \to \infty$: $N_1, D_1, D_2 \to \infty$, but $N_1/D_1 D_2 \to 0$, therefore $M(\omega) \to 0$.

We conclude that with increasing frequency $M(\omega)$ rises from zero, to a maximum value then falls asymptotically to zero. (The shape of the $M(\omega)$ curve is seen in fig. 6.42 as the line profile of $|Y(s)|$ along the $j\omega$-axis.)

Now consider variation of the phase response function $\phi(\omega)$. For all ω, $\psi_1 = 90°$ because the zero is at the origin.

For $\omega = 0$: $\theta_1 = -\theta_2$, therefore $\phi(\omega) = 90°$
For $\omega = \omega_n$: $\theta_2 = 0$, therefore $\phi(\omega) = 90° - \theta_1$
For $\omega \to \infty$: $\theta_1 = \theta_2 \to 90°$, therefore $\phi(\omega) \to -90°$

Fig. 6.45. Amplitude–phase diagrams for the admittance function of the general series branch at three different frequencies.

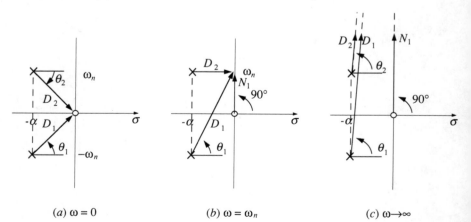

(a) $\omega = 0$ \qquad (b) $\omega = \omega_n$ \qquad (c) $\omega \to \infty$

Pole-zero methods

From the pole zero diagrams shown in fig. 6.45 we can also immediately see the effect of varying one of the circuit parameters. Reducing $\alpha(=R/2L)$, say, will shift the poles nearer to the $j\omega$-axis and increase $M(\omega)$ for values of ω near to ω_n, (that is, near to resonance) because D_2 varies rapidly with α. On the other hand, reducing α will have little effect at very small or very large ω, because vectors D_1 and D_2 then vary slowly with α.

In our study so far of pole-zero methods we have considered only network functions. Excitation functions may also be represented in the s-plane. Pole-zero diagrams for five common excitation functions are shown in fig. 6.46. The impulse has unit value everywhere in the s-plane; the step has a single pole at the origin, while the ramp has a double pole. The sine wave is represented by complex conjugate poles on the $j\omega$-axis; multiplying the sine wave by a damping term e^{-at}, shifts the poles into the left-hand half-plane. The latter two pole zero diagrams and their related transforms illustrate the ideas underlying the concept of a complex frequency. In the

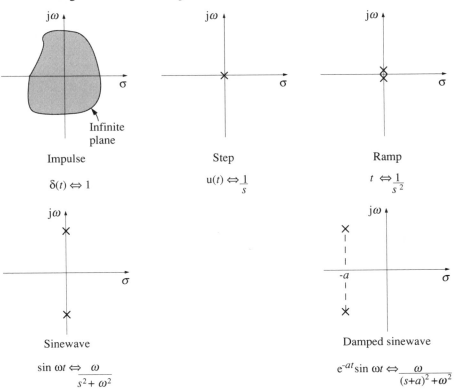

Fig. 6.46. Pole-zero diagrams for five common excitation functions.

Impulse
$\delta(t) \Leftrightarrow 1$

Step
$u(t) \Leftrightarrow \dfrac{1}{s}$

Ramp
$t \Leftrightarrow \dfrac{1}{s^2}$

Sinewave
$\sin \omega t \Leftrightarrow \dfrac{\omega}{s^2+\omega^2}$

Damped sinewave
$e^{-at}\sin \omega t \Leftrightarrow \dfrac{\omega}{(s+a)^2+\omega^2}$

steady-state theory developed in section 3.3 a sine wave of constant amplitude was represented by the complex exponential:

$$V_m \sin(\omega t + \theta) = \mathrm{Im}\, V_m e^{j(\omega t + \theta)} = \mathrm{Im}\, V_m e^{j\theta} e^{j\omega t}$$

The quantity $V_m e^{j\theta} = \mathbf{V}_m$ was termed a phasor.

For a damped sinusoid we have

$$e^{-at} V_m \sin(\omega t + \theta) = \mathrm{Im}\, V_m e^{j\theta} e^{-at} e^{j\omega t} = \mathrm{Im}\, \mathbf{V}_m e^{(-a + j\omega)t}$$

The frequency variable is now $(-a+j\omega)$ instead of $j\omega$. We define this new variable by $s = -a + j\omega$.

In general $s = \sigma + j\omega$ where σ is negative for an exponentially decreasing sinusoid and positive for an exponentially increasing sinusoid.

To complete our study of pole-zero methods let us derive diagrams for the response functions obtained by applying some of the excitation functions illustrated in fig. 6.46 to the general branch. In each case we find the response $I(s)$ assuming that the branch is overdamped, that is

$$I(s) = \frac{1}{L} \frac{s}{(s+m)(s+n)} V(s)$$

We also find the form of the response in the time domain.

Step function $V_m u(t)$

$$I(s) = \frac{1}{L} \frac{s}{(s+m)(s+n)} \cdot \frac{V_m}{s} = \frac{V_m}{L} \cdot \frac{1}{(s+m)(s+n)}$$

In this case the pole associated with the excitation function cancels the zero at the origin (fig. 6.47(a)). The response function is of the form

$$I(s) = \frac{A_1}{s+m} + \frac{A_2}{s+n}$$

$$i(t) = A_1 e^{-mt} + A_2 e^{-nt}$$

Ramp function $V_m t u(t)$

$$I(s) = \frac{1}{L} \frac{s}{(s+m)(s+n)} \cdot \frac{V_m}{s^2} = \frac{V_m}{L} \cdot \frac{1}{s(s+m)(s+n)}$$

Here the excitation function creates a pole at the origin (fig. 6.47(b))

$$I(s) = \frac{A_1}{(s+m)} + \frac{A_2}{(s+n)} + \frac{A_3}{s}$$

$$i(t) = A_1 e^{-mt} + A_2 e^{-nt} + A_3$$

Pole-zero methods

Fig. 6.47. Pole-zero diagrams for the admittance function of the general series branch showing the response $I(s)$ for various excitation functions $V(s)$.

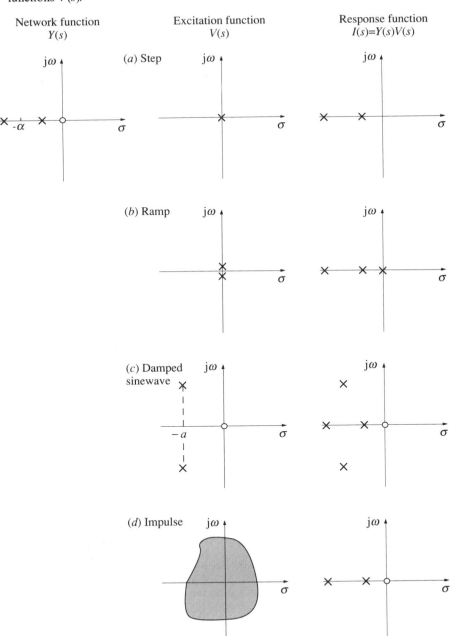

We see that the pole at the origin in the s-plane corresponds to a d.c. component in the time domain.

Damped sine wave $V_m e^{-at}\sin\omega t$

$$I(s) = \frac{1}{L} \frac{s}{(s+m)(s+n)} \cdot \frac{V_m\omega}{(s+a)^2+\omega^2}$$

The excitation function creates additional complex poles (fig. 6.46(c)), and the response is of the form

$$I(s) = \frac{A_1}{(s+m)} + \frac{A_2}{s+n} + \frac{A_3}{(s+a+j\omega)} + \frac{A_4}{(s+a-j\omega)}$$

In the time domain

$$i(t) = A_1 e^{-mt} + A_2 e^{-nt} + Be^{-at}\sin(\omega t + \theta)$$

The last term in this expression is, of course, the forced response.

Impulse function $\delta(t)$

Since the transform of the impulse function is unity, multiplying the network function by the excitation function in this case leaves the network function unchanged, that is

$$I(s) = \frac{1}{L} \frac{s}{(s+m)(s+n)} = Y(s)$$

The pole-zero diagram of the response function is identical to that of the network function (fig. 6.47(d)) and response function is of the form

$$I(s) = \frac{A_1}{(s+m)} + \frac{A_2}{(s+n)}$$

In the time domain

$$i(t) = A_1 e^{-mt} + A_2 e^{-nt}$$

which is the impulse response of the series branch.

6.11 Worked example

The circuit of a fourth-order low-pass Butterworth filter is shown in fig. 6.48. Derive the transfer function $H(s) = V_2(s)/V_1(s)$ for this filter and show that its poles are equi-spaced on a semicircle of unit radius in the left-hand half plane of the pole-zero diagram. Determine graphically the amplitude response function for frequencies $\omega = 0$; $\omega = 0.5$; $\omega = 1$, and $\omega = 1.5$ rad/s.

Worked example

Solution. To obtain the transfer function the ladder method described in section 2.15.3 is used. Assume $V_2 = 1$ volt, then, with the nodes lettered as shown in the diagram:

$$I_{BC} = sC_1 + \frac{1}{R}$$

$$V_{BO} = I_{BC}sL_1 + 1$$
$$I_{BO} = sC_2 V_{BO}$$
$$I_{AB} = I_{BO} + I_{BC} = sC_2 V_{BO} + I_{BC}$$
$$V_{AO} = I_{AB}sL_2 + V_{BO} = (sC_2 V_{BO} + I_{BC})sL_2 + V_{BO}$$
$$= s^2 C_2 L_2 V_{BO} + I_{BC} sL_2 + V_{BO}$$
$$= V_{BO}(s^2 C_2 L_2 + 1) + sL_2 I_{BC}$$
$$= (I_{BC} sL_1 + 1)(s^2 C_2 L_2 + 1) + sL_2 I_{BC}$$

$$= \left(sC_1 + \frac{1}{R}\right)(s^3 L_1 L_2 C_2 + sL_1 + sL_2) + s^2 C_2 L_2 + 1$$

$$= s^4 C_1 C_2 L_1 L_2 + s^3 C_2 L_1 L_2 / R + s^2 (C_1 L_1 + C_1 L_2 + C_2 L_2)$$
$$+ s(L_1 + L_2)/R + 1$$

Now, in the above calculation, V_2 is assumed to be 1 volt therefore, the ratio

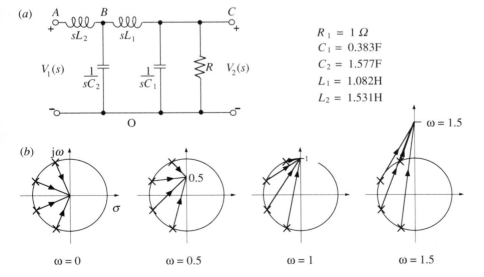

Fig. 6.48. Diagrams for worked example (section 6.11). (*a*) Circuit for Butterworth filter (element values are normalized to an impedance level of $1\,\Omega$ and a frequency of $\omega = 1$ rad/s). (*b*) Pole-zero diagrams with amplitude vectors drawn for four selected frequencies.

$R_1 = 1\,\Omega$
$C_1 = 0.383\mathrm{F}$
$C_2 = 1.577\mathrm{F}$
$L_1 = 1.082\mathrm{H}$
$L_2 = 1.531\mathrm{H}$

V_1/V_2 must be equal to V_{AO}. The transfer function is therefore $H(s) = V_2(s)/V_1(s) = 1/V_{AO}$. Putting in numerical values we obtain

$$H(s) = \frac{1}{s^4 + 2.613s^3 + 3.414s^2 + 2.613s + 1}$$

Using program C4 in Appendix C it is found that the roots (poles) of the denominator polynomial in the above expression are:

$$s_1, s_2 = -0.38268 \pm j0.92388$$
$$s_3, s_4 = -0.92388 \pm j0.38268$$

In polar form these become:

$$1.0\underline{/112.5}; \quad 1.0\underline{/-112.5}; \quad 1.0\underline{/157.5}; \quad 1.0\underline{/-157.5}$$

and in order of increasing positive angle we have

$$1.0\underline{/112.5}; \quad 1.0\underline{/157.5}; \quad 1.0\underline{/202.5}; \quad 1.0\underline{/247.5}$$

Thus, the poles lie on a circle of unit radius in the left-hand half plane with an angular spacing between them of 45°.

The amplitude response function is given by (equation (6.152))

$$M(\omega) = \frac{1}{D_1 D_2 D_3 D_4}$$

where $D_1 \ldots D_4$ are the lengths of the vectors extending from each pole to a particular frequency on the $j\omega$-axis. Fig. 6.48(b) shows vectors drawn for the four selected frequencies. By direct measurement (from a diagram with a scale of unit frequency = 100 mm) we find:

at $\omega = 0$ $\quad M(0) \quad = \dfrac{1}{(1.0)(1.0)(1.0)(1.0)} = 1$

at $\omega = 0.5$ $\quad M(0.5) = \dfrac{1}{(0.58)(0.94)(1.28)(1.47)} = 0.98$

at $\omega = 1$ $\quad M(1) \quad = \dfrac{1}{(0.39)(1.11)(1.67)(1.97)} = 0.70$

at $\omega = 1.5$ $\quad M(1.5) = \dfrac{1}{(0.69)(1.45)(2.11)(2.46)} = 0.19$

It is seen that the amplitude response falls sharply at frequencies above $\omega = 1$, the normalized cut-off frequency. At cut-off frequency the amplitude response is, theoretically, $1/\sqrt{2} = 0.707$, that is, -3 dB referred to the response at $\omega = 0$.

6.12 Pulse and repeated driving functions

The techniques that have been described so far for determining the transient and steady-state conditions in a circuit allow us to deal with only a limited range of driving functions such as the step and sinusoidal functions. However, these techniques may be readily extended to cover single and repeated pulse waveforms of various shapes. Pulse wavetrains of simple shape, such as a repeated series of rectangular pulses, applied to a first order circuit, can be dealt with adequately by elementary methods. For more complicated waveshapes and higher order circuits, the formal methods of the Laplace transform are often to be preferred.

6.12.1 Pulse response of first order circuits

Consider a rectangular pulse of amplitude V and duration a applied to the RC circuit shown in fig. 6.49(a). For the duration of the pulse, $0 \leq t \leq a$, the circuit response is the same as that for an applied step function, which is from the theory presented in section 6.6.2,

$$v(t) = V(1 - e^{-t/RC}) \qquad 0 \leq t \leq a \qquad (6.153)$$

If the time constant RC is very short compared to the duration of the pulse, v will rise to a value substantially equal to the pulse amplitude V (the steady state value) and the output waveform will appear as in fig. 6.49(b). If the time constant is large compared with a, then v will not reach its maximum possible value before the end of the pulse (fig. 6.49(c)).

At the instant $t = a$ the input voltage drops to zero and the capacitance discharges through the resistance via the ideal source supplying the input pulse. The output voltage then decays from some initial value $v(a) = V(1 - e^{-a/RC})$ according to:

$$v(t) = v(a) e^{-(t-a)/RC} \qquad t \geq a \qquad (6.154)$$

Fig. 6.49. Response of an RC circuit to an applied rectangular pulse of amplitude V and duration a for two different time constants RC.

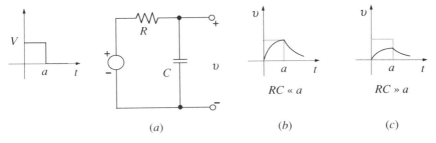

Expressions (6.153) and (6.154) can be used to find the response of the RC circuit to a train of rectangular pulses of the type shown in fig. 6.50(a). In this pulse train each pulse is of duration a, the interval between pulses is of duration b. (The ratio a/b is called the *mark–space ratio*.) We see that the pulse train is periodic with period $T = a + b$.

Fig. 6.50(b) shows the response for the condition $RC \ll a$: in this case the output voltage rises to its full (steady state) value V during each pulse and decays substantially to zero between pulses. For $RC \gg a$, we can encounter the condition shown in fig. 6.50(c) in which the output voltage has time to rise to only a small fraction of its maximum possible value during a pulse, and fails to decay to zero during the intervals between pulses. We now show that, under these circumstances, the output can build up so that it eventually reaches some steady-state value which is less than the pulse amplitude V.

Referring to fig. 6.50(c), let $v(t_1) = v_1$, $v(t_2) = v_2$ etc., then at the end of the first pulse we obtain, by (6.153),

$$v_1 = V(1 - e^{-t_1/RC}) \tag{6.155}$$

and at the end of the first interval, by (6.154),

$$v_2 = v_1 e^{-(t_2 - t_1)/RC} \tag{6.156}$$

During the second pulse, the output will rise from an initial value v_2 to some value v_3, which may be determined in the usual way by considering the

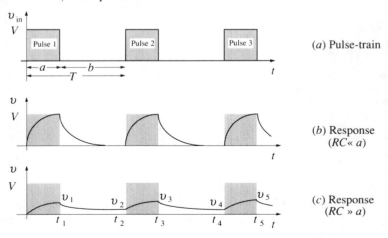

Fig. 6.50. Response of an RC circuit to a train of pulses of mark-space ratio a/b and period $T = a + b$.

(a) Pulse-train

(b) Response ($RC \ll a$)

(c) Response ($RC \gg a$)

Pulse and repeated driving functions

response as a sum of transient and steady-state terms. The steady-state term is obviously V, so that the output during the second pulse will be given by

$$v = Ae^{-(t-t_2)/RC} + V \qquad t_2 \leq t \leq t_3$$

where A is a constant which is determined from the initial condition: $v = v_2$ at $t = t_2$. This gives $A = v_2 - V$, hence,

$$\begin{aligned}v &= (v_2 - V)e^{-(t-t_2)/RC} + V \\ &= V[1 - e^{-(t-t_2)/RC}] + v_2 e^{-(t-t_2)/RC} \qquad t_2 \leq t \leq t_3\end{aligned}$$

Putting $t = t_3$ in this expression gives

$$v_3 = V[1 - e^{-(t_3-t_2)/RC}] + v_2 e^{-(t_3-t_2)/RC}$$

But from (6.156)

$$v_2 = v_1 e^{-(t_2-t_1)/RC}$$

therefore

$$\begin{aligned}v_3 &= V[1 - e^{-(t_3-t_2)/RC}] + v_1 e^{-(t_3-t_1)/RC} \\ &= V[1 - e^{-a/RC}] + v_1 e^{-T/RC} \end{aligned} \qquad (6.157)$$

where a is the pulse duration and T is the period.

Combining (6.155) and (6.157) gives

$$v_3 = v_1 + v_1 e^{-T/RC}$$

which shows that v_3 exceeds v_1 by an amount $v_1 e^{-T/RC}$. By a similar process we can show that $v_5 > v_3$, $v_7 > v_5$ etc. Thus, the output voltage builds up until an equilibrium condition is reached at which the voltage exhibits a cyclic variation about some mean level. This condition is the steady-state response to the pulse train;* which is to be distinguished from the steady-state response (either zero or V) associated with each individual transition in the pulse train.

Fig. 6.51. Steady-state response of RC circuit to a pulse train: $v_{n-1} = v_{n+1}$.

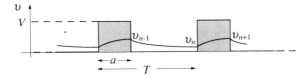

* Some authors prefer to use the term 'quasi-steady state' for this type of equilibrium condition.

We now determine the steady-state response of the RC circuit to an input pulse train under the condition $RC \gg a$ (fig. 6.50(c)). Referring to fig. 6.51, let v_n be the output voltage at the beginning of a particular pulse in the train, and let v_{n+1} and v_{n-1} be the voltages respectively at the end of this pulse and the end of the preceding pulse. Then, by analogy with 6.156 and 6.157,

$$v_n = v_{n-1} e^{-(T-a)/RC}$$

and

$$v_{n+1} = V(1 - e^{-a/RC}) + v_{n-1} e^{-T/RC}$$

But in the equilibrium (steady-state) condition $v_{n-1} = v_{n+1}$, therefore

$$v_{n-1} = \frac{V(1 - e^{-a/RC})}{(1 - e^{-T/RC})}$$

and

$$v_n = \frac{V(1 - e^{-a/RC})}{(1 - e^{-T/RC})} \cdot e^{-(T-a)/RC} \qquad (6.158)$$

The mean level is

$$\frac{v_{n-1} + v_n}{2} = \frac{V}{2} \frac{(1 - e^{-a/RC})}{(1 - e^{-T/RC})} [1 + e^{-(T-a)/RC}]$$

If, for example, the mark–space ratio is unity ($T = 2a$), then, the mean level is

$$\frac{V}{2} \frac{(1 - e^{-a/RC})}{(1 - e^{-2a/RC})} (1 + e^{-a/RC}) = \frac{V}{2}$$

So, for pulses with a mark–space ratio of unity, the output voltage of the RC circuit settles down to a mean level of just half the pulse amplitude. For other mark–space ratios, the steady-state output will be greater or less than $V/2$ depending on the particular value of a/b.

Another example for which elementary theory is well suited is provided by the circuit of fig. 6.52(a). Here an ideal current source drives a sawtooth waveform of current (fig. 6.52(b)) through an RL circuit. Such a circuit might represent, in idealized form, the deflection system of a cathode-ray tube display. L and R are the inductance and resistance of the deflection coils; the slow, positive-going ramp of current in the coils generates the linear sweep of the cathode-ray tube spot, while the fast, negative-going ramp corresponds to the flyback period. The form of the voltage generated across the deflection coils is of some interest to the circuit designer. In the following derivation of the source voltage waveforms, the current is assumed, for simplicity, to have unit amplitude.

Pulse and repeated driving functions

The circuit equation is

$$v = L\frac{di}{dt} + Ri$$

In the interval $0 \le t \le t_1$, the current rises linearly from zero at a rate of k_1 amperes per second, hence,

$$v = L\frac{d}{dt}(k_1 t) + Rk_1 t = Lk_1 + Rk_1 t \qquad 0 \le t \le t_1$$

In this equation it must be remembered that the current is the driving function, and v is the unknown variable. Setting the RHS of the above equation to zero to find the transient response gives $v = 0$; from which we infer that the transient term is zero. The above equation, which is of straight-line form with intercept Lk_1 and slope Rk_1, represents, therefore, the total response to the driving current.

At the instant $t = t_1$, the current starts to fall and it is then expressed by $i = 1 - k_2(t - t_1)$, where k_2 is the new slope of the function. In the interval $t_1 \le t \le t_2$, we have, therefore,

$$v = L\frac{d}{dt}[1 - k_2(t - t_1)] + R[1 - k_2(t - t_1)]$$

or

$$v = -Lk_2 + R[1 - k_2(t - t_1)] \qquad t_1 \le t \le t_2$$

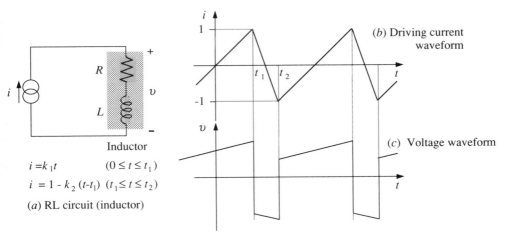

Fig. 6.52. Voltage response of an inductor driven by a sawtooth current waveform.

(a) RL circuit (inductor)

$i = k_1 t \qquad (0 \le t \le t_1)$
$i = 1 - k_2(t - t_1) \quad (t_1 \le t \le t_2)$

(b) Driving current waveform

(c) Voltage waveform

Again this is an equation of straight-line form. At $t=t_2$ the current commences to rise with slope k_1 and the cycle repeats. The complete cyclic voltage waveform is shown in fig. 6.52(c). The reader might find it instructive to consider how this waveform would be modified if either L or R were reduced to zero.

6.12.2 Delayed singularity functions: transforms of recurrent waveforms

In this section we establish methods of describing repeated pulses or recurrent waveforms by means of delayed singularity functions; this in turn will allow us to obtain the Laplace transforms of such waveforms. For the sake of simplicity it will be convenient to assume waveforms of unit amplitude; the results obtained apply equally to waveforms of amplitude V provided that all derived waveforms are multiplied by a factor V.

It will be recalled from our discussion in section 6.8.5 that the delayed unit step $u(t-a)$ provides a convenient way of representing the start of a function at a time $t=a$. This device can be used also to describe pulses and repeated functions. For example, the pulse shown in fig. 6.53 can be resolved into two step functions: a positive step starting at $t=0$ and a negative step starting at $t=a$. The expression for the pulse function is then written:

$$p(t) = u(t) - u(t-a)$$

The transform of the first term is $1/s$ while that for the second term is, according to the shift theorem, e^{-as}/s (transform pair No. 22). Thus, in the s-domain:

$$P(s) = \frac{1}{s} - \frac{e^{-as}}{s} = \frac{1}{s}(1 - e^{-as}) \qquad (6.159)$$

Fig. 6.53. Resolution of a pulse into two step functions.

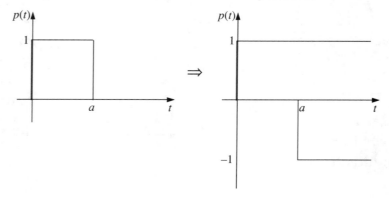

Pulse and repeated driving functions

A pulse train of the type depicted in fig. 6.54(a) may be similarly resolved into a series of step functions:

$$f(t) = u(t) - u(t-a) + u(t-2a) - u(t-3a) + \ldots \tag{6.160}$$

with transform

$$F(s) = \frac{1}{s} - \frac{e^{-as}}{s} + \frac{e^{-2as}}{s} - \frac{e^{-3as}}{s} + \ldots$$

$$= \frac{1}{s}(1 - e^{-as} + e^{-2as} - e^{-3as} + \ldots)$$

The geometric progression within brackets may be expressed in closed form using the following identity:

$$\frac{1}{1+x} = 1 - x + x^2 - x^3 + \ldots \qquad x < 1 \tag{6.161}$$

Thus, the transform of the pulse train of fig. 6.54(a) is

$$F(s) = \frac{1}{s(1 + e^{-as})} \tag{6.162}$$

This technique for finding transforms of repeated functions may be expressed in more general terms as follows. Let $p(t)$ be a pulse of finite duration, as shown in fig. 6.55(a), and let $P(s)$ be its transform. Then, for the repeated function $f(t)$ of fig. 6.55(b) with period T, we have

$$f(t) = p(t)u(t) + p(t-T)u(t-T) + p(t-2T)u(t-2T) + \ldots$$

and its transform is

$$F(s) = P(s) + P(s)e^{-Ts} + P(s)e^{-2Ts} + \ldots$$
$$= P(s)(1 + e^{-Ts} + e^{-2Ts} + \ldots)$$

Fig. 6.54. Resolution of a pulse train (square wave) into an infinite series of step functions.

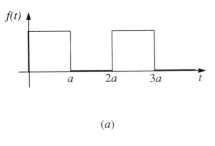
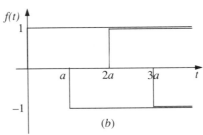

(a)   (b)

Now we use the identity

$$\frac{1}{1-x} = 1 + x + x^2 + x^3 + \ldots \qquad x < 1 \tag{6.163}$$

to obtain

$$F(s) = \frac{P(s)}{1 - e^{-Ts}} \tag{6.164}$$

In a similar fashion it may be shown that the transform of the repeated function shown in fig. 6.55(c) is given by

$$F(s) = \frac{P(s)}{1 + e^{-Ts}} \tag{6.165}$$

The expressions (6.164) and (6.165), which are sometimes referred to as the *periodicity theorem*, enable one to find the transform of a sequence of repeated pulses or a recurrent waveform given the transform of the individual pulse or waveform of which it is composed. For instance, the transform of the single rectangular pulse of duration a shown in fig. 6.53 is $\frac{1}{s}(1 - e^{-as})$ from (6.159). If this pulse is repeated with period T, then, by the periodicity theorem (6.164), the transform of the rectangular pulse train will be

$$F(s) = \frac{P(s)}{1 - e^{-Ts}} = \frac{1}{s} \frac{(1 - e^{-as})}{(1 - e^{-Ts})} \tag{6.166}$$

Fig. 6.55. A single pulse function $p(t)$ and its repeated versions $f(t)$.

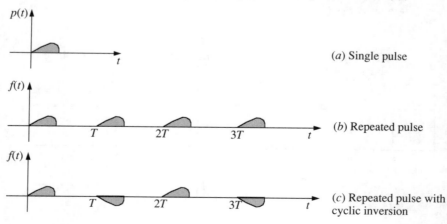

(a) Single pulse

(b) Repeated pulse

(c) Repeated pulse with cyclic inversion

Pulse and repeated driving functions

If the pulse train takes the form of a square wave, that is, if $T = 2a$ as in fig. 6.54, then

$$F(s) = \frac{1}{s} \frac{(1-e^{-as})}{(1-e^{-2as})} = \frac{1}{s(1+e^{-as})} \tag{6.167}$$

which agrees with our previous result (6.162).

As a further example, consider the triangular pulse shown in fig. 6.56. We may resolve this function into three components: a positive ramp starting at $t=0$, a negative ramp starting at $t=1$ and a negative step also starting at $t=1$. For $t \geq 1$ the sum of the three components is zero, thus the function is cut off at $t=1$. Now the first component is the unit ramp $\rho(t)$, and the second component is the (negative) unit delayed ramp $\rho(t-a)$ (see section 6.8.4). So, we may express the function as:

$$P(t) = \rho(t) - \rho(t-1) - u(t-1)$$

Referring to the table of transforms it is found that

$$P(s) = \frac{1}{s^2} - \frac{e^{-s}}{s^2} - \frac{e^{-s}}{s}$$

and the transform of the repeated pulse (fig. 6.57(a)) is by (6.164)

$$F(s) = \frac{P(s)}{1-e^{-Ts}} = \frac{1-e^{-s}-se^{-s}}{s^2(1-e^{-s})}$$

By the use of (6.165) we may also immediately write the transform of the waveform shown in fig. 6.57(b) as

Fig. 6.56. Triangular pulse resolved into two ramp functions and one step function.

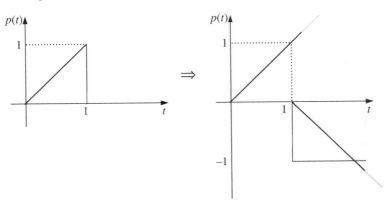

384 Transient and steady-state analysis

$$F(s) = \frac{P(s)}{1+e^{-Ts}} = \frac{1-e^{-s}-se^{-s}}{s^2(1+e^{-s})}$$

When applying the periodicity theorem, it is essential to ensure that the function $p(t)$ is correctly defined in terms of its resolved components. These must sum to zero for all t greater than the specified pulse length. This point is illustrated in our final example: the half-wave rectified sine wave shown in fig. 6.58(a).

Over the interval $0 \le t \le T/2$, $p(t)$ is specified by the function $u(t)\sin\omega t$ in which $u(t)$ indicates that $p(t)$ is initiated at $t=0$. However, as it stands, this function states that $p(t)$ is continuous for all $t > 0$ whereas we require it to be zero for $t > T/2$. This is accomplished by the addition of an identical sinusoidal function, shifted so that it starts at $t = T/2$, which has the effect of cancelling the original function for $t > T/2$ (fig. 6.58(b)). Thus,

$$p(t) = u(t)\sin\omega t + u\left(t - \frac{T}{2}\right)\sin\omega\left(t - \frac{T}{2}\right)$$

Fig. 6.57. Repeated versions of the triangular waveform of fig. 6.55.

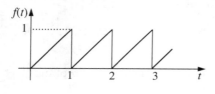

(a) Repeated triangular waveform (sawtooth)

(b) Repeated triangular waveform with cyclic inversion

Fig. 6.58. Formation of a single half-sine-wave pulse by the combination of a continuous function with a shifted function.

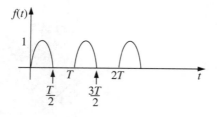

(a) Half-wave rectified sine wave

(b) Single half-sine-wave pulse

The transform of this function is

$$P(s) = \frac{\omega}{s^2 + \omega^2} + \frac{\omega e^{-Ts/2}}{s^2 + \omega^2} = \frac{\omega}{s^2 + \omega^2}(1 + e^{-Ts/2})$$

Now we can apply the periodicity theorem (6.164) to find the transform of the repeated version of this function. Note that the function is repeated with period T, consequently the exponent in the exponential of (6.164) is Ts, not $(T/2)s$. Therefore, the transform of the half-wave rectified sine wave of unit amplitude is:

$$F(s) = \frac{\omega}{(s^2 + \omega^2)} \cdot \frac{(1 + e^{-Ts/2})}{(1 - e^{-Ts})} = \frac{\omega}{(s^2 + \omega^2)(1 - e^{-Ts/2})} \qquad (6.168)$$

6.12.3 Response by the Laplace transform

In order to illustrate the transform method, and for comparison with a more elementary approach adopted in section 6.12.1, we return to a consideration of the RC circuit excited by a single pulse of amplitude V and duration a (fig. 6.59(a)). The transform of a rectangular pulse has already been derived (equation (6.159)) so that the circuit in the s-domain becomes as shown in fig. 6.59(b). The voltage across the capacitance is, therefore,

$$V(s) = \frac{1/sC}{R + 1/sC} \cdot \frac{V}{s}(1 - e^{-as}) = V\left[\frac{1/RC}{s(s + 1/RC)}\right](1 - e^{-as}) \qquad (6.169)$$

Because of the factor $(1 - e^{-as})$, which is not a polynomial, it is not possible to expand the whole of this function in the usual way using partial fractions, but the term within square brackets can be expanded to give:

$$\frac{1/RC}{s(s + 1/RC)} = \frac{1}{s} - \frac{1}{s + 1/RC}$$

Thus,

$$V(s) = V\left[\frac{1}{s} - \frac{1}{s + 1/RC}\right](1 - e^{-as})$$

Fig. 6.59. RC circuit excited by a single rectangular pulse of duration a.

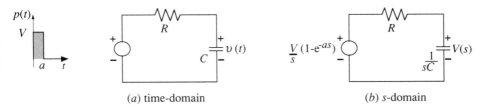

(a) time-domain (b) s-domain

which may be written

$$V(s) = V\left\{\left[\frac{1}{s} - \frac{1}{s+1/RC}\right] - \left[\frac{1}{s} - \frac{1}{s+1/RC}\right]e^{-as}\right\}$$

Upon inversion the first term of this expression yields $(1-e^{-t/RC})$, while the second term yields precisely the same function but delayed in time according to the shift factor e^{-as}. Hence, in the time domain we have:

$$v = V[(1-e^{-t/RC})u(t) - (1-e^{-(t-a)/RC})u(t-a)] \qquad (6.170)$$

Comparing this expression with our previous results we see that, for the duration of the pulse, only the first term of (6.170) is operative (because $u(t-a) = 0$ for $t < a$)) and we obtain:

$$v = V(1-e^{-t/RC}) \qquad 0 \le t \le a$$

which agrees with (6.153).

After the termination of the pulse, *both* terms of (6.170) are operative and

$$\begin{aligned}v &= V[(1-e^{-t/RC}) - (1-e^{-(t-a)/RC})] \qquad t \ge a \\ &= V(e^{-(t-a)/RC} - e^{-t/RC})\end{aligned}$$

which is in accord with (6.153) and (6.154).

Factors of the form $(1-e^{-as})$, like that appearing in (6.169), commonly occur in problems involving pulse driving functions. Such factors play no part in determining coefficients in the expansion of the s-domain circuit equation. Their effect is simply to establish two identical functions in the time domain: one starting at $t=0$, the other at $t=a$.

Continuing with our comparison of elementary and transform methods, we now use the latter to determine the response of the RC circuit to a train of rectangular pulses of unit amplitude (fig. 6.60(a)). The transform of the pulse train is given by (6.166), hence in the s-domain, we have the circuit of fig. 6.60(b). The circuit response is

$$V(s) = \frac{1/sC}{(R+1/sC)} \cdot \frac{1}{s} \frac{(1-e^{-as})}{(1-e^{-Ts})} = \left[\frac{1/RC}{s(s+1/RC)}\right]\frac{(1-e^{-as})}{(1-e^{-Ts})}$$

Expanding the term in square brackets gives

$$V(s) = \left[\frac{1}{s} - \frac{1}{s+1/RC}\right]\frac{(1-e^{-as})}{(1-e^{-Ts})}$$

$$= \left[\frac{1}{s} - \frac{1}{s+1/RC}\right]\frac{1}{(1-e^{-Ts})} - \left[\frac{1}{s} - \frac{1}{s+1/RC}\right]\frac{e^{-as}}{(1-e^{-Ts})}$$

Let $Q(s) = \left[\dfrac{1}{s} - \dfrac{1}{s+1/RC}\right]$,

then

$$V(s) = \frac{Q(s)}{1-e^{-Ts}} - \frac{Q(s)e^{-as}}{1-e^{-Ts}} \qquad (6.171)$$

Now the inverse of $Q(s)$ is $(1-e^{-t/RC})$ therefore, by using the periodicity theorem (6.164) in reverse, we find that the inverse of the first term in (6.171) is

$$f_1(t) = [1-e^{-t/RC}]u(t) + [1-e^{-(t-T)/RC}]u(t-T)$$
$$+ [1-e^{-(t-2T)/RC}]u(t-2T) + \ldots$$

To find the inverse of the second term in (6.171) we expand the factor $1/(1-e^{-Ts})$ using the identity (6.163):

$$Q(s)\frac{e^{-as}}{1-e^{-Ts}} = Q(s)e^{-as}(1+e^{-Ts}+e^{-2Ts}+e^{-3Ts}+\ldots)$$

$$= Q(s)[e^{-as} + e^{-(a+T)s} + e^{-(a+2T)s} + \ldots]$$

The series of shift factors within brackets leads to delayed functions in the time domain of the form

$$u(t-a) + u(t-a-T) + u(t-a-2T) + \ldots$$

Thus, the inverse of the second term in (6.171) is

$$f_2(t) = [1-e^{-(t-a)/RC}]u(t-a) + [1-e^{-(t-a-T)/RC}]u(t-a-T)$$
$$+ [1-e^{-(t-a-2T)/RC}]u(t-a-2T) + \ldots$$

The complete result is

$$v(t) = f_1(t) - f_2(t)$$

Fig. 6.60. RC circuit excited by a train of rectangular pulses of duration a and period T.

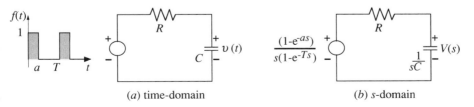

(a) time-domain

(b) s-domain

$$= \{[1-e^{-t/RC}]u(t) + [1-e^{-(t-T)/RC}]u(t-T)$$
$$+ [1-e^{-(t-2T)/RC}]u(t-2T) + \ldots\}$$
$$- \{[1-e^{-(t-a)/RC}]u(t-a) + [1-e^{-(t-a-T)/RC}]u(t-a-T)$$
$$+ [1-e^{-(t-a-2T)/RC}]u(t-a-2T) + \ldots\} \quad (6.172)$$

Let us use this expression to find the output of the circuit at the end of the second pulse, that is, at the instant $t = a + T$. In this case, all terms except the first two in each of the series contained in (6.172) vanish because of the operation of the delayed step functions. This leaves

$$v(t) = \{[1-e^{-(a+T)/RC}] + [1-e^{-a/RC}]\}$$
$$- \{[1-e^{-T/RC}][1-e^0]\}$$

which upon re-arrangement becomes

$$v(t) = (1-e^{-a/RC}) + (1-e^{-a/RC})e^{-T/RC}$$

This result is consistent with the expressions (6.155) and (6.157) obtained previously in section 6.12.1. A comparison of the above method of solution with that of section 6.12.1 will reveal that, although the amount of algebraic manipulation required in the transform method is formidable, it offers a more systematic approach. This is an advantage that becomes more marked the greater the complexity of the network and the excitation. The following example will serve to illustrate this point.

Fig. 6.61 shows a circuit used in a.c.–d.c. power supplies. The rectifier in this type of power supply produces a series of half-sine-wave pulses, as shown in the figure, and the function of the circuit is to smooth out these pulsations so that the supply delivers a constant d.c. output voltage. The steady-state analysis of this type of circuit is best accomplished by the use of Fourier series methods (see chapter 7), but sometimes the circuit designer may wish to determine the behaviour of the circuit under the transient conditions prevailing when the supply is first switched on. The transform method provides such information.

Using the admittance divider principle, we have for the circuit transfer function

Fig. 6.61. *RLC* circuit driven by a half-wave rectified sine-wave.

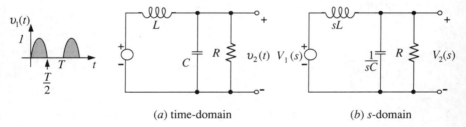

(a) time-domain   (b) s-domain

Worked example

$$H(s) = \frac{V_2(s)}{V_1(s)} = \frac{1/sL}{1/sL + sC + 1/R} = \frac{1}{LC(s^2 + s/RC + 1/LC)}$$

This type of circuit is always heavily damped so that the natural response will be non-oscillatory, and the two roots of the denominator polynomial in the above expression will be real. Let these roots be $-\alpha$ and $-\beta$ where $\alpha + \beta = 1/RC$ and $\alpha\beta = 1/LC$, then

$$H(s) = \frac{\alpha\beta}{(s+\alpha)(s+\beta)}$$

Now the transform of the half-wave rectified sine-wave has already been determined (equation 6.168), hence, the output voltage $V_2(s)$ is

$$V_2(s) = H(s)V_1(s) = \frac{\alpha\beta}{(s+\alpha)(s+\beta)} \cdot \frac{\omega}{(s^2+\omega^2)(1-e^{-Ts/2})}$$

$$= \alpha\beta\omega \left[\frac{1}{(s+\alpha)(s+\beta)(s^2+\omega^2)}\right] \frac{1}{(1-e^{-Ts/2})}$$

Partial fraction expansion of the term within square brackets gives

$$V_2(s) = \alpha\beta\omega \left[\frac{A_1}{s+\alpha} + \frac{A_2}{s+\beta} + \frac{A_3}{s+j\omega} + \frac{A_3^*}{s-j\omega}\right] \frac{1}{(1-e^{-Ts/2})}$$

where A_1, A_2, A_3, which are functions of α, β, ω, are found by the methods described in section 6.9.3. The inverse of the terms within square brackets is of the form

$$f(t) = A_1 e^{-\alpha t} + A_2 e^{-\beta t} + 2A_3 \cos(\omega t - \phi)$$

The function $1/(1-e^{-Ts/2})$ may be expanded using the identity (6.163)

$$\frac{1}{1-e^{-Ts/2}} = 1 + e^{-Ts/2} + e^{-Ts} + e^{-3Ts/2} + \ldots$$

which, as in our previous examples, is interpreted as a series of delayed functions in the time domain. Thus, the output voltage is finally given by

$$v_2(t) = \alpha\beta\omega \left[f(t)u(t) + f\left(t - \frac{T}{2}\right)u\left(t - \frac{T}{2}\right) + f(t-T)u(t-T) + \ldots\right]$$

6.13 Worked example

The TV video pulse train shown in fig. 6.62(a) is applied to the circuit of fig. 6.62(b). When the output voltage $v_2(t)$ of the RC circuit reaches a threshold voltage V_T, the field time base is triggered.

(a) Assuming that the circuit has attained steady-state conditions during the line pulse sequence, calculate the voltage $v_2(t)$ at the instant $t=t_A$.
(b) Calculate the voltage $v_2(t)$ at the end of the equalizing pulse sequence, i.e. at the instant $t=t_B$.
(c) If the field time base triggers at the instant $t=t_C=(t_B+48)\,\mu s$ (middle of second field pulse period), estimate the threshold voltage V_T.

Solution
(a) Under steady-state conditions the output of the RC circuit at the end of a pulse period, for an input pulse train of duration a and period T, (fig. 6.51) is given by

$$v_n = \frac{V(1-e^{-a/RC})}{(1-e^{-T/RC})} \cdot e^{-(T-a)/RC} \qquad (6.158)$$

For the line pulse sequence: $a=4.7\,\mu s$, $T=64\,\mu s$; the circuit time constant $RC=(18 \times 10^3) \times (2 \times 10^{-9}) = 36\,\mu s$. Hence, by 6.158, with $V=1$ V, we have

Fig. 6.62. Diagrams for worked example (section 6.13).

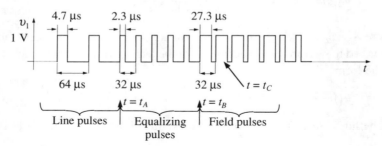

(a) TV video waveform: pulse train for field synchronisation.

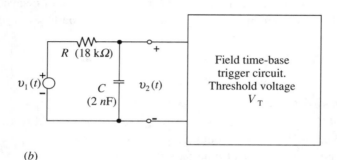

(b)

$$v_2 = \frac{(1-e^{-4.7/36})}{(1-e^{-64/36})} \cdot e^{-(64-4.7)/36} = 0.028 \text{ V}$$

(b) The output voltage of an *RC* circuit subject to an input pulse train of unit magnitude is, during the transient period, expressed by equation 6.172. A simplification of the algebra and a resulting saving of labour is achieved if in this equation and those derived from it, units of time are normalized with respect to the circuit time constant *RC*. In this case 6.172 may be written:

$$\begin{aligned} v_2 = &\{[1-e^{-t}]u(t) + [1-e^{-(t-T)}]u(t-T) \\ &+ [1-e^{-(t-2T)}]u(t-2T) + \ldots\} \\ &- \{[1-e^{-(t-a)}]u(t-a) + [1-e^{-(t-a-T)}]u(t-a-T) \\ &+ [1-e^{-(t-a-2T)}]u(t-a-2T) + \ldots\} \end{aligned}$$

The equalizing pulse sequence in the waveform shown in fig. 6.62(*a*) contains five complete pulse periods, therefore, taking the first five terms in each of the series in the above expression, defining a new time origin at $t = t_A$, and putting $t = 5T$ we obtain

$$\begin{aligned} v_2 = &\{[1-e^{-5T}] + [1-e^{-4T}] + \ldots + [1-e^{-T}]\} \\ &- \{[1-e^{-(5T-a)}] + [1-e^{-(4T-a)}] + \ldots + [1-e^{-(T-a)}]\} \end{aligned}$$

Combining corresponding terms from each of the series in this expression gives

$$\begin{aligned} v_2 &= e^{-5T}(e^a - 1) + \ldots + e^{-T}(e^a - 1) \\ &= (e^a - 1)(e^{-5T} + e^{-4T} + e^{-3T} + e^{-2T} + e^{-T}) \end{aligned}$$

Substituting normalized values: $a = 2.3/36$, $T = 32/36$, yields

$$v_2 = 0.066(0.01 + 0.03 + 0.07 + 0.17 + 0.41) = 0.045 \text{ V}$$

The above calculation assumes that the capacitance is uncharged, and therefore v_2 is zero, at the commencement of the equalizing pulse sequence, whereas it was found in section (a) above that a small but finite voltage (0.028 V) remains at the end of the line pulse sequence. This voltage will decay during the equalizing pulse sequence according to

$$v_2' = 0.028 e^{-5T} = 3.29 \times 10^{-4} \text{ V}$$

By superposition, the net voltage at the end of the equalizing pulse sequence will be the sum

$$v_2 + v_2' = 0.045 + 0.0003$$

It is seen from the above calculation that the initial voltage results in only a very small contribution to the voltage on C at the end of the equalizing pulse sequence.*

(c) In the following calculation for estimating the threshold voltage V_T we neglect the small voltage on C at the beginning of the field pulse sequence. (Its effect could, as in part (b), be calculated by superposition.) Since the field time-base triggers in the middle of the second field pulse period, we have by equation 6.172 (expressed in normalized units of time)

$$V_T = [1-\mathrm{e}^{-t}] + [1-\mathrm{e}^{-(t-T)}] - [1-\mathrm{e}^{-(t-a)}]$$

Substituting normalized values ($a = 27.3/36$; $T = 32/36$; $t = 48/36$) gives

$$V_T = 0.736 + 0.359 - 0.437 = 0.658 \text{ V}$$

This example illustrates the application of the RC circuit as a sync-separator. The output of the circuit remains at a low level during line and equalizing pulse sequences but rises rapidly on inception of the longer field pulses. This action causes the field time base to trigger.

†6.14 Convolution

The methods discussed so far in this chapter for finding the response of a circuit to a given excitation are confined to circumstances in which the excitation is expressible as an analytical function whose Laplace transform can be determined. The convolution method described here is not restricted in this way, and it is applicable also to cases for which the excitation function can be expressed only by a numerical data sequence.

In the convolution method the excitation function in the time-domain is resolved into a sequence of impulses, the response of the circuit to each impulse in the sequence is found and, finally, the responses are superposed to give the overall response function. To establish the basis of the method, we first consider how a function may be expressed in terms of a sequence of impulses.

6.14.1 Representation of a function by an impulse train

Any function $f(t)$ may be represented by a train of step or impulse functions; here we consider only the latter case. Fig. 6.63(a) shows a

---

* The voltage on C at the end of the line pulse sequence is slightly different for odd and even fields of a standard TV video waveform because even fields end with a half line pulse period whereas odd fields end with a full period (the latter is shown in fig. 6.62(a)). The equalizing pulse sequence is included in the video waveform to allow the voltage on C to decay substantially to the same small value for both odd and even fields of the interlaced picture. Without equalizing pulses the field time base would trigger at slightly different times after the start of the field pulse sequence, resulting in line pairing.

Convolution

function sampled at intervals of time $0, \Delta, 2\Delta, 3\Delta. \ldots$ During the first interval $(0 < t < \Delta)$ the function may be approximated by its value at the start of the interval, namely $f(0)$, as shown in the figure. During the second interval $(\Delta < t < 2\Delta)$ the function may be approximated by $f(\Delta)$; likewise during the third and succeeding intervals by $f(2\Delta), f(3\Delta)$ and so on. The function is thus broken up into short pulses of duration Δ with areas $\Delta f(0)$, $\Delta f(\Delta), \Delta f(2\Delta)$ etc. Recalling from our discussion in section 6.8.2 that a short pulse may be represented by an impulse of magnitude equal to its area, we may represent the complete function $f(t)$ by a train of impulses as shown in fig. 6.63(b). (In this figure the impulses are shown located at the start of their corresponding intervals whereas it might appear to be more logical to place them centrally. However, as we shall see later their precise location within the interval is immaterial.) The first impulse in the train will then be $\Delta f(0)\delta(t)$, the second impulse $\Delta f(\Delta)\delta(t-\Delta)$ and so on. Thus the function $f(t)$ may be expressed by the approximation:

$$f(t) \simeq \Delta f(0)\delta(t) + \Delta f(\Delta)\delta(t-\Delta) + \Delta f(2\Delta)\delta(t-2\Delta) + \ldots$$
$$+ \Delta f(n\Delta)\delta(t-n\Delta) + \ldots \Delta f(N\Delta)\delta(t-N\Delta)$$

$$\simeq \sum_{n=0}^{N} \Delta f(n\Delta)\delta(t-n\Delta) \qquad (6.173)$$

where $N = t_N/\Delta$ and t_N is some time at which the function terminates.

It must be remembered that in this summation all terms are zero except the term in which the argument of the impulse function is zero, that is, when the sampling instant $n\Delta$ is equal to t.

Fig. 6.63. Representation of a continuous function $f(t)$ by an impulse train.

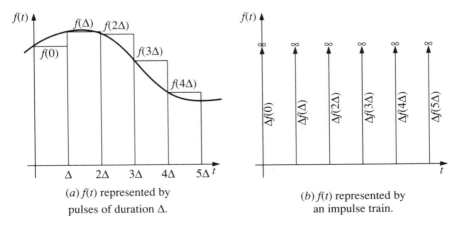

(a) $f(t)$ represented by pulses of duration Δ.

(b) $f(t)$ represented by an impulse train.

Now $f(t)$ can be approximated to any desired degree of accuracy by letting Δ become as small as necessary and by increasing N correspondingly. In the limit, as $\Delta \to 0$ and $N \to \infty$, $n\Delta$ becomes a continuous time τ while Δ becomes the differential $d\tau$. The function is then represented by a continuum of impulses and can be expressed exactly by the integral

$$f(t) = \int_0^{t_N} f(\tau)\delta(t-\tau)d\tau \qquad (6.174)$$

This integral may be interpreted in the following way. As the sampling time τ is varied over the range $0 \leq \tau \leq t_N$, the integrand, and therefore the integral, vanishes everywhere except when the argument of the impulse function is zero, that is, when $\tau = t$. At this instant the integral becomes

$$f(t)\int_0^{t_N} \delta(0)\delta\tau = f(t)$$

since, by definition, $\int \delta(0) = 1$.

Because we are concerned only with sampling times τ extending up to the value t, the upper limit of the integral (6.174) may be replaced by t, thus

$$f(t) = \int_0^t f(\tau)\delta(t-\tau)d\tau \qquad (6.175)$$

The property of the impulse function which enables it to be used to express a continuous function in the above manner is called the *sampling* or *sifting* property.

6.14.2 The convolution integral

Consider a network having an impulse response $h(t)$ (fig. 6.64(a)). An impulse of unit magnitude applied to the input terminals of this network will produce an output $h(t)$ typically of the form shown in the figure.

Next, consider the same network with an excitation function $f_1(t)$,[*] described by the approximate expression (6.173), applied to its terminals (fig. 6.64(b)). The first impulse of the sequence will produce a response $\Delta f_1 \Delta(0) h(t)$, the second impulse a response $\Delta f_1(\Delta) h(t-\Delta)$, and so on. We assume here that Δ is small compared with the effective time constant of the network (see section 6.8.2). Each succeeding impulse produces a response starting at the instant of the impulse and decaying in a fashion determined

---

[*] In previous sections the excitation and response functions in the time-domain have been denoted by $e(t)$ and $r(t)$ respectively. In this section we use $f_1(t)$ and $f_2(t)$, which conforms to the more usual notation found in texts on convolution theory.

Convolution

by the particular impulse characteristics of the network. The output at any particular instant t will be the resultant of all the impulses and responses occurring prior to, and including, that particular instant. By linearity, the output function will be given by the superposition of all the individual responses, that is, by

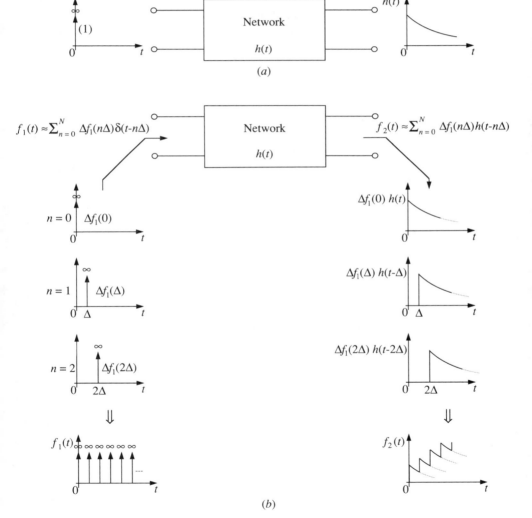

Fig. 6.64. (a) Response of a network to unit impulse excitation. (b) Response to an input function $f_1(t)$ approximated by a sequence of impulses.

$$f_2(t) \simeq \sum_{n=0}^{N} \Delta f_1(n\Delta) h(t - n\Delta) \tag{6.176}$$

The result of superposing the first four responses is indicated in the lower diagram appertaining to the output sequence.

In fig. 6.64 we have chosen for the sake of clarity a large sampling interval Δ, which has resulted in a sawtooth-like output waveform. But it will be appreciated that by taking shorter and shorter intervals, a progressively more faithful representation of the true output waveform will be obtained. The effect of halving the interval Δ is illustrated in figs. 6.65(a) and (b) for both input and output waveforms. Notice that halving Δ also halves the magnitudes of the input impulses and therefore their corresponding output responses. The net effect is a substantial smoothing of the output waveform. Carrying this process to the limit, and letting $\Delta \to 0$, the approximation (6.176) becomes an exact integral, the *convolution integral*:

$$f_2(t) = \int_0^t f_1(\tau) h(t-\tau) \, d\tau \tag{6.177}$$

Fig. 6.65. The effect of sampling interval on response by convolution. (a), (b) The effect of halving the interval. (c) Sampling continuum.

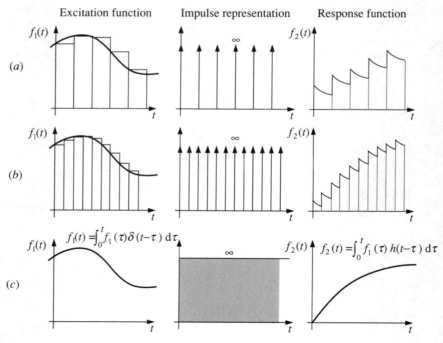

The limiting process used to derive this expression corresponds exactly to that leading to (6.175) and the symbols τ and t have the same meaning. Indeed, the convolution integral (6.177) follows directly from the integral (6.175) since for a network with an impulse response $h(t)$ the response to $\delta(t-\tau)$ is $h(t-\tau)$. These two integrals provide exact expressions for the excitation and response functions respectively (fig. 6.65(c)).

The convolution integral is often written in its most general form with the range of integration extending from $-\infty$ to $+\infty$. This then allows the range in any particular problem to be set in accordance with the constraints imposed by the conditions of that problem. In this book we are concerned almost entirely with functions that are zero for negative time, we shall, therefore, usually write the convolution integral, as may be convenient, either in the form (6.177) or as

$$f_2(t) = \int_0^\infty f_1(\tau) h(t-\tau) \, d\tau \qquad (6.178)$$

Further insight into the convolution integral will be gained from consideration of the graphical interpretation of the integration processes shown in fig. 6.66. The curves in this figure are all plotted with τ as the independent variable while t is a fixed point on the abscissa. The reason for this is that in the evaluation of the convolution integral, t is regarded as a fixed parameter while τ is a variable of integration (a dummy variable) ranging from zero to some upper limit.

The integrand of the convolution integral contains two functions: $f_1(\tau)$ and $h(t-\tau)$. We obtain a graphical plot of the latter in three stages: first, the function $h(\tau)$ is plotted (fig. 6.66(a)), and then this is folded about the vertical axis to form its mirror image $h(-\tau)$ (fig. 6.66(b)). Finally, this function is shifted to the right by an amount t to form the function $h(t-\tau)$ (fig. 6.66(c)). Of course, the form of $h(t-\tau)$ in (c) could be deduced directly without going through the intermediate stages (a) and (b), but we have adopted this procedure to illustrate the folding and shifting process from which the name 'convolution' integral, was historically derived.*

The next step in this graphical integration procedure is to form the product $f_1(\tau) h(t-\tau)$; this is shown in figs. 6.66(d) and (e). Finally, the area under this last curve, between the limits 0 and t, gives the value of the integral, which is equal to the response function $f_2(t)$ at the particular value of t considered. To find $f_2(t)$ at other values of t, the curve shown in fig. 6.66(c) must be shifted along the axis to the appropriate points. We may

---

* In the German language the name given to this important integral is the 'Faltung' (folding) integral.

398 *Transient and steady-state analysis*

then imagine the function $f_2(t)$ being generated by sliding or scanning the $h(t-\tau)$ function along the τ-axis of fig. 6.66(c) whilst performing the subsequent operations illustrated in the following figures (d) and (e). For this reason the $h(t-\tau)$ function is sometimes called the *scanning function*.

We now illustrate the convolution method using as an example the simple *RC* circuit excited by a pulse of unit amplitude and duration one second. This problem is one which has been dealt with at length in sections

Fig. 6.66. Graphical interpretation of convolution.

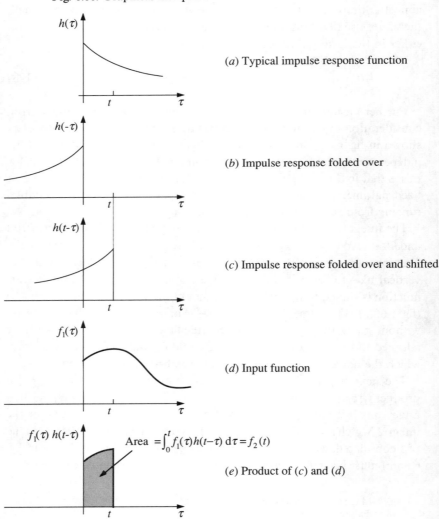

(a) Typical impulse response function

(b) Impulse response folded over

(c) Impulse response folded over and shifted

(d) Input function

Area $= \int_0^t f_1(\tau)h(t-\tau)\,d\tau = f_2(t)$

(e) Product of (c) and (d)

6.12.1 and 6.12.3; the reader should compare the following treatment with the methods described in these sections.

The first step in applying the convolution method is to find the impulse response of the circuit. Usually this is most easily accomplished by first finding the transfer function. For the RC circuit,

$$H(s) = \frac{1/sC}{R + 1/sC} = \frac{1}{RC} \frac{1}{(s + 1/RC)}$$

Thus,

$$h(t) = \mathscr{L}^{-1} H(s) = \frac{1}{RC} e^{-t/RC} u(t) \tag{6.179}$$

Assuming, for simplicity, that the time constant $RC = 1$, then

$$h(t) = e^{-t} u(t) \tag{6.180}$$

In the above expressions for $h(t)$ the step function $u(t)$ indicates that the response is zero for all negative time.

Now a rectangular pulse of unit amplitude may be described by (see section 6.12.2.)

$$v_1(t) = u(t) - u(t-a) \tag{6.181}$$

hence, using the convolution integral (6.178) we obtain

$$v_2(t) = \int_0^\infty \{[u(\tau) - u(\tau - a)] e^{-(t-\tau)} u(t-\tau)\} \, d\tau$$

$$= \int_0^\infty u(\tau) u(t-\tau) e^{-(t-\tau)} \, d\tau - \int_0^\infty u(\tau - a) u(t-\tau) e^{-(t-\tau)} \, d\tau$$

We have chosen here to write the convolution integral with limits 0 and ∞; this has been done because the limits of integration now need to be determined in accordance with the particular parameters of the problem. In the following argument it will be of help to recall the definition of the unit step function, viz.,

$$u(t) = 1 \quad t \geq 0$$
$$ = 0 \quad t < 0$$

Turning our attention to the integrand of the first integral above, $u(\tau)$ is zero for $\tau < 0$ and unity for $\tau \geq 0$, so the lower limit of integration is zero. The function $u(t - \tau)$ is unity for $\tau \leq t$ and zero for $\tau > t$, so the upper limit of integration is t. Similar considerations apply to the second integral: in this case the function $u(\tau - a)$ is zero for $\tau < a$ and unity for $\tau \geq a$ hence integration starts at the lower limit a. As before, the $u(t-\tau)$ defines the upper limit at t.

Using these new limits of integration the response becomes

$$v_2(t) = \int_0^t u(\tau)u(t-\tau)e^{-(t-\tau)}\,d\tau - \int_a^t u(\tau-a)u(t-\tau)e^{-(t-\tau)}\,d\tau$$

Now over the range of integration, all of the step functions are unity, so we obtain on integrating:

$$v_2(t) = [u(\tau)u(t-\tau)e^{-(t-\tau)}]_{\tau=0}^t - [u(\tau-a)u(t-\tau)e^{-(t-\tau)}]_{\tau=a}^t$$
$$v_2(t) = u(t)(1-e^{-t}) - u(t-a)(1-e^{-(t-a)}) \qquad (6.182)$$

(Some care is required when applying limits in the type of expressions encountered above: note particularly that, with $\tau=t$, $u(t-\tau) = u(t-t) = u(0) = 1$.)

Equation (6.182) is in agreement with (6.170) obtained previously. Over $0 \leq t \leq a$ only the first term is operative and the output rises exponentially. For $t \geq a$, both terms are operative and the output falls.

It will now be instructive to solve this same problem using a graphical method of convolution based on the approximate equation (6.176). We divide the pulse (assumed to be of 1 second duration) into n equal sub-intervals. Each sub-interval will be of duration $1/n$, and the area under the curve contained within each sub-interval will also be $1/n$. For the sake of clarity in the diagrams we choose a small number of sub-intervals, say $n=4$ (fig. 6.67(a)). The four pulses thus obtained are now replaced by four impulse functions of magnitude $\frac{1}{4}$ (fig. 6.67(b)). The response to each of these impulses will be $(\frac{1}{4})e^{-t}$ as shown in fig. 6.67(c). The total response, obtained by summing the individual responses, is shown by the dotted

Fig. 6.67. Response of the RC circuit excited by a single rectangular pulse: graphical convolution. (a) Pulse divided into four short pulses. (b) Each short pulse replaced by an impulse. (c) Response obtained by convolution (dotted curve) compared with theoretical curve (dashed curve).

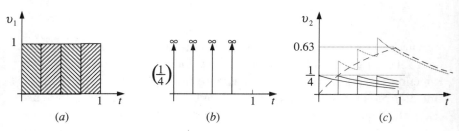

curve. This may be compared with the exact solution indicated by the dashed curve.

The above example is intended to help the reader new to the theory of convolution to gain familiarity with the concepts involved; for this particular problem, however, we should stress that one of the elementary methods of solution discussed in earlier sections would be more appropriate. Moreover, for problems not amenable to an analytical solution, it would be more usual to employ a numerical method for evaluating the convolution integral (see section 6.14.4) rather than the graphical procedure, used in this example.

The convolution integral is encountered frequently in many branches of engineering and physics. In general terms the convolution of two functions, say $\phi_1(t)$ and $\phi_2(t)$, is denoted symbolically by

$$\phi_1 * \phi_2 \equiv \int_{-\infty}^{\infty} \phi_1(\tau)\phi_2(t-\tau)\,d\tau \qquad (6.183)$$

where $*$ is read as *'convolved with'*. We can show very easily that ϕ^1 convolved with ϕ_2 is identical to ϕ_2 convolved with ϕ_1.

Let $z = t - \tau$ then $dz = -d\tau$ and (6.183) becomes

$$\phi_1 * \phi_2 = -\int_{\infty}^{-\infty} \phi_1(t-z)\phi_2(z)\,dz$$

$$= \int_{-\infty}^{\infty} \phi_2(z)\phi_1(t-z)\,dz = \phi_2 * \phi_1 \qquad (6.184)$$

6.14.3 The convolution theorem

In preceding sections of this chapter we have developed a method of finding the response $f_2(t)$ of a circuit to a given excitation $f_1(t)$ using the Laplace transform. The method involves: (1) finding the transform of the excitation, $F_1(s)$; (2) finding the transfer function of the circuit, $H(s)$; and (3) finding the inverse transform of the product $F_1(s) \times H(s)$.

The convolution integral, on the other hand, allows one to find the response directly in the time domain. The two approaches to the problem, and the way in which they are related, are illustrated in fig. 6.68. It will be evident from this diagram that convolution in the time domain corresponds to multiplication in the transform domain; a relationship expressed by the *convolution theorem*:

$$f_2(t) = \mathscr{L}^{-1}[F_1(s)H(s)] = \int_0^{\infty} f_1(\tau)h(t-\tau)\,d\tau \qquad (6.185)$$

402 Transient and steady-state analysis

This theorem may be derived directly from the definition of the Laplace transform. Taking the transform of the RHS of (6.185), and calling this $F(s)$, we have

$$F(s) = \mathscr{L}\left[\int_0^\infty f_1(t)h(t-\tau)\,d\tau\right] = \int_0^\infty \left[\int_0^\infty f_1(\tau)h(t-\tau)\,d\tau\right]e^{-st}\,dt$$

Changing the order of integration:

$$F(s) = \int_0^\infty f_1(\tau)\left[\int_0^\infty h(t-\tau)e^{-st}\,dt\right]d\tau$$

The integral within brackets is recognized as the Laplace transform of the delayed function $h(t-\tau)$: which is, from the shift theorem (6.141), $e^{-s\tau}H(s)$. Therefore,

$$F(s) = \int_0^\infty f_1(\tau)e^{-s\tau}H(s)\,d\tau = H(s)F_1(s)$$

which proves (6.185).

The commutative property of convolution (6.184) also follows from the convolution theorem since

$$\mathscr{L}^{-1}[\Phi_1(s)\Phi_2(s)] = \phi_1(t) * \phi_2(t)$$

and

$$\mathscr{L}^{-1}[\Phi_2(s)\Phi_1(s)] = \phi_2(t) * \phi_1(t)$$

Clearly, these two expressions are identical.

For the majority of the analytical excitation functions encountered in circuit theory, convolution in the time domain involves a more difficult

Fig. 6.68. Illustrating the convolution theorem.

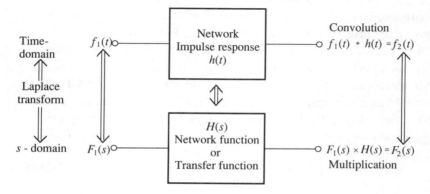

Convolution

mathematical procedure than multiplication in the transform domain. However, when the excitation is arbitrary and cannot be expressed or modelled by an analytical function, or if the Laplace transforms of the functions involved cannot be readily found, then numerical evaluation of the convolution integral provides a powerful technique for obtaining the network response.

6.14.4 Worked example

The 'phase advance' circuit shown in fig. 6.69 is often used in electronic control system networks. Show that the voltage transfer function for this circuit is

$$H(s) = \frac{V_2(s)}{V_1(s)} = \frac{s + 1/T}{s + 1/GT}$$

Fig. 6.69. Circuit and waveforms for worked example (section 6.14.4).

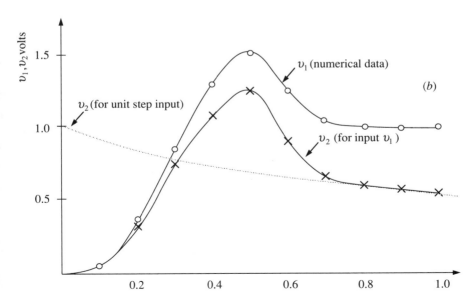

404 Transient and steady-state analysis

where $T=CR_1$ and $G=R_2/(R_1+R_2)$. Hence determine the impulse response $h(t)$ if $T=1$ second and $G=0.5$.

Find a general expression for the output $v_2(t)$ in terms of the input $v_1(t)$ and $h(t)$. Hence, determine the output if the input is: (a) the unit step function; (b) the waveform specified by the following numerical data:

t seconds)	0	0.1	0.2	0.3	0.4	0.5	0.6	0.7	0.8	0.9	1.0
v_1 (volt)	0	0.07	0.38	0.83	1.3	1.5	1.23	1.04	1.0	1.0	1.0

Solution

By the method established in section 6.9.6 we obtain

$$H(s) = \frac{sC + 1/R_1}{sC + 1/R_1 + 1/R_2} = \frac{s + 1/CR_1}{s + (1/C)(R_1 + R_2)/R_1 R_2} = \frac{s + 1/T}{s + 1/GT}$$

For $T=1$ second and $G=0.5$:

$$H(s) = \frac{s+1}{s+2}$$

Expressing $H(s)$ as a partial fraction (see section 6.9.3)

$$H(s) = 1 - \frac{1}{s+2}$$

hence, the impulse response is

$$h(t) = \mathscr{L}^{-1}[H(s)] = [\delta(t) - e^{-2t}]u(t)$$

Assuming that the input voltage is zero for $t<0$, then the input is described by $v_1(t)u(t)$ and by convolution we obtain

$$v_2(t) = \int_0^\infty v_1(\tau) h(t-\tau) \, d\tau$$

$$= \int_0^\infty v_1(\tau) u(\tau) \{[\delta(t-\tau) - e^{-2(t-\tau)}]u(t-\tau)\} \, d\tau$$

$$= \int_0^\infty v_1(\tau) u(\tau) \delta(t-\tau) u(t-\tau) \, d\tau$$

$$- \int_0^\infty v_1(\tau) u(\tau) e^{-2(t-\tau)} u(t-\tau) \, d\tau$$

In the above integrals the step functions define the range of integration as $0 \le \tau \le t$, and are unity over this range, therefore,

$$v_2(t) = \int_0^t v_1(\tau) \delta(t-\tau) \, d\tau - \int_0^t v_1(\tau) e^{-2(t-\tau)} \, d\tau$$

The first of these integrals is, from (6.175), simply equal to the input function itself, hence a general expression for the output voltage is

$$v_2(t) = v_1(t) - e^{-2t} \int_0^t v_1(\tau) e^{2\tau} d\tau \qquad (6.186)$$

(a) *Step function input*
Putting $v_1(t) = u(t)$ in (6.186) gives

$$v_2(t) = u(t) - e^{-2t} \int_0^t u(\tau) e^{2\tau} d\tau$$

$$= u(t) - \frac{e^{-2t}}{2} u(t)[e^{2\tau}]_0^t$$

$$v_2(t) = \tfrac{1}{2}(1 + e^{-2t})u(t)$$

This expression is shown plotted in fig. 6.69(b).

(b) *Numerical input data*
The input waveform, defined by the numerical data, is shown in fig. 6.69(b). Numerical integration is used to evaluate (6.186).

τ or t	0.1	0.2	0.3	0.4	0.5	0.6	0.7	0.8	0.9	1.0
$v_1(\tau)e^{2\tau}$	0.085	0.57	1.51	2.89	4.07	4.08	4.21	4.95	6.04	7.38
$I = \int_0^t v_1(\tau)e^{2\tau}d\tau$	0.004	0.04	0.14	0.36	0.71	1.12	1.53	1.99	2.54	3.21
$v_2 = v_1 - e^{-2t} I$	0.067	0.36	0.75	1.14	1.24	0.89	0.66	0.60	0.58	0.57

In the above table the integral I has been evaluated using the trapezoidal rule and a pocket calculator. The results are plotted in fig. 6.69(b). As to be expected, the output initially follows the input quite closely because of the capacitive coupling between input and output; with increasing time the output diverges from the input and decays towards its d.c. level of 0.5 V.

6.15 Summary

In this chapter a variety of methods have been developed for finding the response of a linear network to various forms of excitation. Broadly, these fall into three categories: time-domain techniques in which the network differential equations are set up and solved; Laplace transform techniques in which the network equations are formulated and solved in the transform domain; and convolution techniques in which the network is characterized by an impulse response and the network response is found by means of a convolution integral.

Transient and steady-state analysis

For any linear network the response $r(t)$ and excitation $e(t)$ are related by a differential equation of the form:

$$a_n \frac{d^n r}{dt^n} + a_{n-1} \frac{d^{n-1} r}{dt^{n-1}} + \ldots + a_1 \frac{dr}{dt} + a_0$$

$$= b_m \frac{d^m e}{dt^m} + b_{m-1} \frac{d^{m-1} e}{dt^{m-1}} + \ldots + b_1 \frac{de}{dt} + b_0 \quad (6.136)$$

The solution of this equation can be written:

$$\text{total response } r(t) = r_n(t) + r_{ss}(t)$$

where $r_n(t)$ is the transient or natural response, and $r_{ss}(t)$ is the steady-state or forced response.

The natural response contains terms of the form $A_i e^{-t/\tau_k}$ which die away with increasing time according to the network time constants τ_k. The constants A_i depend on both the initial energy states of the network and upon the form of the excitation; they can be evaluated only after the form of the complete solution of the network differential equation has been found. For high-order networks, the evaluation of these constants can be troublesome.

The forced response depends only on the excitation; for d.c. or steady sinusoidal excitation it is most easily found by using the standard techniques of d.c. and a.c. network analysis developed in chapters 2 and 3.

For many problems, it is often convenient to find the natural and forced responses separately, and then combine them to find the complete response. This approach is particularly advantageous for first and second order networks and where the excitation consists of step, sinusoidal or other simple functions.

The D-operator method facilitates the process of setting up the network differential equation and, for certain types of excitation, of finding its solution. In this method the network differential equation (6.136) is replaced by an equation of the form:

$$(a_n D^n + a_{n-1} D^{n-1} + \ldots + a_1 D + a_0) r(t)$$
$$= (b_m D^m + b_{m-1} D^{m-1} + \ldots + b_1 D + b_0) e(t)$$

in which D may be treated as an algebraic quantity.

The Laplace transform provides the most powerful and comprehensive means of analysing the transient and steady-state behaviour of linear networks. By taking the Laplace transform of the general network differential equation (6.136) we obtain

Summary

$$(a_n s^n + a_{n-1} s^{n-1} + \ldots + a_1 s + a_0) R(s)$$
$$= (b_m s^m + b_{m-1} s^{m-1} + \ldots + b_1 s + b_0) E(s) \tag{6.139}$$

where s is the complex frequency, and $R(s)$ (the response function) and $E(s)$ (the excitation function) are the Laplace transforms of $r(t)$ and $e(t)$ respectively. The ratio of $R(s)$ and $E(s)$, defines a network function:

$$H(s) = \frac{R(s)}{E(s)} \tag{6.134}$$

In the practical application of the method, each element of the network is expressed as a generalized reactance or impedance function in s, which process allows the network function $H(s)$ to be formulated. The excitation function $E(s)$ is conveniently obtained from $e(t)$ using a table of Laplace transform pairs. The response function $R(s)$ is then formed from the product $H(s) \times E(s)$, and finally the inverse of $R(s)$ is found, again using a table of transform pairs, which yields the response $r(t)$ in the time domain.

An important advantage of the Laplace transform method is that information concerning the initial energy states of the network can be incorporated into the excitation function; thus, the necessity of evaluating the arbitrary constants, A_i, as required in the solution of the network differential equation, is obviated. Also, because the complex frequency s may be manipulated in the s-domain as an algebraic identity, the method retains, in this respect, the advantage of the D-operator method. The pole-zero diagram provides an important adjunct to the Laplace transform method. It affords the network designer with a ready means for the appreciation and understanding of network behaviour under both transient and steady-state conditions. No such pictorial device exists for the other methods of network analysis.

In general, the use of the Laplace transform method is advantageous when dealing with networks of order three or higher, and for excitations of complex form. For uncomplicated circuits and excitations of simple form, its use can involve unnecessary algebraic complexity.

The response of a network can be found by the Laplace transform method only if the excitation is expressible as an analytical function. The same applies to the use of the network differential equation. The convolution method, on the other hand, suffers from no such restriction; the network response may be found even in cases where the excitation is described in terms of a numerical data sequence. In the convolution method, $e(t)$ is expressed as a summation of delayed impulse functions; the network is characterized by an impulse function $h(t)$ (the inverse transform of $H(s)$), and the overall response is found by convolving $e(t)$ with $h(t)$ using the convolution integral. We have shown, by means of the convolution theorem, that this procedure corresponds to multiplication of $E(s)$ by $H(s)$ in the transform domain. The convolution method is particularly well suited to the numerical integration procedures available on most calculators and small computers.

6.16 Problems

1. A voltage having a step waveform of amplitude 50 V is applied to a circuit formed by a 1 MΩ resistor in series with a 1 µF capacitor. What is the time constant of the circuit? How long does the capacitor take to charge to: (a) 25 V; (b) 47.5 V?

2. In the circuit of fig. 6.70 the switch has been in position 1 for a long time. At $t=0$ it is thrown to position 2.
(a) What are i_c and v_c at $t=0^+$?
(b) Determine an expression for $i_c(t)$ for $t>0$.
(c) What are i_c and v_c at $t=\infty$?
(d) How much energy is stored in the capacitor at $t=\infty$?
(e) How much energy has been supplied by the battery during the charging process?
(f) Show that the energy of part (e) is twice that stored in the capacitance, regardless of the size of the resistance.

3. For the circuit of fig. 6.71, derive equations for i, i_1 and i_2 valid from the instant $t=0$ when switch S is closed. (C is uncharged initially.) S is reopened

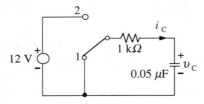

Fig. 6.70. Circuit for problem 2.

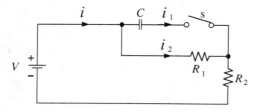

Fig. 6.71. Circuit for problem 3.

after one time constant has elapsed. If $V=1\,\text{V}$, $R_1=R_2=2\,\text{k}\Omega$ and $C=1\,\mu\text{F}$, what will be the values of i immediately before and immediately after opening S?
Sketch the variation of current i from $t=0$.
(Manchester University)

4. In the circuit shown in fig. 6.72 switches S_1 and S_2 are open.
(a) At $t=0$, S_2 is closed. Obtain expressions for the current in and the voltage across C for time $t=0.01\,\text{s}$. C is initially uncharged.
(b) At $t=0.05\,\text{s}$, S_2 is opened and at the same time S_1 is closed. Obtain an expression for the current in C subsequent to this operation and determine the voltage across C for $t=0.1\,\text{s}$.
(c) With S_1 remaining closed, write down the circuit equations and the initial conditions for the sudden reclosing of S_2.
(Wales Science and Technology)

5. The operating coil of a relay working on a 20 V d.c. supply and activated by opening and closing switch S is represented by L, R in the circuit of fig. 6.73. The coil is shunted by a non-inductive resistor R_1 and the parameters of the circuit are as shown.

The relay closes when the current through its coils reaches 180 mA and opens when the current falls to 60 mA. Calculate the time lags for opening and closing respectively.
(Newcastle University)

Fig. 6.72. Circuit for problem 4.

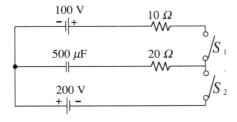

Fig. 6.73. Circuit for problem 5.

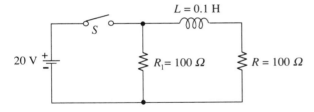

6. In the circuit of fig. 6.74 the initial current in the inductance is 2 mA.
(a) With $e_s(t) = 4\cos 10^3 t$, the switch is closed at $t=0$. Find an expression for $i_L(t)$ for $t>0$.
(b) If $e_s(t) = 4\cos(10^3 t + \phi)$, determine the value of ϕ such that there will be no transient when the switch is closed at $t=0$.

7. The current through the deflecting coils of a cathode-ray tube is required to rise linearly with time from zero at the rate of 3 A/s. What must be the form of the applied voltage if the coils have an inductance of 0.1 H and a resistance of 5 Ω? What would be the form of the voltage if the resistance were zero?

8. The following problem relating to the circuit shown in fig. 6.75, is an exercise in determining initial and final conditions. The switch has been open for a long time.
(a) What are the values of i_1, i_2 and v_c? At $t=0$, the switch is closed.
(b) Calculate: i_1, i_2, v_c, di_1/dt, di_2/dt and dv_c/dt all at $t=0^+$.
(c) Calculate: i_1, i_2 and v_c at $t=\infty$.

9. The circuit between two terminals consists of two branches in parallel. One branch contains an inductance L and a resistance R_1 in series; the other a capacitance C and a resistance R_2 in series. By considering the

Fig. 6.74. Circuit for problem 6.

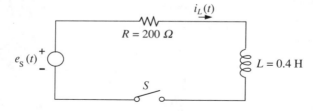

Fig. 6.75. Circuit for problem 8.

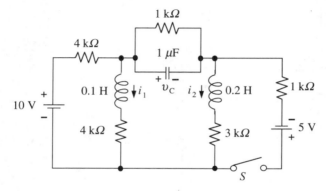

currents which flow in the branches when a generator of e.m.f. E and zero internal impedance is suddenly connected to the terminals, find the relations between L, C, R_1 and R_2 such that the circuit behaves as a pure resistance.

10. In the circuit of fig. 6.76 determine the value of R in terms of L, C and r so that the potential difference between A and B will be non-oscillating when switch S is opened.
(Manchester University)

11. For the circuit of fig. 6.77; derive the differential equation, expressed in terms of the D-operator, relating $v_2(t)$ and $v_1(t)$. Assuming that the circuit is over-damped, write the form for the natural response for $v_2(t)$ and derive the time constants in the expression.

12. Two coils each having inductance L have mutual inductance M. One coil has a resistance R connected in parallel with it, and the other has a voltage E suddenly applied through a resistance R to its terminals. Find an expression for the subsequent voltage between the terminals of the first coil.

13. In the circuit of fig. 6.78 the initial conditions are: $i(0)=2$ A, $v_c(0)=1$ V. The switch is closed at $t=0$.
(a) If $e_s(t)=1$, find $v_c(t)$ for $t>0$.
(b) If $e_s(t)=2\cos 2t$, find $v_c(t)$ and $i(t)$ for $t>0$.

Fig. 6.76. Circuit for problem 10.

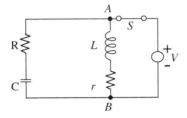

Fig. 6.77. Circuit for problem 11.

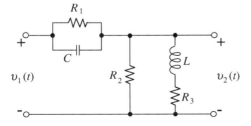

14. When a 250 pF capacitor charged to 100 V is connected to an inductance coil, it is found that the discharge is oscillatory, and that the peak voltage falls to 10 V after 150 μs, corresponding to 150 cycles of oscillation.

It is also found that if the experiment is repeated with a pure resistor R_1 connected across the coil, the peak voltage falls from 100 V to 10 V after 90 μs, corresponding to 90 cycles of oscillation. Determine:
(a) the inductance and Q factor of the coil;
(b) the ohmic value of the resistance R_1.

The expression $v = V_0 \exp(-Rt/2L)\cos\omega t$ for the transient voltage v may be assumed.
(London University)

15. In the circuit of fig. 6.79 the switch is closed at $t = 0$. Show that

$$i_1 = \frac{V}{R_1}\left[1 - e^{-\alpha t}\left\{\cosh\beta t - \frac{L_2 R_1 - L_1 R_2 \pm 2MR_1}{2(L_1 L_2 - M^2)\beta}\sinh\beta t\right\}\right]$$

where

$$\alpha = \frac{L_1 R_2 + L_2 R_1}{2(L_1 L_2 - M^2)}; \qquad \beta^2 = \alpha^2 - \frac{R_1 R_2}{L_1 L_2 - M^2}$$

(Manchester University)

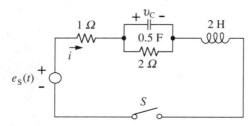

Fig. 6.78. Circuit for problem 13.

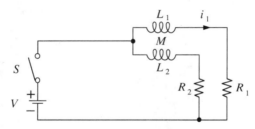

Fig. 6.79. Circuit for problem 15.

Problems

16. In the circuit of fig. 6.80 the switch is initially open and the capacitor is charged so that $v_c(0) = 10$ V. At $t=0$, the switch is closed. Use the Laplace transform method to find $i(t)$ and $v_c(t)$ for $t > 0$.

17. Discuss the benefits of using the Laplace transform method in network analysis.

In the network of fig. 6.81, $v_1(t)$ and $v_2(t)$ are the voltages on C_1 and C_2 respectively, and $i_L(t)$ is the current through L.

(a) If the components have the values indicated, show that the Laplace transform of the output voltage $v_2(t)$ is of the form:

$$V_2(s) = \frac{I(s) + as^2 + bs + c}{(s+1)(s^2+s+1)}$$

where $I(s)$ is the Laplace transform of the input $i(t)$. Derive expressions for the coefficients a, b and c in terms of the initial conditions $v_1(0)$, $v_2(0)$, and $i_L(0)$.

(b) If $i(t)$ is a unit step function of current applied at $t=0$, what initial conditions are required to produce an output of 1 V for $0 < t < \infty$? Show that your answer is consistent with the *initial value* and *final value* theorems.

(c) If $i(t)$ is a unit step function of current and the circuit is initially quiescent, find an expression for the time-varying voltage for $0 < t < \infty$.

(Cambridge University)

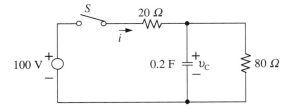

Fig. 6.80. Circuit for problem 16.

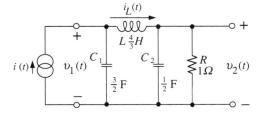

Fig. 6.81. Circuit for problem 17.

414   Transient and steady-state analysis

18. A voltage waveform which starts at $t=0$, rises linearly at the rate of 10 V/s until it reaches an amplitude of 1 V and then instantaneously falls to zero, is applied to a resonant circuit consisting of resistance R, inductance L, and capacitance C, all connected in series. Determine the Laplace transform of the voltage waveform and the current $i(t)$ flowing in the series resonant circuit for $t>0$. Assume the Q factor of the circuit to be greater than 100 and that it is at rest when the waveform is applied.
(Newcastle University)

19. In the circuit of fig. 6.82 the switch is opened at $t=0$. Show that for $t>0$:

$$i = \frac{j\omega CI}{1 - (RC\omega)^2 - j3\omega CR}$$

$$\times \left[e^{j\omega t} - e^{-\alpha t} \left\{ \cosh\beta t - \frac{2 - j3\omega CR}{j\omega CR\sqrt{5}} \sinh\beta t \right\} \right]$$

where

$$\alpha = \frac{3}{2RC} \qquad \beta = \frac{\sqrt{5}}{2RC}$$

(Manchester University)

20. (a) Find the Laplace transform of each of the following time functions:

$$e^{\alpha t}, \ te^{\alpha t}, \ \sin\omega t, \ \sin(\omega t + \phi)$$

Fig. 6.82. Circuit for problem 19.

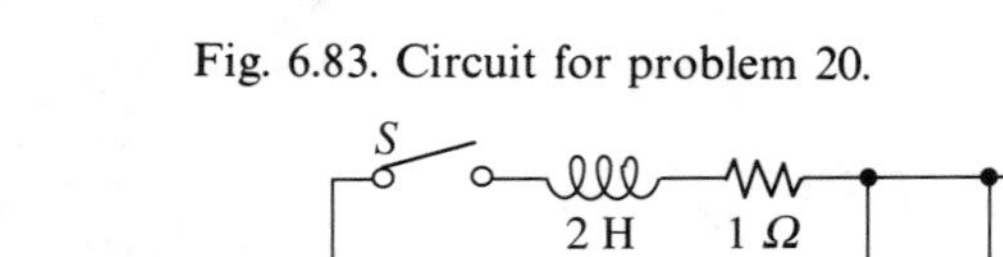

Fig. 6.83. Circuit for problem 20.

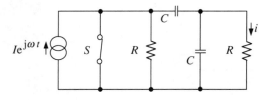

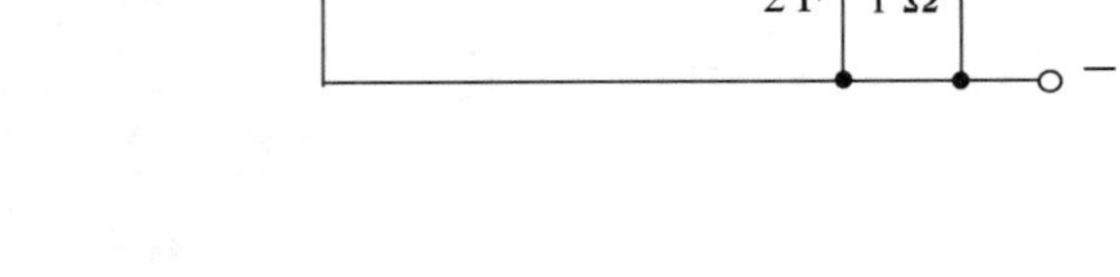

(b) The switch S in the circuit shown in fig. 6.83 is closed at $t=0$. If all the initial conditions in the circuit are zero, find the output voltage v_2 as a function of the complex frequency s and as a function of time t.
(University of Kent)

21. Calculate the output voltage $v_0(t)$ in the circuit of fig. 6.84 if the current source delivers a pulse of 1 mA lasting 50 μsec. $R_1 = R_2 = 10\,\text{k}\Omega$, $C_1 = 100\,\text{pF}$, $C_2 = 0.01\,\mu\text{F}$.
(Oxford University)

22. Show that the voltage transfer function $G(s)$ of the network of fig. 6.85 may be expressed as:

$$G(s) = \frac{E_0(s)}{E_{in}(s)} = \frac{1 + 2s\tau + \alpha s^2 \tau^2}{1 + 2\beta + s\tau(2 + \alpha + \alpha\beta) + \alpha s^2 \tau^2}$$

where $\tau = CR$, $\alpha = C_1/C$, $\beta = R/R_L$, and $E_{in}(s)$ and $E_0(s)$ are the Laplace Transforms of the input voltage $e_{in}(t)$ and output voltage $e_0(t)$ respectively.

Sketch the frequency response of the network for the case where R_L is infinite and $C_1 > C$. Indicate the asymptotic values of the response, and any maxima or minima, with their corresponding frequencies.

Show how this network might be used as the frequency-determining element of a sinusoidal oscillator.
(University of Kent)

23. Find the transfer function $H(s) = V_2(s)/V_1(s)$ of the third-order Butterworth filter shown in fig. 6.86. Verify that the amplitude response

Fig. 6.84. Circuit for problem 21.

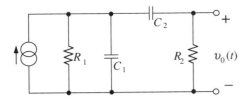

Fig. 6.85. Circuit for problem 22.

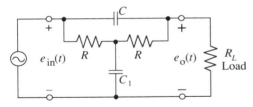

function for a sinusoidal input at frequency ω is given by

$$|H(j\omega)| = \frac{1}{\sqrt{(1+\omega^6)}}$$

Show that the poles of the transfer function lie on a circle of unity radius.
(Cambridge University)

24. Obtain an expression for the input impedance (in operational form) of the circuit shown in fig. 6.87. Find the zeros of this expression and hence find the minimum resistance required if the excitation of current oscillations by a step voltage input is to be avoided.
(Oxford University)

25. The impulse response of a potential divider is

$$f(t) = e^{-\alpha t}[\alpha \cosh \beta t - \beta \sinh \beta t], \qquad \alpha > \beta$$

Find the frequency response function of the divider, and from it devise a possible circuit.
(Manchester University)

26. The equivalent circuit of an L–C surge absorber interposed at the junction between two transmission lines is shown in fig. 6.88.

An 'impulse' voltage (VT) volt-secs, expressible as $e_1 = (VT)s.1$, is applied at time $t=0$ to the input terminals as shown. Show that the resultant output voltage e_2 may be expressed as

Fig. 6.86. Circuit for problem 23.

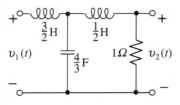

Fig. 6.87. Circuit for problem 24.

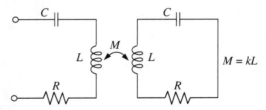

$$e_2 = (VT)R_2 \frac{s}{LCR_2 s^2 + (L + R_1 R_2 C)s + (R_1 + R_2)} \cdot 1$$

Show also that, if e_2 is to be non-oscillatory, the necessary condition which must be fulfilled is

$$(Z^2 - R_1 R_2) > 2ZR_2$$

where

$$Z = \sqrt{\frac{L}{C}}$$

In a particular case, with numerical values inserted, the foregoing expression for e_2 reduces to

$$e_2 = 10^6 \left[\frac{s}{s+1} - \frac{s}{s+10} \right] \cdot 1$$

where t is measured in micro-seconds.

Sketch the resultant wave-form of e_2, and show that it attains its maximum value of approximately 698 000 V at time $t \doteqdot 0.26\,\mu s$. (Newcastle University)

Fig. 6.88. Circuit for problem 26.

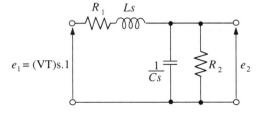

7
Non-linear circuit analysis

7.1 Introduction: linear and non-linear elements

In the preceding chapters, discussions have been confined to circuits that could be modelled by linear elements. The property that characterizes a linear element is independent of current or voltage. For example, the voltage–current characteristic of a resistor, modelled by a pure, linear resistance R is a straight line passing through the origin and having a slope equal to $1/R$ (fig. 7.1(a)). A linear resistor is also *bilateral*, that is, its voltage–current relationship is the same in the first and third quadrants of the characteristic. This property implies that the component can be connected into a circuit without regard to the polarity of the voltage to which it is subjected.

An example of a non-linear resistor is the incandescent lamp whose v–i characteristic is shown in fig. 7.1(b). The non-linearity in this case results from the great increase in temperature of the filament as it becomes incandescent. The characteristics of the lamp are the same whichever way round it is connected to the supply, and the device, although non-linear, is bilateral. On the other hand, the diode, whose v–i characteristic is shown in fig. 7.1(c), is neither linear nor bilateral.

Linear inductors and linear capacitors present a behaviour similar to that of the linear resistor. The response, $v_L = L\dfrac{\mathrm{d}}{\mathrm{d}t}[i(t)]$, of a linear inductor to a changing current is independent of the magnitude of the current in the inductor. Likewise, $i_C = C\dfrac{\mathrm{d}}{\mathrm{d}t}[v(t)]$ is independent of the magnitude of the voltage across the capacitor. An ordinary capacitor rarely exhibits appreciable non-linear behaviour unless it is driven beyond the voltage range for which it was designed. An inductor that contains no ferromagnetic material, or which has an appreciable air gap in its ferromagnetic core if it is present, exhibits linear behaviour. If, however, the path for the flux is entirely contained within ferromagnetic material, then, except for vanishingly small amplitudes of currents, the device will be non-linear and will exhibit the phenomena of saturation and hysteresis that were discussed in chapter 4.

In chapter 2 theorems were developed that depended upon the property of linearity, and we saw in later chapters how one of the most important of these – superposition – allowed us to find the total response of a circuit to an excitation by adding or superposing the transient and steady-state responses found separately. This technique cannot be applied to circuits containing non-linear devices, such as the diode, since in this case the incremental response will no longer be proportional to the excitation. Similar remarks may be made concerning all of the linear circuit theorems. Consequently the theory and techniques appertaining to the analysis of non-linear circuits are very much more restricted than is the case for linear circuits.

In contrast with the linear circuit theorems, Kirchhoff's Laws, being essentially expressions of the laws of conservation of energy and charge, are universally applicable.

7.2 Graphical analysis

A non-linear device may always be represented by its experimentally determined v–i characteristic. A single v–i (or i–v) plot is sufficient to completely characterize a simple two-terminal device such as a resistor. When a device has more than two terminals, its behaviour must be described by either a family of curves or, more generally, several families of curves.

Fig. 7.1. Voltage–current characteristics of: (*a*) linear and, ((*b*) and (*c*)), non-linear devices.

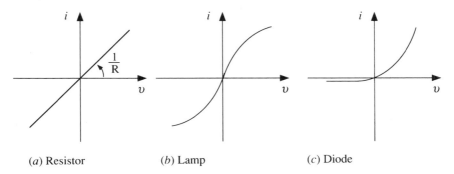

(*a*) Resistor (*b*) Lamp (*c*) Diode

Non-linear circuit analysis

The graphical method is very often used to find the currents and/or voltages when two circuit elements whose characteristics are given are connected in series or in parallel. The graphical construction solves whichever of Kirchhoff's laws is appropriate to the given problem. The method will be introduced with the example, shown in fig. 7.2(a), of a series circuit consisting of a non-linear resistor, a linear resistor, and a voltage source. We wish to find the current I and the voltage V_2 across the non-linear resistor. The characteristic of the non-linear resistor, shown in fig. 7.2(b), gives one relation between i and v_2. A second relation is obtained by application of KVL to the circuit of fig. 7.2(a).

$$V_0 = v_1 + v_2 = iR + v_2$$

So

$$i = \frac{V_0}{R} - \frac{v_2}{R} \tag{7.1}$$

This is the equation of a straight line with slope $-1/R$ and intercept V_0/R. Also, when $i = 0$, $v_2 = V_0$. In fig. 7.2(c) (7.1) is plotted on the same axes as the characteristic of the non-linear resistor. The intersection of the two lines provides the required combination of I and V_2. The voltage across R is $(V_0 - V_2)$. The construction of the straight line is simple. One locates V_0 on the voltage axis and V_0/R on the current axis and joins these points by a straight line.

Examination of fig. 7.2(c) shows that the straight line represents the characteristic of the linear resistor R in a coordinate system having its origin at V_0 and having voltage increasing to the left. Use of this fact makes possible the extension of the graphical method to the case of two non-linear resistors in series as illustrated in fig. 7.3.

Fig. 7.2. Graphical solution for a circuit containing a non-linear resistor (NLR): (a) circuit; (b) non-linear resistor characteristic; (c) graphical construction for solution.

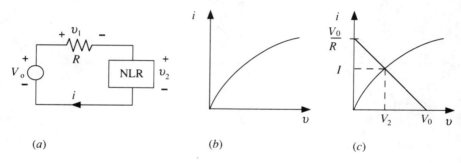

Graphical analysis

The same graphical approach is applicable when circuit elements are connected in parallel. Fig. 7.4(a) shows a parallel combination of a non-linear resistor and a linear resistor supplied from a current source of magnitude I_0. The problem is to find the current in each circuit element and the voltage across the combination. Since voltage now is the common quantity, we draw the non-linear characteristic with v as the dependent variable. This is shown in fig. 7.4(b). Now KCL gives

$$I_0 = i_1 + i_2 = i_1 + \frac{v}{R}$$

So

$$v = I_0 R - i_1 R \qquad (7.2)$$

This is the equation of a straight line with slope $-R$ and intercept $I_0 R$. Furthermore, when $v = 0$, $i_1 = I_0$. In fig. 7.4(c) the line represented by (7.2) is drawn on the characteristic of the non-linear resistor. The intersection of

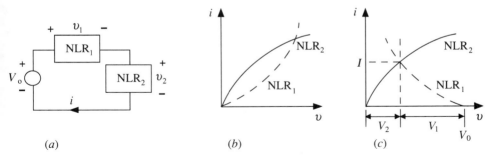

Fig. 7.3. Graphical solution for two non-linear resistors in series.

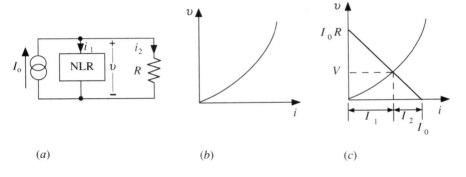

Fig. 7.4. Graphical solution for linear and non-linear resistors in parallel.

the two lines gives the combination I_1 and V, representing the current in and the voltage across the non-linear resistor. The current in R is $I_2 = (I_0 - I_1)$.

In section 1.7.3 formulae were derived for the equivalent resistance of a series/parallel combination of several linear resistances. Non-linear resistances cannot be combined using these formulae and it is necessary to adopt other means; a graphical approach is often appropriate. In fig. 7.5 are drawn the characteristics of two non-linear resistors. If these elements are connected in series, then the current must be the same in both. For example, a current I_1 requires a voltage V_a across device number 1 and a voltage V_b across device number 2. The total voltage required to maintain I_1 in the series combination is therefore $V_a + V_b$. We thus determine one point on a new non-linear (composite) characteristic that represents the series combination of the two non-linear resistors. By assuming other currents, we may obtain other points on the composite characteristic.

If the circuit elements are connected in parallel then the composite curve is found by finding the total current required to maintain a given voltage across the combination. Details are shown in fig. 7.6.

Fig. 7.5. Composite v–i characteristic for two non-linear resistors in series.

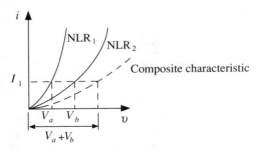

Fig. 7.6. Composite v–i characteristic for two non-linear resistors in parallel.

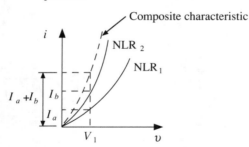

We examine next a three-terminal device as shown schematically in fig. 7.7(a). (This obviously is the special case of a two-terminal pair device (or two-port) in which there is a common connection between input and output.) Now there are two current–voltage pairs to be considered. Subscript 1 refers to the input and subscript 2 refers to the output. If the device is of any practical use, there will be interaction between input and output and two sets of characteristic curves will be required to describe the device behaviour.

Fig. 7.7(a) is an appropriate representation of a transistor. In what follows we shall assume that we are dealing with an npn transistor in the common–emitter connection.* The notation generally employed with transistors is shown in fig. 7.7(b).

The characteristics of a typical silicon transistor are shown in fig. 7.8. It is important to observe that i_B is almost independent of v_{CE} while i_C is strongly dependent upon i_B.

Suppose now that we make external connections to the transistor as shown in fig. 7.9. We wish to determine the currents I_B and I_C. Following the procedure already described, we construct *load lines* on the input and output characteristics as shown in fig. 7.10. (Note that the spacing between the curves in fig. 7.10(a) has been exaggerated for the sake of clarity.) From the intersections of the input load line with the family of input characteristics, we obtain pairs of values of I_B and V_{CE} that satisfy the constraints imposed by the combination of supply voltage V_1 and resistance R_1. When transferred to the output characteristic, these pairs of values determine points that establish a *transfer characteristic* indicated by the broken line in fig. 7.10(b). The intersection of the transfer characteristic with the output

Fig. 7.7. Schematic representation of three-terminal devices.

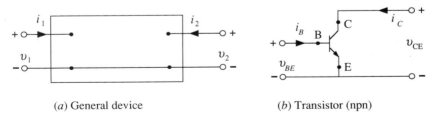

(a) General device (b) Transistor (npn)

* The three terminals of a transistor are called, respectively, the base, the emitter and the collector. Any one of these may be the terminal that is common to input and output. The common–emitter connection is most often employed. Schematically, the transistor in the common–emitter connection is shown in fig. 7.7(b). For an npn device, the collector and the base normally are maintained positive with respect to the emitter. For further details see reference 5.

Fig. 7.8. Silicon transistor characteristics for common–emitter connection.

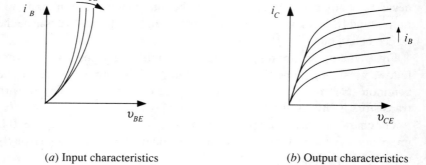

(a) Input characteristics

(b) Output characteristics

Fig. 7.9. Transistor common–emitter connections to input and output.

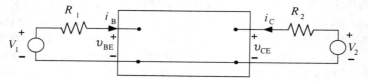

Fig. 7.10. Load-line method for determining operating point of a transistor.

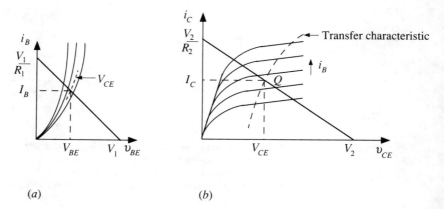

(a)

(b)

Small-signal models

load line (determined by V_2 and R_2) represents the *operating point Q* and determines the current I_C and the corresponding voltage V_{CE}. We then use V_{CE} and the input characteristic to determine I_B and V_{BE}. The application of this method to bipolar transistors is simplified by the fact that the curves on the input characteristic are very close together. Consequently, the effect of v_{CE} upon i_B may be ignored. One simply finds I_B from the input characteristic (assuming that the device may be represented by a single line) and then uses the appropriate line on the output characteristic to establish the operating point.

7.3 Small-signal models

7.3.1 Non-linear resistor model

In Section 7.2 we described how to find the current in a series circuit consisting of a non-linear resistor (NLR), a linear resistor R and a voltage source V_0. The circuit and the graphical construction are repeated in fig. 7.11. The current I and the voltage V_{NLR} determine the *quiescent* or *operating* point, commonly designated by the letter Q. Now assume that V_0 increases by a small amount ΔV. There will then be a new quiescent point Q', found by a new construction as shown in fig. 7.11(c). A small decrease in V_0 will result in a shift of the operating point to Q''.

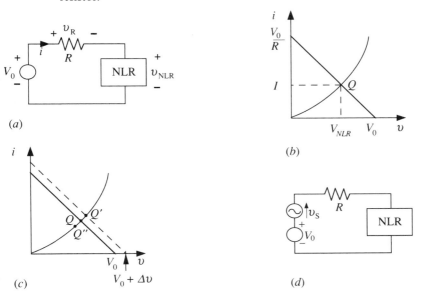

Fig. 7.11. Development of the small-signal model for a non-linear resistor.

Let us now revise the circuit of fig. 7.11(a) to include a sinusoidal voltage source $v_s = \Delta V \sin \omega t$, as shown in fig. 7.11(d). Now the operating point in fig. 7.11(c) will move periodically along the characteristic curve between limits Q' and Q''. As ΔV decreases, the relevant portion of the characteristic curve decreases in length. In the limit, as $\Delta V \to 0$, this portion of the curve may be approximated by a straight line whose slope is di/dv at point Q. Let $di/dv = 1/r_n$. Then, *as far as the voltage v_s is concerned*, the circuit is equivalent to the series combination of two linear resistors and the source v_s as shown in fig. 7.12. Then the alternating component of the voltage across the NLR is

$$v_n = \frac{r_n}{r_n + R} v_s \tag{7.3}$$

In fig. 7.11(d), the circuit is said to be *biased* by the source V_0 at the point Q whose coordinates are I and V_{NLR}. I is the *bias current* and the resistance r_n in fig. 7.12 is determined by the slope of the NLR characteristic at the Q point. The resistance r_n is referred to variously as the *slope*, *dynamic*, or *incremental* resistance, and once it has been determined, the circuit of fig. 7.12 is sufficient for calculating the alternating current behaviour of the circuit. Fig. 7.12 is truly valid only for alternating voltages of vanishingly small amplitudes. It is called the *small-signal* model. It is apparent that this model utilizes a linear representation of the NLR – a representation whose validity depends upon the amplitude of the driving alternating voltage and upon the shape of the NLR characteristic.

7.3.2 Transistor model

Let us now examine the possibility of obtaining a small-signal model of the transistor that may be used in circuit calculations without recourse to the characteristic curves.

Referring again to fig. 7.10 we assume that the transistor is biased at an appropriate point by means of batteries and resistors. Consider first the input characteristic, a change in v_{BE} can be caused *either* by a change in i_B or by a change in v_{CE}. If v_{CE} is held constant, then the combination (i_B, v_{BE})

Fig. 7.12. Small signal model for a non-linear resistor.

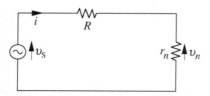

Small-signal models

must lie on the appropriate v_{CE} line. Let $i_b = I_m \sin\omega t$ represent an alternating current with $I_m \ll I_B$. Then, just as in the case of the two-terminal device the characteristic may be represented by a straight line having the slope of the characteristic at the operating point. Then we may write

$$v_{be} = \left.\frac{\partial v_{be}}{\partial i_b}\right|_{v_{CE}} i_b = h_{ie} i_b \qquad (7.4)$$

where $h_{ie} = \partial v_{be}/\partial i_b$ is the slope of the characteristic. Now let i_B remain constant at I_B and let v_{CE} change. Then the corresponding change in v_{BE} is

$$v_{be} = \frac{\Delta v_{be}}{\Delta v_{ce}} v_{ce}$$

In the limit then

$$v_{be} = \left.\frac{\partial v_{be}}{\partial v_{ce}}\right|_{i_B} v_{ce} = h_{re} v_{ce} \qquad (7.5)$$

If both i_B and v_{CE} may change, (7.4) and (7.5) give

$$v_{be} = h_{ie} i_b + h_{re} v_{ce} \qquad (7.6)$$

In fig. 7.10(b), i_C depends upon both v_{CE} and i_B. Suppose, with the transistor biased at the point Q, i_B is held constant and v_{CE} changes. Then the combination (i_C, v_{CE}) must move along the line that represents a constant value of I_B. So, again considering small amplitude a.c. quantities

$$i_c = \left.\frac{\partial i_c}{\partial v_{ce}}\right|_{i_B} v_{ce} = h_{oe} v_{ce} \qquad (7.7)$$

Finally, if v_{CE} is held constant and if i_B changes the corresponding change in i_C is

$$i_c = \left.\frac{\partial i_c}{\partial i_b}\right|_{v_{CE}} i_b = h_{fe} i_b \qquad (7.8)$$

So, if both i_B and v_{CE} change, (7.7) and (7.8) give

$$i_c = h_{fe} i_b + h_{oe} v_{ce} \qquad (7.9)$$

It is a straightforward matter to devise a linear, two-port model that represents equations 7.6 and 7.9. This *hybrid parameter* model (so called because the *h*s do not all have the same dimensions) is shown in fig. 7.13. Observe that this model contains two dependent sources. In many transistors the parameter h_{re} is so small as to be negligible in most applications. (This corresponds to the situation where the curves on the

428 Non-linear circuit analysis

input characteristic are very close together.) Also, h_{oe} is frequently very small. (This corresponds to the curves in the output characteristic being almost horizontal and thus having zero slope.) So, in many applications the appropriate small-signal model of the transistor is as shown in fig. 7.14. It must be remembered that the models of figs 7.13 and 7.14 apply only to currents and voltages of small amplitude, and that both models are based upon the assumption that d.c. voltages have been applied to bias the transistor at the appropriate operating point.

7.4 Piecewise-linear circuits

7.4.1 Piecewise-linear approximation

The small-signal model of a device uses a linear approximation that is valid over a narrow region of the device characteristic. We examine next a model that may represent a non-linear device over an arbitrarily wide region of its characteristic.

In general, a curve describing the characteristic of a real device may be

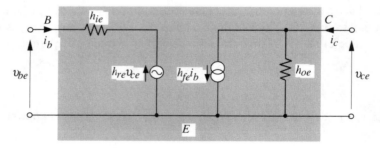

Fig. 7.13. Hybrid-parameter model of a transistor in the common–emitter connection.

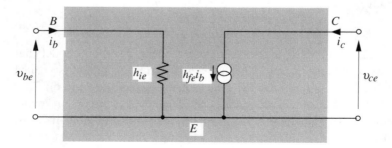

Fig. 7.14. Simplified hybrid-parameter model in the common emitter connection.

Piecewise-linear circuits

represented to any required accuracy by a broken line consisting of many short straight segments. Many real non-linear devices are represented adequately by two or three such segments. Fig. 7.15 shows one example of a characteristic and a three-segment approximation. If we can devise a model that has this piecewise-linear property, the device may be represented by the model and methods of linear circuit analysis may be used.

Two kinds of problems are of interest in applications of piecewise-linear models. The first type involves *synthesis* of a circuit that will reproduce, to whatever accuracy is required and over a specific range of operation, the non-linear behaviour of a device. The second type of problem is concerned with *analysis* of a given circuit to determine the slopes of the linear segments and the coordinates of the *break points* i.e. those points at which the slope of the characteristic changes. When slopes and break points are known, the piecewise-linear characteristic of the circuit may be drawn. In synthesis we are given a characteristic and we must find a circuit to represent it. In analysis we are given a circuit and are required to find the corresponding piecewise-linear representation of its characteristic. It is possible to construct piecewise-linear circuit models using resistances and *ideal diodes*.

7.4.2 The ideal diode

An ideal diode is a voltage-controlled two-terminal device that has the characteristics of a switch. If voltage of one polarity is applied, the diode is a short circuit (i.e., a closed switch) while if the voltage polarity is reversed the diode is an open circuit (i.e., an open switch). The diode is represented as shown in fig. 7.16(*a*) where the arrow represents the direction of current when the diode is conducting. The i–v characteristic of a diode is as shown in fig. 7.16(*b*), with positive values of i and v defined in fig. 7.16(*a*). Except where real diodes are specified, all diodes in the circuits that follow are ideal.

Fig. 7.15. Piecewise-linear approximation.

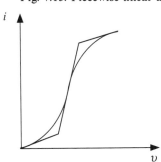

7.4.3 Combinations of resistances and ideal diodes

Before discussing synthesis and analysis, we examine the v–i characteristics of some resistance–diode combinations. These simple circuits may then be used as 'building blocks' in either the synthesis or the analysis of more complex circuits.

Fig. 7.17 shows four diode–resistance combinations and the v–i characteristic of each. For these circuits the characteristics are easily determined. For example, in fig. 7.17(a), when v is positive the diode conducts. Since a conducting diode has no resistance, there is no voltage drop across it and so the v–i characteristic is simply the straight line of slope $1/R$ that represents the resistance R. When v is negative, the diode does not conduct and so the current is zero.

The combinations shown in fig. 7.17 all have a single break point at the origin. By including a voltage source in the series circuits of figs. 7.17(a) and (b), we can shift the break point along the voltage axis (see fig. 7.18). A current source in parallel with the resistance–diode combinations of figs. 7.17(c) and (d) shifts the break point along the current axis (see fig. 7.19). Again, it is a simple matter to sketch the i–v characteristic. For example, in fig. 7.19(a), as long as the input current exceeds $(-I)$ there is a net forward current through the diode and so the voltage across it is zero. When i is less than $(-I)$ there is current through R producing a voltage drop that turns the diode off. As i decreases further, a current $(i-I)$ flows in R and the characteristic is a straight line having slope $1/R$ and passing through the point $(0, -I)$.

Consider next the effect of adding a second resistance to the circuit of fig. 7.17(a). The resulting circuit and its characteristic are shown in fig. 7.20(a).

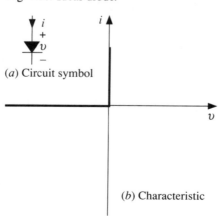

Fig. 7.16. Ideal diode.

(a) Circuit symbol

(b) Characteristic

Piecewise-linear circuits

Fig. 7.17. Diode–resistance combinations with break point at the origin.

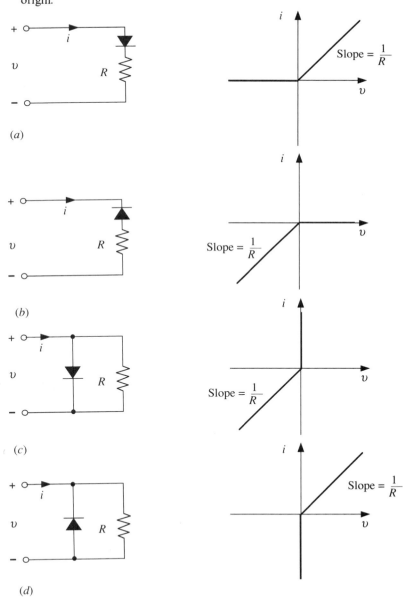

432 *Non-linear circuit analysis*

Fig. 7.18. Diode–resistance combinations with the break point on the v-axis.

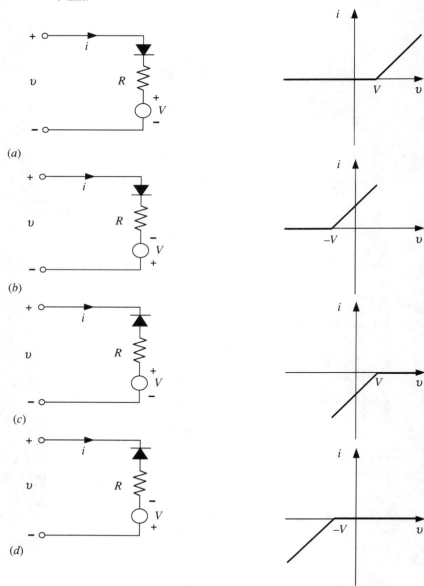

Piecewise-linear circuits

Fig. 7.19. Diode–resistance combinations with break point on *i*-axis.

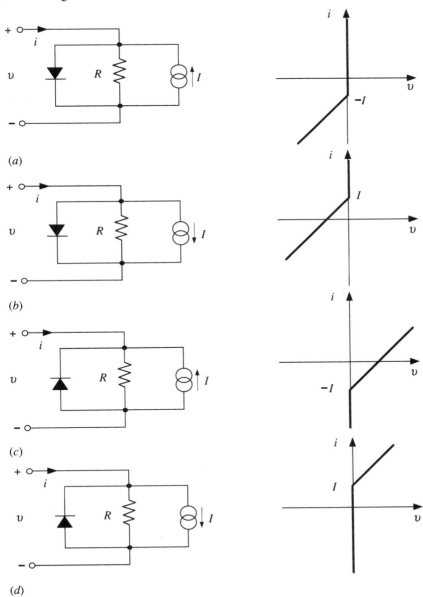

(a)

(b)

(c)

(d)

434   Non-linear circuit analysis

A resistance added to the circuit of fig. 7.17(c) yields the result shown in fig. 7.20(b).

Addition of a resistance to the circuits of figs. 7.18 and 7.19 causes the break point to move off the axis. Consider, for example, the circuit of fig. 7.18(a) shown again as fig. 7.21(a). Let R_2 be added as in fig. 7.21(b). We determine the break point by making use of the fact that *when the diode switches from the conducting state to the non-conducting state both the voltage v_d across the diode and the current i_d through the diode must be identically zero.* For the circuit of fig. 7.21(b)

$$v - v_d - i_d R_1 - V = 0 \tag{7.10}$$

and when $v_d = 0$ and $i_d = 0$, then $v = V$. The condition $i_d = 0$ means that there is no current in the diode branch. Therefore, at the break point, $i = V/R_2$.

Above the break point when the diode is conducting the slope of the characteristic is $1/(R_1//R_2) = (R_1 + R_2)/R_1 R_2$, while below the break point (diode not conducting) the slope is $1/R_2$. Observe that at $v = 0$, $i = 0$, so the characteristic passes through the origin.

Consider next the circuit of fig. 7.19(b), redrawn in fig. 7.22(a). Let R_2 be added as in fig. 7.22(b). Now to find the break point we impose the condition, $i_d = 0$ and $v_d = 0$. When $v_d = 0$, the current through R_1 must be

Fig. 7.20. Diode resistance combinations with two different slopes.

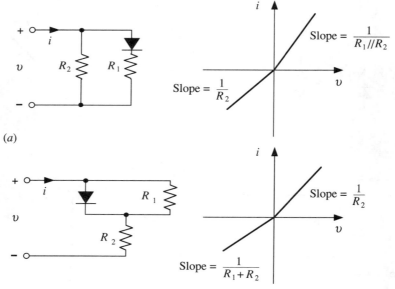

(a)

(b)

Piecewise-linear circuits

Fig. 7.21. Effect of added parallel resistance to circuit of fig. 7.18(a). (a) Original circuit and characteristic; (b) Circuit and characteristic with addition of R_2.

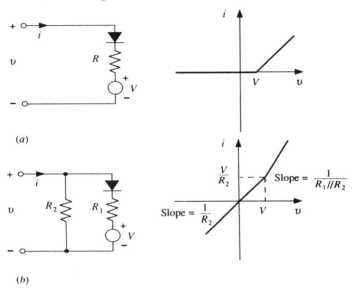

Fig. 7.22. Effect of added series resistance to circuit of fig. 7.19(b). (a) Original circuit characteristic; (b) Circuit and characteristic with addition of R_2.

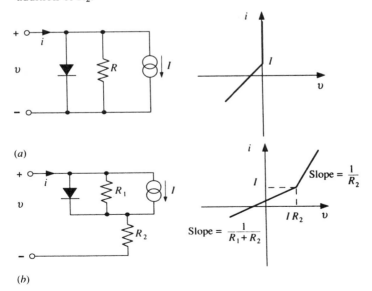

zero. Since at the same time $i_d = 0$, it follows that at the break point $i = I$. But if $i = I$ and the voltage across R_1 is zero, then at the break point $v = IR_2$.

Above the break point, with the diode conducting, the slope of the characteristic is simply $1/R_2$. When the diode is not conducting, KVL gives

$$v - iR_1 + IR_1 - iR_2 = 0$$

and so

$$i = \frac{v + IR_1}{R_1 + R_2} = \frac{v}{R_1 + R_2} + \frac{R_1}{R_1 + R_2} I \qquad (7.11)$$

From this expression we see that the slope of the characteristic below the

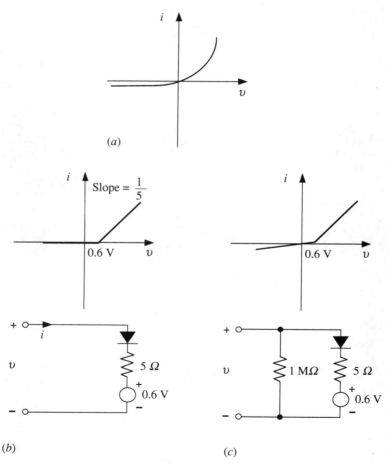

Fig. 7.23. Development of circuit model for the forward characteristic of a real diode.

break point is $1/(R_1+R_2)$. For this characteristic to pass through the origin, it is necessary that $R_1=0$. Then the break point disappears because there is just the resistance R_2 across v.

7.4.4 The real diode

In contrast to the ideal diode that we have used so far, a real diode has a characteristic as shown in fig. 7.23(a). Typical semiconductor diodes have resistance of the order of a megohm for reverse voltage so the characteristic in the third quadrant is nearly horizontal. When forward voltage is applied, current rises exponentially with voltage.

For many purposes the semiconductor diode may be represented by a piecewise-linear approximation having a single break point on the voltage axis as shown in fig. 7.23(b). An approximation that accounts for the small *reverse leakage current* that exists under reverse voltage conditions is shown in fig. 7.23(c).

If the reverse voltage is made sufficiently large, a real diode exhibits *breakdown*, that is, it begins conducting heavily in the reverse direction. Its resistance then is of the order of a few ohms. Fig. 7.24 shows a model that accounts for breakdown. For most purposes R_3 may be omitted. Then both break points will lie on the voltage axis.

7.4.5 The Zener diode

The voltage at which reverse breakdown occurs in a semiconductor diode is termed the *Zener voltage*. In rectifier or other applications based upon the unidirectional conducting properties of the diode, the Zener voltage is of interest simply because it specifies the maximum peak inverse voltage that the diode will withstand. The *Zener diode*, on the other hand, is a device whose normal mode of operation lies within the reverse breakdown

Fig. 7.24. Circuit model for the complete characteristic of a real diode.

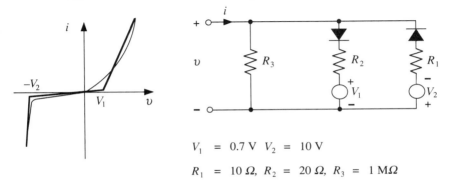

$V_1 = 0.7 \text{ V}$ $V_2 = 10 \text{ V}$

$R_1 = 10 \, \Omega, \; R_2 = 20 \, \Omega, \; R_3 = 1 \, \text{M}\Omega$

region. Special care is taken in the manufacture of this device to ensure that the slope of the characteristic in this region is very steep, that is, that the effective resistance is very small (see fig. 7.25). Thus, the voltage remains substantially constant over a wide range of operating currents. Inspection of fig. 7.25 shows that there is a minimum reverse current, in the region of the knee, that must exist in the diode in order for it to operate in the constant voltage region. The maximum current is determined by the heat dissipating properties of the diode. Zener diodes are available with operating voltages between 2 V and 200 V and with power ratings up to the order of 50 W.

The primary application of the Zener diode is that of maintaining a constant voltage across a load regardless of fluctuations in supply voltage or load currents. A basic circuit used for this purpose is shown in fig. 7.26(a). The supply voltage V_0 is large in relation to the load voltage V_L so that the current through R_1 is substantially constant; thus, changes in load current I_L are reflected in equal and opposite changes in diode current I_Z. The diode current, therefore, swings over a range of values equal to the load current variation and, since the characteristic of the Zener diode has a finite slope, this produces small fluctuation in the load voltage. Variations in supply voltage will also cause the load voltage to vary. We now establish an expression, using piecewise-linear analysis, for the variation in V_L that results from small variations in I_L and V_0.

Consider the circuit of fig. 7.26(b) in which the Zener diode has been modelled by the circuit of fig. 7.18(a). By KVL we obtain

$$(I_L + I_Z)R_1 + I_Z R_Z = V_0 - V_Z$$

and

$$I_Z R_Z = V_L - V_Z$$

Fig. 7.25. Zener diode characteristic.

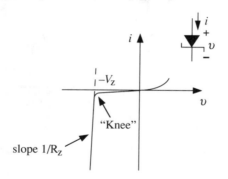

Piecewise-linear circuits

Eliminating I_Z from these equations gives

$$\left(\frac{R_1}{R_Z}+1\right)V_L - \frac{R_1}{R_Z}V_Z + R_1 I_L = V_0 \qquad (7.12)$$

Now, from the total differential, the incremental change in V_L is given by:

$$\Delta V_L = \frac{\partial V_L}{\partial I_L}\Delta I_L + \frac{\partial V_L}{\partial V_0}\Delta V_0 \qquad (7.13)$$

where ΔI_L and ΔV_0 are the incremental changes in load current and supply voltage.

Differentiating (7.12) we obtain

$$\left(\frac{R_1}{R_Z}+1\right)\frac{\partial V_L}{\partial I_L} + R_1 = 0$$

and

$$\left(\frac{R_1}{R_Z}+1\right)\frac{\partial V_L}{\partial V_0} = 1$$

Hence, by substitution in (7.13)

$$\Delta V_L = -\left(\frac{R_1 R_Z}{R_1 + R_Z}\right)\Delta I_L + \frac{R_Z}{R_1 + R_Z}\Delta V_0 \qquad (7.14)$$

The coefficient of the first term in (7.14) is recognied as the parallel combination of R_1 *and* R_Z; the coefficient of the second term derives from the voltage-divider principle. Clearly, ΔV_L is reduced by making $R_1 \gg R_Z$, in which case (7.14) reduces to

$$\Delta V_L \simeq -R_Z \Delta I_L + \frac{R_Z}{R_1}\Delta V_0 \qquad (7.15)$$

Fig. 7.26. Voltage stabilizer circuit incorporating a Zener diode.

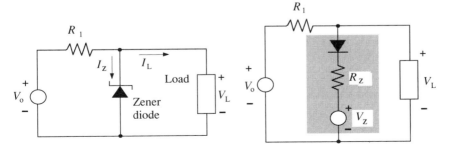

(*a*) Circuit (*b*) Piecewise-linear circuit model

440 Non-linear circuit analysis

Expressions (7.14) and (7.15) indicate that R_1 should be as large as possible in order to achieve good immunity from supply voltage variation. The value of R_1 determines, however, the quiescent operating point on the characteristic, and its upper limit will be set by the required load-current swing.

7.4.6 Analysis of piecewise linear circuits

To analyze a circuit consisting of ideal diodes and resistors, we find break points by determining the input-voltage/input-current combination that exists as each diode changes from the non-conducting state to the conducting state. We then locate these break points on an i–v plane. By joining adjacent break points with straight lines we have the desired piecewise-linear characteristic of the given circuit.

Sometimes we may be interested in the output voltage v. input voltage characteristic. Once the break points have been determined, this characteristic may readily be found. The procedure is illustrated by the following example.

7.4.7 Worked example

For the circuit of fig. 7.27(a), find the break points and plot
(a) the i_{in} v. v_{in} characteristic;
(b) the v_{out} v. v_{in} characteristic.

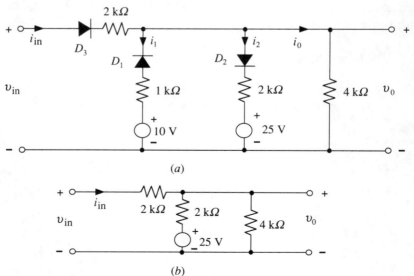

Fig. 7.27. Circuits for worked example.

Piecewise-linear circuits

Solution
Because there are three diodes, there are three break points. We find these by considering one diode at a time in the order of their subscripts.

Diode D_1: When $v_0 < 10$ V, D_1 conducts. When $v_0 > 10$ V, D_1 is reverse biased and so is not conducting. The first break point occurs at $v_0 = 10$ V. When $v_0 = 10$ V, D_2 is not conducting, so $i_2 = 0$. Also, at the first break point, $i_1 = 0$. Therefore, at the first break point, $v_0 = 10$ V, and

$$i_{in} = i_0 = 10/4000 = 2.5 \text{ mA}$$
$$v_{in} = 10 + (2.5 \times 10^{-3} \times 2000) = 15 \text{ V}$$

Diode D_2: When $v_0 < 25$ V, D_2 is reverse biased and does not conduct. When $v_0 > 25$ V, D_2 conducts. When $v_0 = 25$ V, D_1 is not conducting, so $i_1 = 0$. The second break point occurs then at $v_0 = 25$ V and $i_2 = 0$. So

$$i_{in} = i_0 = 25/4000 = 6.25 \text{ mA},$$
$$v_{in} = 25 + (6.25 \times 10^{-3} \times 2000) = 37.5 \text{ V}$$

Diode D_3: Assume D_3 is not conducting. Then v_0 is provided by the 10 V source in series with D_1 and $v_0 = 10(4/5) = 8$ V. For this value of v_0, D_2 is not conducting so $i_2 = 0$. If v_{in} exceeds 8 V, D_3 conducts, but if v_{in} is less than 8 V, D_3 cannot conduct. Therefore, the third break point is at $v_{in} = 8$ V, $i_{in} = 0$.

We now must determine what happens above the second break point, that is, when v_{in} exceeds 37.5 V. For this condition, D_1 is not conducting and D_2 conducts. We then have the circuit of fig. 7.27(b). By nodal analysis

$$\frac{v_0 - v_{in}}{2} + \frac{v_0}{4} + \frac{v_0 - 25}{2} = 0$$

which reduces to

$$v_0 = (0.4 v_{in} + 10)$$

Then,

$$i_{in} = \frac{v_{in} - v_0}{2} = \frac{v_{in} - (0.4 v_{in} + 10)}{2} = (0.3 v_{in} - 5) \text{ mA}$$

Therefore, for $v_{in} > 37.5$ V, the curve v_0 v. v_{in} has slope 0.4. The curve i_{in} v. v_{in} has slope 0.3×10^{-3}, corresponding to a resistance of 3.3. kΩ.

We now have information from which we may draw the two required characteristics. They are shown in fig. 7.28.

7.4.8 Synthesis of piecewise-linear circuits

The first requirement for synthesizing a piecewise-linear circuit is to decide upon an appropriate linear approximation for the given

442 Non-linear circuit analysis

characteristic of the device or circuit. Then the slopes of the straight line segments may be found. The break points are the intersections of these segments. By referring to the 'building blocks' whose characteristics have already been found (figs. 7.17–7.22) we then choose the proper circuits and the appropriate values of resistors and sources to provide the required break points and slopes. The procedure is illustrated in the following example.

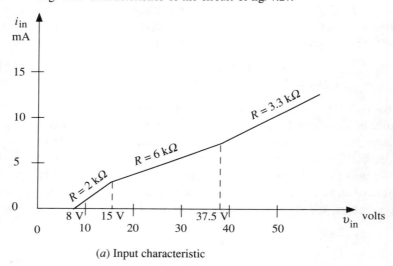

Fig. 7.28. Characteristics of the circuit of fig. 7.27.

(a) Input characteristic

(b) Transfer characteristic

7.4.9 Worked example

Given the continuous characteristic of fig. 7.29(a), determine an appropriate three-segment piecewise-linear approximation and then design a circuit that has the desired characteristic.

Solution

The dashed line in fig. 7.29(a) indicates a three-segment linear approximation that may be taken as an adequate representation of the original characteristic. The coordinates of the break points are (0, 0), V_a, I_a) and V_b, I_b). The synthesis is accomplished by starting at the lowest values of v and i and working through successive break points.

In region I of fig. 7.29(a) the characteristic is a straight line through the origin. This is represented as in fig. 7.29(b) by a diode D_1 in series with a resistor R_1 of value $1/M_1 \, \Omega$.

At break-point a, the slope increases, indicating that the total circuit resistance has decreased. Therefore, we add in parallel with our original D_1–R_1 combination a series combination of D_2, R_2 and V_2. The appropriate values are:

$$V_2 = V_a \text{ and } R_1 R_2/(R_1 + R_2) = 1/M_2$$

Since R_1 has already been found, R_2 may be calculated. The circuit now is as shown in fig. 7.29(c).

At break-point b, the slope decreases. This means that the circuit resistance in region III must be greater than in region II. To accomplish this resistance change we use a parallel diode/resistor/current-source combination as shown in fig. 7.29(d). Here, $I_3 = I_b$ and R_3 is chosen so that

$$R_3 + R_1 R_2/(R_1 + R_2) = 1/M_3$$

The justification for the parallel resistance combination is as follows. As long as $i < I_3$, there is no current through R_3, and diode D_3 carries current $(i - I_3)$. For $i > I_3$, some current must go through R_3. The resulting voltage drop across R_3 provides reverse bias for D_3 and so the diode does not conduct and its current is zero. Therefore, for $i > I_3$, R_3 is effectively added in series with the parallel combination of R_1 and R_2.

For design of a practical circuit, we would use the Thévenin–Norton transformation to convert the current source I_3 to a voltage source V_3, as shown in fig. 7.29(e).

7.5 Analytical methods

The characteristics of many non-linear devices may be expressed by means of analytical functions or approximated by power series. In such cases the circuit of which they form part may be solved either analytically or

444 Non-linear circuit analysis

Fig. 7.29. Diagrams for worked example: (*a*) original characteristic with three-segment linear approximation; (*b*) and (*c*) steps in circuit synthesis; (*d*) and (*e*) two forms of final circuit.

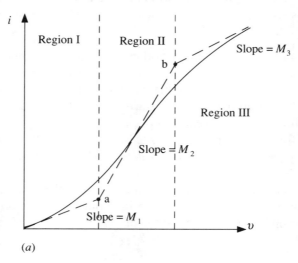

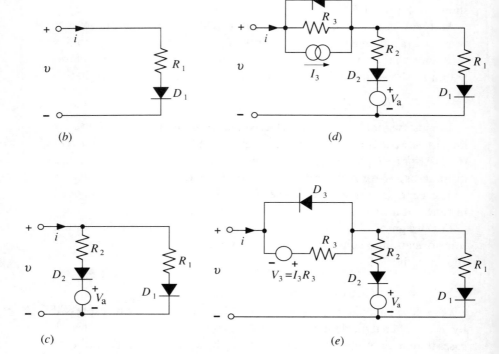

Analytical methods

numerically. For example, certain non-linear resistors, consisting of crystals of silicon carbide bonded together and fired at high temperature, have a voltage–current relationship of the form

$$i = kv^p \qquad (7.16)$$

where k and p depend on the nature and physical state of the material. The index p usually lies in the range 3–5. If a device of this type is connected in a series circuit such as that shown in fig. 7.2, we have by KVL

$$V_0 = iR + v_2$$

But from (7.16) $v_2 = v = (i/k)^{1/p}$ hence,

$$V_0 = iR + \left(\frac{i}{k}\right)^{1/p}$$

This equation may be solved by numerical iteration using the recurrence relationship

$$i_{n+1} = \frac{V_0}{R} - \frac{1}{R}\left(\frac{i_n}{k}\right)^{1/p}$$

The voltage–current characteristic of a diode may also be expressed analytically by the equation

$$i = I_s(e^{Kv} - 1) \qquad (7.17)$$

where I_s and K are constants dependent upon temperature. (A similar equation (the Ebers–Moll equation) relates the collector current of a bipolar transistor to its base–emitter voltage.) By expanding the exponential term in (7.17) as a power series we obtain

$$i = I_s\left[\left(1 + Kv + \frac{(Kv)^2}{2!} + \frac{(Kv)^3}{3!} + \ldots\right) - 1\right]$$

or

$$i = I_s\left[Kv + \frac{(Kv)^2}{2!} + \frac{(Kv)^3}{3!} + \ldots\right] \qquad (7.18)$$

The polynomial formed by taking the first few terms of this expression may then be used to obtain an algebraic or numerical solution for a circuit incorporating the diode. The procedure is similar to that adopted above for the case of the non-linear resistor.

The characterization of a non-linear device by means of a power series provides a powerful tool for the analysis of modulators and frequency changer circuits used extensively in communication networks. We conclude this section with an examination of the simple circuit shown in fig. 7.30(a), which may be used both as a modulator and as a frequency changer.

When used as a modulator, two sinusoidal signals v_c and v_m, of widely differing frequencies ($\omega_c \gg \omega_m$), are applied to the input. The output is an *amplitude-modulated* wave of the form shown in fig. 7.30(b). The signal at the higher frequency is called the *carrier*, while that at the lower frequency is the *modulating* signal. From fig. 7.30(b) it is evident that the instantaneous value of the modulated wave is

$$v_0 = V_c' \sin\omega_c t + (V_m' \sin\omega_m t)\sin\omega_c t \tag{7.19}$$

We now show that this form of output arises directly as a result of the non-linearity of the diode characteristic.

Fig. 7.30. Modulator (frequency changer) circuit.

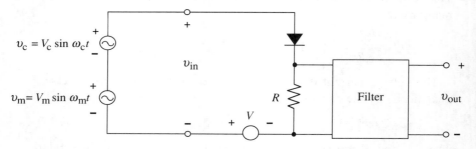

(a) Diode modulator circuit and filter

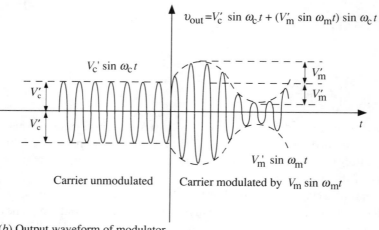

(b) Output waveform of modulator

Analytical methods

The two input signals, together with a d.c. bias voltage V, which ensures that the diode operates in an appropriate region of its characteristic, are applied to the diode and resistance connected in series. If R is made sufficiently small, substantially the whole of the voltage $(V+v_m+v_c)$ appears across the diode, and the current in the circuit may be described by an expression of the form (7.18). Thus, the voltage developed across R may be written:

$$v_R = a(V+v_m+v_c)+b(V+v_m+v_c)^2 + \ldots \tag{7.20}$$

where a and b are constants. For our present purposes the first two terms of (7.20) are of interest; when these are expanded we obtain:

$$v_R = aV + av_m + av_c + bV^2 + 2bVv_m + 2bVv_c + bv_m^2 + 2bv_mv_c + bv_c^2 \tag{7.21}$$

If

$$v_m = V_m \sin\omega_m t \text{ and } v_c = V_c \sin\omega_c t$$

then

$$v_m^2 = V_m^2 \sin^2\omega_m t = \frac{V_m^2}{2}(1-\cos2\omega_m t)$$

and

$$v_c^2 = V_c^2 \sin^2\omega_c t = \frac{V_c^2}{2}(1-\cos2\omega_c t)$$

Also

$$v_m v_c = V_m V_c \sin\omega_m t \sin\omega_c t$$

$$= \frac{V_m V_c}{2}[\cos(\omega_c-\omega_m)t - \cos(\omega_c+\omega_m)t] \tag{7.22}$$

Substitution of these expressions in (7.21) followed by regrouping of terms gives

$$v_R = aV + bV^2 + \frac{bV_m^2}{2} + \frac{bV_c^2}{2}$$

$$+ (a+2bV)V_m\sin\omega_m t + (a+2bV)V_c\sin\omega_c t$$

$$- \frac{bV_m^2}{2}\cos2\omega_m t - \frac{bV_c^2}{2}\cos2\omega_c t$$

$$+ bV_m V_c \cos(\omega_c-\omega_m)t - bV_m V_c \cos(\omega_c+\omega_m)t \tag{7.23}$$

Now, the voltage v_R is applied to the input of a filter which takes the form of a resonant circuit tuned to the carrier frequency ω_c. This tuned circuit is designed to pass frequencies at, or close to, ω_c and reject all others. The first four terms of (7.23) constitute a d.c. component and are rejected, as are the terms in ω_m and $2\omega_m$, which lie well below the pass band. The terms in $2\omega_c$ lie above the pass band and are likewise rejected. The sum and difference frequencies, however, lie close to ω_c (because $\omega_c \gg \omega_m$) and are passed. The output voltage (assuming that amplitudes are unmodified by the filter) is then

$$v_{\text{out}} = (a + 2bV)V_c\sin\omega_c t + bV_m V_c\cos(\omega_c - \omega_m)t + bV_m V_c\cos(\omega_c + \omega_m)t$$

But by (7.22) this may be written as

$$v_{\text{out}} = (a + 2bV)V_c\sin\omega_c t + bV_m V_c\sin\omega_m t\sin\omega_c t$$

which is an amplitude modulated wave of the form (7.19).

As mentioned above, the circuit of fig. 7.30(a) may also be used as a frequency changer. A common application of frequency changing, is to be found in the superheterodyne receiver. In this case the received radio frequency signal is applied to the circuit together with a signal generated by a 'local' oscillator within the receiver. The frequency of the local oscillator is arranged to be close to that of the received signal so that their *difference* is much lower than the frequency of either. If in fig. 7.30(a), v_c is the received signal and v_m is the local oscillator signal, then $(\omega_c - \omega_m) \ll \omega_c$. The filter is tuned to the frequency $(\omega_c - \omega_m)$, and all terms in (7.23) other than the term in $(\omega_c - \omega_m)$ are rejected. Thus the frequency ω_c of the received signal is changed by the circuit to the lower frequency $(\omega_c - \omega_m)$. It may be shown that if the received signal is amplitude modulated, then the difference frequency is similarly amplitude modulated.

Practical modulator and frequency changing circuits utilize transistors rather than diodes, but the underlying principles of operation remain the same.

7.6 Rectifier circuits

7.6.1 Half-wave rectifier

A diode is often used as a *rectifier* to convert alternating voltage to unidirectional voltage. In fig. 7.31(a), the alternating voltage $v = V_m\sin\omega t$ is applied to the series combination of a diode and a resistor. If the diode is ideal, then during each positive half cycle of v the diode conducts and there is current in the circuit. The input voltage, the voltage across each element and the current are shown in fig. 7.31(b) as functions of time. The single

Rectifier circuits

Fig. 7.31. Half-wave rectifier: (*a*) circuit using ideal diode; (*b*) waveforms for circuit (*a*); (*c*) circuit using real diode modelled by ideal diode and a resistance; (*d*) waveforms for circuit (*c*).

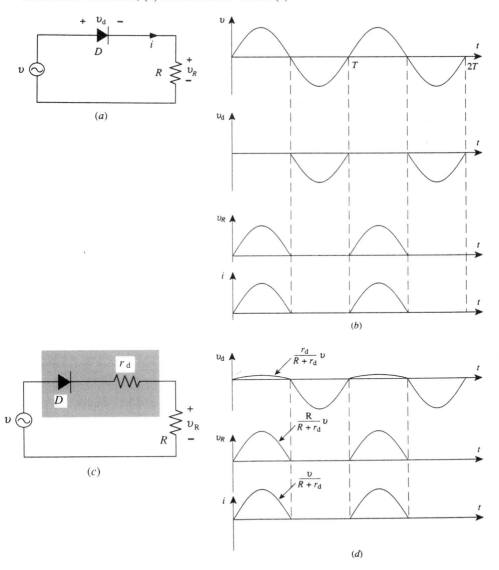

diode is a *half-wave* rectifier, providing current in the load R that is always in the same direction and that exists for half of each cycle of the input voltage.

We have seen that a real diode may be modelled in several ways depending upon the importance of such factors as forward voltage drop, forward resistance, reverse current and reverse breakdown voltage. In most rectifier applications we need consider only the forward resistance and the reverse breakdown voltage. (If the amplitude V_m of the input voltage is very small, the forward voltage drop may cause a significant decrease in the fraction of a cycle during which the diode conducts, because the diode current will be small until the applied voltage exceeds the diode forward voltage drop.)

The forward resistance of the diode often must be included when calculating the load current. If the diode is to act effectively as a rectifier, its reverse breakdown voltage must exceed by a safe margin the maximum signal voltage that appears periodically as a reverse voltage across the diode. In specifying diodes for rectifier service it is appropriate to state the required *power dissipation* (which may be calculated from the r.m.s. current and the forward resistance) and the peak inverse voltage (PIV) that one expects to apply to the diode.

An appropriate model for the half-wave rectifier is as shown in fig. 7.31(*c*). Voltage and current waveforms are shown in fig. 7.31(*d*). The instantaneous current is

$$i = I_m \sin\omega t \quad v > 0$$
$$i = 0 \quad v < 0$$

where

$$I_m = V_m/(r_d + R)$$

The average value of this current is of interest because it is what a d'Arsonval type ammeter would indicate if it were included in the circuit.

$$I_{av} = \frac{1}{T}\int_0^{T/2} I_m \sin\omega t \, dt = \frac{I_m}{\pi} \tag{7.24}$$

To calculate the power dissipated in the diode resistance and the power delivered to the load we must find the effective (r.m.s.) current

$$I = \left[\frac{1}{T}\int_0^{T/2} I_m^2 \sin^2\omega t \, dt\right]^{\frac{1}{2}} = \frac{I_m}{2} \tag{7.25}$$

The power rating of the diode then must be at least

$$P = I^2 r_d = \frac{I_m^2 r_d}{4} = \frac{V_m^2 r_d}{4(R + r_d)^2} \tag{7.26}$$

Rectifier circuits

A simple half-wave rectifier may be used to charge a storage battery. The fact that the current amplitude is not constant is immaterial; we are interested only in the total charge that passes through the battery. This total charge and the battery voltage together determine the amount of energy supplied and stored as chemical energy in the battery.

7.6.2 Worked example

It is necessary to charge ten 12-volt automobile batteries. Design a half-wave circuit using a diode with $1\,\Omega$ forward resistance that will provide an average charging current of $1\,A$. (Assume the battery has negligible internal resistance.)

Solution: The batteries must be connected in series in order to ensure that all have the same charging current. We must determine:
(1) the amplitude of the supply voltage;
(2) the maximum current;
(3) the power dissipated in the diode;
(4) the peak inverse voltage (PIV) rating of the diode.

The circuit is shown in fig. 7.32(*a*). Calculations are simplified if we use $v = V_m \cos\omega t$. We saw that for a resistive load there is current during the

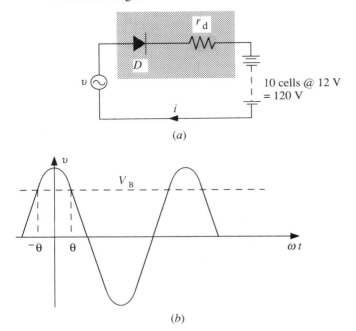

Fig. 7.32. Diagrams for worked example: (*a*) circuit; (*b*) definition of conduction angle.

whole positive half cycle. Now, however, the bank of batteries provides a reverse bias of 120 volts and there will be current only when $v > V_B = 120$ V. So

$$i = \frac{1}{r_d}(V_m \cos\omega t - V_B) \tag{7.27}$$

It is convenient to define a *conduction angle* 2θ that represents the fraction of each positive half cycle during which the diode conducts; this is shown in fig. 7.32(b). We use the given information, $I_{av} = 1$ A, to calculate V_m. From (7.27), using $r_d = 1\,\Omega$,

$$I_{av} = \frac{1}{2\pi} \int_{-\theta}^{\theta} (V_m \cos\omega t - 120) \mathrm{d}\omega t$$

$$= \frac{1}{\pi}(V_m \sin\theta - 120\theta) = 1\text{ A} \tag{7.28}$$

Also from (7.27) the equation for the angle θ is

$$V_m \cos\theta - V_B = 0$$

or,

$$V_m = \frac{V_B}{\cos\theta}$$

Hence, (7.28) becomes

$$\frac{120}{\pi}(\tan\theta - \theta) = 1$$

To find θ we use the approximation, valid for small angles,

$$\tan\theta = \theta + (1/3)\theta^3$$

These two equations give,

$$\theta = 0.43 \text{ radians} = 24.6°$$

Then, $\cos\theta = 0.91$ and $V_m = 120/0.91 = 132$ V.
The maximum current is $I_m = (132 - 120)/1 = 12$ A.

To find the power dissipated in the diode we must find the effective current.

$$I = \left[\frac{1}{2\pi}\int_{-\theta}^{\theta}\left(\frac{V_m \cos\omega t - V_B}{r_d}\right)^2 \mathrm{d}\omega t\right]^{\frac{1}{2}}$$

Rectifier circuits

This integral may be evaluated using the approximation $\cos\phi = 1 - \frac{1}{2}\phi^2$, and the relation $V_B = V_m \cos\theta$. Then

$$I = \left[\frac{V_m^2}{2\pi r_d^2} \int_{-\theta}^{\theta} \left(\frac{\theta^2}{2} - \frac{(\omega t)^2}{2}\right)^2 d\omega t\right]^{\frac{1}{2}}$$

When the integration is performed and the limits substituted, the result is

$$I = \left[\left(\frac{2}{15\pi}\right)\left(\frac{V_m \theta^5}{r_d^2}\right)\right]^{\frac{1}{2}}$$

The power dissipated in the diode is $I^2 r_d$, so,

$$P_{\text{diode}} = (2/15\pi)(132)^2(0.43)^5 = 10.87 \text{ W}$$

The maximum reverse voltage across the diode is $V_m + V_B = 252$ V. Therefore, the PIV rating of the diode must be greater than 252 V.

7.6.3 Full-wave rectifier

There are a few applications, such as the battery charger of the previous example, where the pulsating unidirectional current supplied by the half-wave rectifier is satisfactory. We shall see in a later section (7.9) how filters may be used to modify the output waveform of the half-wave rectifier and make the voltage more nearly constant. Filtering of the output voltage to get a constant value is easier if the output voltage does not remain zero for half of each input cycle. A *full-wave* rectifier utilizes both the positive and the negative halves of the alternating input voltage.

There are two common full-wave rectifier circuits. The first uses two diodes and requires a transformer with a centre-tap (see fig. 7.33(a)). During half of the a.c. cycle, v_1 and v_2 are positive. Then diode D_1 is forward biased and conducts while D_2 is reverse biased and is off. During the other half cycle, D_2 conducts and D_1 is off. Regardless of which diode is conducting, the voltage drop across R is always of the polarity shown in fig. 7.33(a). The total secondary transformer voltage and the current in the load are shown in fig. 7.33(b). In this circuit only one diode is conducting at any instant, so the resistance of only one diode must be considered in calculating the current. The peak inverse voltage is V_m.

The bridge rectifier shown in fig. 7.34 gets its name from the fact that the arrangement of circuit elements resembles the bridge used for measuring resistance or impedance. It requires four diodes but does not need a centre-tapped transformer. In fig. 7.34, D_1 and D_3 conduct during one half cycle while D_2 and D_4 conduct during the other half cycle.

The output voltage waveform of the bridge circuit is identical with that of the full-wave rectifier. An advantage of the bridge circuit is that the inverse

Fig. 7.33. Full-wave rectifier: (a) circuit; (b) waveforms.

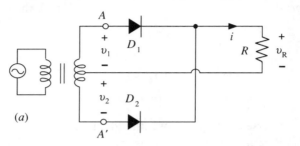

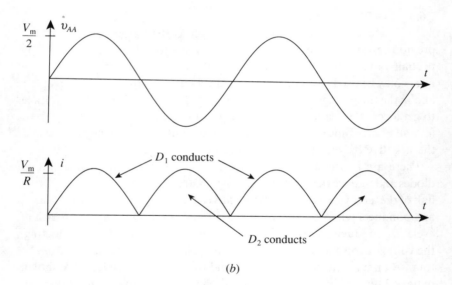

Fig. 7.34. Bridge rectifier.

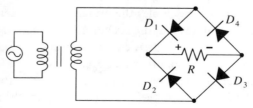

voltage is across two diodes in series so the PIV rating of each diode is half that required for the standard full-wave arrangement. The bridge circuit can be connected directly across the power line. Such a connection can be dangerous, however, since neither side of the load can then be grounded (because one side of the power line normally is grounded). Because the bridge circuit uses two diodes in series, an additional voltage drop appears in the circuit. This usually is not significant. The effective value of the load current is the same as has already been calculated for sinusoidal alternating current, that is, $I = I_m/\sqrt{2}$. The average value of the full-rectified current wave is

$$I_{av} = \frac{1}{2\pi} \int_{-\pi}^{\pi} I_m \cos\omega t \, d\omega t = \frac{2I_m}{\pi}$$

7.7 Thyristor circuits

With either a half-wave or a full-wave rectifier one may vary the average current supplied to a given load by (1) controlling the amplitude of the alternating voltage supply or (2) including a variable resistor in series with the load. The first method requires a variable transformer and the second is wasteful of energy. Another method of controlling the load current employs the controlled rectifier. The controlled rectifier is usually fabricated of silicon and so is often called, particularly in the U.S.A., a *silicon controlled rectifier*, abbreviated SCR. In the U.K. the term *thyristor* is more frequently used.

The two-terminal diode conducts whenever the anode is positive with respect to the cathode. The thyristor is a diode in which a voltage applied between a third terminal (the *gate*) and the cathode can affect the forward conducting characteristic (fig. 7.35(a)). The characteristic between anode

Fig. 7.35. Characteristic for a thyristor.

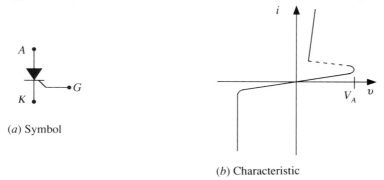

(a) Symbol

(b) Characteristic

and cathode of the SCR when the gate–cathode voltage is negative has the form shown in fig. 7.35(b). Unless the anode–cathode voltage exceeds V_A the diode current is insignificant. The dashed line represents an unstable region. When the anode–cathode voltage exceeds V_A, the diode suddenly becomes fully conducting, the voltage drops sharply, and the current increases, being limited by whatever external resistance there is in series with the diode. The device now behaves like an ordinary two-terminal diode. Current continues until the anode–cathode voltage is reduced to zero.

For most practical applications, however, the thyristor is chosen so that the anode–cathode voltage will not exceed V_A. The diode is instead made to conduct by driving the gate positive with respect to the cathode by an amount sufficient to cause a *trigger current* of a few milliamps to flow in the gate circuit. This trigger current initiates forward conduction in the diode. In contrast with the turn-on process, the traditional thyristor can be turned off only if there is a *negative* gate current comparable in magnitude to the forward diode current. So it has been customary to say that, practically, once the diode conducts the gate loses control and conduction ceases only when the anode–cathode voltage drops to zero. Recently, however, there has appeared a controlled diode which can be turned off by a negative gate current of the order of 1% or less of the diode current. Such a device increases the possible applications of the thyristor.

Fig. 7.36(a) shows a thyristor circuit in which the gate is used to control the average current in the load of a half-wave rectifier. As shown there, the gate is connected through resistor R to the same side of the power line as the anode. During the negative half of the input voltage cycle the diode D_2 in the gate circuit is reverse biased and cannot conduct. As the input voltage enters its positive half cycle, D_2 conducts, the gate–cathode voltage becomes positive and the gate current increases. For a particular value of R the input voltage must reach a specific amplitude in order to furnish the required trigger current that initiates conduction in the diode. The larger R, the later in the positive half cycle conduction begins. It is apparent that if conduction has not occurred by the time the input voltage reaches its maximum value, conduction cannot occur at all. Thus, by varying R we may reduce the average load current smoothly from its maximum value to one-half the maximum value, but no lower. The circuit of fig. 7.36 is said to provide a 'retard angle' that can lie between zero and 90 degrees. Fig. 7.36(b) shows the currents and voltages when R is chosen for an angle of 45°.

Smooth control of the load current over almost the full range from zero to its maximum value may be achieved by *phase control* of the gate voltage.

Thyristor circuits

Fig. 7.37(a) shows a half-wave rectifier using a thyristor. Now the gate voltage comes from the power line through an RC phase-shift circuit. When the thyristor is not conducting, the voltage $V_{G'K}$ depends upon R and C:

$$V_{G'K} = \frac{1/j\omega C}{1/j\omega C + R} V_{in} = \frac{V_{in}}{1 + j\omega CR} = \frac{V_{in}}{\sqrt{[1+(\omega CR)^2]}} \underline{/\theta}$$

where $\theta = -\tan^{-1}(R\omega C)$. Thus, $V_{G'K}$ lags V_{in} and there can be no gate current until $V_{G'K}$ becomes positive. The maximum angle of lag (90°) is approached as (ωCR) becomes much greater than 1.0.

Now R both adjusts the phase of $V_{G'K}$ and limits the gate current. Fig. 7.37(b) shows the case where R is set for $\theta = -45°$. There is an additional delay in onset of thyristor conduction as the gate current rises to the 'trigger' level. Even after R has reached a value such that $\theta \approx -90°$, further delay in 'firing' is achieved by further increase in R. In a practical case, R may consist of a small fixed resistor in series with a variable resistor, so that the gate current is always held to a safe value.

Fig. 7.36. Gate current control of a thyristor.

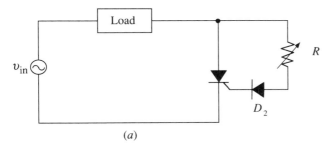

(a)

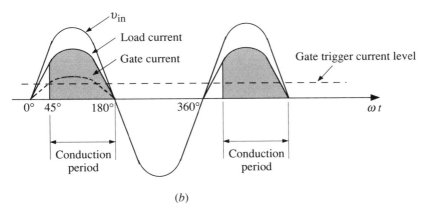

(b)

458 Non-linear circuit analysis

Fig. 7.37. Phase control of thyristor: (a) circuit; (b) waveform for 45° phase angle; (c) waveform for 90° phase angle.

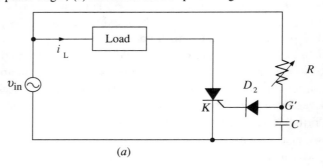

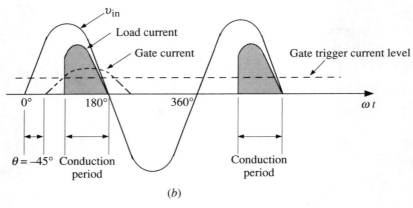

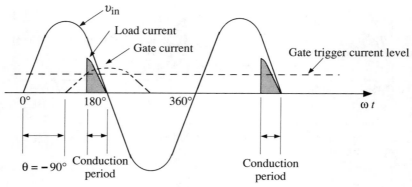

Thyristor circuits

In addition to providing control of average rectified current in a load, the circuit of fig. 7.37(*a*) finds application in dimming circuits for incandescent lamps and in the control of the speed of fractional horsepower motors (those used in hand-held electric drills, for example). Wider range of control in these applications is achieved by use of the *triac*, a three terminal device that exhibits identical bidirectional conduction and control characteristics. Its characteristic in either direction is the same as that of the thyristor. Fig. 7.38 shows the triac connected in a full-wave control circuit. The drive for the common gate is obtained from a phase shift circuit as in fig. 7.37(*a*). Note that diode D_2 in fig. 7.37(*a*) has been replaced by resistor R_2 because now the gate must carry current in both directions.

Because it is a bidirectional device, the triac does *not* function as a full-wave rectifier. However, this may be achieved by using two thyristors in place of the conventional diodes of fig. 7.33(*a*). Such a circuit is shown in fig. 7.39 where each gate is supplied from a separate phase-shift circuit. The two phase-shift resistors may be identical and may be mounted on the same shaft so that a single knob controls both. Usually it is desirable to have the two diodes conduct for equal fractions of the positive and negative half cycles.

Fig. 7.38. Triac with phase control.

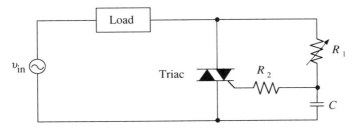

Fig. 7.39. Full-wave rectifier using phase-controlled SCRs.

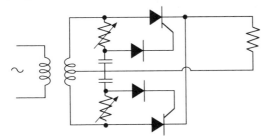

7.8 Fourier analysis of periodic waves

7.8.1 Fourier expansion

Periodic non-sinusoidal waveforms are encountered in many branches of science and engineering. The rectifier circuits that have been discussed in the preceding sections, for example, give output voltages that are periodic but non-sinusoidal. The techniques of a.c. circuit analysis, introduced in chapter 3, apply only to sinusoidal waveforms and we cannot employ them directly to find the response of a circuit to a non-sinusoidal waveform. We can, however, transform the problem so that such techniques are applicable.

By means of Fourier analysis a periodic, non-sinusoidal function may be expressed as a series containing a constant term together with a number of sinusoidal terms. The response to each component term of the series may be found using standard techniques. Provided the circuit is linear, the response to the original function may then be obtained from the superposition of the separate responses. This approach to the analysis of circuits is of great power and generality.

As an introduction to Fourier analysis consider fig. 7.40(a), which shows a constant and several sinusoids whose frequencies are integral multiples of the lowest frequency shown. In fig. 7.40(b) we have plotted the algebraic sum of the functions of time shown individually in fig. 7.40(a). The sum is a periodic but non-sinusoidal wave whose period is the period of the lowest frequency in fig. 7.40(a). The waveform of fig. 7.40(b) may be written as the sum of its components

$$f(\omega t) = f_1 + f_2 + f_3 + f_4 = 1 + 3\sin\omega t + 1\sin 2\omega t - \tfrac{1}{2}\sin 3\omega t \qquad (7.29)$$

This expression illustrates the following general statement: within certain limits of finiteness and continuity that always are met in practice, periodic functions may be expressed as infinite series of the form:

$$f(\omega t) = \frac{a_0}{2} + (a_1\cos\omega t + b_1\sin\omega t) + (a_2\cos 2\omega t + b_2\sin 2\omega t) + \ldots$$
$$+ (a_n\cos n\omega t + b_n\sin n\omega t) + \ldots \qquad (7.30)$$

The series in (7.30) is the *Fourier series*, or the *Fourier expansion*, for $f(\omega t)$, and the a_n and b_n are the *Fourier coefficients*.

Equation (7.30) may also be written more compactly as

$$f(\omega t) = \frac{a_0}{2} + \sum_{n=1}^{\infty} (a_n\cos n\omega t + b_n\sin n\omega t) \qquad (7.31)$$

Fourier analysis of periodic waves

or in the alternative form

$$f(\omega t) = \frac{a_0}{2} + \sum_{n=1}^{\infty} K_n \cos(n\omega t + \phi_n) \quad (7.32)$$

where

$$K_n = \sqrt{(a_n^2 + b_n^2)} \text{ and } \phi_n = \tan^{-1}\left(\frac{-b_n}{a_n}\right)$$

Since the average value of each sine or cosine term in the above equations is zero, it follows that the term $a_0/2$ must be the average value of $f(\omega t)$. The term $K_1 \cos(\omega t + \phi_1)$ in (7.32) is called the *fundamental* or *first harmonic* and has the same period as the original function. The nth term in the summation is the nth harmonic; its frequency is n times the fundamental frequency.

Interpreted as a periodic function of voltage, (7.32) may be written as

$$v(\omega t) = V_0 + \hat{V}_1 \cos(\omega t + \phi_1) + \hat{V}_2 \cos(2\omega t + \phi_2) + \ldots$$
$$+ \hat{V}_n \cos(n\omega t + \phi_n) + \ldots$$

Fig. 7.40. Addition of sinusoids.

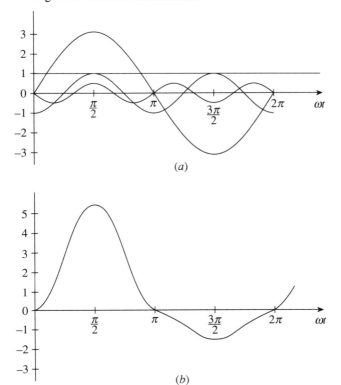

In this expression $V_0(=a_0/2)$ is the average value or *d.c. component* of the waveform; $\hat{V}_n(=K_n)$ is the *amplitude* of the nth harmonic ($\hat{V}_n = V_n\sqrt{2}$, where V_n is the r.m.s. magnitude), and ϕ_n is the phase of the nth harmonic.*

The Fourier expansions represented by (7.30) and (7.32) are infinite series. In many cases of practical interest in electrical engineering the amplitudes of the terms in the series decrease rapidly as the order of the harmonics increases, and often the first three or four terms will provide a satisfactory representation of the original function.

The process of Fourier analysis consists essentially in finding the coefficients a_n and b_n. This process is somewhat simplified if we express the variable of the given function in terms of angle rather than time. In the following theory we shall therefore, where appropriate, put $\omega t = \theta$. In this case (7.31) becomes

$$f(\theta) = \frac{a_0}{2} + \sum_{n=1}^{\infty} (a_n \cos n\theta + b_n \sin n\theta) \qquad (7.33)$$

To obtain an expression for a_n, let the function $f(\theta)$ be defined over one complete period, conveniently over the range $-\pi < \theta < \pi$. Multiply both sides of (7.33) by $\cos m\theta \, d\theta$ (where m is an integer not equal to zero) and integrate between the limits $-\pi$ to π. The resulting expression will contain, on the right-hand side, integrals of the form:

$$\int_{-\pi}^{\pi} \cos n\theta \cos m\theta \, d\theta = \pi \quad \text{for } m=n$$
$$= 0 \quad \text{for } m \neq n$$

and

$$\int_{-\pi}^{\pi} \sin n\theta \cos m\theta \, d\theta = 0 \quad \text{for all } m \text{ and } n$$

(The values of these integrals for the given m and n may be confirmed by simple integration.) We see that this procedure eliminates all terms except the one containing a_n, hence,

$$\int_{-\pi}^{\pi} f(\theta) \cos n\theta \, d\theta = a_n \pi$$

or

$$a_n = \frac{1}{\pi} \int_{-\pi}^{\pi} f(\theta) \cos n\theta \, d\theta \qquad (7.34)$$

* The notation $\hat{V}$ (rather than V_m) to denote amplitude avoids the confusion of subscripts in Fourier analysis.

Fourier analysis of periodic waves

We have previously noted that $a_0/2$ is the average value of $f(\theta)$, therefore,

$$\frac{a_0}{2} = \frac{1}{2\pi} \int_{-\pi}^{\pi} f(\theta) \, d\theta)$$

or

$$a_0 = \frac{1}{\pi} \int_{-\pi}^{\pi} f(\theta) \, d\theta$$

From this we see that a_0 also can be found from (7.34) for the special case $n=0$. (Note that this result is obtained because we have expressed the first term of the series as $a_0/2$, not simply as a_0.)

A similar procedure allows us to find b_n: multiply (7.33) by $\sin m\theta \, d\theta$ and integrate from $-\pi$ to π. Again all terms on the right-hand side of the resulting expression vanish except that containing b_n. The result is:

$$b_n = \frac{1}{\pi} \int_{-\pi}^{\pi} f(\theta) \sin n\theta \, d\theta \tag{7.35}$$

For purposes of evaluating a_n and b_n, any point on the periodic function $f(\theta)$ may be taken as the origin of time or angle. The amplitudes of the harmonics (represented by $K_n = \sqrt{(b_n^2 + a_n^2)}$) depend upon the form of the original function and are independent of the choice of origin. The phase angle ϕ_n, however, will depend upon the location of the origin. The above statements are equivalent to saying that it is sometimes possible to choose the origin so that either all the a_n or all the b_n in (7.31) are identically zero.

Equations (7.34) and (7.35) are useful only when: (1) we have an explicit expression for the original function $f(\theta)$, and (2) we can perform the integrations indicated by the equations. These conditions are met for a wide variety of waveforms encountered in electrical engineering practice.

7.8.2 Worked example

Derive the Fourier expansion for the square wave (fig. 7.41(a)) defined by

$$v(\omega t) = -V \quad -\pi < \omega t < 0$$
$$= +V \quad 0 < \omega t < \pi$$

Solution

Let $\omega t = \theta$, then, by (7.34),

$$a_n = \frac{1}{\pi} \int_{-\pi}^{0} (-V) \cos n\theta \, d\theta + \frac{1}{\pi} \int_{0}^{\pi} (+V) \cos n\theta \, d\theta$$

Fig. 7.41. Diagrams for worked example: (a) square wave; (b) Fourier components; (c) line frequency spectrum.

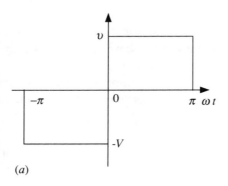

(a)

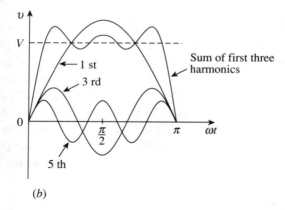

(b)

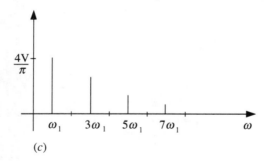

(c)

Fourier analysis of periodic waves

$$= \frac{V}{n\pi}[-\sin n\theta]^0_{-\pi} + \frac{V}{n\pi}[\sin n\theta]^\pi_0$$

$$= 0$$

and, by (7.35)

$$b_n = \frac{1}{\pi}\int_{-\pi}^0 (-V)\sin n\theta \, d\theta + \frac{1}{\pi}\int_0^\pi (+V)\sin n\theta \, d\theta$$

$$= \frac{V}{n\pi}[\cos n\theta]^0_{-\pi} + \frac{V}{n\pi}[-\cos n\theta]^\pi_0$$

$$= \frac{2V}{n\pi}(1 - \cos n\pi)$$

$$b_n = \begin{cases} 0 & n \text{ even} \\ \dfrac{4V}{n\pi} & n \text{ odd} \end{cases}$$

Hence, using (7.33)

$$v(\theta) = \frac{4V}{\pi}\sum_{n=1}^\infty \frac{1}{n}\sin n\theta \qquad n \text{ odd}$$

In expanded form, the representation of the square wave is (with $\theta = \omega t$)

$$v(\omega t) = \frac{4V}{\pi}[\sin\omega t + \tfrac{1}{3}\sin 3\omega t + \tfrac{1}{5}\sin 5\omega t + \ldots] \tag{7.36}$$

We observe that there are no even harmonics. It is true in general that there are no even harmonics when the function has identical positive and negative parts, that is, when $f(\omega t) = -f(\omega t + \tfrac{1}{2}\omega T)$.

Fig. 7.41(b) shows a partial sum of the first three terms of the expansion (7.36); the resulting wave is already a good approximation to the original function. Increasing the number of terms improves the approximation; in particular, the slope of the transitions between positive and negative values is increased.* But it must be remembered that real pulse waveforms, for which the square wave is an idealized representation, possess finite transition times and it is, therefore, unnecessary to include more terms in

---

* In the case of the square wave an infinite number of terms will not produce a perfect representation. A detailed mathematical treatment shows that the infinite Fourier series (7.36) is not uniformly convergent and that there is appreciable overshoot (about 18%) at the discontinuities between positive and negative values. This result is known as the *Gibbs phenomenon* (see reference 14).

the Fourier representation than is warranted by the transition times occurring in the real waveforms under consideration.

An informative way of displaying graphically the relative amplitudes of the harmonics in a waveform is by means of the *line frequency spectrum*. This is shown for the first few terms of the square wave in fig. 7.41(c).

7.8.3 Odd and even functions

It is often possible to choose the origin so that the wave being analyzed is symmetrical about $\omega t = \theta = 0$. If the symmetry is such that $f(\theta)$ is an *even* function (that is, $f(\theta) = f(-\theta)$), all the b_n are zero. Furthermore, integration from $-\pi$ to 0 gives identical results to integration from 0 to π. To find a_n in this case, therefore, it is convenient to change the lower limit in (7.34) to zero and double the result. When this is done, we obtain *for an even function*:

$$a_n = \frac{2}{\pi} \int_0^\pi f(\theta) \cos n\theta \, d\theta \qquad (7.37)$$

and

$$f(\theta) = \frac{a_0}{2} + \sum_{n=1}^\infty a_n \cos n\theta \qquad (7.38)$$

If the wave is an odd function ($f(\theta) = -f(-\theta)$), then all the a_n are zero and, again, similar remarks apply concerning the interval of integration. In this case we obtain *for an odd function*:

$$b_n = \frac{2}{\pi} \int_0^\pi f(\theta) \sin n\theta \, d\theta \qquad (7.39)$$

and

$$f(\theta) = \sum_{n=1}^\infty b_n \sin n\theta \qquad (7.40)$$

The square wave of section 7.8.2 was an example of an odd function resulting in an expansion containing only sine terms. It should be noted that if $f(\theta)$ is even, the amplitude of the nth harmonic is $K_n = a_n$; if $f(\theta)$ is odd, then $K_n = b_n$.

7.8.4 Worked example

Find the Fourier cosine series and Fourier sine series for a triangular waveform of current having peak values $\pm I$ and period T.
Solution: The function of current $i(\omega t)$ is shown in figs. 7.42(a) and (b) where $\omega = 2\pi/T$. In fig. 7.42(a) the origin has been chosen so that the function is

Fourier analysis of periodic waves

even; thus, all b_n in the expansion are zero and we obtain the Fourier cosine series.

Let $\omega t = \theta$, then, in the range $0 < \theta < \pi$, the slope of the function $i(\theta)$ is $2I/\pi$ and the intercept on the vertical axis is $-I$. The function may therefore be described by

$$i(\theta) = \frac{2I}{\pi}\theta - I = I\left(\frac{2\theta}{\pi} - 1\right) \quad 0 < \omega t < \pi$$

The coefficients a_n are given for an even function by (7.37), which in this case becomes:

$$a_n = \frac{2I}{\pi}\int_0^\pi \left(\frac{2\theta}{\pi} - 1\right)\cos n\theta \, d\theta$$

$$= \frac{2I}{\pi}\left[\frac{2}{\pi}\int_0^\pi \theta\cos n\theta \, d\theta - \int_0^\pi \cos n\theta \, d\theta\right]$$

Because of symmetry about the horizontal axis the second integral in this expression is obviously zero. The first integral is readily evaluated using integration by parts (putting $u = \theta$; $dv = \cos n\theta$). This gives

$$a_n = \frac{4I}{\pi^2}\left[\frac{1}{n^2}(\cos n\pi - 1)\right]$$

Therefore,

$$a_n = \begin{cases} 0 & n \text{ even} \\ \dfrac{-8I}{\pi^2 n^2} & n \text{ odd} \end{cases}$$

Fig. 7.42. Diagrams for worked example: triangular wave.

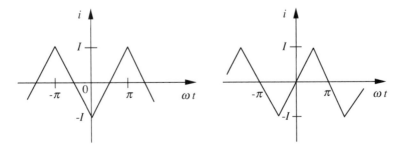

(a) Even function (b) Odd function

468 Non-linear circuit analysis

So the Fourier cosine series for the triangular wave is (with $\theta = \omega t$)

$$i(\omega t) = -\frac{8I}{\pi^2}\left[\cos\omega t + \frac{1}{9}\cos 3\omega t + \frac{1}{25}\cos 5\omega t + \ldots\right] \tag{7.41}$$

To find the Fourier sine series we choose the origin as shown in fig. 7.42(b). In the range $0 < \theta < \frac{\pi}{2}$ the slope of the function is $2I/\pi$, while in the range $\frac{\pi}{2} < \theta < \pi$ it is $-2I/\pi$, hence, the current is described by:

$$i(\theta) = \frac{2I}{\pi}\theta \qquad 0 < \theta < \frac{\pi}{2}$$

$$i(\theta) = I - \frac{2I}{\pi}\left(\theta - \frac{\pi}{2}\right) = \frac{2I}{\pi}(\pi - \theta) \qquad \frac{\pi}{2} < \theta < \pi$$

The coefficients b_n are given for an odd function by (7.39):

$$b_n = \frac{2I}{\pi}\left[\frac{2}{\pi}\int_0^{\pi/2}\theta\sin n\theta\, d\theta + \frac{2}{\pi}\int_{\pi/2}^{\pi}(\pi - \theta)\sin n\theta\, d\theta\right]$$

Integration by parts (putting $u = (\pi - \theta)$; $dv = \sin n\theta$, in the second integral) gives

$$b_n = \frac{4I}{\pi^2}\left[\frac{2}{n^2}\sin n\frac{\pi}{2}\right]$$

$$b_n = \begin{cases} 0 & n \text{ even} \\ +\dfrac{8I}{\pi^2 n^2} & n = 1, 5, 9 \ldots \\ -\dfrac{8I}{\pi^2 n^2} & n = 3, 7, 11 \ldots \end{cases}$$

Hence, the Fourier sine series for the triangular wave is

$$i(\omega t) = \frac{8I}{\pi^2}\left[\sin\omega t - \frac{1}{9}\sin 3\omega t + \frac{1}{25}\sin 5\omega t + \ldots\right] \tag{7.42}$$

Comparing (7.41) and (7.42) we see that in the cosine series all terms are of the same sign whereas in the sine series terms alternate in sign. The reason for this will become apparent if the first three terms of the series for this wave are sketched out for a full period of the fundamental (in a fashion similar to that shown in fig. 7.41(b) for the square wave). It will be seen that a shift in the origin of $\pi/2$ (fig. 7.42) changes the terms in the summation

Fourier analysis of periodic waves

from cosine to sine and at the same time changes the signs of the first and fifth harmonics leaving the third harmonic unchanged.

Mathematically (7.42) may be derived from (7.41) using the relationship $\cos n\theta = \sin\left(\frac{\pi}{2} - n\theta\right) = -\sin\left(n\theta - \frac{\pi}{2}\right)$. A phase shift of $\pi/2$ (corresponding to a shift in the origin of $\pi/2$) referred to the fundamental, corresponds to a phase shift of $n\pi/2$ referred to the nth harmonic. Hence the nth term of the cosine series transforms to

$$-\sin\left[n\theta - \frac{\pi}{2} + n\frac{\pi}{2}\right] = -\sin\left[n\theta + (n-1)\frac{\pi}{2}\right]$$

For

$$\begin{aligned}
n &= 1 & \cos\theta &\to -\sin\theta \\
n &= 3 & \cos 3\theta &\to -\sin(3\theta + \pi) = \sin 3\theta \\
n &= 5 & \cos 5\theta &\to -\sin(5\theta + 2\pi) = -\sin 5\theta \text{ etc.}
\end{aligned}$$

The series for the triangular wave converges more rapidly (as $1/n^2$) than does the series for the square wave. The difference in the rates of convergence is related to the fact that the square wave is discontinuous whereas the triangular wave is continuous but has discontinuous derivatives. In general, the smoother the original wave the more accurately it can be represented by a few terms in the series. The square wave requires higher harmonics to fill in the corners.

Observe that in both examples the constant term is zero. This could have been predicted from the fact that in both cases the positive and negative half cycles are identical and so the average value of the wave is zero.

7.8.5 Fourier expansion for rectifier output

Of particular interest are the Fourier expansions for the output voltages of half-wave and full-wave rectifiers.

(a) *Half-wave rectifier.* For convenience the amplitude of the output waveform (fig. 7.43) is taken to be unity. We choose the origin where the voltage is maximum, which results in an even function so that all the b_n are zero. The a_n are derived from (7.37).

$$a_n = \frac{2}{\pi}\int_0^\pi f(\theta)\cos n\theta \, d\theta = \frac{2}{\pi}\int_0^{\pi/2}\cos\theta\cos n\theta \, d\theta \tag{7.43}$$

Integration of (7.43) is accomplished by making use of the identity $\cos A \cos B = \frac{1}{2}[\cos(A+B) + \cos(A-B)]$. The result is:

$$a_n = \frac{2}{\pi(1-n^2)}\cos\frac{n\pi}{2} \tag{7.44}$$

Now

$$\cos\frac{n\pi}{2} = \begin{cases} 0 & n \text{ odd } (n \neq 1) \\ -1 & n = 2, 6, 10 \ldots \\ 1 & n = 4, 8, 12 \ldots \end{cases}$$

For $n=1$, (7.44) is indeterminate; therefore, we return to (7.43) which, with $n=1$, becomes

$$a_1 = \frac{2}{\pi} \int_0^{\pi/2} \cos^2\theta \, d\theta$$

This, after integration and substitution of limits, gives $a_1 = \frac{1}{2}$. The other coefficients, including a_0, are evaluated using (7.44). The Fourier expansion for the half-wave rectifier output is then (with $\theta = \omega t$)

$$f(\omega t) = \frac{1}{\pi}\left[1 + \frac{\pi}{2}\cos\omega t + \frac{2}{3}\cos 2\omega t - \frac{2}{15}\cos 4\omega t + \ldots\right] \quad (7.45)$$

(b) *Full-wave rectifier.* We choose the origin as shown in fig. 7.44. Here again the wave is an even function and so there are no sine terms. From (7.37)

$$a_n = \frac{2}{\pi}\int_0^\pi \sin\theta\cos(n\theta)\,d\theta$$

$$= \frac{1}{\pi}\left[\int_0^\pi \sin(1+n)\theta \, d\theta + \int_0^\pi \sin(1-n)\theta \, d\theta\right]$$

$$= \frac{-1}{\pi(1+n)}[\cos(1+n)\pi - 1] + \frac{-1}{\pi(1-n)}[\cos(1-n)\pi - 1]$$

When n is odd, both terms in the brackets are zero. When n is even, each bracket has the value -2. Therefore,

$$a_n = \frac{4}{\pi(1-n^2)}$$

Fig. 7.43. Half-wave rectifier output represented as an even function.

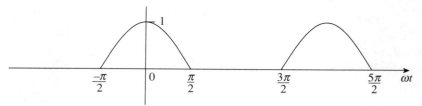

Fourier analysis of periodic waves

The Fourier expansion for the full-wave output is then

$$f(\omega t)=\frac{2}{\pi}\left[1-\frac{2}{3}\cos2\omega t-\frac{2}{15}\cos4\omega t-\frac{2}{35}\cos6\omega t\ldots\right] \quad (7.46)$$

7.8.6 Expansion of functions of time

Sometimes it is convenient to specify waveforms as functions of time rather than of angle. In this case the expressions for the Fourier expansion take a slightly different form.

For a periodic function $f(t)=f(t+T)$ we may put $\omega t = 2\pi t/T$ in (7.31) to give

$$f(t)=\frac{a_0}{2}+\sum_{n=1}^{\infty}\left(a_n\cos\frac{2n\pi t}{T}+b_n\sin\frac{2n\pi t}{T}\right) \quad (7.47)$$

Since π, the half-period in angle, corresponds to $T/2$ in time, (7.34) and (7.35) become

$$a_n=\frac{2}{T}\int_{-T/2}^{T/2}f(t)\cos\frac{2n\pi t}{T}dt \quad (7.48)$$

and

$$b_n=\frac{2}{T}\int_{-T/2}^{T/2}f(t)\sin\frac{2n\pi t}{T}dt \quad (7.49)$$

These equations allow one to evaluate the Fourier coefficients directly in terms of time; however, they are usually more cumbersome to use than the corresponding equations (7.34) and (7.35) in terms of angle. For this reason it is usually preferable in a practical problem to specify functions in terms of angle.

7.8.7 Complex exponential form of Fourier series

The Fourier series expansion (7.31) or (7.33) may also be written in terms of the complex exponential. This approach to Fourier expansion is of particular importance in the theory of communication networks.

Fig. 7.44. Full-wave rectifier output represented as an even function.

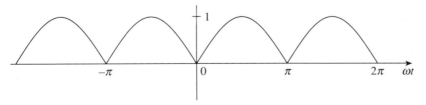

The pair of general terms in (7.33) may be written as

$$a_n\cos n\theta + b_n\sin n\theta = a_n\frac{e^{jn\theta}+e^{-jn\theta}}{2} + b_n\frac{e^{jn\theta}-e^{-jn\theta}}{j2}$$

$$= \frac{a_n-jb_n}{2}e^{jn\theta} + \frac{a_n+jb_n}{2}e^{-jn\theta}$$

Now if we define *complex* coefficients

$$c_n = \frac{a_n-jb_n}{2} \quad \text{and} \quad c_{-n} = \frac{a_n+jb_n}{2} \tag{7.50}$$

then the expansion (7.33) may be written

$$f(\theta) = c_0 + \sum_{n=1}^{\infty}(c_n e^{jn\theta} + c_{-n}e^{-jn\theta})$$

where $c_0 = a_0/2$.

This expression may, by allowing n to have all integral values from $-\infty$ to $+\infty$, including zero, be written in the compact form:

$$f(\theta) = \sum_{n=-\infty}^{\infty} c_n e^{jn\theta} \tag{7.51}$$

By combining (7.50) with (7.34) and (7.35) we obtain

$$c_n = \frac{a_n-jb_n}{2} = \frac{1}{2\pi}\int_{-\pi}^{\pi} f(\theta)\cos n\theta\, d\theta - \frac{j}{2\pi}\int_{-\pi}^{\pi} f(\theta)\sin n\theta\, d\theta$$

$$= \frac{1}{2\pi}\int_{-\pi}^{\pi} f(\theta)(\cos n\theta - j\sin n\theta)\, d\theta$$

or

$$c_n = \frac{1}{2\pi}\int_{-\pi}^{\pi} f(\theta)e^{-jn\theta}\, d\theta \tag{7.52}$$

It may be shown that this equation holds for all n, positive, negative and zero.

Although (7.52) provides a direct method for deriving the coefficients c_n, in practice it is usually easier to evaluate a_n and b_n and then to derive c_n from (7.50).

The Fourier series expansion in the form (7.51) may be interpreted in a mathematical rather than a physical sense, as a series containing terms of both positive and negative frequency. The individual harmonic components of the trigonometric Fourier series are, however, composed of pairs of terms from the positive and negative sequences, as will be appreciated if the coefficients c_n are expressed in terms of the amplitudes and phases of the harmonics. In the form (7.32) (with $\omega t = \theta$) the Fourier expansion is

$$f(\theta) = \frac{a_0}{2} + \sum_{n=1}^{\infty} K_n \cos(n\theta + \phi_n) \qquad (7.32)$$

where:

$$\text{amplitude of } n\text{th harmonic } K_n = \sqrt{(a_n^2 + b_n^2)}$$

and

$$\text{phase of } n\text{th harmonic } \phi_n = \tan^{-1}\left(\frac{-b_n}{a_n}\right)$$

From (7.50) and (7.32) we have

$$c_n = \frac{a_n - jb_n}{2} = \tfrac{1}{2}\sqrt{(a_n^2 + b_n^2)}\, e^{j\phi_n} = \frac{K_n}{2} e^{j\phi_n}$$

$$c_{-n} = \frac{a_n + jb_n}{2} = \tfrac{1}{2}\sqrt{(a_n^2 + b_n^2)}\, e^{-j\phi_n} = \frac{K_n}{2} e^{-j\phi_n} \qquad (7.53)$$

The sum of the two terms corresponding to the nth harmonic in the series is, therefore

$$c_n e^{jn\theta} + c_{-n} e^{-jn\theta} = \frac{K_n}{2}\left(e^{j(n\theta + \phi_n)} + e^{-j(n\theta + \phi_n)}\right)$$

$$= K_n \cos(n\theta + \phi_n)$$

Thus, the sum of pairs of negative and positive frequency components contained in the complex exponential series corresponds to one harmonic component in the trigonometric series. We observe also, from (7.53) that the amplitude of each coefficient of the pair of terms is half that of the corresponding coefficient in the trigonometric series, that is,

$$|c_n| = |c_{-n}| = \frac{K_n}{2} \qquad (7.54)$$

The magnitude of the phase angle is the same in both complex and trigonometric series.

As an example of the application of the complex form of Fourier series, let us consider again the square wave of amplitude $\pm V$ shown in fig. 7.41 and analyzed in section 7.8.2. The origin was chosen so that the wave was an odd function thus giving $a_n = 0$. The coefficients b_n were found to be

$$b_n = \begin{cases} \dfrac{4V}{n\pi} & n \text{ odd} \\ 0 & n \text{ even} \end{cases}$$

Therefore, from (7.50),

$$\left. \begin{aligned} c_n &= \dfrac{-\mathrm{j}}{2} \dfrac{4V}{n\pi} = -\mathrm{j}\dfrac{2V}{n\pi} \\ c_{-n} &= \dfrac{\mathrm{j}}{2} \dfrac{4V}{n\pi} = +\mathrm{j}\dfrac{2V}{n\pi} \end{aligned} \right\} n \text{ odd}$$

The complex Fourier expansion is then, from (7.51),

$$f(\omega t) = \sum_{n=-\infty}^{\infty} -\mathrm{j}\dfrac{2V}{n\pi} \mathrm{e}^{\mathrm{j}n\omega t}$$

$$= \ldots + \mathrm{j}\dfrac{2V}{5\pi}\mathrm{e}^{-\mathrm{j}5\omega t} + \mathrm{j}\dfrac{2V}{3\pi}\mathrm{e}^{-\mathrm{j}3\omega t} + \mathrm{j}\dfrac{2V}{\pi}\mathrm{e}^{-\mathrm{j}\omega t}$$

$$- \mathrm{j}\dfrac{2V}{\pi}\mathrm{e}^{\mathrm{j}\omega t} - \mathrm{j}\dfrac{2V}{3\pi}\mathrm{e}^{\mathrm{j}3\omega t} - \mathrm{j}\dfrac{2V}{5\pi}\mathrm{e}^{\mathrm{j}5\omega t} \ldots \quad (7.55)$$

It is seen that the phases of the complex coefficients c_n and c_{-n} are $-90°$ and $+90°$ respectively. Amplitude and phase spectra are plotted in fig. 7.45. Plots of this type find application in the theory of communication networks. In fig. 7.45(c) amplitude and phase information has been combined in a single diagram, the phase change of $+90°$ to $-90°$ being indicated by oppositely directed amplitude components. This form of representation is possible only if the original function is either even or odd, otherwise the c_n coefficients possess both real and imaginary parts and the phase changes continuously throughout the spectrum.

†7.8.8 Expansions for r.m.s. values and power

It is frequently of interest to know the r.m.s. or effective value of a periodic, non-sinusoidal waveform of current or voltage. This may be calculated (from the definition given in section 3.1) using the analytical function describing the waveform. Such calculation yields an r.m.s. value that includes the effect of all of the harmonics contained in the waveform. It may, however, be of interest to know the r.m.s. value of just a portion of the spectrum and for this a series expansion is required. We may also wish to know the power developed at a terminal pair in a circuit where voltage and current are periodic at the same fundamental frequency but which possess different waveforms. This may be found by taking the instantaneous

Fig. 7.45. Line spectra for the square wave of fig. 7.41.

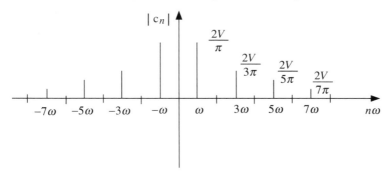

(a) Amplitude

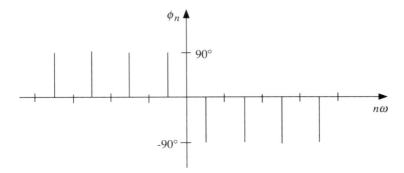

(b) Phase

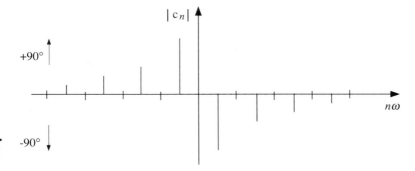

(c) Combined amplitude and phase

476 Non-linear circuit analysis

product of the two waveforms and integrating to find the average value of the product over one period of the waveform. But, again, it may be of interest to know the power contributed by one or more harmonics present in either of the waveforms.

We now proceed to establish expressions that will allow us to calculate the r.m.s. values and powers associated with one or more harmonics in non-sinusoidal waveforms.

Consider two functions, $f(\omega t)$ and $g(\omega t)$, represented by their Fourier expansions in the form (7.32):

$$f(\omega t) = \frac{a_{0f}}{2} + \sum_{n=1}^{\infty} K_{nf}\cos(n\omega t + \phi_{nf}) \tag{7.56}$$

$$g(\omega t) = \frac{a_{0g}}{2} + \sum_{m=1}^{\infty} K_{mg}\cos(m\omega t + \phi_{mg}) \tag{7.57}$$

In these expressions $a_{0f}/2$ and $a_{0g}/2$ are the average values or d.c. components of the two functions; K_{nf} and K_{mg} are the amplitudes, and ϕ_{nf} and ϕ_{mg} are the phase angles of the nth and mth harmonic components of the two functions respectively.

The average value of the product of the two functions is given by

$$\overline{f(\omega t)g(\omega t)} = \frac{1}{2\pi} \int_0^{2\pi} f(\omega t)g(\omega t)\, d\omega t \tag{7.58}$$

When we multiply (7.56) by (7.57), the right-hand side of (7.58) is seen to contain:

(a) self-product terms of the same frequency ($m = n$);
(b) cross-product terms of different frequency ($m \neq n$);
(c) terms consisting of products of the d.c. components with harmonics;
(d) the product of the two d.c. components.

Type (a) terms are of the form:

$$\frac{1}{2\pi} \int_0^{2\pi} K_{nf}\cos(n\omega t + \phi_{nf}) K_{ng}\cos(n\omega t + \phi_{ng})\, d\omega t$$

$$= \frac{K_{nf}K_{ng}}{2\pi} \int_0^{2\pi} \tfrac{1}{2}[\cos(2n\omega t + \phi_{nf} + \phi_{ng}) + \cos(\phi_{nf} - \phi_{ng})]\, d\omega t$$

(using the identity $\cos A \cos B = \tfrac{1}{2}[\cos(A+B) + \cos(A-B)]$). The first term within square brackets is a cosine function of time whose average value is zero; the second term is a constant. Since the average value of a constant is the constant itself, the average value of the nth self-product term is

$$\frac{K_{nf}K_{ng}}{2}\cos(\phi_{nf} - \phi_{ng}) \tag{7.59}$$

Fourier analysis of periodic waves

Type (b) terms are of the form:

$$\frac{1}{2\pi}\int_0^{2\pi} K_{nf}\cos(n\omega t+\phi_{nf})K_{mg}\cos(m\omega t+\phi_{mg})\,d\omega t$$

$$=\frac{K_{nf}K_{mg}}{2\pi}\int_0^{2\pi}\tfrac{1}{2}[\cos\{(n+m)\omega t+\phi_{nf}+\phi_{mg}\}$$

$$+\cos\{(n-m)\omega t+\phi_{nf}-\phi_{mg}\}]\,d\omega t=0$$

since for $n\neq m\neq 0$, each of the terms within the square brackets is a cosine function of time whose average is zero.

Type (c) terms also give rise to cosine functions of time with zero average. With regard to type (d) terms, clearly, the average value of the product of the two d.c. components is the product itself, namely,

$$\frac{a_{0f}a_{0g}}{4} \tag{7.60}$$

Combining (7.59) and (7.60) we obtain a series representing the average value of the product of two functions:

$$\overline{f(\omega t)g(\omega t)}=\frac{a_{0f}a_{0g}}{4}+\sum_{n=1}^{\infty}\frac{K_{nf}K_{ng}}{2}\cos(\phi_{nf}-\phi_{ng}) \tag{7.61}$$

This expression may be used to find the r.m.s. value of a current $i(\omega t)$ in terms of its harmonic components.

Let $i(\omega t)=f(\omega t)$, then, from (7.56), with $a_{0f}/2=I_0$ and $K_{nf}=\hat{I}_n$ (where $\hat{I}_n$ is the *amplitude* of the nth harmonic), the expansion of $i(\omega t)$ may be written as

$$i(\omega t)=I_0+\sum_{n=1}^{\infty}\hat{I}_n\cos(n\omega t+\phi_n)$$

Equation (7.61), with $i(\omega t)=f(\omega t)=g(\omega t)$, then gives:

$$\overline{i(\omega t)^2}=I_0^2+\sum_{n=1}^{\infty}\frac{\hat{I}_n^2}{2} \tag{7.62}$$

By definition, the r.m.s. value of a current $i(\omega t)$ is

$$i(\omega t)_{\text{r.m.s.}}=\sqrt{[\text{average value of } i(\omega t)^2]}$$

$$=\sqrt{\left[I_0^2+\sum_{n=1}^{\infty}\frac{\hat{I}_n^2}{2}\right]} \tag{7.63}$$

Expressing the harmonic components in terms of their r.m.s. values $I_n=\hat{I}_n/\sqrt{2}$, (7.63) becomes

$$i(\omega t)_{\text{r.m.s.}} = \sqrt{\left[\sum_{n=0}^{\infty} I_n^2\right]}$$

or

$$i(\omega t)_{\text{r.m.s.}} = \sqrt{[I_0^2 + I_1^2 + I_2^2 + \ldots]} \qquad (7.64)^*$$

A similar expression holds for the r.m.s. value of a voltage function.

Equation (7.61) enables us to find also an expansion for the average power P at a terminal pair in a circuit, given expressions for the instantaneous voltage and current $v(\omega t)$ and $i(\omega t)$.

Using (7.56) and (7.57), voltage and current functions may be expressed as

$$v(\omega t) = V_0 + \sum_{n=1}^{\infty} \hat{V}_n \cos(n\omega t + \phi_{nv})$$

and

$$i(\omega t) = I_0 + \sum_{m=1}^{\infty} \hat{I}_m \cos(m\omega t + \phi_{mi})$$

Then from (7.61)

$$P = \overline{v(\omega t)i(\omega t)} = V_0 I_0 + \sum_{n=1}^{\infty} \frac{\hat{V}_n \hat{I}_n}{2} \cos(\phi_{nv} - \phi_{ni}) \qquad (7.65)$$

If we let $\phi_n = \phi_{nv} - \phi_{ni}$ be the phase angle of the nth harmonic of voltage with respect to current, and if we let $V_n = \hat{V}_n/\sqrt{2}$ and $I_n = \hat{I}_n/\sqrt{2}$, then (7.65) becomes

$$P = V_0 I_0 + \sum_{n=1}^{\infty} V_n I_n \cos\phi_n$$

or

$$P = V_0 I_0 = V_1 I_1 \cos\phi_1 + V_2 I_2 \cos\phi_2 + \ldots$$
$$= \text{Sum of powers for each harmonic component alone} \qquad (7.66)$$

If voltage and current are pure sinusoids, then there are no harmonics and this expression reduces to $VI\cos\phi$, which is the expression for power obtained by other means in section 4.2.

An important conclusion which may be drawn from the foregoing theory

---

* Equations (7.64) is a special case of *Parseval's theorem* (see, for example, references 2, 8, 10)

Fourier analysis of periodic waves

is that only voltages and currents of the same frequency can interact to produce finite average power. In particular, d.c. components do not interact with a.c. components. Products of quantities of different frequency average over a period of time to zero.

7.8.9 Summary of formulae
Basic forms:

$$f(\omega t) = \frac{a_0}{2} + \sum_{n=1}^{\infty} (a_n \cos n\omega t + b_n \sin n\omega t) \quad (7.31)$$

$$= \frac{a_0}{2} + \sum_{n=1}^{\infty} K_n \cos(n\omega t + \phi_n) \quad (7.32)$$

where

$$K_n = \sqrt{(a_n^2 + b_n^2)} \text{ and } \phi_n = \tan^{-1}\left(\frac{-b_n}{a_n}\right)$$

and

$$a_n = \frac{1}{\pi} \int_{-\pi}^{\pi} f(\omega t) \cos n\omega t \, d\omega t \quad (7.34)$$

$$b_n = \frac{1}{\pi} \int_{-\pi}^{\pi} f(\omega t) \sin n\omega t \, d\omega t \quad (7.35)$$

For an even function

$$a_n = \frac{2}{\pi} \int_0^{\pi} f(\omega t) \cos n\omega t \, d\omega t; \; b_n = 0 \quad (7.37)$$

For an odd function

$$b_n = \frac{2}{\pi} \int_0^{\pi} f(\omega t) \sin n\omega t \, d\omega t; \; a_n = 0 \quad (7.39)$$

Complex form:

$$f(\omega t) = \sum_{n=-\infty}^{\infty} c_n e^{jn\omega t} \quad (7.51)$$

where

$$\begin{aligned} c_n &= \tfrac{1}{2}(a_n - jb_n) & n > 0 \\ &= \tfrac{1}{2}(a_n + jb_n) & n < 0 \\ &= a_0/2 & n = 0 \end{aligned}$$

or

$$c_n = \frac{1}{2\pi}\int_{-\pi}^{\pi} f(\omega t)e^{-jn\omega t}\,d\omega t \tag{7.52}$$

$$|c_n| = K_n/2 \tag{7.54}$$

Square wave (unit amplitude; odd function)

$$f(\omega t) = \frac{4}{\pi}\left[\sin\omega t + \frac{1}{3}\sin 3\omega t + \frac{1}{5}\sin 5\omega t + \ldots\right] \tag{7.36}$$

Triangular wave (unit amplitude; odd function)

$$f(\omega t) = \frac{8}{\pi^2}\left[\sin\omega t - \frac{1}{9}\sin 3\omega t + \frac{1}{25}\sin 5\omega t - \ldots\right] \tag{7.42}$$

Half-wave rectified sine wave (unit amplitude; even function)

$$f(\omega t) = \frac{1}{\pi}\left[1 + \frac{\pi}{2}\cos\omega t + \frac{2}{3}\cos 2\omega t - \frac{2}{15}\cos 4\omega t + \ldots\right] \tag{7.45}$$

Full-wave rectified sine wave (unit amplitude; even function)

$$f(\omega t) = \frac{2}{\pi}\left[1 - \frac{2}{3}\cos 2\omega t - \frac{2}{15}\cos 4\omega t - \frac{2}{35}\cos 6\omega t - \ldots\right] \tag{7.46}$$

Average value of product of two functions. If

$$f(\omega t) = \frac{a_{0f}}{2} + \sum_{n=1}^{\infty} K_{nf}\cos(n\omega t + \phi_{nf}) \tag{7.56}$$

and

$$g(\omega t) = \frac{a_{0g}}{2} + \sum_{m=1}^{\infty} K_{mg}\cos(m\omega t + \phi_{mg}) \tag{7.57}$$

then

$$\overline{f(\omega t)g(\omega t)} = \frac{a_{0f}a_{0g}}{4} + \sum_{n=1}^{\infty} \frac{K_{nf}K_{ng}}{2}\cos(\phi_{nf} - \phi_{ng}) \tag{7.61}$$

R.M.S. value of a function in terms of harmonic components.
  If $i(\omega t)$ is a function of current with r.m.s. value I, then

$$I = [\overline{i(\omega t)^2}]^{\frac{1}{2}} = \sqrt{(I_0^2 + I_1^2 + I_2^2 + \ldots)} \tag{7.64}$$

where I_0, I_1, I_2, etc., are the r.m.s. values of harmonic components.
  Power in terms of harmonic components.

If, at a terminal pair, $v(\omega t)$ and $i(\omega t)$ are functions of voltage and current with the same periodicity, then the average power is given by

$$P = \overline{v(\omega t)i(\omega t)} = V_0 I_0 + V_1 I_1 \cos\phi_1 + V_2 I_2 \cos\phi_2 \ldots \quad (7.66)$$

where $(V_1, I_1); (V_2, I_2)$; etc. are the r.m.s. values of harmonic components of like frequency.

†7.9 Filter circuits for rectifiers

Half-wave or full-wave rectifier circuits of the type discussed in section 7.6 provide the basis for most d.c. power supplies used in electronic equipment, but for many applications the unidirectional but fluctuating voltages provided by such circuits is not sufficiently constant, and provision must be made for 'smoothing' their outputs. This is accomplished by means of a filter circuit designed to attenuate the fluctuating harmonic components of the output voltage waveform whilst passing the constant, d.c. component.

Referring to fig. 7.46, each harmonic component in the Fourier expansion of the rectifier output may be regarded as deriving from a separate ideal voltage source. The attenuation of the filter may then be calculated separately for each input and, assuming that filter and load are composed of linear circuit elements, the overall effect of all sources acting together may be obtained by superposition. In practice it is sufficient to know that the amplitude of a particular component is below a certain specified level. In the type of circuit considered here only the first two or three harmonic components are of interest since higher harmonics, firstly,

Fig. 7.46. Analysis of a rectifier/filter circuit using Fourier component representation of rectifier output waveform.

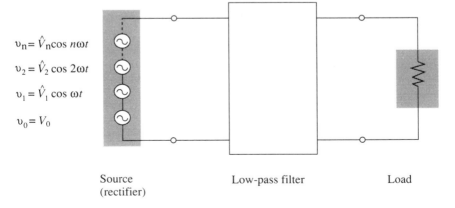

have relatively small amplitudes (see equations (7.45) and (7.46)); and, secondly, they suffer greater attenuation by the filter.

Various combinations of inductive and capacitive elements are used in rectifier filter circuits; some of the more common arrangements are considered below.

7.9.1 Inductor

One of the simplest filter circuits consists of a single inductor in series with the load, assumed to be a pure resistance, as shown in fig. 7.47. We analyze first the smoothing effect of this circuit upon the output of the full-wave rectifier, making use of the series expansion (7.46) and considering each term separately.

The first term represents the d.c. component of the rectified wave with amplitude $2V_m/\pi$, where V_m is the peak value of the rectified wave. Since this is a direct voltage all of it appears across R (assuming that the inductor has negligible resistance).

The next term in the expansion is the second harmonic component with amplitude $4V_m/3\pi$. If the supply frequency is ω, then the second harmonic is at frequency 2ω and the factor by which this component is reduced by the filter is $R/\sqrt{[R^2+(2\omega)^2 L^2]}$. Thus, the amplitude V_{2m} of the ripple across R is given by

$$V_{2m} = \frac{4V_m}{3\pi} \cdot \frac{R}{\sqrt{(R^2+(2\omega)^2 L^2)}} = \frac{4V_m}{3\pi} \cdot \frac{1}{\sqrt{[1+(2\omega L/R)^2]}} \quad (7.67)$$

Usually, with this type of circuit, the reactance of the inductor $(2\omega L)$ is arranged to be large in comparison with R, in which case V_{2m} is given to a sufficiently good approximation by

$$V_{2m} \approx \frac{4V_m}{3\pi} \cdot \frac{R}{2\omega L} = \frac{2V_m R}{3\pi \omega L} \quad (2\omega L \gg R)$$

Fig. 7.47. Inductor smoothing of rectifier output.

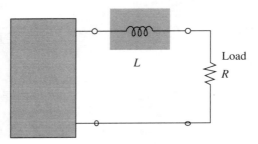

Filter circuits for rectifiers

The fourth harmonic component has amplitude $4V_m/15\pi$, and the amplitude of the voltage appearing across R due to this component is approximately

$$V_{4m} \approx \frac{4V_m}{15\pi} \cdot \frac{R}{4\omega L} = \frac{V_m R}{15\pi \omega L}$$

From the above expressions we see that the amplitude of the fourth harmonic component across the load is only about one tenth that of the second harmonic component and can usually be ignored.

It is customary to describe the effectiveness of the smoothing action of a filter circuit by stating the peak-to-peak value of the most significant ripple component as a percentage of the d.c. component.

$$\text{percentage ripple} = \frac{\text{peak-to-peak ripple amplitude}}{\text{magnitude of direct component}} \times 100 \quad (7.68)$$

For the full-wave rectifier with series inductor smoothing the percentage ripple is

$$2 \times \frac{2V_m R}{3\pi \omega L} \times \frac{\pi}{2V_m} \times 100 = \frac{2R}{3\omega L} \times 100\%$$

As an example, consider a full-wave rectifier designed to supply a 50 Ω load. If $L = 10$ H and the supply frequency is 50 Hz, then the percentage ripple will be

$$\frac{2 \times 50 \times 100}{3 \times 2\pi \times 50 \times 10} = 1\%$$

Note that the full-wave rectifier with inductor smoothing produces a lower ripple than the half-wave rectifier because, for the former, the fundamental component of ripple is at twice supply frequency, while for the latter it is at the same frequency as supply frequency (compare series expansions (7.46) and (7.45)).

Carrying out a similar analysis for the half-wave rectifier, using the expansion (7.45), gives the following results:

D.C. component $\qquad V_0 = V_m/\pi$
Amplitude of first harmonic $\qquad V_{1m} = V_m/2$

Reduction factor
(for first harmonic) $\qquad = \dfrac{R}{\omega L}$

$$\text{Percentage ripple} = 2 \times \frac{V_m}{2} \times \frac{R}{\omega L} \times \frac{\pi}{V_m} \times 100 = \frac{\pi R}{\omega L} \times 100\%$$

$$= 5\% \text{ (for } R = 50\,\Omega;\ L = 10\text{ H)}$$

7.9.2 L-section

Considering again the full-wave rectifier, a further degree of smoothing is obtained by the addition of a capacitor to the circuit of fig. 7.47, as shown in fig. 7.48. It is usual to make the reactance of this capacitor ($1/2\omega C$) small in comparison with R so that substantially the whole of the second harmonic component of current goes through the capacitor. The inductor and capacitor then constitute a voltage divider and the factor by which the second-harmonic voltage is reduced is, approximately,

$$\frac{1/2\omega C}{\sqrt{[(1/2\omega C)^2 + (2\omega L)^2]}} \approx \frac{1}{4\omega^2 LC} \tag{7.69}$$

7.9.3 Capacitor

The output of a rectifier circuit may be smoothed by placing a capacitor across the load, as shown in fig. 7.49(a). With this type of circuit the diode conducts for only a fraction of each half-cycle of the supply voltage, and the output waveform can no longer be represented by the Fourier series expansions given in section 7.8.5. The action of the circuit is illustrated in fig. 7.49(b). On positive half cycles of the input voltage the capacitor charges to the peak value V_m of the supply. As the supply voltage falls, the diode ceases to conduct and the capacitor discharges through the load R with a time constant RC. If this time constant is large compared with the period T of the input voltage, then the capacitor will lose only a small fraction of its initial charge before the next half cycle of the voltage appears and raises the capacitor voltage back to V_m. During the period in which the diode is non-conducting the capacitor supplies the whole of the current to the load, and for this reason it is often referred to as a 'reservoir' capacitor.

Fig. 7.48. L-section filter.

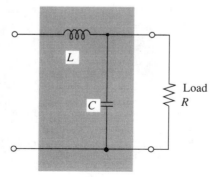

Filter circuits for rectifiers

The capacitor voltage during the discharge period is

$$v_c = V_m e^{-t/RC}$$

If $RC \gg T$, then we may approximate the exponential by the first two terms of its series expansion to give

$$v_c \approx V_m\left(1 - \frac{t}{RC}\right)$$

Now, if we further assume that the charging time t_1 is short compared with T, the decrease in capacitor voltage ΔV is approximately

$$\Delta V \approx \frac{V_m T}{RC}$$

and the average voltage (d.c. component) across the load is

$$V_0 \approx V_m - \frac{\Delta V}{2} = V_m\left(1 - \frac{T}{2RC}\right) \tag{7.70}$$

Using the definition (7.68) the percentage ripple for the half-wave rectifier with capacitor smoothing is given by

Fig. 7.49. Half-wave rectifier with capacitor smoothing: (a) circuit (b) waveforms.

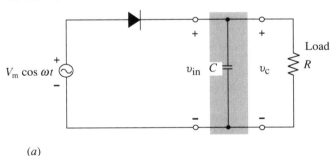

(a)

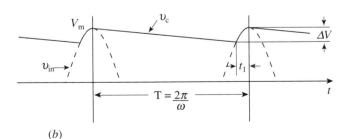

(b)

Non-linear circuit analysis

$$\text{Percentage ripple} = \frac{\Delta V}{V_0} \times 100 \approx \frac{(V_m T/RC) \times 100}{V_m} \approx \frac{T}{RC} \times 100\%$$

This approximate expression is usually sufficiently accurate to allow one to choose a suitable value for the filter capacitor. However, if a more detailed and accurate treatment is required, in particular if the power in the diode is to be calculated, then one may use methods similar to those employed in section 7.6.2 in relation to the battery charger problem. These methods allow one to calculate the conduction period t_1 and, given the diode resistance, to derive the instantaneous function of current during the charging period. The power dissipated in the diode may then be determined.

7.9.4 π-section

If a high degree of smoothing is required, the single capacitor of fig. 7.49(a) may be combined with the L-section filter of fig. 7.48 to produce the π-section filter circuit shown in fig. 7.50. In this circuit the voltage across C_1 will resemble the sawtooth voltage of fig. 7.49(b). In order to estimate the ripple across the load it is usual to make the simplifying assumption that the fundamental component of ripple across C_1, (for the half-wave rectifier, this is the first harmonic Fourier component of the sawtooth at supply frequency) has an amplitude equal to half the peak-to-peak voltage of the sawtooth. (Such an assumption gives, of course, an overestimate of the ripple.) Then, making also the assumptions discussed in section 7.9.2, we may deduce that the factor by which this ripple component is reduced is $1/(\omega^2 LC)$. (For a full-wave rectifier the fundamental component of ripple is at twice supply frequency and the corresponding factor is $1/(4\omega^2 LC)$.)

In order to reduce weight and cost of components, π-section filters are often designed with a resistor R_F in place of the inductor. In this case the ripple is reduced by the factor $1/\omega CR_F$ for the half-wave rectifier, or

Fig. 7.50. π-section filter.

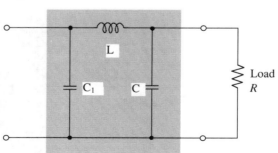

$1/2\omega CR_F$ for the full-wave rectifier. A disadvantage of this circuit is that R_F and R now form a voltage divider for the d.c. component of the rectifier output and so the d.c. voltage across the load may be reduced significantly if a low ripple, and therefore a high value of R_F, is required.

7.10 Summary

The analysis of circuits containing non-linear elements requires special techniques. These fall into four major categories: graphical analysis, small-signal models, piecewise-linear techniques, and analytical techniques.

By means of graphical analysis, voltages and currents can be found in circuits containing simple series and parallel combinations of linear and non-linear elements; this method depends upon a knowledge of the complete voltage–current characteristic of the non-linear devices in the circuit. For a two-terminal non-linear device, such as a diode, a single characteristic curve relating voltage and current is all that is required. For a three-terminal device, such as a bipolar transistor, two sets of characteristic curves are required. Graphical analysis is commonly used to determine the *bias* and *operating point* in circuits containing transistors, the procedure in this case being often referred to as the *load line* method.

If the incremental voltage and current swings in a non-linear device are small, then over a limited region, its voltage–current characteristic may be considered as being of straight-line form and a linear relationship may be assumed between voltage and current. This is the basis of the small-signal approach to the analysis of non-linear circuits in which the non-linear devices in the circuit are described by one or more linear parameters. This approach is of particular importance in the case of transistor circuits. The *hybrid-parameter* (small-signal) models of figs. 7.13 and 7.14 are commonly employed in this context.

The small-signal model of a device uses a linear approximation that is valid over a narrow region of the device characteristic. By this means circuits including such devices may be treated as linear and all the techniques of linear circuit analysis become applicable. An extension of this approach is to approximate the voltage–current characteristic of a device by a series of straight-line segments extending over an arbitrarily wide region of the characteristic. Within each segment the device is modelled by appropriate linear parameters. This *piecewise-linear* approximation then allows linear circuit analysis to be applied over any desired range of operating voltages and currents. The analysis of circuits containing diodes is one important area of application for this technique, in particular rectifier circuits. The piecewise-linear approach is also useful when we wish to synthesize circuits that will reproduce given non-linear characteristics by means of combinations of resistors and diodes.

488 Non-linear circuit analysis

The characteristics of many non-linear devices may be expressed by means of analytical functions or approximated by power series. In such cases the circuits of which they form part may be solved using either algebraic or numerical techniques. The characterization of a non-linear device by means of a power series provides a method for the analysis of modulators and frequency changer circuits of the type used in communication networks.

Rectifiers, employing diodes and thyristors for a.c.–d.c. power conversion, comprise a broad class of circuits requiring non-linear techniques of analysis. In such circuits it is often sufficiently accurate to approximate a diode or thyristor by a two-segment piecewise-linear approximation, which in its simplest form is equivalent to treating the device as a switch.

Many types of non-linear circuit produce waveforms that are periodic but non-sinusoidal; rectifier circuits being but one example. The response of linear circuits to such waveforms cannot be determined directly using standard techniques of d.c. or a.c. circuit analysis. However, by means of Fourier series analysis a periodic waveform may be resolved into a d.c. component plus a series of harmonic components of sinusoidal form. The response to such a waveform may then be conveniently found by determining the response to each Fourier component separately and then combining these individual responses by superposition. The utility of the technique lies in the fact that it is usually necessary to consider only the first few harmonic components in order to obtain a sufficiently accurate knowledge of circuit behaviour for design purposes. This approach is particularly useful in the design of filter circuits for rectifiers.

Finally, Fourier theory may be extended to allow the calculation of the r.m.s. value of an arbitrary number of harmonic components, and the power associated with the harmonic components of periodic non-sinusoidal voltage and current waveforms.

7.11 Problems

1. A source of e.m.f. 4.5 V is connected in series with a 225 Ω resistor and a diode operating in the forward conducting region. The diode has the voltage–current characteristic given in table 7.1. Determine by a graphical method the current in the circuit and the voltage across the diode.

Current (mA)	0	1	2	3	4	5	6	7	8	9	10	11	12
Voltage (V)	0	0.45	0.78	1.02	1.23	1.38	1.56	1.74	1.92	2.04	2.16	2.28	2.36

Table 7.1 for problems 1 and 2

Problems

2. Each of the diodes in the circuit of fig. 7.51 has the voltage–current characteristic given in table 7.1. Use a graphical method to determine the combined v–i characteristic for the two diodes and the 1.5 V battery. Hence determine the value of the current I and the voltage V. Determine also the dissipation in each diode.
(London University)

3.

Current (mA)	0	1	2	3	4	5	6	7	8	9	10
Voltage D_1	0	0.7	1.3	1.7	2.1	2.3	2.6	2.9	3.2	3.4	3.6
Voltage D_2	0	1.8	2.8	3.6	4.3	4.9	5.3	5.8	6.2	6.6	6.9

Table 7.2 for problem 3

The diodes in the circuit of fig. 7.52 have the characteristics given in table 7.2. Using a graphical method to combine the diode characteristic, determine the current I, the voltage V and the voltage across each diode. Find also for this bias condition the effective d.c. resistance and the incremental a.c. resistance of the diode combination.
(London University)

Fig. 7.51. Circuit for problem 2.

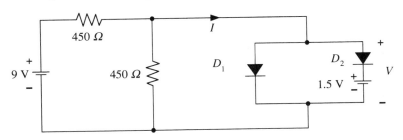

Fig. 7.52. Circuit for problem 3.

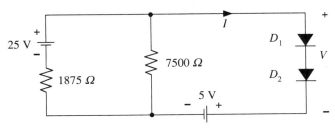

490   Non-linear circuit analysis

4. A tunnel diode having the characteristic shown in fig. 7.53 is connected in series with a voltage source V_0 and a resistor R.
   (a) If $V_0 = 0.75$ V, what range of values of resistance R allows the circuit to have two stable states?
   (b) If $R = 1$ kΩ, what change in V_0 is necessary to change the state of the circuit?
5. The Zener diode in the circuit of fig. 7.54(a) has the characteristic shown in fig. 7.54(b).
   (a) What is the voltage across the diode if $R_L = \infty$?
   (b) What value of R_L will reduce the diode voltage to 90% of the open circuit value?
6. Determine the slopes and break points of the v–i relationship for the piecewise-linear circuit shown in fig. 7.55. (Assume ideal components.)
7. Determine the slopes and break points of the v–i relationship for the piecewise-linear circuit shown in fig. 7.56. (Assume ideal components.)
8. Design a circuit consisting of ideal sources, diodes and resistors that will have the piecewise-linear characteristic shown in fig. 7.57.
9. Fig. 7.58(a) shows two ideal voltage generators V_1 and V_2 connected to a load resistance R_3 through R_1 and R_2 plus a perfect diode. The voltages are shown in fig. 7.58(b).

   Draw the waveform of the voltage across R_2 and the currents in R_1 and R_2, when $R_1 = R_2 = R_3 = R$. What power is consumed in R_3?
   (Oxford University)

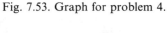

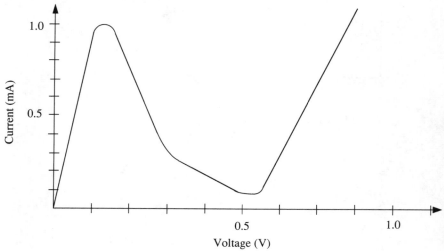

Fig. 7.53. Graph for problem 4.

Problems

Fig. 7.54. Circuit and graph for problem 5.

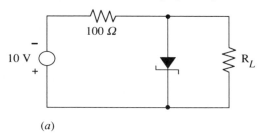

(a)

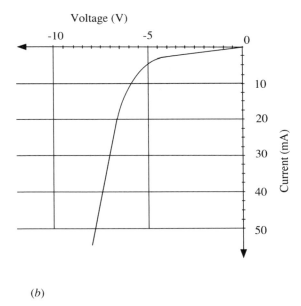

(b)

Fig. 7.55. Circuit for problem 6.

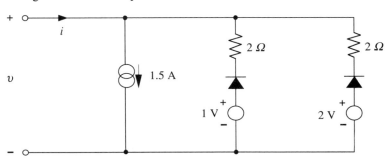

492   Non-linear circuit analysis

10. A diode whose forward characteristic is given in fig. 7.59(a) is connected as shown in the circuit of fig. 7.59(b).
(a) What will be the diode voltage, current and power dissipation?
(b) Approximate the diode characteristic, over the full range shown, by three straight-line segments, keeping the errors in the diode current to within 0.4 mA. Hence draw a piecewise-linear equivalent circuit for the diode when conducting in the forward direction. If the diode is replaced by the equivalent circuit, what will be the power dissipated in the equivalent circuit?
(Cambridge University: Second year)

Fig. 7.56. Circuit for problem 7.

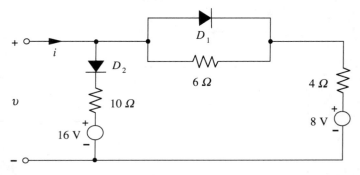

Fig. 7.57. Graph for problem 8.

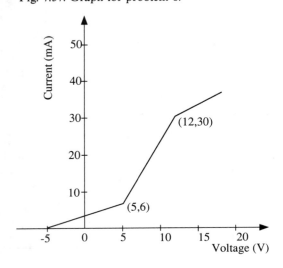

Problems

11. The circuit shown in fig. 7.60 is employed to approximate the characteristic $i = v^2/10^4$ amperes over the range $0 \leqslant v \leqslant 10$ V by a piecewise-linear approximation. Assuming that the diodes are ideal and that D_2 is to conduct at $10/3$ V and D_3 at $20/3$ V, calculate suitable values for the resistors.
(Cambridge University: Second year)

12. In the circuit of fig. 7.61 the non-linear resistor r has a characteristic described by $v = i(1+i^2)^{-\frac{1}{2}}$. Determine the current I.

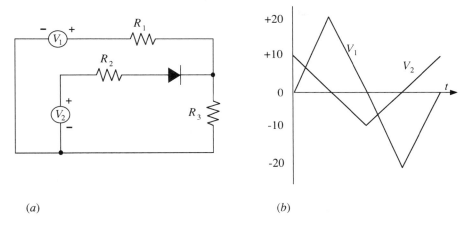

Fig. 7.58. Circuit and graph for problem 9.

(a) (b)

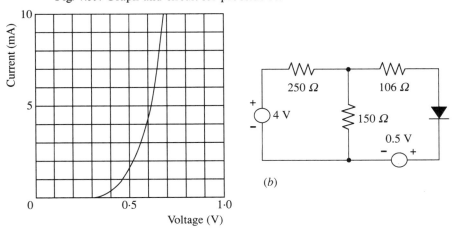

Fig. 7.59. Graph and circuit for problem 10.

(a) (b)

13. Two non-linear devices d_1, d_2 have the following v–i characteristics:

$d_1:$ $i_1 = 0.3(e^{v_1} - 1)$
$d_2:$ $i_2 = v_2 + 0.4v_2^2$

The devices are connected in series across a voltage source.
(a) If the voltage source is ideal with e.m.f. 3 V, find the current in the circuit and the voltage across each device.
(b) If the voltage source has an e.m.f. of 3 V and internal resistance 0.5 Ω, find the current in the circuit and the voltage across each device.

14. In the circuit shown in fig. 7.62, if the output load resistance R_L is constant, show that the incremental variation of output voltage V_o with a change of input voltage V_{in} is equal to

$$x\left(x + R + \frac{Rx}{R_L}\right)^{-1}$$

Fig. 7.60. Circuit for problem 11.

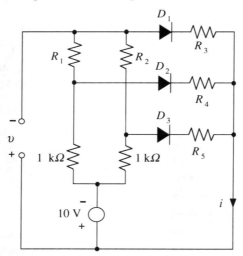

Fig. 7.61. Circuit for problem 12.

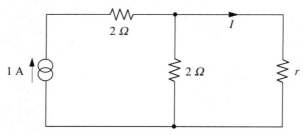

where x is the incremental impedance of the Zener diode.
(Sheffield University: Second year)

15. (a) Derive a two-slope circuit model for the diode whose characteristic is given in fig. 7.63(a).
(b) The diode is used in the simple rectifying circuit of fig. 7.63(b). What is the average current in the circuit if the supply voltage $v(t)$ is:
(i) $1.0\cos\omega t$ volts;
(ii) $100\cos\omega t$ volts?

Fig. 7.62. Circuit for problem 14.

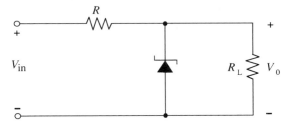

Fig. 7.63. Graph and circuit for problem 15.

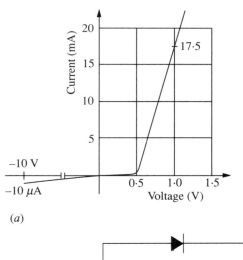

(a)

(b)

16. Switch contacts may be damaged by the opening of a switch that supplies current to a coil. Protection to the contacts is afforded by installation of a diode as shown in fig. 7.64.

(a) Without the diode, the switch contacts would 'arc' when the switch is opened. Why?

(b) Explain how the diode solves the 'arcing' problem but does not interfere with normal operation of the coil.

(c) What is the steady state current in the coil with the switch closed?

(d) How long after the switch is opened will be required for the coil current to reach 3 mA?

17. In the circuit shown in fig. 7.65 the two voltage sources are equal, the diode is perfect and the resistance is small compared to $\frac{1}{2}\sqrt{(LC)}$. Initially there is no charge on the capacitor. Calculate and sketch the variation of capacitor voltage with time after the switch is closed.
(Oxford University)

18. (a) A half-wave rectifier consists of a silicon diode in series with a load resistance of 2 Ω and is supplied from the secondary of a 50 Hz transformer which has an open-circuit voltage of 5.0 V r.m.s. and an effective resistance of 0.1 Ω. The diode characteristic may be represented by an ideal diode in series with a resistance of 0.04 Ω. Determine the mean current flowing in the load.

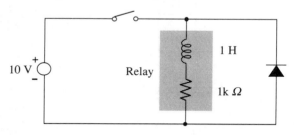

Fig. 7.64. Circuit for problem 16.

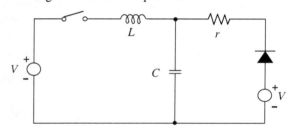

Fig. 7.65. Circuit for problem 17.

(b) A smoothing capacitor, connected across the load, is needed to reduce the peak-to-peak ripple voltage to 10% of the mean voltage at the load. Determine the capacitance to be used and comment on the result. Show that the mean voltage at the load is now about 5 V.
(Cambridge University: Second year)

19. A half-wave rectifier supplies a load resistor $R = 5\,\text{k}\Omega$ in parallel with a filter capacitor $C = 32\,\mu\text{F}$. The a.c. supply is $300\sin(100\pi t)\,\text{V}$ and the combined forward resistance of the rectifier diode and the resistance of the supply is $30\,\Omega$. Estimate:
(a) the average current in the load resistor;
(b) the percent amplitude of the ripple voltage;
(c) the peak current in the rectifier;
(d) the r.m.s. current in the rectifier;
(e) the percent ripple if a second 32 microfarad capacitor is added in parallel;
(f) the percent ripple if a second 32 microfarad capacitor and a 20 henry choke are used in the circuit. (The two capacitors and the choke are arranged to make an L-section filter.)

20. (a) A transformer with centre-tapped secondary each half of which has an e.m.f. of 400 V r.m.s. at 50 Hz is used with a 20 henry inductor and two diodes of zero forward resistance to provide d.c. power to a $500\,\Omega$ load. Calculate the average voltage across the load and the amplitude of the lowest frequency component of the ripple voltage. It may be assumed that one or the other rectifier is conducting at all times.

(b) A filter capacitor is used instead of the inductor and the transformer e.m.f. is altered to give approximately the same average voltage and percent ripple across the $500\,\Omega$ load. What is the new e.m.f. and what capacitance is required?

21. (a) For the general bridge circuit of fig. 7.66(a) show that balance occurs when $Z_1 Z_4 = Z_2 Z_3$. The bridge is unbalanced by changing Z_4 by a small increment δZ_4. Show that the unbalance voltage $|V_{AB}|$ is, to a good approximation, proportional to $|\delta Z_4|$. If $Z_1 = Z_2 = Z_3 = (1000 + j0)\,\Omega$, $Z_4 = (990 + j0)\,\Omega$, and V is a sinusoidal voltage of r.m.s. magnitude 10 V, what is the unbalance voltage?

(b) In the phase-sensitive detector circuit of fig. 7.66(b) the signal and reference voltages are sinusoidal and of the same frequency but differ in phase by an angle ϕ. V_s and V_r are the r.m.s. magnitudes of the corresponding transformer secondary voltages ($V_r > V_s$). The time constant CR is long compared with the period of these waveforms. Derive an expression for the d.c. output voltage V_d in terms of V_s, V_r and ϕ. It may be assumed that the diodes are ideal and that no current is drawn from the output terminals.

(c) Explain the advantages of using the circuit of fig. 7.66(b) as a detector for the bridge circuit of fig. 7.66(a). Indicate how the circuits are connected, and suggest one possible application of such an arrangement.
(Cambridge University: Second year)

22. In the circuit of fig. 7.67 the resistance R controls the mean power to the load. The gate current required to trigger the thyristor is 20 mA. What approximately is the maximum angle of delay achievable with this circuit, and to what value must R be set to obtain this angle?

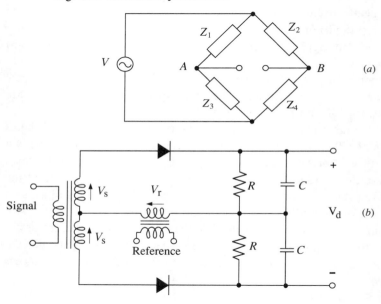

Fig. 7.66. Circuits for problem 21.

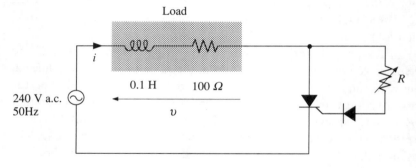

Fig. 7.67. Circuit for problem 22.

Problems

With the angle of delay set to 45°, derive an expression for the load current i during the period of conduction. Sketch the waveforms of i and v for one complete period of the a.c. mains supply. Indicate on your sketch the approximate angle at which the current reaches a maximum and the angle at which conduction ceases.
(Cambridge University: Second year)

23. The trapezoidal wave shown in fig. 7.68 has period T seconds, amplitude A and rise and fall times between zero and A of p seconds. Derive a Fourier expansion for this waveform choosing a time axis that will give rise to sine terms only. Hence find the sine series for:
(a) a square wave;
(b) a triangular wave.
Find values of p for which:
(c) there is no third harmonic;
(d) there is no fifth harmonic.

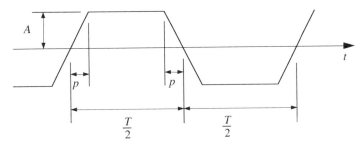

Fig. 7.68. Waveform for problem 23.

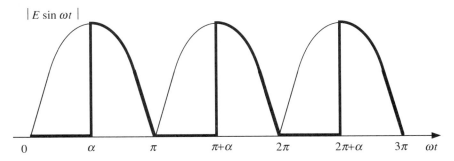

Fig. 7.69. Waveform for problem 24.

24. The voltage waveform indicated by the heavy line in fig. 7.69 is produced by a thyristor circuit. Show that the amplitude of the fundamental ripple component is

$$c_2 = \frac{E}{3\pi}\sqrt{[(14+12\cos\alpha - 4\cos 3\alpha - 6\cos 2\alpha)]}$$

Hence calculate the angle α for which c_2 is a maximum and the ratio of c_2 to the mean value of the voltage at that angle.
(Cambridge University: Second year)

8
Two-port networks

8.1 Introduction

Networks are frequently encountered in electronics, control and communication systems in which an input signal is impressed at one pair of terminals and an output signal is taken from another pair of terminals; such networks are called *two-terminal pair* networks or, more frequently *two-port* networks.

The resistance voltage divider, first introduced in section 2.2, is an example of an elementary two-port network, and many other examples have occurred in the intervening chapters. In particular, it was demonstrated in section 7.3.2 that a non-linear device such as the transistor could, for small-signal conditions, be modelled by linear, two-port network. The concepts relating to such networks are therefore of considerable generality, and in this chapter we examine the theory of two-ports in greater detail and introduce further applications.

The theory contained in this chapter is concerned with the functional relationships among the voltage and current variables at the two ports of the network as defined in fig. 8.1. If the variables are steady-state a.c. quantities, then these relationships are formulated in terms of impedances or admittances. These same relationships will, of course, apply if the variables are transformed variables; in this case impedances and admittances are generalized functions of complex frequency as defined in chapter 6. In the case of purely resistive networks the functional relationships among variables are identical in form for instantaneous and d.c. quantities as well as for a.c. quantities. It should also be noted that the same relationships apply for incremental (small-signal) a.c. quantities as defined in section 7.3.1.*

> * In view of the variety of meanings that may be attached to the current and voltage variables and to impedance, the use of bold face type to indicate phasors and complex quantities is discontinued in this chapter.

502 Two-port networks

A network having a single pair of terminals, a one-port network has associated with it a single current and a single voltage. The ratio of voltage to current, or current to voltage, is the impedance (admittance), which parameters completely characterize the circuit so far as any external network is concerned. For a two-port network there are four variables: an input voltage–current pair and an output voltage–current pair. The possible combinations of four variables taken two at a time is six; it is therefore possible to define six sets of parameters that characterize the two-port, one set for each pair of variables that are chosen to be independent. Which of the parameter sets chosen to characterize a given two-port depends on the application. In the following sections the parameter sets and their interrelationships are derived, and their areas of application are indicated.

Two-port networks are often categorized according to the degree of electrical symmetry which they possess. If a network is symmetrical about the axis XX' (fig. 8.2(a)) then it is said to be *balanced*; if it possesses symmetry about YY', then it is said to be *symmetrical*. Some typical network configurations with various types of symmetry are illustrated in figs. 8.2(b)–(e).

It will be noted that the networks shown in figs. 8.2(b) and (d) possess four distinct terminals while the networks shown in figs. 8.2(c) and (e) possess only three, terminals 1' and 2' being common to both input and output ports in both cases. However, in the theory which follows, no distinction is drawn between these two different configurations, the theory being concerned only with a description of a two-port with respect to the defined variables. So, although in the case of fig. 8.2(b), a potential difference may, in general, exist between terminals 1' and 2' while for fig. 8.2(c) terminals 1' and 2' must be at the same potential, as far as the theory here is concerned both networks possess identical characteristics. From a practical point of view this implies that connections between a two-port and its external circuits must be made in such a way as to prevent any external mesh current from flowing through any internal impedance other than the defined

Fig. 8.1. Voltage and current variables appertaining to the theory of two-port networks.

currents I_1 and I_2. If such currents flow through any of the impedances within the two-port, then the established relationships among the defined variables will no longer apply. In this respect the three-terminal networks of figs. 8.2(c) and (e) provide somewhat greater flexibility since external mesh currents can flow in the common connection without the predicted relationships among the defined variables being affected.

8.2 Admittance, impedance and hybrid parameters

8.2.1 Admittance parameters

The reference directions of currents and voltages at the two ports of the networks are indicated in fig. 8.1. Let us first consider the voltages as the excitations and the currents as the responses, then by linearity:

Fig. 8.2. Balance and symmetry in two-port networks.

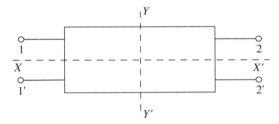

(a) Reference axes

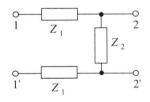

(b) Balanced - unsymmetrical

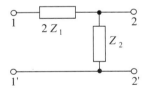

(c) Unbalanced - unsymmetrical

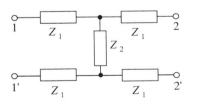

(d) Balanced - symmetrical

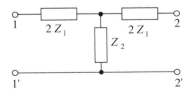

(e) Unbalanced - symmetrical

$$I_1 = y_{11}V_1 + y_{12}V_2 \atop I_2 = y_{21}V_1 + y_{22}V_2 \} \qquad (8.1)$$

We may interpret the coefficients in (8.1), which have dimensions of admittance, in terms of measurements made at the terminals. For example, if the output terminals are short circuited, so that $V_2=0$, and if the responses I_1 and I_2 are measured when an excitation V_1 is applied, then

$$y_{11} = \frac{I_1}{V_1}\bigg|_{V_2=0} \qquad y_{21} = \frac{I_2}{V_1}\bigg|_{V_2=0} \qquad (8.2)$$

y_{11} is clearly an *input admittance*. y_{21} represents the output current response to an input voltage: it is called the *forward transfer admittance*. If the input terminals are shorted so that $V_1=0$, and if the responses I_1 and I_2 are measured when an excitation V_2 is applied, then

$$y_{12} = \frac{I_1}{V_2}\bigg|_{V_1=0} \qquad y_{22} = \frac{I_2}{V_2}\bigg|_{V_1=0} \qquad (8.3)$$

Now y_{22} is the *output admittance* and y_{12} is the *reverse transfer admittance*.

The four admittances completely characterize the two-port. They are commonly referred to as the short-circuit admittance parameters. The input admittance y_{11} and the output admittance y_{22} are frequently called *driving-point* admittances because each describes the response at a port when a driving function is applied. y_{12} and y_{21} are *transfer* admittances, describing the response at one port to excitation at the other.

As an example let us find the y-parameters for the π-network shown in fig. 8.3 in which the elements are specified in terms of their admittances. With $V_2=0$, the input admittance is simply the parallel combination of Y_1 and Y_3. Therefore,

$$y_{11} = Y_1 + Y_3$$

Also, with $V_2=0$, the current I_2 is the current in Y_3. Since with $V_2=0$ the voltage V_1 appears directly across Y_3 it follows that

$$y_{21} = -Y_3$$

The negative sign appears because with $V_2=0$ the current that results from V_1 is opposite in direction to the positive direction of I_2. Now we short circuit the input port making $V_1=0$. Then we find

$$y_{22} = Y_2 + Y_3 \qquad \text{and} \qquad y_{12} = -Y_3$$

where the negative sign indicates that when $V_1=0$, V_2 produces a short-circuit input current that is opposite in direction to the positive direction of I_1.

Admittance, impedance and hybrid parameters

We see that for this example $y_{12} = y_{21}$ and we conclude that with equal voltage excitations the current responses will be equal. This is an example of *reciprocity*. As long as the circuit elements are linear and bilateral and the reciprocity principle applies then $y_{12} = y_{21}$ no matter how complicated the circuit that joins the two ports. Two-port networks of this type, which include all circuits containing only passive elements, are said to be *reciprocal*; those that have unequal transfer admittances, which include many types of active circuit, are said to be *non-reciprocal*.

8.2.2 Impedance parameters

If in the circuit of fig. 8.1 we assume the currents are the excitations, then the voltages are the responses and

$$\left.\begin{array}{l} V_1 = z_{11}I_1 + z_{12}I_2 \\ V_2 = z_{21}I_1 + z_{22}I_2 \end{array}\right\} \quad (8.4)$$

Now the parameters may be evaluated by making measurements of the three remaining variables when first I_1 and then I_2 is required to be zero. Then

$$z_{11} = \left.\frac{V_1}{I_1}\right|_{I_2=0} \quad z_{21} = \left.\frac{V_2}{I_1}\right|_{I_2=0}$$
$$z_{12} = \left.\frac{V_1}{I_2}\right|_{I_1=0} \quad z_{22} = \left.\frac{V_2}{I_2}\right|_{I_1=0} \quad (8.5)$$

The zs all have dimensions of impedance. z_{11} and z_{22} are the *driving point impedances* with, respectively, the output and the input open circuited. z_{21} is the *forward transfer impedance* and z_{12} is the *reverse transfer impedance*. As is the case for the admittance parameters if the circuit elements are linear and bilateral, $z_{12} = z_{21}$.

It should be evident that there is no direct correspondence between the ys and the zs, that is, z_{11} is *not* the reciprocal of y_{11}. However, because they describe the same network there must be definite relationships between the

Fig. 8.3. π-network.

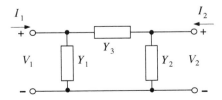

impedance parameters and the admittance parameters. We may find these interrelationships by solving (8.1) for V_1 and V_2 and then comparing the results with (8.4). We find:

$$z_{11}=\frac{y_{22}}{\Delta_y} \quad z_{12}=\frac{-y_{12}}{\Delta_y} \quad z_{21}=\frac{-y_{21}}{\Delta_y} \quad z_{22}=\frac{y_{11}}{\Delta_y} \tag{8.6}$$

where

$$\Delta_y = y_{11}y_{22} - y_{12}y_{21}$$

By solving (8.4) for I_1 and I_2 and comparing ith (8.2), we obtain

$$y_{11}=\frac{z_{22}}{\Delta_z} \quad y_{12}=\frac{-z_{12}}{\Delta_z} \quad y_{21}=\frac{-z_{21}}{\Delta_z} \quad y_{22}=\frac{z_{11}}{\Delta_z} \tag{8.7}$$

where

$$\Delta_z = z_{11}z_{22} - z_{12}z_{21}$$

Let us now find the z-parameters of the T-network of fig. 8.4. First apply a current source I_1 with the output port open so that $I_2 = 0$. The current flows through Z_1 and Z_3 in series, so

$$z_{11} = Z_1 + Z_3 \quad \text{and} \quad z_{21} = Z_3$$

Next apply current source I_2 with $I_1 = 0$. Now the current flows through Z_2 and Z_3 in series, so

$$z_{12} = Z_3 \quad \text{and} \quad z_{22} = Z_2 + Z_3$$

8.2.3 Hybrid and inverse hybrid parameters

These parameters derive their names from the fact that in neither set do all the parameters have the same dimensions. The *hybrid* parameters result when I_1 and V_2 are chosen as independent variables. Then

$$\left.\begin{array}{l}V_1 = h_{11}I_1 + h_{12}V_2 \\ I_2 = h_{21}I_1 + h_{22}V_2\end{array}\right\} \tag{8.8}$$

Fig. 8.4. T-network.

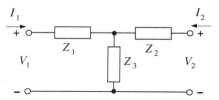

Equivalent circuits and circuit models

Hybrid parameters are defined operationally in a manner similar to the admittance and impedance parameters:

$$h_{11} = \frac{V_1}{I_1}\bigg|_{V_2=0} \qquad h_{21} = \frac{I_2}{I_1}\bigg|_{V_2=0}$$
$$h_{12} = \frac{V_1}{V_2}\bigg|_{I_1=0} \qquad h_{22} = \frac{I_2}{V_2}\bigg|_{I_1=0} \tag{8.9}$$

It is apparent that h_{11} has dimensions of impedance, h_{22} has dimensions of admittance, and h_{12} and h_{21} are dimensionless.

When V_1 and I_2 are independent variables, I_1 and V_2 are written in terms of the *inverse hybrid* parameters:

$$\left.\begin{array}{l} I_1 = g_{11}V_1 + g_{12}I_2 \\ V_2 = g_{21}V_1 + g_{22}I_2 \end{array}\right\} \tag{8.10}$$

These parameters may be found from appropriate open- and short-circuit measurements. g_{11} is an admittance, g_{22} is an impedance, and g_{12} and g_{21} are dimensionless.

8.3 Equivalent circuits and circuit models

For any one of the six possible sets of 2-port parameters, equivalent circuits may be derived which conform to the voltage–current relations expressed by those parameters. For example, the circuits shown in fig. 8.5 are two possible configurations conforming to the relations (8.1) in which the coefficients are the admittance parameters. In the case of fig. 8.5(*a*) we see that the first expression in (8.1) is simply the nodal equation for the left-hand half of the circuit, while the second expression represents the nodal equation for the right-hand half. (It is left as an exercise for the reader to show that the circuit of fig. 8.5(*b*) also conforms to the relations (8.1).)

The circuits shown in fig. 8.5 are not unique in expressing the relations (8.1), but they are two of the simplest and are generally the most convenient to use for the purposes of practical circuit analysis. It should be observed that the sources in these equivalent circuits are of the *controlled* or *dependent* type; the methods of analysis of circuits containing such sources have been treated in section 2.13. It is also worthy of note that the single dependent current source in the circuit of fig. 8.5(*b*) becomes zero (an open circuit) if $y_{12} = y_{21}$, that is, if the circuit is reciprocal. From this we may infer that a passive 2-port network, which must be reciprocal, may be represented by three immittances connected in a π-configuration.

Equivalent circuits derived from the admittance, impedance or hybrid parameter sets are commonly used as a basis for the modelling of non-linear devices such as the transistor. In section 7.3.2 we saw how an h-parameter model for the transistor, operating under small-signal conditions, could be derived by considering its input and output characteristics. This model (shown in fig. 7.13) is repeated in fig. 8.6 together with an equivalent circuit conforming to the relations (8.8). The first expression in (8.8) derives from the mesh equation for the left-hand half of the equivalent circuit; the second expression from the nodal equation for the right-hand half. The h-parameter subscripts differ in the two circuits of fig. 8.6, but the correspondence, element for element, will be obvious.*

The behaviour of a transistor operating at low or medium frequencies may be adequately represented by a linear model containing only resistive elements; consequently, in fig. 8.6, the impedance h_{11} corresponds to resistance h_{ie} and the admittance h_{22} corresponds to conductance h_{oe}. The parameters h_{re} and h_{fe} are, of course, dimensionless.

Although the h-parameter equivalent circuit shown in fig. 8.6 is the one commonly used for modelling the transistor, other models may for particular applications be more convenient or appropriate. For example, the model shown in fig. 8.7 is of interest because each of the elements relates directly to some part of the two-junction physical model for the transistor

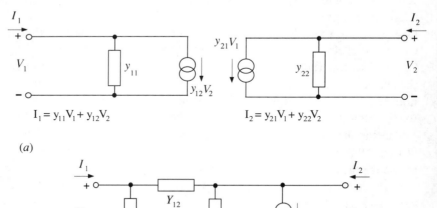

Fig. 8.5. Admittance parameter equivalent circuits.

---

* In the context of transistor circuits the h-parameters are designated by the letters i, o, r and f; standing for, respectively, input, output, reverse and forward. The second letter indicates which of the three terminals of the transistor (emitter, base, collector) is common to the input and output ports.

Equivalent circuits and circuit models

(see reference 5). It is possible to show that this model is related to an equivalent circuit defined in terms of the z-parameters (see problem 8.3).

The 2-port parameters chosen to model a device such as a transistor are determined in practice by making direct, physical measurements at its terminals. At low or medium frequencies such measurements present no special difficulty and the hybrid parameters, or indeed any of the 2-port parameters sets, may be readily determined. At high or very high

Fig. 8.6. Hybrid parameter circuits.

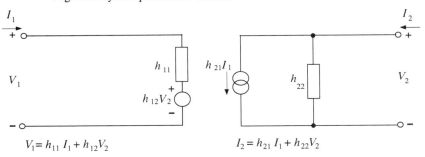

$V_1 = h_{11} I_1 + h_{12} V_2$ $I_2 = h_{21} I_1 + h_{22} V_2$

(a) h - parameter equivalent

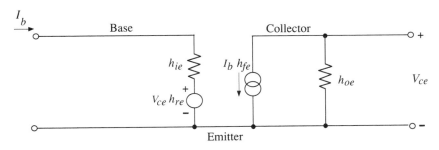

(b) Transistor model

Fig. 8.7. T-equivalent network for the bipolar transistor.

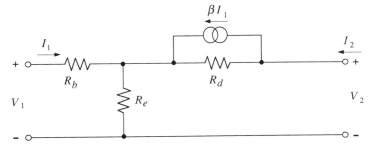

510 Two-port networks

frequencies, however, sophisticated measurement techniques become necessary and in this respect a model based on the admittance parameters is advantageous. The reason for this will become apparent if we compare the definitions of the h- and y-parameter sets. From (8.9) it is seen that determination of h_{21} and h_{22} necessitates making measurements with terminals open circuit, but the creation of a valid open circuit at high frequencies is complicated by the presence of stray capacitance. The y-parameters, on the other hand, can all be determined by making admittance measurements with terminals short circuit (see definitions (8.2) and (8.3)), and it is easier to produce a valid short circuit in the presence of stray capacitance than it is an open circuit. Thus the admittance parameters may be related directly to the most appropriate set of physical measurements and for this reason are preferred as a basis for modelling transistors intended for high-frequency operation.

As a final example of device modelling by means of the two-port parameters we consider the circuit of fig. 8.8(a), which is commonly used to model an operational amplifier at low or medium frequencies. (The use of this model was illustrated in section 2.14.) The model is based on the z-parameter equivalent circuit shown in fig. 8.8(b). The mesh equations for the left- and right-hand halves of this circuit conform to (8.4). At low or medium frequencies, only the resistive components of the impedances

Fig. 8.8. Circuit model for an operational amplifier based on an impedance-parameter equivalent circuit.

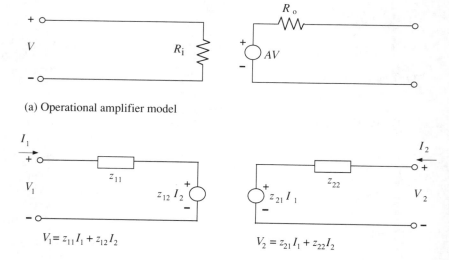

(a) Operational amplifier model

$V_1 = z_{11}I_1 + z_{12}I_2$ $V_2 = z_{21}I_1 + z_{22}I_2$

(b) z-parameter equivalent circuit

associated with the amplifier need be considered, hence the impedances z_{11} and z_{22} are identified respectively with the input resistance R_i and output resistance R_o of the amplifier.

For a practical operational amplifier the voltage and current conditions at its output have an insignificant effect on conditions at its input, in other words, the transfer impedance z_{12} is negligibly small, consequently a voltage source corresponding to $(z_{12}I_2)$ is absent from the model. If z_{12} is negligible, then for fig. 8.8(b) the input current is $I_1 = V_1/z_{11}$ and the magnitude of the voltage source in the output circuit is $z_{21}I_1 = (z_{21}/z_{11})V_1$. The ratio (z_{21}/z_{11}) is a dimensionless quantity called the *forward voltage transfer coefficient*; it may be identified with A, the gain of the amplifier.

8.4 Transmission, inverse transmission and *ABCD* parameters

When a 2-port is used to transfer a signal from one port to another, it is appropriate to express the voltage and current at one port in terms of the voltage and current at the other. If V_2 and I_2 are the independent variables,

$$\left. \begin{array}{l} V_1 = a_{11}V_2 + a_{12}I_2 \\ I_1 = a_{21}V_2 + a_{22}I_2 \end{array} \right\} \tag{8.11}$$

where the as are the *transmission* parameters. When V_1 and I_1 are the independent variables,

$$\left. \begin{array}{l} V_2 = b_{11}V_1 + b_{12}I_1 \\ I_2 = b_{21}V_1 + b_{22}I_1 \end{array} \right\} \tag{8.12}$$

where the bs are the *inverse transmission* parameters.* Here again the parameters do not all have the same dimensions. a_{12} and b_{12} are impedances, a_{21} and b_{21} are admittances and the other four parameters are dimensionless.

At this point we encounter a problem in terminology. Although the definition of the a and b parameters in (8.11) and (8.12) follow directly from what has gone before, these are not the parameters commonly used in circuit analysis. Transmission parameters were defined originally for power system calculations, and the positive direction of the current at port number 2 was taken to be *out* of the positive end of that port. The parameters A, B, C and D are then defined by

---

* The terminology here is not altogether logical. The *transmission* parameters express *input* in terms of *output*, while the *inverse* parameters express *output* in terms of *input*.

512 Two-port networks

$$V_s = AV_r + BI_r \brace I_s = CV_r + DI_r$$ (8.13)

where the currents and voltages are as shown in fig. 8.9. The subscripts r and s may be taken to stand, respectively, for receiving and sending. In order to avoid confusion with the transmission parameters defined by (8.11) we shall, in all that follows, refer to the parameters defined by (8.13) as the *ABCD* parameters.

In terms of short-circuit and open-circuit measurements:

$$A = \frac{V_s}{V_r}\bigg|_{I_r=0} \qquad B = \frac{V_s}{I_r}\bigg|_{V_r=0}$$
$$C = \frac{I_s}{V_r}\bigg|_{I_r=0} \qquad D = \frac{I_s}{I_r}\bigg|_{V_r=0}$$ (8.14)

The *inverse ABCD parameters* are defined by the following relations:

$$V_r = A'V_s + B'I_s \brace I_r = C'V_s + D'I_s$$ (8.15)

In the analysis of two-port networks it is frequently useful to be able to convert from one set of parameters to another set. The relationships between the *ABCD* parameters and the admittance and impedance parameters already defined are readily determined.

Using (8.1) but with the currents and voltages as defined in fig. 8.9 we obtain:

$$I_s = y_{11}V_s + y_{12}V_r$$
$$-I_r = y_{21}V_s + y_{22}V_r$$

Solving these equations for V_s and I_s and comparing the results with (8.13) we find:

$$A = -\frac{y_{22}}{y_{21}} \qquad B = -\frac{1}{y_{21}}$$
$$C = y_{12} - \frac{y_{11}y_{22}}{y_{21}} \qquad D = -\frac{y_{11}}{y_{12}}$$ (8.16)

By using (8.4) with currents and voltages as defined in fig. 8.9, we obtain:

$$A = \frac{z_{11}}{z_{21}} \qquad B = \frac{z_{11}z_{22}}{z_{21}} - z_{12}$$
$$C = \frac{1}{z_{21}} \qquad D = \frac{z_{22}}{z_{21}}$$ (8.17)

Matrix notation

We have already shown that for passive, linear circuit elements $y_{12} = y_{21}$ and $z_{12} = z_{21}$. It follows then from either (8.16) or (8.17) that

$$AD - BC = 1 \quad \text{or} \quad \begin{vmatrix} A & B \\ C & D \end{vmatrix} = 1 \tag{8.18}$$

If the network is symmetrical, then $z_{11} = z_{22}$ and $y_{11} = y_{22}$ and therefore $A = D$. It follows then that *for a symmetrical 2-port*

$$\begin{vmatrix} A & B \\ C & A \end{vmatrix} = 1 \tag{8.19}$$

8.5 Matrix notation

It is often convenient to write the various sets of parameters for the two-port in matrix form. Thus, the relations (8.1) for the admittance parameter may be written:

$$\begin{bmatrix} I_1 \\ I_2 \end{bmatrix} = \begin{bmatrix} y_{11} & y_{12} \\ y_{21} & y_{22} \end{bmatrix} \begin{bmatrix} V_1 \\ V_2 \end{bmatrix} \tag{8.20}$$

or

$$[I] = [Y)][V]$$

The matrix

$$[Y] = \begin{bmatrix} y_{11} & y_{12} \\ y_{21} & y_{22} \end{bmatrix}$$

is often referred to as the *short-circuit admittance matrix*. The term 'short circuit' refers to the fact that we may determine numerical values of the matrix elements by making short-circuit measurements on the two-port. (Or by making short-circuit calculations if the details of the two-port are known.)

In similar fashion one may, from (8.4), write the *open-circuit impedance matrix*:

Fig. 8.9. Reference directions for current and voltage variables for ABCD parameters.

$$\begin{bmatrix} z_{11} & z_{12} \\ z_{21} & z_{22} \end{bmatrix}$$

where 'open circuit' refers to the conditions under which measurements (or calculations) are made on the two-port to determine the matrix elements.

It is apparent that for each of the six possible sets of equations that may be used to characterize the two-port we may write an appropriate matrix and express the behaviour of the two-port by means of matrix equations.

8.6 Worked example

For the simple unsymmetrical two-port shown in fig. 8.10, calculate the impedance, admittance and *ABCD* parameters and express the appropriate equations in matrix form.

Solution. From (8.4) and (8.5), it follows that the impedance parameters are

$$z_{11} = 2 + 4 = 6\,\Omega \qquad z_{22} = 8 + 4 = 12\,\Omega$$
$$z_{12} = 4\,\Omega \qquad z_{21} = 4\,\Omega$$

Then

$$\Delta_z = z_{11}z_{22} - z_{21}z_{12} = 6 \times 12 - 4 \times 4 = 56$$

Now we use (8.7) to calculate the *y*-parameters

$$y_{11} = \frac{z_{22}}{\Delta_z} = \frac{3}{14} \qquad y_{22} = \frac{z_{11}}{\Delta_z} = \frac{3}{28}$$

$$y_{12} = -\frac{z_{12}}{\Delta_z} = -\frac{1}{14} \qquad y_{21} = -\frac{z_{21}}{\Delta_z} = -\frac{1}{14}$$

To find the transmission parameters we use (8.17)

$$A = \frac{z_{11}}{z_{21}} = \frac{3}{2} \qquad B = \frac{z_{11}z_{22}}{z_{21}} - z_{12} = 18 - 4 = 14$$

$$C = \frac{1}{z_{21}} = \frac{1}{4} \qquad D = \frac{z_{22}}{z_{21}} = 3$$

Fig. 8.10. Circuit for worked example (section 8.6).

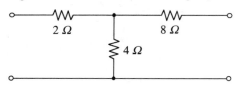

We may use the relation (8.18) to check our results.

$$(AD - BC) = (3/2) \cdot 3 - 14 \cdot (1/4) = (9/2) - (7/2) = 1$$

Now in matrix form the equations for the two-port of fig. 8.10 are

$$\begin{bmatrix} I_1 \\ I_2 \end{bmatrix} = \begin{bmatrix} 3/14 & -1/14 \\ -1/14 & 3/28 \end{bmatrix} \begin{bmatrix} V_1 \\ V_2 \end{bmatrix}$$

$$\begin{bmatrix} V_1 \\ V_2 \end{bmatrix} = \begin{bmatrix} 6 & 4 \\ 4 & 12 \end{bmatrix} \begin{bmatrix} I_1 \\ I_2 \end{bmatrix}$$

$$\begin{bmatrix} V_s \\ I_s \end{bmatrix} = \begin{bmatrix} 3/2 & 14 \\ 1/4 & 3 \end{bmatrix} \begin{bmatrix} V_r \\ I_r \end{bmatrix}$$

8.7 Relationships between direct and inverse ABCD parameters

In matrix form the relations (8.15) for the inverse $ABCD$ parameters are

$$\begin{bmatrix} V_r \\ I_r \end{bmatrix} = \begin{bmatrix} A' & B' \\ C' & D' \end{bmatrix} \begin{bmatrix} V_s \\ I_s \end{bmatrix} \qquad (8.21)$$

We now employ matrix algebra to find the relations among the parameters A, B, C, D and the parameters A', B', C', D'. From (8.13) we have

$$\begin{bmatrix} V_s \\ I_s \end{bmatrix} = \begin{bmatrix} A & B \\ C & D \end{bmatrix} \begin{bmatrix} V_r \\ I_r \end{bmatrix} \qquad (8.22)$$

or

$$[S] = [M][R] \qquad (8.23)$$

Now we multiply (8.23) by $[M]^{-1}$. Then, since $[M][M]^{-1}$ is the unit matrix, we have

$$[R] = [M]^{-1}[S] \qquad (8.24)$$

Now, from the theory of matrix algebra, a matrix defined by

$$[L] = \begin{bmatrix} L_{11} & L_{12} \\ L_{21} & L_{22} \end{bmatrix}$$

has an inverse

$$[L]^{-1} = \frac{1}{\Delta} \begin{bmatrix} L_{22} & L_{12} \\ -L_{21} & L_{11} \end{bmatrix}$$

where

$$\Delta = (L_{11}L_{22} - L_{12}L_{21})$$

Applying this result in (8.24) gives

$$[R] = \frac{1}{(AD-BC)} \begin{bmatrix} D & -B \\ -C & A \end{bmatrix} [S]$$

But $(AD - BC) = 1$. Hence

$$\begin{bmatrix} V_r \\ I_r \end{bmatrix} = \begin{bmatrix} D & -B \\ -C & A \end{bmatrix} \begin{bmatrix} V_s \\ I_s \end{bmatrix} \qquad (8.25)$$

Comparing the matrices (8.21) and (8.25) we find:

$$A' = D, \quad B' = -B, \quad C' = -C, \quad \text{and } D' = A$$

Then in terms of the direct $ABCD$ parameters, V_r and I_r are given by

$$V_r = DV_s - BI_s$$
$$I_r = -CV_s + AI_s$$

8.8 Parameter relationships for π- and T-networks

The networks shown in fig. 8.11 are commonly employed in both power and communications branches of electrical engineering, and reference has already been made to them in sections 2.9.2 and 5.3.6 where they were referred to as star–delta or wye–delta networks. As we have already observed in connection with power systems, the analysis of circuits containing such configurations is sometimes considerably facilitated by the transformation of one configuration into the other. One technique for effecting this transformation has been given in section 2.9.2 and the sets of transformation relations are expressed in (2.23) for the case of a purely resistive network.

An alternative method for deriving the relations that must hold at the two ports of the T- and π-networks if they are to be equivalent is to

Fig. 8.11. T–π transformation.

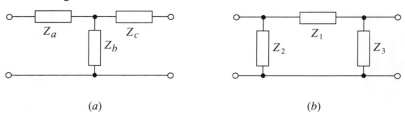

(a)   (b)

Worked example

determine, say, the transmission parameters for the two networks and then set corresponding parameters equal. It is left as an exercise for the reader to prove in this way the following relations, which are similar in form to those in (2.23):

$$Z_1 = Z_a + Z_c + \frac{Z_a Z_c}{Z_b}; \quad Z_a = \frac{Z_1 Z_2}{Z_1 + Z_2 + Z_3}$$

$$Z_2 = Z_b + Z_a + \frac{Z_b Z_a}{Z_c}; \quad Z_b = \frac{Z_2 Z_3}{Z_1 + Z_1 + Z_3} \qquad (8.25)$$

$$Z_3 = Z_c + Z_b + \frac{Z_c Z_b}{Z_a}; \quad Z_c = \frac{Z_3 Z_1}{Z_1 + Z_1 + Z_3}$$

8.9 Worked example

The resistors in the circuit of fig. 8.12(a) are used as part of a bridge circuit in which the effective resistances in parallel with C_1, C_2 can be varied. Calculate the maximum and minimum value of the resistances appearing in parallel with C_1, C_2.

Solution

The resistive T-network in fig. 8.12(a) can be replaced exactly by the π-network of fig. 8.12(b). The resistors of interest are R_2, R_3. Using the notation of fig. 8.11 we have

$$Z_a = [(1-\alpha) + 1] \times 10^4 \, \Omega = (2-\alpha) \times 10^4 \, \Omega$$
$$Z_c = (1+\alpha) \times 10^4 \, \Omega$$
$$Z_b = 10^5 \, \Omega$$

Fig. 8.12. Circuit for worked example (section 8.9).

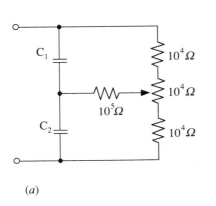
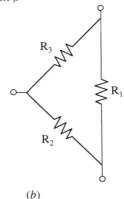

(a) (b)

in which α represents the fractional position of the slider on the potentiometer. Equations (8.25) can then be applied to give

$$R_1 = 3 \times 10^4 + (2-\alpha)(1+\alpha) \times 10^3$$

$$R_2 = 10^5 + (2-\alpha) \times 10^4 + \frac{2-\alpha}{1+\alpha} \times 10^5$$

$$R_3 = 10^5 + (1+\alpha) \times 10^4 + \frac{1+\alpha}{2-\alpha} \times 10^5$$

From the nature of the circuit it is clear that the extremes of R_2, R_3 will occur for the slider at an end of the potentiometer, i.e. $\alpha = 0$ or 1.

Putting $\alpha = 0$ we find

$$R_2 = 3.2 \times 10^5 \, \Omega$$
$$R_3 = 1.6 \times 10^5 \, \Omega$$

In the middle for $\alpha = 0.5$

$$R_2 = R_3 = 2.15 \times 10^5 \, \Omega$$

8.10 Cascaded two-ports and chain matrices

Of special interest is the situation where two or more two-ports are connected in *cascade* with the output of the first providing the input to the second and so on. We are interested then in calculating the output of the final two-port in terms of the input to the first two-port. For such calculations the *ABCD* parameters are appropriate. The matrix that represents the *ABCD* parameters is frequently referred to as the *chain matrix*.

Consider the cascade connection shown in fig. 8.13, where the parameters of the two circuits are identified by the subscripts '1' and '2'. The subscript 's' identifies the input to the first two-port and the subscript 'r' identifies the output of the second two-port. Then in fig. 8.13

$$\begin{bmatrix} V_n \\ I_n \end{bmatrix} = \begin{bmatrix} A_2 & B_2 \\ C_2 & D_2 \end{bmatrix} \begin{bmatrix} V_r \\ I_r \end{bmatrix}$$

Fig. 8.13. Cascade connection of two-ports.

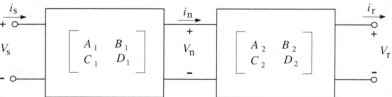

Cascaded two-ports and chain matrices

and

$$\begin{bmatrix} V_s \\ I_s \end{bmatrix} = \begin{bmatrix} A_1 & B_1 \\ C_1 & D_1 \end{bmatrix} \begin{bmatrix} V_n \\ I_n \end{bmatrix}$$

Then

$$\begin{bmatrix} V_s \\ I_s \end{bmatrix} = \begin{bmatrix} A_1 & B_1 \\ C_1 & D_1 \end{bmatrix} \begin{bmatrix} A_2 & B_2 \\ C_2 & D_2 \end{bmatrix} \begin{bmatrix} V_r \\ I_r \end{bmatrix} \tag{8.26}$$

We now show how the *ABCD* parameters of a two-port network may be derived by combining the chain matrices for the individual elements or sections of which the two-port is comprised. Consider first the two simple two-port networks of fig. 8.14.

Clearly in fig. 8.14(a)

$$V_s = V_r + ZI_r$$
$$I_s = I_r$$

Thus the *ABCD* matrix for this network is

$$\begin{bmatrix} 1 & Z \\ 0 & 1 \end{bmatrix} \tag{8.27}$$

For the circuit of fig. 8.14(b) (in which *Y* represents the shunt admittance)

$$V_s = V_r$$
$$I_s = YV_r + I_r$$

Thus the *ABCD* matrix is

$$\begin{bmatrix} 1 & 0 \\ Y & 1 \end{bmatrix} \tag{8.28}$$

For both of these we may check the validity of $AD - BC = 1$. Applying (8.26) the *ABCD* matrix for the circuit of fig. 8.15(a) is found as

Fig. 8.14. Single element two-ports.

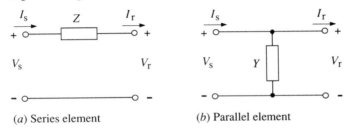

(a) Series element   (b) Parallel element

$$\begin{bmatrix} 1 & Z_1 \\ 0 & 1 \end{bmatrix} \begin{bmatrix} 1 & 0 \\ 1/Z_2 & 1 \end{bmatrix} = \begin{bmatrix} 1+Z_1/Z_2 & Z_1 \\ 1/Z_2 & 1 \end{bmatrix} \tag{8.29}$$

For the circuit of fig. 8.15(b)

$$\begin{bmatrix} 1 & 0 \\ 1/Z_2 & 1 \end{bmatrix} \begin{bmatrix} 1 & Z_1 \\ 0 & 1 \end{bmatrix} = \begin{bmatrix} 1 & Z_1 \\ 1/Z_2 & 1+Z_1/Z_2 \end{bmatrix} \tag{8.30}$$

By combining these results with (8.27, 8.28) we may then find the ABCD matrices for the T- and π-sections of fig. 8.11. For the T-section of fig. 8.11(a) we have, using (8.27, 8.29),

$$\begin{bmatrix} 1+Z_a/Z_b & Z_a \\ 1/Z_b & 1 \end{bmatrix} \begin{bmatrix} 1 & Z_c \\ 0 & 1 \end{bmatrix} = \begin{bmatrix} 1+Z_a/Z_b & (Z_aZ_b+Z_bZ_c+Z_cZ_a)/Z_b \\ 1/Z_b & 1+Z_c/Z_b \end{bmatrix} \tag{8.31}$$

The same result can be obtained by pre-multiplication of equation (8.30) by (8.27).

For the π-section of fig. 8.11(b) we similarly find for the ABCD matrix

$$\begin{bmatrix} 1 & Z_1 \\ 1/Z_2 & 1+Z_1/Z_2 \end{bmatrix} \begin{bmatrix} 1 & 0 \\ 1/Z_3 & 1 \end{bmatrix}$$
$$= \begin{bmatrix} 1+Z_1/Z_3 & Z_1 \\ (Z_1+Z_2+Z_3)/Z_2Z_3 & 1+Z_1/Z_2 \end{bmatrix} \tag{8.32}$$

These results illustrate the formal procedure for evaluating the ABCD matrix of any ladder network (consisting solely of series or shunt elements).

For other circuits, equations similar to those by which (8.27) and (8.28) were derived may be set, or the formal expression of equations (8.14) may be used. This latter process will be illustrated for the balanced lattice of fig. 8.16(a). To apply equations (8.14) we need to analyse the circuit for the two cases $I_r = 0$ and $V_r = 0$. When $I_r = 0$ the relevant variables are shown in fig. 8.16(b) whence

Fig. 8.15. Cascade connections of the two-ports of fig. 8.14.

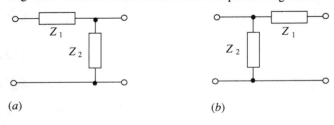

(a) (b)

Cascaded two-ports and chain matrices

$$V_r = \left(\frac{Z_2}{Z_1+Z_2} - \frac{Z_1}{Z_1+Z_2}\right)V_s = \left(\frac{Z_2-Z_1}{Z_2+Z_1}\right)V_s$$

and

$$I_s = 2V_s/(Z_1+Z_2)$$

These two results used in equations (8.14) give

$$A = \frac{Z_2+Z_1}{Z_2-Z_1}$$

$$C = \frac{2}{Z_2-Z_1}$$

The case $V_r = 0$ is shown in fig. 8.16(c), whence

$$B = \frac{2Z_1Z_2}{Z_2-Z_1}$$

$$D = \frac{Z_2+Z_1}{Z_2-Z_1}$$

Thus the *ABCD* matrix for the balanced lattice is given by

$$\frac{1}{Z_2-Z_1}\begin{bmatrix} Z_2+Z_1 & 2Z_1Z_2 \\ 2 & Z_2+Z_1 \end{bmatrix} \qquad (8.33)$$

Fig. 8.16. Balanced lattice network.

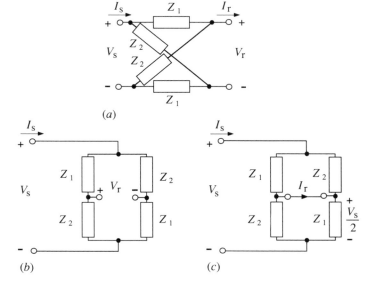

As a check the symmetry is observed and the relation $AD - BC = 1$ may be verified.

Another common component is the ideal transformer, shown in fig. 8.17. The action of this ideal component is defined by the relations

$$V_r = \frac{1}{n} V_s$$

$$I_r = n I_s$$

whence the $ABCD$ matrix is given by

$$\begin{bmatrix} n & 0 \\ 0 & 1/n \end{bmatrix} \tag{8.34}$$

The simplest equivalent circuit for a real transformer takes the form shown in fig. 8.18, in which Z_1, Z_2 will be inductances. Following previous examples the overall matrix may be calculated as

$$\begin{bmatrix} 1 & 0 \\ 1/Z_1 & 1 \end{bmatrix} \begin{bmatrix} n & 0 \\ 0 & 1/n \end{bmatrix} \begin{bmatrix} 1 & Z_2 \\ 0 & 1 \end{bmatrix}$$

$$= \begin{bmatrix} n & nZ_2 \\ n/Z_1 & \frac{1}{n} + \frac{nZ_2}{Z_1} \end{bmatrix} \tag{8.35}$$

The above results are summarized in table 8.1. Two other two-ports are also listed which cannot be constructed from bilateral components since

Fig. 8.17. Ideal transformer.

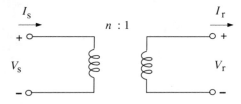

Fig. 8.18. Non-ideal transformer.

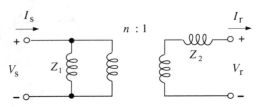

Table 8.1

Circuit	Chain matrix
	$\begin{bmatrix} A & B \\ C & D \end{bmatrix}$
	$\begin{bmatrix} 1 & Z \\ 0 & 1 \end{bmatrix}$
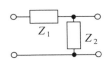	$\begin{bmatrix} 1 & 0 \\ Y & 1 \end{bmatrix}$
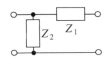	$\begin{bmatrix} 1 + Z_1/Z_2 & Z_1 \\ 1/Z_2 & 1 \end{bmatrix}$
	$\begin{bmatrix} 1 & Z_1 \\ 1/Z_2 & 1 + Z_1/Z_2 \end{bmatrix}$
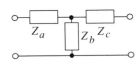	$\begin{bmatrix} 1 + Z_a/Z_b & (Z_a Z_b + Z_b Z_c + Z_c Z_a)/Z_b \\ 1/Z_b & 1 + Z_c/Z_b \end{bmatrix}$
	$\begin{bmatrix} 1 + Z_1/Z_3 & Z_1 \\ (Z_1 + Z_2 + Z_3)/Z_2 Z_3 & 1 + Z_1/Z_2 \end{bmatrix}$
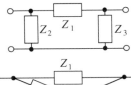	$\dfrac{1}{Z_2 - Z_1} \begin{bmatrix} Z_2 + Z_1 & 2Z_1 Z_2 \\ 2 & Z_2 + Z_1 \end{bmatrix}$

Table 8.1 (cont.)

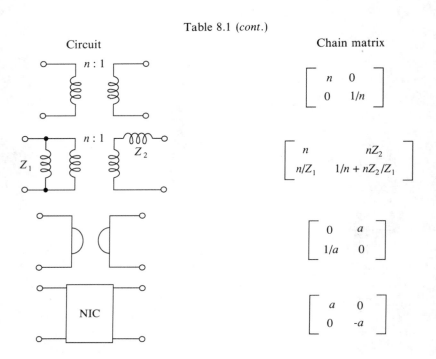

both are non-reciprocal. These are the *gyrator* and the *negative-impedance-converter*. Using the standard conventions for V_s, I_s, V_r and I_r the power dissipated in the two-ports is $V_s I_s - V_r I_r$. For a passive lossless component therefore

$$V_s I_s = V_r I_r$$

Two simple sets of relations satisfy this equation:

either:

$$\frac{V_s}{V_r} = \frac{I_r}{I_s} = n$$

or

$$\frac{V_s}{I_r} = \frac{V_r}{I_s} = a$$

The first of these is applied to the ideal transformer, which was considered above, the second defines the gyrator. Hence the *ABCD* matrix for the gyrator is

$$\begin{bmatrix} 0 & a \\ 1/a & 0 \end{bmatrix} \tag{8.36}$$

for which $AD - BC = -1$ and so is non-reciprocal. The conventional symbol is shown in fig. 8.19(a). The input impedance resulting from a termination in impedance Z, fig. 8.19(b), will be seen to be

a^2/Z

The negative-impedance-converter (NIC) is defined by the relations

$V_s = aV_s$
$I_s = -aI_r$

corresponding to the $ABCD$ matrix

$$\begin{bmatrix} a & 0 \\ 0 & -a \end{bmatrix} \tag{8.37}$$

Such a two-port takes its name because the input impedance is the negative of the terminating impedance, as indicated in fig. 8.20. As might be expected by considering a resistive termination this two-port is active as well as non-reciprocal.

Both these two-ports can be realized approximately at low frequencies by use of operational amplifiers.

Fig. 8.19. Gyrator.

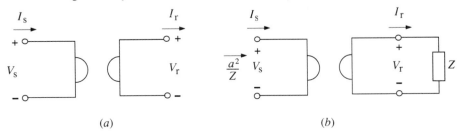

Fig. 8.20. Negative-impedance-converter.

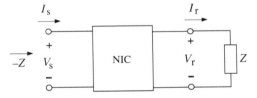

8.11 Worked example

A two-wire high voltage, 50 Hz, transmission line has the following parameters:

Series inductance	2.2 mH/km
Series resistance	0.2 Ω/km
Shunt capacity	15 nF/km

Suggest a T-network which will represent a 50 km length of such a line, and calculate the chain matrix at 50 Hz.

Solution

Total series elements are $10\,\Omega$ in series with $110\,\text{mH}$; shunt capacitance is $0.75\,\mu\text{F}$. The T-network which suggests itself is shown in fig. 8.21(a); it is not quite accurate since it neglects the fact that the components are distributed along the length of the wire. However for this length at 50 Hz the error is not significant.

The value for the chain-matrix can be obtained by use of table 8.1. We find

$$A = D = 1 + (R + j\omega L)j\omega C$$
$$C = j\omega C$$
$$B = 2(R + j\omega L) + j\omega C(R + j\omega L)^2$$

Using the values in the circuit, we find for 50 Hz

$$A = D = 0.996 + j10^{-3}$$
$$C = j2.4\,10^{-4}$$
$$B = 10 + j34.6$$

In this case the shunt impedance is very high compared with the series impedance (by some $4000\times$); this also implies that the π-network of fig. 8.21(b) is an acceptable equivalent.

Fig. 8.21. Circuit for worked example (section 8.11).

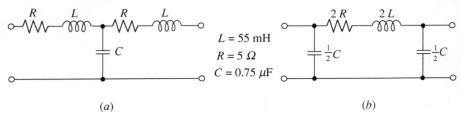

(a) $L = 55\,\text{mH}$ $R = 5\,\Omega$ $C = 0.75\,\mu\text{F}$ (b)

†8.12 Series and parallel connections of two-ports

Two ways of interconnecting two two-ports are shown in fig. 8.22: that in fig. 8.22(a) is the series connection, that in fig. 8.22(b) the parallel connection. Each way results in a composite two-port. In order to deduce the two-port parameters of the composite in terms of the individual two-port parameters before interconnection, it is important that the current flow is not affected by the interconnection.

In the series connection, fig. 8.22(a), describing the individual two-ports by their impedance matrices as indicated in the figure, we have from (8.4)

$$V'_1 = z'_{11} I'_1 + z'_{12} I'_2$$
$$V'_2 = z'_{21} I'_1 + z'_{22} I'_2$$
$$V''_1 = z''_{11} I''_1 + z''_{12} I''_2$$
$$V''_2 = z''_{21} I''_1 + z''_{22} I''_2$$

Fig. 8.22. Interconnection of two-port networks.

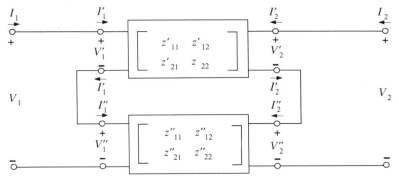

(a) Series connection

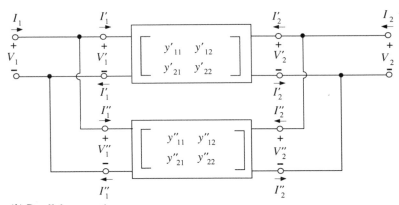

(b) Parallel connection

528 Two-port networks

The interconnection imposes the following relations

$$V_1 = V'_1 + V''_1$$
$$V_2 = V'_2 + V''_2$$
$$I_1 = I'_1 = I''_1$$
$$I_2 = I'_2 = I''_2$$

Hence

$$V_1 = (z'_{11} + z''_{11})I_1 + (z'_{12} + z''_{12})I_2$$
$$V_2 = (z'_{21} + z''_{21})I_1 + (z'_{22} + z''_{22})I_2$$

showing that the impedance matrix of the composite two-port is the sum of the individual impedance matrices.

For the parallel connections the following relations are imposed

$$V_1 = V'_1 = V''_1$$
$$V_2 = V'_2 = V''_2$$
$$I_1 = I'_1 + I''_1$$
$$I_2 = I'_2 + I''_2$$

The description of the two-ports by the admittance parameters, equation (8.1), combined with these relations show that the admittance matrix of the composite is the sum of the individual admittance matrices

$$\begin{bmatrix} y'_{11} + y''_{11} & y'_{12} + y''_{12} \\ y'_{21} + y''_{21} & y'_{22} + y''_{22} \end{bmatrix}$$

†8.13 Worked example

Find the conditions in the twin-T circuit of fig. 8.23(a) so that an input signal does not give rise to any output.

Solution

From (8.1) the conditions for no output from a two-port is $y_{12} = y_{21} = 0$, effectively disconnecting the input from the output. The circuit of fig. 8.23(a) is reciprocal, so that a physical interpretation of y_{12} is the negative of the admittance of the series element in the equivalent π-circuit. Accordingly the easiest way to proceed is to use the T-π equivalence in section 8.8 to find π-networks for each of the T-components of fig. 8.23(a). These are shown in figs. 8.23(b) and (c), and combined in fig. 8.23(d) to give the π-equivalent of the twin T-circuit. Note that this circuit could not simply be made since it involves a resistance which is both negative and varies with frequency.

The condition $y_{12} = 0$ then implies

$$\left(-\frac{2}{\omega^2 C^2 R} + \frac{2}{j\omega C} \right)^{-1} + (2R + j\omega 2R^2 C)^{-1} = 0$$

or

$$-\frac{1}{\omega^2 C^2 R}+\frac{1}{j\omega C}=-(R+j\omega R^2 C)$$

Both real and imaginary parts of this equation are satisfied for $\omega CR = 1$, which is the required condition.

Fig. 8.23. Circuits for worked example (section 8.13).

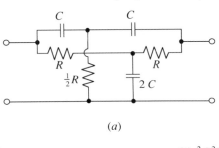

(a)

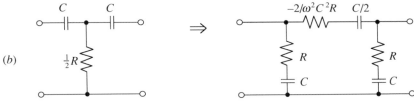

(b)

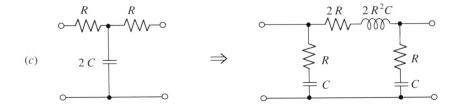

(c)

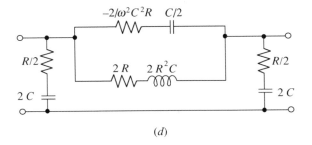

(d)

†8.14 Iterative and image impedances

Although equation (8.26) provides a way of determining the overall *ABCD* matrix of any number of two-ports in cascade the process is clearly laborious. The process can be made simpler in certain cases, leading to the concept of iterative and image impedances.

8.14.1 Iterative impedances

This impedance relates to a cascade of identical two-ports, fig. 8.24(*a*): if a value of Z_L exists for which the input impedance to a two-port is also Z_L then clearly we may consider each two-port in the cascade separately. The impedance if this condition is true is known as an iterative impedance, of which in general there are two, one for each direction. The two possibilities are indicated in fig. 8.24(*b*), (*c*).

Consider the situation in fig. 8.24(*b*). We have

$$V_s = AV_r + BI_r$$
$$I_s = CV_r + DI_r$$
$$V_r = Z_{it1} I_r$$

Hence we must have

$$Z_{it1} = \frac{V_s}{I_s} = \frac{AZ_{it1} + B}{CZ_{it1} + D}$$

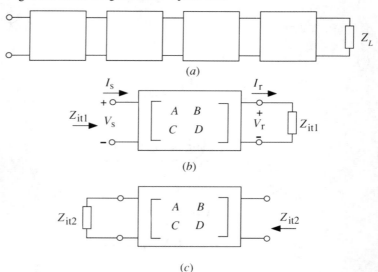

Fig. 8.24. Illustrating iterative impedances.

Iterative and image impedances

or

$$CZ_{it1}^2 + (D-A)Z_{it1} - B = 0$$

whence

$$Z_{it1} = \frac{1}{2C}[A - D \pm \sqrt{\{(D-A)^2 + 4BC\}}] \qquad (8.38)$$

Carrying out a similar process for the situation in fig. 8.24(c) we find

$$Z_{it2} = \frac{1}{2C}[D - A \pm \sqrt{\{(D-A)^2 + 4BC\}}] \qquad (8.39)$$

In each case we shall choose a value with positive resistive part. If the network is symmetrical, $A = D$ and

$$Z_{it1} = Z_{it2} = \sqrt{(B/C)} \qquad (8.40)$$

8.14.2 Image impedances

The concept of iterative impedance outlined in the last section is of use only for cascades of identical two-ports. The situation may be broadened slightly by use of image impedances. The two image impedances of a two-port, Z_{I1}, Z_{I2}, are defined by the circuit configurations of fig. 8.25. For the case of fig. 8.25(a) we have

$$V_s = AV_r + BI_r$$
$$I_s = CV_r + DI_r$$
$$V_r = Z_{12}I_r$$

We require

$$V_s = Z_{11}I_s$$

Hence

$$Z_{11} = \frac{AZ_{12} + B}{CZ_{12} + D} \qquad (8.41)$$

Fig. 8.25. Illustrating image impedances.

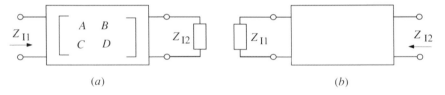

(a)   (b)

532 Two-port networks

In the case of fig. 8.25(b)

$$V_s = -Z_{11}I_s$$

and we require

$$V_r = -Z_{12}I_r$$

Hence

$$-Z_{11} = \frac{-AZ_{12} + B}{-CZ_{12} + D} \tag{8.42}$$

Adding equations (8.41, 8.42) we find

$$Z_{12} = \sqrt{\left(\frac{BD}{AC}\right)} \tag{8.43}$$

and

$$Z_{11} = \sqrt{\left(\frac{BA}{DC}\right)} \tag{8.44}$$

For a symmetrical network $A = D$ and we see that

$$Z_{11} = Z_{12} = Z_{it1} = Z_{it2} = \sqrt{\left(\frac{B}{C}\right)}$$

This common value applying to a symmetrical network is frequently referred to as the *characteristic impedance*.

Equations (8.43, 8.44) may be more meaningfully expressed in terms of the open- and short-circuit impedances of the two-port, defined in fig. 8.26. Straightforward application of the *ABCD* matrix equation shows

$$Z_{oc1} = \frac{A}{C}; \; Z_{sc1} = \frac{B}{D}$$

Fig. 8.26. Definition of open- and short-circuit impedances.

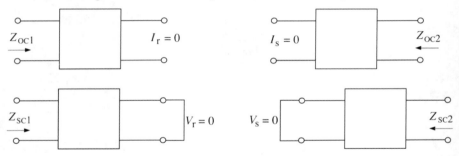

Attenuators

$$Z_{oc2} = \frac{D}{C}; \quad Z_{sc2} = \frac{B}{A}$$

We see that for equation (8.44) we may write

$$Z_{11} = \sqrt{(Z_{oc1}Z_{sc1})} \tag{8.45}$$

and for equation (8.43)

$$Z_{12} = \sqrt{(Z_{oc2}Z_{sc2})}$$

The most straightforward application of these concepts is to attenuator circuits composed only of resistors.

†8.15 Attenuators

Consider the symmetrical T-network shown in fig. 8.27(a). For this a characteristic impedance, Z_0, exists given by

$$Z_0 = \sqrt{(Z_{oc}Z_{sc})}$$

We have

$$Z_{oc} = R_1 + R_2$$

$$Z_{sc} = R_1 + \frac{R_1 R_2}{R_1 + R_2} = \frac{R_1^2 + 2R_1 R_2}{R_1 + R_2}$$

Hence

$$Z_0 = \sqrt{(R_1^2 + 2R_1 R_2)} \tag{8.46}$$

In use the attenuator section will be connected as in fig. 8.27(b), terminated in Z_0. Since by definition of Z_0, the impedance seen by the source V_s is also Z_0

$$I_s = \frac{V_s}{Z_0}$$

Thus the potential across R_2 is

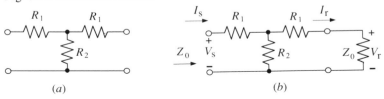

Fig. 8.27. Attenuator section.

$$V_s - \frac{R_1 V_s}{Z_0}$$

and finally

$$V_r = \frac{Z_0}{R_1 + Z_0} V_s \left(1 - \frac{R_1}{Z_0}\right)$$

or

$$\frac{V_s}{V_r} = \frac{Z_0 + R_1}{Z_0 - R_1} \qquad (8.47)$$

The ratio

$$\frac{\text{power from source}}{\text{power to load}}$$

is usually measured in decibels and termed the attenuation A of the section

$$A = 10 \log_{10} \frac{V_s^2/Z_0}{V_r^2/Z_0}$$

$$= 20 \log_{10} \frac{Z_0 + R_1}{Z_0 - R_1} \qquad (8.48)$$

†8.16 Worked example

Design an attenuator section having a characteristic impedance of $600\,\Omega$ with attenuation of 10 dB.

Solution
From equation (8.48)

$$\frac{Z_0 + R_1}{Z_0 - R_1} = 10^{\frac{1}{2}}$$

whence

$$\frac{R_1}{Z_0} = 0.519$$

or

$$R_1 = 311\,\Omega$$

Using equation (8.46)

$$R_2 = \frac{Z_0^2 - R_1^2}{2 R_1} = 423\,\Omega$$

Worked example

Consider the L-section of fig. 8.28. The T-section just designed can be regarded as two of these sections back-to-back. We now determine the image and iterative impedances for this section. From (8.45):

$$Z_{11}^2 = (311 + 846)311 \qquad Z_{11} = 600$$

$$Z_{12}^2 = 846\left(\frac{1}{311} + \frac{1}{846}\right)^{-1} \qquad Z_{12} = 439$$

The complete T-section is shown in fig. 8.28(b) and illustrates the use of image impedances: the right-hand section is terminated in its correct image impedance of 600 Ω and its input has impedance 439 Ω. This is the correct image impedance with which to terminate the left-hand section so that its input impedance is 600 Ω.

To evaluate the iterative impedance the *ABCD* parameters may be determined using table 8.1, but it is simpler to proceed directly from the circuit definition of fig. 8.24(b), (c). We have

$$Z_{it1} = 311 + \left(\frac{1}{Z_{it1}} + \frac{1}{846}\right)^{-1}$$

whence

$$Z_{it1}^2 - 311 Z_{it1} - 311 \times 846 = 0$$

Choosing the positive root

$$Z_{it1} = 691 \, \Omega$$

Similarly

$$\frac{1}{Z_{it2}} = \frac{1}{846} + \frac{1}{Z_{it2} + 311}$$

whence

$$Z_{it2}^2 + 311 Z_{it2} - 311 \times 846 = 0$$

giving

$$Z_{it2} = 380 \, \Omega$$

Fig. 8.28. Circuits for worked example (section 8.16).

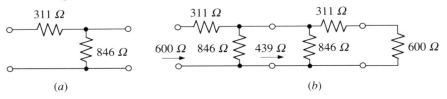

(a)   (b)

†8.17 Insertion loss

A principal procedure in electrical circuits is feeding power from a source to a load as indicated in fig. 8.29(a), in which the source and load impedances have been taken to be resistive. Frequently a two-port may be interposed between source and load, perhaps as a filter, and the voltage across the load will be a new value V_L (which may be calculated if parameters of the two-port are known). The ratio

$$\frac{V_{L0}}{V_L}$$

is known as the *insertion loss*. It is usually measured in decibels

$$I = 10 \log_{10}\left(\frac{V_{L0}^2/R_2}{V_L^2/R_2}\right) = 20 \log_{10}\left(\frac{V_{L0}}{V_L}\right) \tag{8.49}$$

It is clear that the attenuator section designed in section 8.16 has, when operated between source and load impedances of 600 Ω, an insertion loss of 10 dB.

†8.18 Worked example

The attenuator section designed in section 8.16 is mistakenly worked between source and load of 50 Ω. Determine the insertion loss.

The circuit is shown in fig. 8.30. The simplest way to determine the output voltage is to proceed as follows: 423 Ω in parallel with (311 + 50) Ω is 195 Ω.

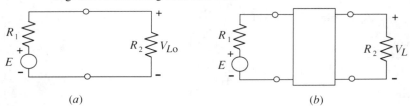

Fig. 8.29. Illustrating insertion loss.

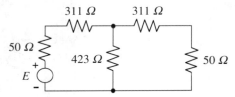

Fig. 8.30. Circuit for worked example (section 8.18).

Thus

$$V_L = \frac{50}{(311+50)} \times E \times \frac{195}{50+311+195} = 0.0486\,E$$

Since $V_{L0} = 0.5\,E$, the insertion loss is

$$\frac{0.5}{0.0486} = 10.3$$

or

$$I = 20.2\,\text{dB}$$

8.19 Summary

The two-port network, or two-terminal pair network, is one of the most frequently used circuit configurations found in electrical and electronic engineering practice. The theory of such networks relates the voltage and current variables at the two ports of the network; depending upon which two of these four variables is chosen to be the independent variables, six possible sets of parameters may be defined, any one of which completely characterizes the behaviour of the network so far as externally impressed voltages and currents are concerned. Any of the sets of parameters may be expressed in terms of any other set, however, for a particular application, one of the six sets of parameters is often found to be the most convenient and is chosen in preference to the others.

For each parameter set, equivalent circuits may be derived which conform to the two circuit equations corresponding to that particular parameter set. The elements (sources and impedances) of these equivalent circuits are functions of the parameters and of the terminal variables. Such equivalent circuits are of great practical value in the analysis of circuits part of which consists of a two-port network. They are also commonly used as a basis for modelling the small-signal characteristics of devices such as transistors.

A recurring problem in circuit analysis is to determine the voltage–current relationships between input and output of several two-port networks connected in cascade. This may be achieved by finding the $ABCD$ parameters (modified transmission parameters) of each individual two-port and expressing these in matrix form – the so-called chain matrix. Multiplication of the individual chain matrices yields the overall matrix for the cascade from which voltage–current relationships between input and output may be found. Series and parallel interconnections of two-ports also arise (fig. 8.22) and for these cases matrix addition of, respectively, the impedance and admittance parameters (expressed in matrix form) is an appropriate technique for determining the overall properties of the interconnected networks.

Finally, the concepts of iterative and image impedances are useful for circuits consisting of identical two-ports connected in cascade, and when it is desired to achieve matching between a source and a load by means of a two-port network.

8.20 Problems

1. Find the hybrid parameters of the circuit shown in fig. 8.10.
2. Show that if a two-port network, described by its hybrid parameters, is reciprocal then $h_{12} = -h_{21}$. (Hint: apply the Reciprocity theorem to the relations (8.8).)
3. Find the z-parameters of the circuit shown in fig. 8.7; hence, express the elements of this circuit in terms of its z-parameters.
4. Determine the y-parameters for the circuit shown in fig. 8.31.
5. (a) Obtain the transmission ($ABCD$) matrices for the networks shown in fig. 8.32. If the two are connected in cascade, obtain the overall $ABCD$ matrix.

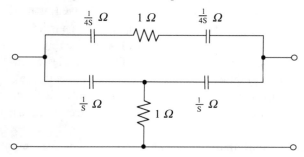

Fig. 8.31. Circuit for problem 4.

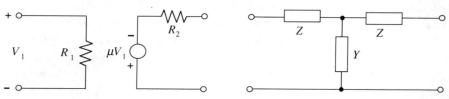

Fig. 8.32. Circuit for problem 5.

(b) If $R_1 = 100 \,\Omega$; $R_2 = 1 \,\text{k}\Omega$; $Y = j100\,S$; $Z = j20 \,\Omega$ and $\mu = 20$, calculate the overall current gain when the output is short-circuited.
(University of Kent)

6. Derive the transfer ($ABCD$) matrix of:
(a) a simple series impedance;
(b) a simple shunt admittance.

Using these transfer matrices derive any one matrix representation of the parallel twin-tee network shown in fig. 8.33. Prove any matrix conversion you require.

What is the condition for infinite attenuation for this network?
(Sheffield University)

7. (i) For the cascade connection of the two-port networks shown in fig. 8.34(a), show that the short-circuit transfer admittance of the overall circuit is given by

$$y_{12} = -\frac{(y_{12})_A (y_{12})_B}{(y_{11})_B + (y_{22})_A}$$

(ii) Using this result or otherwise, find the transfer admittances of the network shown in fig. 8.34(b) ($T = RC$).
(University of Wales)

8. Find the iterative and image impedances of the circuit shown in fig. 8.35.

9. Show that when the symmetrical two-port described by the matrix

$$\begin{bmatrix} A & B \\ C & A \end{bmatrix}$$

is terminated in its image impedance, the ratio V_{in}/V_{out} is given by

$$A + \sqrt{(A^2 - 1)}$$

Fig. 8.33. Circuit for problem 6.

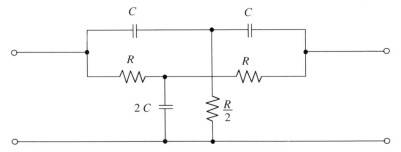

Use this result to find the ratio V_{in}/V_{out} for the circuit shown in fig. 8.36 when it is terminated in its image impedance. Evaluate this ratio and the image impedance if $R = 1\,\text{k}\Omega$, $L = 1\,\text{H}$, and $\omega = 100^3\,\text{rad/s}$.

10. The insertion loss of a two-port is defined as:

$$\frac{V_{out} \text{ with network removed}}{V_{out} \text{ with network inserted}}$$

Show that for the circuit given in fig. 8.37 the insertion loss is:

$$\frac{(z_{11} + Z_s)(z_{22} + Z_r) - z_{12}^2}{z_{12}(Z_r + Z_s)}$$

(Manchester University)

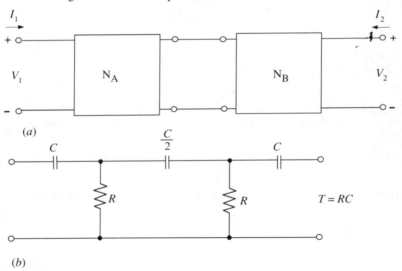

Fig. 8.34. Circuit for problem 7.

Fig. 8.35. Circuit for problem 8.

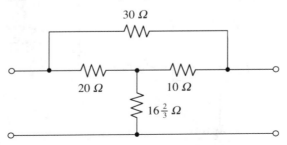

Problems

Fig. 8.36. Circuit for problem 9.

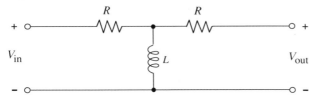

Fig. 8.37. Circuit for problem 10.

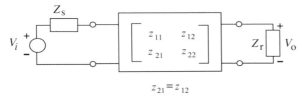

$z_{21} = z_{12}$

Appendix A

Units, symbols and abbreviations

The International System of Units (SI) comprises seven base units and two supplementary units. The ampere is a base unit and is defined as follows: that constant current which, if maintained in two straight parallel conductors of infinite length, of negligible cross-section, and placed 1 metre apart in vacuum, would produce between these conductors a force equal to 2×10^{-7} newton per metre of length. (For definitions of other SI units, see reference 16.)

SI base units

Quantity	Name	Symbol
length	metre	m
mass	kilogram	kg
time	second	s
electric current	ampere	A
thermodynamic temperature	kelvin	K
amount of substance	mole	mol
luminous intensity	candela	cd

SI supplementary units

plane angle	radian	rad
solid angle	steradian	sr

SI derived units

Units for other quantities commonly used in the electrical and other sciences are expressed in terms of base and supplementary units. The most important of these, which have been given special names and symbols, are listed below:

Units, symbols and abbreviations

Quantity	Name	Symbol	Derivation
force	newton	N	kg m/s^2
energy	joule	J	N m
electric charge	coulomb	C	A s
electric potential	volt	V	J/C
power	watt	W	J/s
apparent power	volt-ampere	VA	J/s
reactive power	var	VAr	J/s
resistance	ohm	Ω	V/A
conductance	siemens	S	A/V
capacitance	farad	F	C/V
inductance	henry	H	V s/A
magnetic flux	weber	Wb	V s
magnetic flux density	tesla	T	Wb/m^2
frequency	hertz	Hz	s^{-1}
pressure, stress	pascal	Pa	N/m^2
luminous flux	lumen	lm	cd sr
illuminance	lux	lx	lm/m^2

SI compound units

The following quantities have not been given special unit names or symbols, being expressed directly in terms of base, supplementary and derived units.

electric field strength	volt per metre	V/m
magnetic field strength	ampere per metre	A/m
electric flux density	coulomb per square metre	C/m^2
permittivity	farad per metre	F/m
permeability	henry per metre	H/m
mass density	kilogram per cubic metre	kg/m^3
thermal conductivity	watt per metre Kelvin	W/(m K)
torque	newton metre	N m
rotational frequency	radian per second	rad/s
luminance	candela per square metre	cd/m^2

Quantity and unit symbols

(a) Quantity symbols are generally expressed by capital or small letters of the Latin or Greek alphabet in italic (sloping) type. Vector, phasor or complex quantities are expressed, where necessary, by bold-face type (in typescript etc. an underline or an overline may be used). Subscripts are used to differentiate maximum values from magnitudes.

Examples: potential difference, electromotive force, current.

V, E, I	d.c. or a.c. (r.m.s.) magnitudes;
V_m, E_m, I_m	a.c. maximum values;
V, E, I	phasors (complex quantities);
$v(t), e(t), i(t)$ or v, e, i	instantaneous values of time-varying functions.

(b) Unit symbols are abbreviated unit names (e.g. A, Ω, m) expressed in roman (upright) characters and are used only after numerical values. A space is set between the number and its unit symbol.

Decimal multiple and submultiple indicators

The following indicators are prefixed to the unit symbols without a space (e.g. 5.5 kV); powers in steps of 3 are preferred:

10^{18}	exa	E				10^{-3}	milli	m
10^{15}	peta	P	10^{2}	hecto	h	10^{-6}	micro	μ
10^{12}	tera	T	10^{1}	deca	da	10^{-9}	nano	n
10^{9}	giga	G	10^{-1}	deci	d	10^{-12}	pico	p
10^{6}	mega	M	10^{-2}	centi	c	10^{-15}	femto	f
10^{3}	kilo	k				10^{-18}	atto	a

Abbreviations

These are set in small roman (lower-case upright) letters except at the beginning of a sentence where capitals are preferred (e.g. A.C. not A.c.). The following well-known abbreviations are used in electrical engineering:

alternating current	a.c	phase	ph
direct current	d.c.	(e.g. 3-ph supply)	
electromotive force	e.m.f.	potential difference	p.d.
magnetomotive force	m.m.f.	power factor	p.f.
per-unit	p.u.	revolution	r (rev)
		(e.g. r/s or r/min)	
		root-mean-square	r.m.s.

Appendix B

The general mesh equations and proofs of the network theorems

Mesh equations

The mesh equations for a general M-mesh network are:

$$Z_{11}I_1 + Z_{12}I_2 \ldots + Z_{1M}I_M = V_{11}$$
$$Z_{21}I_1 + Z_{22}I_2 \ldots + Z_{2M}I_M = V_{22}$$
$$\vdots$$
$$Z_{m1}I_1 + Z_{m2}I_2 \ldots + Z_{mM}I_M = V_{mm} \quad \text{(B.1)}$$
$$\vdots$$
$$Z_{M1}I_1 + Z_{M2}I_2 \ldots + Z_{MM}I_M = V_{MM}$$

where V_{mm} is the net e.m.f. in the mth mesh, $I_1 \ldots I_M$ are the M dependent mesh currents, and the coefficients Z are the network self and mutual impedances (all quantities complex).

The network determinant is then

$$\Delta = \begin{vmatrix} Z_{11} & Z_{12} & \ldots & Z_{1M} \\ Z_{21} & Z_{22} & \ldots & Z_{2M} \\ \vdots & \vdots & & \vdots \\ Z_{M1} & Z_{M2} & \ldots & Z_{MM} \end{vmatrix} \quad \text{(B.2)}$$

and the solution for the current I_m in the mth mesh is

$$I_m = \frac{1}{\Delta} \begin{vmatrix} Z_{11} & Z_{12} & \ldots & Z_{1m-1} & V_{11} & Z_{1m+1} & \ldots & Z_{1M} \\ Z_{21} & Z_{22} & \ldots & Z_{2m-1} & V_{22} & Z_{2m+1} & \ldots & Z_{2M} \\ \vdots & \vdots & & \vdots & \vdots & \vdots & & \vdots \\ Z_{M1} & Z_{M2} & \ldots & Z_{Mm-1} & V_{MM} & Z_{Mm+1} & \ldots & Z_{MM} \end{vmatrix} \quad \text{(B.3)}$$

The network theorems may be deduced directly from this solution.

The Superposition theorem

For a discussion of this theorem see section 2.6.1. Expanding the numerator of (B.3) about the mth column we obtain

$$(-1)^{m-1}I_m = V_{11}\frac{\Delta_{1m}}{\Delta} - V_{22}\frac{\Delta_{2m}}{\Delta} + V_{33}\frac{\Delta_{3m}}{\Delta} \ldots + (-1)^{M-1}V_{MM}\frac{\Delta_{Mm}}{\Delta}$$

(B.4)

where Δ_{1m} denotes the determinant remaining when the first row and mth column are deleted from (B.2). All the determinants in this expression are functions only of the complex impedances in the network and, for a given linear network, are therefore constants. Furthermore, each term contains only a single net e.m.f. (there are, for example, no squared or cross-product terms). The superposition theorem is therefore proved.

The Reciprocity theorem

For a discussion of this theorem see section 2.6.2. Consider two branches in the general network and let us choose our meshes such that one branch occurs only in one mesh which we may label mesh (1) and the other only in some other mesh which we may label mesh (2). (A little thought will show that this is always possible.) Under these conditions the current in branch (1) will be I_1 only, and that in branch (2) will be I_2 only. Furthermore, an e.m.f. V_1 in branch (1) will form part of the V_{11} only, and an e.m.f. V_2 in branch (2) will form part of V_{22} only. The superposition theorem tells us that currents caused by one e.m.f. are independent of all other e.m.f.s, so that without loss of generality we may set all e.m.f.s except V_1 and V_2 to zero. Using (B.4) the solutions for the mesh current I_1 and I_2 are then:

$$I_1 = V_1\frac{\Delta_{11}}{\Delta} - V_2\frac{\Delta_{21}}{\Delta}$$

$$I_2 = -V_1\frac{\Delta_{12}}{\Delta} + V_2\frac{\Delta_{22}}{\Delta}$$

Now consider identical e.m.f.s, V say, acting in each of the branches. The current in branch (1) due to V acting *alone* in branch (2) will be

$$I_1' = -V\frac{\Delta_{21}}{\Delta}$$

and the current in branch (2) due to V acting alone in branch (1) will be

$$I_2' = -V\frac{\Delta_{12}}{\Delta}$$

General mesh equations and proofs of theorems

Now, in the general mesh equations, for any pair of subscript values p and q, we have $Z_{pq} = Z_{qp}$, and recalling (from the theory of determinants) that rows and columns of a determinant may be interchanged without affecting its value, it may be readily seen that $\Delta_{pq} = \Delta_{qp}$, hence $I_1' = I_2'$. Since this same result may be obtained for any two meshes chosen arbitrarily, the theorem is proved.

Thévenin's theorem

For a statement and discussion of this theorem see section 2.7. Let the open circuit e.m.f. between the terminals of the network be V_T, and let the impedance measured between these terminals with all internal voltage sources short circuited be Z_T. (We assume that any current sources will have been transformed to voltage sources.) To prove the theorem we have to find expressions for V_T, Z_T and the current that will flow in an external load impedance Z connected between the terminals.

We first connect an impedance Z in series with a source of e.m.f. E between the terminals. This operation will create an additional mesh in the network, which we take as mesh (1), the total number of meshes in the network then being M. The current in Z is then I_1 which, by (B.3), is

$$I_1 = \frac{1}{\Delta} \begin{vmatrix} V_{11} & Z_{12} & Z_{13} & \cdots & Z_{1M} \\ V_{22} & Z_{22} & Z_{23} & \cdots & Z_{2M} \\ \cdot & \cdot & \cdot & & \cdot \\ \cdot & \cdot & \cdot & & \cdot \\ V_{MM} & Z_{M2} & Z_{M3} & \cdots & Z_{MM} \end{vmatrix} = \frac{\Delta'}{\Delta} \quad (B.5)$$

In the present case, E will form part of V_{11} only, and Z a part of Z_{11} only. Let

Δ_0' be the value of Δ' when E is zero;
Δ_0 be the value of Δ when Z is zero;
Δ_{11}' be the minor of Δ'

then,

$$\Delta' = \Delta_0' + E\Delta_{11}' = \Delta_0' + E\Delta_{11} \quad (B.6)$$

and

$$\Delta = \Delta_0 + Z\Delta_{11} \quad (B.7)$$

The voltage between the terminals of the network is now:

$$I_1 Z + E = \frac{\Delta'}{\Delta} Z + E = \left[\frac{\Delta_0' + E\Delta_{11}}{\Delta_0 + Z\Delta_{11}} \right] Z + E$$

$$= \frac{\Delta_0' + E\Delta_{11}}{\Delta_0/Z + \Delta_{11}} + E$$

The open circuit e.m.f. between the terminals is the value of this expression when Z becomes infinite and E is zero, so that

$$V_T = \frac{\Delta_0'}{\Delta_{11}} \tag{B.8}$$

The impedance between the terminals with all internal voltage sources short-circuited is the value of E/I_1 with Z equal to zero. Under these conditions Δ_0' is zero and from (B.6) we obtain

$$E = \frac{\Delta'}{\Delta_{11}}$$

Also, from (B.5),

$$I_1 = \frac{\Delta'}{\Delta_0}$$

hence,

$$Z_T = \frac{E}{I_1} = \frac{\Delta_0}{\Delta_{11}} \tag{B.9}$$

Considering again the solution (B.5) for the current, we have

$$I_1 = \frac{\Delta'}{\Delta}$$

which, from (B.6) and (B.7), becomes

$$I_1 = \frac{\Delta_0' + E\Delta_{11}}{\Delta_0 + Z\Delta_{11}} = \frac{\Delta_0'/\Delta_{11} + E}{\Delta_0/\Delta_{11} + Z}$$

Substituting (B.8) and (B.9) in this expression yields

$$I_1 = \frac{V_T + E}{Z_T + Z} \tag{B.10}$$

But (B.10) is precisely the equation which applies to a single-mesh network containing total impedance $Z_T + Z$ and total e.m.f. $V_T + E$. The theorem is therefore proved.

Appendix C

Computer programs

Many versatile computer programs are now available to the circuit designer; one such program, SPICE (Simulation Program for Integrated Circuit Electronics), written at the University of California, is freely available on the computers of most universities and polytechnics, and at the later stages of a degree course students, particularly those intending to specialise in electrical subjects, will benefit from an acquaintance with such a program.

In the introductory stages of an electrical circuits course, however, assimilation of principles is of paramount importance; the programs described and listed in this appendix have been written with this in mind – they are not intended as circuit design tools but rather to assist the student in working through the examples and problems given in the text. In particular, the program for solving simultaneous equations and the circuit calculator program will remove much of the drudgery associated with the manipulation of complex quantities in a.c. circuit problems.

The programs will run on any personal computer supporting GW BASIC or similar versions of the BASIC language allowing the use of defined functions.*

C1   Simultaneous equations

Program SIMUL solves the set of equations:

$$(a_{11}+jb_{11})x_1 + (a_{12}+jb_{12})x_2 + \ldots + (a_{1n}+jb_{1n})x_n = c_1\underline{/\theta_1}$$
$$(a_{21}+jb_{21})x_1 + (a_{22}+jb_{22})x_2 + \ldots + (a_{2n}+jb_{2n})x_n = c_2\underline{/\theta_2}$$
$$\ldots \quad \ldots \quad \ldots \quad \ldots$$
$$(a_{n1}+jb_{n1})x_1 + (a_{n2}+jb_{n2})x_2 + \ldots + (a_{nn}+jb_{nn})x_n = c_n\underline{/\theta_n}$$

* Minor variations of the BASIC language have, as far as possible, been taken into account in writing these programs. Users of BBC BASIC should, however, note that the pseudo variable INKEY$ takes an argument, viz: INKEY$(10).

This set corresponds to the general mesh equations (B.1) given in Appendix B, or to the nodal equations of similar form. The independent variables in the right-hand column represent, correspondingly, the magnitudes and phases of the net mesh voltages in the case of the mesh equations, or the net injected nodal currents in the case of the nodal equations. The coefficients $(a_{ik}+jb_{ik})$ correspond to impedances or admittances as appropriate.

The program uses the method of Gaussian elimination and back substitution, with partial pivoting.* Real and imaginary parts of coefficients are stored in arrays AA(R,S), BB(R,S), R and S running from 1 to N where N is the number of equations. The independent variables are also stored in these arrays as AA(R,N+1), BB(R,N+1); for brevity in the program instructions these variables are also referred to as coefficients. To avoid corruption of the original data, arrays AA and BB are copied to working arrays A and B.

The dependent variables are stored in the arrays XA(R), XB(R). The program proceeds by eliminating in turn each of the variables $X(1)\ldots X(N-1)$ to produce a solution for $X(N)$. This is back-substituted into the preceding pivotal equation to yield $X(N-1)$, the procedure being repeated for all pivotal equations.

When using the program to solve the standard mesh or nodal equations, it must be remembered that mutual terms are negative (see section 2.5); this can lead to changes of sign which may cause confusion when entering data, particularly if the circuit described by the equations contains both inductive and capacitive elements. It is recommended, therefore, that the equations are written out with coefficients enclosed within brackets, the real and imaginary parts being given the appropriate sign. This is illustrated in the example given below.

An option in the program allows correction of data if a mistake is made during entry. This option also enables changes to be made to one or more circuit parameters and for the program to be rerun so that the effect of such changes may be ascertained.

Example. Use program SIMUL to solve the following set of equations (see section 5.3.5 and Fig. 5.9).

$$(24+j18)I_1+(-2-j4)I_2+(-20-j10)I_3=415\underline{/30}$$
$$(-2-j4)I_1+(24+j18)I_2+(-20-j10)I_3=415\underline{/150}$$
$$(-20-j10)I_1+(-20-j10)I_2+(60+j30)I_3=0$$

---

* A discussion of the underlying principles of the method can be found in: *Basic Numerical Methods* by R.E. Scraton (Edward Arnold, 1984).

Program printout:
Entry complete; check coefficients

Row 1	24	18
	−2	−4
	−20	−10
	415	30
Row 2	−2	−4
	24	18
	−20	−10
	415	150
Row 3	−20	−10
	−20	−10
	60	30
	0	0

Are coefficients correct (Y/N)? Y
Solutions are:

	MAGNITUDE	PHASE (DEG)
X(1)	21.1047266	19.7636417
X(2)	21.1047266	79.7636416
X(3)	12.1848196	49.7636416

Do you wish to change parameters (Y/N)? N

C2 Circuit calculator

The circuit calculator (program CALC) contains two separate but linked programs: program RLC, which computes the impedance of a series or parallel branch, and program CMPLX, which enables one to perform chained complex arithmetic. On entry to CMPLX, results from RLC are automatically transferred. Program RCL extends from line 10 and calls a subroutine at line 2000; program COMPLX extends from line 1010 and calls subroutines at lines 1500, 2000, 3000 and 4000. Linking between RLC and CMPLX is effected by lines 3–8. Error trapping is provided at line 5. The two programs are self-contained and may be entered in the computer and run as separate programs, in which case it will be necessary to include the dimension statement at line 4 early on in CMPLX. It is also advisable to incorporate error trapping in CMPLX to avoid loss of data on error. Other minor modifications required will be obvious to the programmer. (Note: to escape from CALC it is necessary to type ESC followed by Q.)

(a) Program RLC

This program computes the complex impedance (R_S+jX) and complex admittance $(G+jB)$ of any combination of the elements R,L,C connected either all in series or all in parallel. Additionally, if both L and C are present, the frequency at which the circuit resonates is computed together with the Q-factor and half-power bandwidth. The definition of resonant frequency is that given by (3.90). When entering data, a zero (0) is used to indicate that a particular element is absent; e.g., if only L and C are present in a parallel circuit their numerical values are entered but a zero is placed against R (the program then takes into account the fact that the circuit is lossless and R infinite). Values for R,L,C and f are entered in units of ohm, henry, farad and hertz respectively. Note that for a series circuit containing resistance, the value of the series resistance R_S printed out by the program will be identical to the input value of R. However, for a parallel circuit R_S will be a function of the values of all elements present (see section 3.8).

(b) Program CMPLX

In this program complex numbers are held in seven registers designated Z1–Z7. Registers Z1 and Z2 are used for manual entry of data in either Cartesian or polar form. When transferring from program RLC, results are automatically placed in Z1. These registers may also be used for temporary storage of numbers during calculations. Registers Z3–Z5 are used for permanent storage, while the results of arithmetical operations are displayed in registers Z6 or Z7, the default being Z6. Operations may be carried out using any two registers as operands, e.g., the instruction 6*7 multiplies the number in Z6 by that in Z7 placing the result in Z6; the instruction 6*77 places the result in Z7. Data is, of course, overwritten in Z6 or Z7 so that these registers should not be used as operands if their contents are required for further calculation. A number in any register may be placed in any other register using the equality operator, e.g., the instruction 3=1 sets the number in Z3 equal to the number in Z1. Reciprocals are obtained by means of the operator R: instruction 4R47, for example, takes the inverse of the number in Z4 and places it in Z7.

Example. The printouts given below illustrate the use of CALC for solving the a.c. circuit problem given in section 3.7. The three impedances in the circuit (designated Z_1,Z_2,Z_3 in section 3.7) have been calculated using RLC, and these are displayed in registers Z3, Z4 and Z5. The impedance of the two branches connected in parallel $(Z_1//Z_2)$ has been computed (using the 'product-over-sum rule') and added to the series impedance Z_3 to give the total impedance of the circuit; this is displayed in register Z7. The applied circuit voltage has been entered manually via register Z1 and, finally, the operation 1/7 has produced in Z6 the value of the main circuit current.

Program printout:

PROGRAM RLC
Enter R,L,C,F (if element is ABSENT enter a ZERO)

R(ohm) = ?2
L(H) = ?15.9E—3
C(F) = ?0
F(Hz) = ?50

Select circuit: series(S) or parallel(P) S

Series circuit parameters are:

Rs(ohm) = 2
X(ohm) = 4.99513232
Z(ohm) = 5.38064558
AZ(deg) = 68.17934
G(S) = 6.90814147E−2
B(S) = −0.172535404
Y(S) = 0.185851304
AY(deg) = −68.17934

Product LC = 0: fo,Q,Bw undefined

Select prog. RLC(R) or prog. CMPLX(X) X

PROGRAM CMPLX
Result of $Z(1) - (/) - Z(7) - > Z(6)$ is:

	REAL	IMAGINARY	MAGNITUDE	ANGLE(DEG)
ENTER & STORE				
Z(1)	240	0	240	0
Z(2)	0	0	0	0
STORE				
Z(3)	6	20.9858389	21.8267138	74.0443918
Z(4)	30	−7.99773583	31.0477661	−14.9273799
Z(5)	2	4.99513232	5.38064558	68.17934

RESULT
Z6	7.4013709	−7.63636498	10.6345833	−45.8952859
Z7	15.7066099	16.205296	22.5678801	45.8952859

Select: Operation(O);Entry(E);Program RLC(R)

C3   Tee-Pi transformation

Given the impedances ZA,ZB,ZC connected in a T (or star) configuration, or the impedances Z1,Z2,Z3 connected in a π (or delta) configuration, program TEE-PI computes the transformation given by the relations (8.25). These relations are printed out on the VDU together with schematics of the circuit configurations. The operation of this program is self-explanatory.

C4   Roots of polynomials

Program 'ROOTS' finds the roots of a polynomial of the form

$$A(m+1)x^m + A(m)x^{m-1} + \ldots + A(2)x + A(1) = 0$$

where m is the degree of the polynomial and the coefficients A may be complex.* The real and imaginary parts of the coefficients are retained in arrays A(K) and B(K) respectively (K running from 1 to M+1). Values held in these arrays are transferred to working arrays AR(K) and AI(K) as necessary. Roots are found by Laguerre's method which operates iteratively. For a trial value of X, a correction term DX is computed which is a function of the value of the polynomial and its first and second derivative; $X-DX$ then becomes the next trial value, the procedure being repeated until DX is sufficiently small. After each root is found, forward deflation is used to reduce the degree of the polynomial by one, the next root is then found for this new polynomial. Finally, when all roots have been found, their values are refined (polished), again using Laguerre's method, by insertion, in turn, into the unmodified polynomial.

The coefficients of polynomials arising in electrical circuit theory are always real, consequently, the imaginary part of each coefficient must be set to zero. Alternatively, for the purposes of the examples in this book, reference to array B(K) in line 80 of the program may be omitted.

* This program is a modified version of a program, written in FORTRAN, given in *Numerical Recipes* by W.H. Press, B.P. Flannery, S.A. Teukolsky and W.T. Vetterling, (Cambridge University Press, 1986).

Example. Denominator polynomial for a fourth-order Butterworth filter (see section 6.11).

$$s^4 + 2.613s^3 + 3.414s^2 + 2.613s + 1 = 0$$

Program printout:

PROGRAM ROOTS

Degree of polynomial: M? 4

Enter coefficients starting with $A(M+1)$

A(5)(Real,Imag) = ?1,0
A(4)(Real,Imag) = ?2.613,0
A(3)(Real, Imag) = ?3.414,0
A(2)(Real,Imag) = ?2.613,0
A(1)(Real,Imag) = ?1,0

Entry complete; coefficients OK(Y/N)? Y

Roots of polynomial are:

REAL	IMAGINARY	MODULUS	ARGUMENT(DEG)
−0.382629302	0.923901953	1	112.496643
−0.382629302	−0.923901952	1	−112.496643
−0.923870697	−0.382704758	0.999999998	−157.498677
−0.9238707	0.382704758	1	157.498677

Program SIMUL

```
10 CLS:PRINT "PROGRAM SIMUL":PRINT
20 REM SOLUTION OF SIMULTANEOUS EQUATIONS
30 REM BY GAUSSIAN ELIMINATION
40 REM*DATA INPUT*
50 INPUT "Number of equations: N ";N: PRINT
60 DIM AA(N,N+1),BB(N,N+1),A(N,N+1),B(N,N+1),XA(N),XB(N):PRINT
70 PRINT "Enter coefficients":PRINT
80 FOR R=1 TO N
90 FOR S=1 TO N
100 PRINT"A(";R;",";S;") = ";:INPUT AA(R,S)
110 PRINT"B(";R;",";S;") = ";:INPUT BB(R,S)
120 NEXT S
130 PRINT"C(";R;") = ";:INPUT AA(R,N+1)
140 PRINT"THETA(";R;") = ";:INPUT BB(R,N+1): PRINT
150 NEXT R
160 CLS:PRINT:PRINT"Entry complete; check coefficients":PRINT
170 FOR R=1 TO N
180 PRINT"Row";R;
190 FOR S=1 TO N+1
200 PRINT TAB(6); AA(R,S) TAB(20); BB(R,S)
210 NEXT S: PRINT
220 NEXT R
230 PRINT:PRINT"Are coefficients correct (Y/N)? ";:GOSUB 1500
240 IF AN<>1 AND AN<>2 GOTO 230
250 IF AN=1 GOTO 300
```

Appendix C

```
260 PRINT:INPUT"Enter Row,Column,Values(A,B or C,THETA)";R,S,VA,VB
270 IF R<=N AND R>=1 AND S<=N+1 AND S>=1 GOTO 290
280 PRINT:PRINT"Incorrect row and/or column numbers":GOTO 260
290 AA(R,S)=VA:BB(R,S)=VB:GOTO 170
300 REM*FUNCTIONS FOR COMPLEX ARITHMETIC*
310 DEF FNCABS(D,E)=SQR(D*D+E*E)
320 DEF FNPA(D,E,F,G)=D*F-E*G
330 DEF FNPB(D,E,F,G)=D*G+E*F
340 DEF FNQA(D,E,F,G)=(D*F+E*G)/(F*F+G*G)
350 DEF FNQB(D,E,F,G)=(E*F-D*G)/(F*F+G*G)
360 REM*COPY COEFFS TO WORKING ARRAYS*
370 FOR R=1 TO N
380 FOR S=1 TO N+1
390 A(R,S)=AA(R,S):B(R,S)=BB(R,S)
400 NEXT S
410 NEXT R
420 REM*CONVERT (C,THETA) VALUES TO CARTESIAN
430 PIE=3.14159265#
440 FOR R=1 TO N
450 AC=(B(R,N+1))*PIE/180
460 A=A(R,N+1)*COS(AC)
470 B(R,N+1)=A(R,N+1)*SIN(AC)
480 A(R,N+1)=A
490 NEXT R
500 REM*ELIMINATION WITH PARTIAL PIVOTING*
510 FOR Z=1 TO N-1
520 W=0
530 FOR R=Z TO N
540 MA=FNCABS(A(R,Z),B(R,Z))
550 IF MA>W THEN U=R:W=MA
560 NEXT R
570 FOR S=Z TO N+1
580 P=A(U,S):A(U,S)=A(Z,S):A(Z,S)=P
590 P=B(U,S):B(U,S)=B(Z,S):B(Z,S)=P
600 NEXT S
610 FOR R=Z+1 TO N
620 PA=FNQA(A(R,Z),B(R,Z),A(Z,Z),B(Z,Z))
630 PB=FNQB(A(R,Z),B(R,Z),A(Z,Z),B(Z,Z))
640 FOR S=Z+1 TO N+1
650 A(R,S)=A(R,S)-FNPA(PA,PB,A(Z,S),B(Z,S))
660 B(R,S)=B(R,S)-FNPB(PA,PB,A(Z,S),B(Z,S))
670 NEXT S
680 NEXT R
690 NEXT Z
700 REM*BACK SUBSTITUTION*
710 FOR R=N TO 1 STEP-1
720 PA=A(R,N+1)
730 PB=B(R,N+1)
740 IF R=N GOTO 790
750 FOR S=R+1 TO N
760 PA=PA-FNPA(A(R,S),B(R,S),XA(S),XB(S))
770 PB=PB-FNPB(A(R,S),B(R,S),XA(S),XB(S))
780 NEXT S
790 XA(R)=FNQA(PA,PB,A(R,R),B(R,R))
800 XB(R)=FNQB(PA,PB,A(R,R),B(R,R))
810 NEXT R
820 REM* CONVERT RESULTS TO POLAR FORM AND PRINT*
830 PRINT:PRINT"Solutions are:":PRINT
840 PRINT TAB(8)"MAGNITUDE" TAB(23)"PHASE(DEG)":PRINT
850 FOR R=1 TO N
860 U=XA(R):V=XB(R):GOSUB 2000
870 PRINT"X(";R;")"TAB(8);MZ TAB(23);AZ
880 NEXT R
890 PRINT:PRINT"Do you wish to change parameters (Y/N)? ";:GOSUB 1500
900 IF AN<>1 AND AN<>2 GOTO 890
910 IF AN=1 GOTO 170
920 END
1500 REM SUBROUTINE: OPTIONS
1510 Q$=INKEY$:IF Q$<>"" GOTO 1510
1520 Q$=INKEY$:IF Q$="" GOTO 1520
1530 IF Q$>"a" THEN Q$=CHR$(ASC(Q$)-32)
1540 PRINT Q$:AN=0
1550 IF Q$="Y" THEN AN=1
1560 IF Q$="N" THEN AN=2
1570 RETURN
2000 REM SUBROUTINE: POLAR CONVERSION
2010 MZ=FNCABS(U,V)
2020 IF ABS(U)<1E-36 GOTO 2080
```

Computer programs

```
2030 AZ=(ATN(V/U))*180/PIE
2040 IF AZ=0 AND U<0 THEN AZ=180
2050 IF U<0 AND V>0 THEN AZ=180+AZ
2060 IF U<0 AND V<0 THEN AZ=AZ-180
2070 RETURN
2080 IF V>0 THEN AZ=90 ELSE AZ=-90
2090 IF U=0 AND V=0 THEN AZ=0
2100 RETURN
```

Program CALC

```
3 CLS:PRINT"PROGRAM CALC"
4 DIM R(7),X(7),M(7),A(7)
5 ON ERROR GOTO 999
6 PRINT:PRINT"Select: prog.RLC(R); prog.CMPLX(X) or quit(Q) ";:GOSUB 1500
7 IF AN=1 GOTO 1000
8 IF AN<>2 GOTO 6
10 CLS:PRINT"PROGRAM RLC":PRINT
20 REM*PARAMETERS OF RLC SERIES/PARALLEL CIRCUITS*
30 PRINT"Enter R,L,C,F (if element is ABSENT enter a ZERO)":PRINT
40 INPUT"R(ohm)= ";R
50 INPUT"L(H)= "; L
60 INPUT"C(F)= ";C
70 INPUT"F(Hz)= ";F
80 PIE=3.14159265#
90 W=2*PIE*F
100 F0=0
110 IF L>0 AND C>0 THEN F0=1/(2*PIE*SQR(L*C))
120 W0=2*PIE*F0
130 REM*TEST FOR RESONANCE*
140 RE=0
150 IF F0>0 GOTO 160 ELSE GOTO 170
160 IF ABS((F-F0)/F0)<.000001 THEN RE=1
170 DEF FNR(U,V)=U/(U*U+V*V)
180 DEF FNI(U,V)=-V/(U*U+V*V)
190 PRINT:PRINT"Select circuit: series(S) or parallel(P) ";
200 QU$=INKEY$:IF QU$<>"" GOTO 200
210 QU$=INKEY$:IF QU$="" GOTO 210
220 PRINT QU$:IF QU$>="a"THEN QU$=CHR$(ASC(QU$)-32)
230 IF QU$<>"S" AND QU$<>"P" GOTO 190
240 IF QU$="P" GOTO 450
250 REM*COMPUTE PARAMETERS FOR SERIES CIRCUIT*
260 IF C=0 GOTO 380
270 IF RE=1 AND R=0 GOTO 280 ELSE GOTO 290
280 PRINT"Resonance(f=fo):Q infinite,Z zero for Rs=0":END
290 X=W*L-1/(W*C)
300 U=R:V=X:GOSUB 2000
310 Z=M:AZ=A
320 G=FNR(R,X):B=FNI(R,X)
330 U=G:V=B:GOSUB 2000
340 Y=M:AY=A
350 IF R=0 OR F0=0 GOTO 670
360 Q=W0*L/R:BW=F0/Q
370 GOTO 670
380 X=W*L
390 U=R:V=X:GOSUB 2000
400 Z=M:AZ=A
410 G=FNR(R,X):B=FNI(R,X)
420 U=G:V=B:GOSUB 2000
430 Y=M:AY=A
440 GOTO 670
450 REM*COMPUTE PARAMETERS FOR PARALLEL CIRCUIT*
460 IF RE=1 AND R=0 GOTO 470 ELSE GOTO 490
470 PRINT:PRINT"Resonance(f=fo):Q and Z infinite for G=1/R=0"
480 END
490 IF R=0 THEN G=0 ELSE G=1/R
500 IF L=0 GOTO 600
510 B=W*C-1/(W*L)
520 U=G:V=B:GOSUB 2000
530 Y=M:AY=A
540 RS=FNR(G,B):X=FNI(G,B)
550 U=RS:V=X:GOSUB 2000
560 Z=M:AZ=A
570 IF R=0 OR F0=0 GOTO 670
580 Q=R/(W0*L):BW=F0/Q
590 GOTO 670
600 B=W*C
```

```
610 U=G:V=B:GOSUB 2000
620 Y=M:AY=A
630 RS=FNR(G,B):X=FNI(G,B)
640 U=RS:V=X:GOSUB 2000
650 Z=M:AZ=A
660 REM*PRINT RESULTS*
670 PRINT:IF QU$="S" THEN PRINT"Series circuit parameters are:":PRINT
680 IF QU$="P" THEN PRINT"Parallel circuit parameters are:":PRINT
690 IF QU$="S" THEN RS=R
700 PRINT"Rs(ohm)= ";RS
710 PRINT"X(ohm)= ";X
720 PRINT"Z(ohm)= ";Z
730 PRINT"AZ(deg)= ";AZ
740 PRINT"G(S)= ";G
750 PRINT"B(S)= ";B
760 PRINT"Y(S)= ";Y
770 PRINT"AY(deg)= ";AY
780 IF F0>0 AND R>0 GOTO 790 ELSE GOTO 820
790 PRINT"fo(Hz)= ";F0
800 PRINT "Q= ";Q
810 PRINT "Bw= ";BW:GOTO 860
820 IF F0>0 AND R=0 GOTO 830 ELSE GOTO 850
830 PRINT"fo(Hz)= ";F0
840 PRINT:PRINT"Lossless  circuit:Q infinite,Bw zero":GOTO 860
850 PRINT:PRINT"Product LC=0:fo,Q,Bw undefined"
860 R(1)=RS:X(1)=X:M(1)=Z:A(1)=AZ
870 GOTO 6
880 END
999 CLS:PRINT:PRINT"ERROR NUMBER ";ERR:RESUME 6
1000 REM*COMPLEX ARITHMETIC*
1010 CLS:PRINT"PROGRAM CMPLX"
1020 PIE=3.14159265#
1030 GOSUB 4000
1040 PRINT:PRINT"Select:Operation(O);Entry(E);Prog.RLC(R);Quit(Q)";:GOSUB 1500
1050 IF AN<2 OR AN>4 GOTO 1040
1060 ON AN-1 GOTO 10,1120,1070
1070 CLS:GOSUB 4000:GOSUB 3000
1080 IF I<>7 THEN I=6
1090 IF O$="=" THEN I=G
1100 CLS:PRINT"Result of Z(";G;")-(";O$;")-Z(";H;")->Z(";I;") is:":PRINT
1110 GOSUB 4000:GOTO 1040
1120 CLS:GOSUB 4000
1130 PRINT:PRINT"Entry in Cartesian(C) or Polar(P)?";:GOSUB 1500
1140 IF AN<>5 AND AN<>6 THEN GOTO 1130
1150 IF AN=6 GOTO 1200
1160 PRINT:FOR J=1 TO 2
1170 PRINT"Enter Z";J;"(Real,Imag)";:INPUT R(J),X(J)
1180 U=R(J):V=X(J):GOSUB 2000:M(J)=M:A(J)=A
1190 NEXT J:GOTO 1250
1200 PRINT:FOR J=1 TO 2
1210 PRINT"Enter Z";J;"(Mag,Angle(deg))";:INPUT M(J),A(J)
1220 AA=A(J)*PIE/180
1230 R(J)=M(J)*COS(AA):X(J)=M(J)*SIN(AA)
1240 NEXT J
1250 CLS:GOSUB 4000
1260 GOTO 1040
1500 REM SUBROUTINE: OPTIONS
1510 Q$=INKEY$:IF Q$<>"" GOTO 1510
1520 Q$=INKEY$:IF Q$="" GOTO 1520
1530 IF Q$>="a" THEN Q$=CHR$(ASC(Q$)-32)
1540 PRINT Q$:AN=0
1550 IF Q$="X" THEN AN=1
1560 IF Q$="R" THEN AN=2
1570 IF Q$="E" THEN AN=3
1580 IF Q$="O" THEN AN=4
1590 IF Q$="C" THEN AN=5
1600 IF Q$="P" THEN AN=6
1610 IF Q$="Q" GOTO 9999
1620 RETURN
2000 REM SUBROUTINE: POLAR CONVERSION
2010 M=SQR(U*U+V*V)
2020 IF ABS(U)<1E-36 GOTO 2080
2030 A=(ATN(V/U))*180/PIE
2040 IF A=0 AND U<0 THEN A=180
2050 IF U<0 AND V>0 THEN A=180+A
2060 IF U<0 AND V<0 THEN A=A-180
2070 RETURN
```

Computer programs

```
2080 IF V>0 THEN A=90 ELSE A=-90
2090 IF U=0 AND V=0 THEN A=0
2100 RETURN
3000 REM SUBROUTINE: OPERATIONS
3010 PRINT:PRINT"Operations:";
3020 PRINT"+-*/= and R"
3030 I=0:PRINT
3040 INPUT"Z(?)-(OP?)-Z(?)->Z(6-7)";X$
3050 IF LEN(X$)<3 OR LEN(X$)>4 GOTO 3040
3060 IF LEN(X$)=3 GOTO 3090
3070 I=VAL(RIGHT$(X$,1)):IF I<>7 GOTO 3040
3080 X$=LEFT$(X$,3)
3090 G=VAL(LEFT$(X$,1)):H=VAL(RIGHT$(X$,1))
3100 O$=MID$(X$,2,1)
3110 IF G<1 OR G>7 GOTO 3040
3120 IF H<1 OR H>7 GOTO 3040
3130 IF O$<>"+" AND O$<>"-" GOTO 3140 ELSE GOTO 3160
3140 IF O$<>"*" AND O$<>"/" GOTO 3150 ELSE GOTO 3160
3150 IF O$<>"R" AND O$<>"=" GOTO 3040 ELSE GOTO 3160
3160 REM*SELECT OPERANDS*
3170 FOR J=1 TO 7
3180 IF G=J THEN P=R(J)
3190 IF G=J THEN Q=X(J)
3200 IF H=J THEN S=R(J)
3210 IF H=J THEN T=X(J)
3220 NEXT J
3230 REM*EXECUTE OPERATOR*
3240 IF O$="+" GOTO 3250 ELSE GOTO 3260
3250 U=P+S:V=Q+T:GOSUB 2000:GOTO 3430
3260 IF O$="-" GOTO 3270 ELSE GOTO 3280
3270 U=P-S:V=Q-T: GOSUB 2000:GOTO 3430
3280 IF O$="*" GOTO 3290 ELSE GOTO 3300
3290 U=P*S-Q*T:V=P*T+Q*S: GOSUB 2000:GOTO 3430
3300 IF O$="/" GOTO 3310 ELSE GOTO 3330
3310 U=(P*S+Q*T)/(S*S+T*T)
3320 V=(Q*S-P*T)/(S*S+T*T):GOSUB 2000:GOTO 3430
3330 IF O$="R" GOTO 3340 ELSE GOTO 3360
3340 U=P/(P*P+Q*Q)
3350 V=-Q/(P*P+Q*Q):GOSUB 2000:GOTO 3430
3360 IF O$<>"=" GOTO 3430
3370 FOR J=1 TO 7
3380 IF G<>J THEN GOTO 3420
3390 R(J)=S:X(J)=T
3400 U=S:V=T:GOSUB 2000
3410 M(J)=M:A(J)=A
3420 NEXT J:RETURN
3430 IF I=7 GOTO 3450
3440 R(6)=U:X(6)=V:M(6)=M:A(6)=A:RETURN
3450 R(7)=U:X(7)=V:M(7)=M:A(7)=A
3460 RETURN
4000 REM SUBROUTINE:DISPLAY
4010 PRINT TAB(5)"REAL" TAB(18)"IMAGINARY";
4020 PRINT TAB(31)"MAGNITUDE" TAB(44)"ANGLE(DEG)"
4030 FOR J%=1 TO 7
4040 IF J%=1 THEN PRINT"ENTER & STORE"
4050 IF J%=3 THEN PRINT"STORE"
4060 IF J%=6 THEN PRINT"RESULT"
4070 PRINT"Z";J% TAB(5);R(J%) TAB(18);X(J%) TAB(31);M(J%) TAB(44);A(J%)
4080 NEXT
4090 RETURN
9999 PRINT"Quitting prog.CALC"
```

Program TEE-PI

```
10 CLS:PRINT"PROGRAM TEE-PI":PRINT
20 DIM Z$(6),R(6),X(6),M(6),A(6)
30 Z$(1)="Z1":Z$(2)="Z2":Z$(3)="Z3"
40 Z$(4)="ZA":Z$(5)="ZB":Z$(6)="ZC"
50 DEF FNR(P,Q,R,S,U,V)=(U*P*R-U*Q*S+V*P*S+V*Q*R)/(U*U+V*V)
60 DEF FNI(P,Q,R,S,U,V)=(U*P*S+U*Q*R-V*P*R+V*Q*S)/(U*U+V*V)
70 GOSUB 1000
80 PRINT "ZA=Z1*Z2/(Z1+Z2+Z3)"TAB(21)"Z1=ZA+ZC+ZA*ZC/ZB"
90 PRINT "ZB=Z2*Z3/(Z1+Z2+Z3)"TAB(21)"Z2=ZB+ZA+ZB*ZA/ZC"
100 PRINT "ZC=Z3*Z1/(Z1+Z2+Z3)"TAB(21)"Z3=ZC+ZB+ZC*ZB/ZA":PRINT
110 PRINT "Select:TEE-to-PI(T) or PI-to-TEE(P) ";
120 Q$=INKEY$:IF Q$<>"" GOTO 120
130 Q$=INKEY$:IF Q$="" GOTO 130
```

Appendix C

```
140 PRINT Q$:IF Q$>="a" THEN Q$=CHR$(ASC(Q$)-32)
150 IF Q$<>"T" AND Q$<>"P" GOTO 110
160 IF Q$="P" GOTO 300
170 PRINT:PRINT"Enter TEE-circuit values":PRINT
180 FOR J=4 TO 6
190 PRINT Z$(J);"(Real,Imag)";:INPUT R(J),X(J)
200 NEXT J
210 REM*COMPUTE PI-IMPEDANCES*
220 R(1)=R(4)+R(6)+FNR(R(4),X(4),R(6),X(6),R(5),X(5))
230 X(1)=X(4)+X(6)+FNI(R(4),X(4),X(6),X(6),R(5),X(5))
240 R(2)=R(5)+R(4)+FNR(R(5),X(5),R(4),X(4),R(6),X(6))
250 X(2)=X(5)+X(4)+FNI(R(5),X(5),R(4),X(4),R(6),X(6))
260 R(3)=R(6)+R(5)+FNR(R(6),X(6),R(5),X(5),R(4),X(4))
270 X(3)=X(6)+X(5)+FNI(R(6),X(6),R(5),X(5),R(4),X(4))
280 GOSUB 2000
290 CLS:GOSUB 1000:GOTO 440
300 PRINT:PRINT"Enter PI-circuit values":PRINT
310 FOR J=1 TO 3
320 PRINT Z$(J);"(Real,Imag)";:INPUT R(J),X(J)
330 NEXT J
340 REM*COMPUTE TEE-IMPEDANCES*
350 D=R(1)+R(2)+R(3):E=X(1)+X(2)+X(3)
360 R(4)=FNR(R(1),X(1),R(2),X(2),D,E)
370 X(4)=FNI(R(1),X(1),R(2),X(2),D,E)
380 R(5)=FNR(R(2),X(2),R(3),X(3),D,E)
390 X(5)=FNI(R(2),X(2),R(3),X(3),D,E)
400 R(6)=FNR(R(3),X(3),R(1),X(1),D,E)
410 X(6)=FNI(R(3),X(3),R(1),X(1),D,E)
420 GOSUB 2000
430 CLS:GOSUB 1000
440 REM*PRINT RESULTS*
450 PRINT TAB(5)"REAL" TAB(18)"IMAGINARY";
460 PRINT TAB(31)"MAGNITUDE" TAB(44)"ANGLE(DEG)"
470 IF Q$="T" GOTO 540
480 PRINT:PRINT"FOR PI-IMPEDANCES:":PRINT
490 FOR J=1 TO 6
500 IF J=4 THEN PRINT:PRINT"TEE-IMPEDANCES ARE:":PRINT
510 GOSUB 900
520 NEXT J
530 END
540 PRINT:PRINT"FOR TEE-IMPEDANCES:":PRINT
550 FOR J=4 TO 6
560 GOSUB 900
570 NEXT J
580 PRINT:PRINT"PI-IMPEDANCES ARE:":PRINT
590 FOR J=1 TO 3
600 GOSUB 900
610 NEXT J
620 END
900 REM SUBROUTINE: PRINT VALUES
910 PRINT Z$(J) TAB(5);R(J) TAB(18);X(J);
920 PRINT TAB(31);M(J) TAB(44);A(J)
930 RETURN
1000 REM SUBROUTINE: DISPLAY CIRCUIT SCHEMATICS
1010 PRINT TAB(5)"TEE" TAB(26)"PI":PRINT
1020 PRINT"**ZA*****ZC**" TAB(21)"*****ZI*****"
1030 PRINT TAB(6)"*"TAB(23)"*"TAB(30)"*"
1040 PRINT TAB(6)"*"TAB(23)"*"TAB(30)"*"
1050 PRINT TAB(6)"ZB"TAB(13)"<------>"TAB(23)"Z2"TAB(30)"Z3"
1060 PRINT TAB(6)"*"TAB(23)"*"TAB(30)"*"
1070 PRINT TAB(6)"*"TAB(23)"*"TAB(30)"*":PRINT
1080 RETURN
2000 REM SUBROUTINE: CARTESIAN/POLAR CONVERSION
2010 PIE=3.14159265#:FOR J=1 TO 6
2020 M(J)=SQR(R(J)*R(J)+X(J)*X(J))
2030 IF ABS(R(J))<1E-36 GOTO 2050
2040 A(J)=(ATN(X(J)/R(J)))*180/PIE:GOTO 2060
2050 IF X(J)>0 THEN A(J)=90 ELSE A(J)=-90
2060 NEXT J
2070 RETURN
```

Computer programs

Program ROOTS

```
10 CLS:PRINT"PROGRAM ROOTS":PRINT
20 REM*EVALUATES ROOTS OF POLYNOMIAL OF DEGREE M*
30 REM*INPUT DATA*
40 INPUT"Degree of polynomial: M";M
50 DIM A(M+1),B(M+1),AR(M+1),AI(M+1),ROOTR(M),ROOTI(M)
60 PRINT:PRINT"Enter coefficients starting with A(M+1)":PRINT
70 FOR K=M+1 TO 1 STEP-1
80 PRINT "A(";K;")(Real,Imag) = ";:INPUT A(K),B(K)
90 NEXT K
100 PRINT:PRINT"Entry complete; coefficients OK(Y/N)? ";
110 QU$=INKEY$:IF QU$<>"" GOTO 110
120 QU$=INKEY$:IF QU$="" GOTO 120
130 PRINT QU$:IF QU$>="a" THEN QU$=CHR$(ASC(QU$)-32)
140 IF QU$<>"Y" AND QU$<>"N" GOTO 100
150 IF QU$="N" GOTO 60
160 REM*COPY COEFFS TO WORKING ARRAYS*
170 FOR K=1 TO M+1
180 AR(K)=A(K): AI(K)=B(K)
190 NEXT K
200 REM*FUNCTIONS FOR COMPLEX ARITHMETIC*
210 DEF FNCABS(P,Q)=SQR(P*P+Q*Q)
220 DEF FNPR(P,Q,R,S)=P*R-Q*S
230 DEF FNPI(P,Q,R,S)=P*S+Q*R
240 DEF FNQR(P,Q,U,V)=(P*U+Q*V)/(U*U+V*V)
250 DEF FNQI(P,Q,U,V)=(Q*U-P*V)/(U*U+V*V)
260 REM*ROOT FINDING DRIVER ROUTINE*
270 EPS=.000001:REM*ACCURACY*
280 FOR J=M TO 1 STEP -1
290 PLSH$="FALSE"
300 XR=0:XI=0:REM*START ROOT FINDING AT ZERO*
310 GOSUB 1000:REM*CALL SUBROUTINE LAGUERRE*
320 IF ABS(XI)<=2*EPS*EPS*ABS(XR) THEN XI=0
330 ROOTR(J)=XR:ROOTI(J)=XI
340 REM*FORWARD DEFLATION*
350 BR=AR(J+1):BI=AI(J+1)
360 FOR K=J TO 1 STEP-1
370 CR=AR(K):CI=AI(K)
380 AR(K)=BR:AI(K)=BI
390 BRR=FNPR(XR,XI,BR,BI)+CR
400 BI=FNPI(XR,XI,BR,BI)+CI:BR=BRR
410 NEXT K
420 NEXT J
430 REM*POLISH ROOTS USING UNDEFLATED COEFFS*
440 FOR K=1 TO M+1
450 AR(K)=A(K):AI(K)=B(K)
460 NEXT K
470 FOR L=1 TO M
480 J=M
490 PLSH$="TRUE"
500 XR=ROOTR(L):XI=ROOTI(L)
510 GOSUB 1000:REM*CALL LAGUERRE*
520 IF ABS(XI)<=.1*EPS*ABS(XR) THEN XI=0
530 ROOTR(L)=XR:ROOTI(L)=XI
540 NEXT L
550 REM*SORT ROOTS BY REAL PARTS*
560 FOR J=2 TO M
570 XR=ROOTR(J):XI=ROOTI(J)
580 FOR K=J-1 TO 1 STEP-1
590 IF ROOTR(K)<=XR GOTO 630
600 ROOTR(K+1)=ROOTR(K):ROOTI(K+1)=ROOTI(K)
610 NEXT K
620 K=0
630 ROOTR(K+1)=XR:ROOTI(K+1)=XI
640 NEXT J
650 REM*OUTPUT RESULTS*
660 PRINT:PRINT"Roots of polynomial are:":PRINT
670 PRINT "REAL" TAB(14)"IMAGINARY";
680 PRINT TAB(28)"MODULUS" TAB(42)"ARGUMENT(DEG)"
690 FOR K=M TO 1 STEP-1
700 ZR=ROOTR(K):ZI=ROOTI(K)
710 GOSUB 2000
720 MO=FNCABS(ZR,ZI)
730 AZ=AZ*180/PIE
740 PRINT ROOTR(K) TAB(14);ROOTI(K) TAB(28);MO TAB(42);AZ
750 NEXT K
760 END
```

```
1000 REM SUBROUTINE: LAGUERRE
1010 EPSS=6E-09:REM*ACCURACY*
1020 MAXIT=100:REM*NUMBER OF ITERATIONS*
1030 DXOLD=FNCABS(XR,XI)
1040 FOR I=1 TO MAXIT
1050 BR=AR(J+1):BI=AI(J+1)
1060 ERO=FNCABS(BR,BI)
1070 DR=0:DI=0
1080 FR=0:FI=0
1090 ABX=FNCABS(XR,XI)
1100 REM*EVALUATION OF POLYNOMIAL AND FIRST TWO DERIVATIVES*
1110 FOR K=J TO 1 STEP-1
1120 FRR=FNPR(XR,XI,FR,FI)+DR
1130 FI=FNPI(XR,XI,FR,FI)+DI:FR=FRR
1140 DRR=FNPR(XR,XI,DR,DI)+BR
1150 DI=FNPI(XR,XI,DR,DI)+BI:DR=DRR
1160 BRR=FNPR(XR,XI,BR,BI)+AR(K)
1170 BI=FNPI(XR,XI,BR,BI)+AI(K):BR=BRR
1180 ERO=FNCABS(BR,BI)+ABX*ERO
1190 NEXT K
1200 ERO=EPSS*ERO
1210 IF FNCABS(BR,BI)<=ERO THEN GOTO 1220 ELSE GOTO 1250
1220 DXR=0:DXI=0
1230 RETURN
1240 REM*COMPUTE CORRECTION TERM*
1250 GR=FNQR(DR,DI,BR,BI)
1260 GI=FNQI(DR,DI,BR,BI)
1270 G2R=FNPR(GR,GI,GR,GI)
1280 G2I=FNPI(GR,GI,GR,GI)
1290 HR=G2R-2*FNQR(FR,FI,BR,BI)
1300 HI=G2I-2*FNQI(FR,FI,BR,BI)
1310 ZR=(J-1)*(J*HR-G2R)
1320 ZI=(J-1)*(J*HI-G2I)
1330 GOSUB 2000:REM*COMPUTES ARG(Z)=AZ*
1340 SRQ=SQR(FNCABS(ZR,ZI))*COS(AZ/2)
1350 SIQ=SQR(FNCABS(ZR,ZI))*SIN(AZ/2)
1360 GPR=GR+SRQ:GPI=GI+SIQ
1370 GMR=GR-SRQ:GMI=GI-SIQ
1380 IF FNCABS(GPR,GPI)<FNCABS(GMR,GMI) GOTO 1390 ELSE GOTO 1400
1390 GPR=GMR:GPI=GMI
1400 DXR=FNQR(J,0,GPR,GPI)
1410 DXI=FNQI(J,0,GPR,GPI)
1420 X1R=XR-DXR:X1I=XI-DXI
1430 IF XR=X1R AND XI=X1I THEN RETURN
1440 XR=X1R:XI=X1I
1450 CDX=FNCABS(DXR,DXI)
1460 IF I>6 AND CDX>=DXOLD THEN RETURN
1470 DXOLD=CDX
1480 IF PLSH$="TRUE" GOTO 1500
1490 IF FNCABS(DXR,DXI)<=EPS*FNCABS(XR,XI) THEN RETURN
1500 NEXT I
1510 PRINT "TOO MANY ITERATIONS"
1520 RETURN
2000 REM SUBRTN 4-QUADRANT ARCTAN ROUTINE
2010 PIE=3.14159265#
2020 IF ABS(ZR)<1E-36 GOTO 2080
2030 AZ=ATN(ZI/ZR)
2040 IF AZ=0 AND ZR<C THEN AZ=PIE
2050 IF ZR<0 AND ZI>0 THEN AZ=PIE+AZ
2060 IF ZR<0 AND ZI<0 THEN AZ=AZ-PIE
2070 RETURN
2080 IF ZI>0 THEN AZ=PIE/2 ELSE AZ=-PIE/2
2090 IF ZI=0 AND ZR=0 THEN AZ=0
2100 RETURN
```

Appendix D

Laplace transform pairs

$f(t)$ $F(s) = \mathcal{L}\{f(t)\}$

1. t^n $\dfrac{n!}{s^{n+1}}$

2. e^{-at} $\dfrac{1}{s+a}$

3. $\dfrac{1}{a}(1 - e^{-at})$ $\dfrac{1}{s(s+a)}$

4. te^{-at} $\dfrac{1}{(s+a)^2}$

5. $\dfrac{1}{b-a}(e^{-at} - e^{-bt})$ $\dfrac{1}{(s+a)(s+b)}$

6. $\sin \omega t$ $\dfrac{\omega}{s^2 + \omega^2}$

7. $\sin(\omega t + \lambda)$ $\dfrac{\omega \cos \lambda}{s^2 + \omega^2} + \dfrac{s \sin \lambda}{s^2 + \omega^2}$

8. $\cos \omega t$ $\dfrac{s}{s^2 + \omega^2}$

9. $\cos(\omega t + \lambda)$ $\dfrac{s \cos \lambda}{s^2 + \omega^2} - \dfrac{\omega \sin \lambda}{s^2 + \omega^2}$

10. $e^{-at} \sin \omega t$ $\dfrac{\omega}{(s+a)^2 + \omega^2}$

11. $e^{-at} \cos \omega t$ $\dfrac{s+a}{(s+a)^2 + \omega^2}$

12.	$e^{-at}\left(\cos\omega t - \dfrac{a}{\omega}\sin\omega t\right)$	$\dfrac{s}{(s+a)^2+\omega^2}$
13.	$\sinh \omega t$	$\dfrac{\omega}{s^2-\omega^2}$
14.	$\cosh \omega t$	$\dfrac{s}{s^2-\omega^2}$
15.	c (const.)	$\dfrac{c}{s}$
16.	$u(t)$ (step)	$\dfrac{1}{s}$
17.	$u(t-a)$	$\dfrac{1}{s}e^{-as}$
18.	$\delta(t)$ (impulse)	1
19.	$\delta(t-a)$	e^{-as}
20.	$\rho(t)$ (ramp)	$\dfrac{1}{s^2}$
21.	$\rho(t-a)$	$\dfrac{1}{s^2}e^{-as}$
22.	$f(t-a)u(t-a)$	$e^{-as}F(s)$ (shift in t)
23.	$e^{-at}f(t)$	$F(s+a)$ (shift in s)

Operational transforms

In the following,

$$f_0 = f(0^+); \quad f_n = \dfrac{d^n}{dt^n}f(0^+)$$

24.	$\dfrac{df}{dt}$	$sF(s)-f_0$
25.	$\dfrac{d^2f}{dt^2}$	$s^2F(s)-sf_0-f_1$
26.	$\dfrac{d^nf}{dt^n}$	$s^nF(s)-s^{n-1}f_0-s^{n-2}f_1 \cdots -f_{n-1}$

Laplace transform pairs

27. $\int f(t)dt$ $\dfrac{1}{s}F(s)+\dfrac{1}{s}\left[\int f(t)dt\right]_{t=0}$

28. $\int_0^t f(\tau)d\tau$ $\dfrac{1}{s}F(s)$

29. $\int_0^t f_1(\tau)f_2(t-\tau)d\tau$ $F_1(s)F_2(s)$ (convolution)

Bibliography

1. G.W. Carter and A. Richardson: *Techniques of Circuit Analysis*, Cambridge University Press, 1972.
2. H.H. Skilling: *Electrical Engineering Circuits*, (2nd edition). Wiley, 1965.
3. E.A. Guillemin: *Introductory Circuit Theory*, Wiley, 1953.
4. C.A. Desoer and E.S. Kuh: *Basic Circuit Theory*, McGraw-Hill, 1979.
5. A. Ahmed and P.J. Spreadbury: *Analogue and Digital Electronics for Engineers*, (2nd edition), Cambridge University Press, 1984.
6. C.W. Oatley: *Electric and Magnetic Fields*, Cambridge University Press, 1976.
7. R.L. Ferrari: *An Introduction to Electromagnetic Fields*, Van Nostrand Reinhold, 1975.
8. F.F. Kuo: *Network Analysis and Synthesis*, (2nd edition), Wiley, 1965.
9. J.G. Holbrook: *Laplace Transforms for Electronic Engineers*, (2nd edition), Pergamon, 1966.
10. R.E. Scott: *Linear Circuits*, Addison-Wesley, 1960.
11. A.H. Morton: *Advanced Electrical Engineering*, Pitman, 1966.
12. P.G. McLaren: *Elementary Electric Power and Machines*, Ellis Horwood, 1984.
13. G.E. Williams and B.J. Prigmore: *Electrical Engineering*, Heinemann, 1963.
14. K.F. Riley: *Mathematical Methods for the Physical Sciences*, Cambridge University Press, 1984.
15. A.G. Warren: *Mathematics Applied to Electrical Engineering*, Chapman and Hall, 1958.
16. R.J. Bell and David T. Goldman (editors): *The International System of Units*, National Physical Laboratory, H.M.S.O., 1986.

Answers to problems

Chapter 1
1. (a) Source; 6 W. (b) 110 C; 330 J.
2. Sources: A and E. Sinks: B, C, and D. 27 W.
3. $9\,\Omega$; $1/2$ S.
4. $V_{AE}=1$ V; $V_{BE}=1$ V; $V_{CE}=-0.5$ V; $I_{AB}=0$; $I_{CB}=-0.3$ A.
5. 0.624 J; 2.63 kV.
6. $+1/84\,\mu\text{F}$.
7. 1 H; 2 A; $-1/2$ V; 8 J.

Chapter 2
1. 5 V; 40.8 mΩ.
2. (a) 700 V; 250 Ω. (b) 2.8 A; 4 mS.
3. 0.347 V; 0.207 V.
4. 104 V.
5. $(21V_1 - V_2)/109$ A; $(12V_1 + 15V_2)/109$ A; $(13V_1 - 11V_2)/109$ A
   1/109 A; 21/109 A; 18/109 A
   2/21 V; 109/21 Ω.
6. 11 V; $5\frac{1}{2}$ V.
7. Four; two; -5.37 V.
8. 1/2 A.
9. $18\frac{1}{3}$ A; $71\frac{2}{3}$ V.
10. $5r/4\,\mu\text{A}$.
11. 10 kΩ; 30 kΩ.
12. 2 Ω; 4.5 Ω.
13. 5/6 Ω.
14. 40 mΩ (approx); 4 A (approx).
15. Minimum at third load point; $1.32 \times 10^{-4}\,\text{m}^2$

Chapter 3
1. $(E^2 + \frac{1}{2}E_m^2)/R$; $(E^2 + \frac{1}{2}E_m^2)^{1/2}$.
2. (a) $1/\sqrt{3}$; $2/\sqrt{3}$ (b) $1/\sqrt{2}$; $\pi/2\sqrt{2}$.
3. (a) $10 + j10 = 14.1\underline{/45}$; $0.05 - j0.05 = 0.07\underline{/-45}$.
   (b) $10 - j10 = 14.1\underline{/-45}$; $0.05 + j0.05 = 0.07\underline{/45}$.
   (c) $25.9 + j18.55 = 31.9\underline{/35.7}$; $0.25 - j0.18 = 0.031\underline{/-35.7}$.
4. $460 \cos \omega t$.
5. $144\underline{/33.7}$; $17.14\,\Omega$; $109\,\text{mH}$; $139\,\mu\text{F}$.
6. $288\,\text{V}$.
7. $45°$; $45°$.
8. $\omega = [(1/C_1 + 1/C_2 + 1/C_3)/L]^{1/2}$.
12. $4.26\,\text{k}\Omega$; $213\,\text{pF}$; $39.2\,\text{k}\Omega$; 2.62×10^{-3}.
13. $1/(\omega_0^2 C_1 C_2 R) + j(1/\omega_0 C_1 + 1/\omega_0 C_2)$
15. $R + j(\omega L - 1/\omega C - 1/\omega C_1)$; $-j(\omega M - 1/\omega C_1)$.
16. $C_0/(1 + 1/Q^2)$.
17. $35.6\,\text{kHz}$; $0.15\,\text{V}$.
18. (a) $20 + j20$; $2.83 \cos(10t - 45)$. (c) $\omega_1 = \sqrt{50}$; $\omega_2 = \sqrt{150}$.
19. $15.9\,\text{kHz}$

Chapter 4
1. (a) $43.6\,\text{W}$; $129.2\,\text{W}$; $61.7\,\text{W}$. (b) $234\,\text{W}$; $+207\,\text{VAr}$. (c) 0.75.
2. $-j1.77\,\Omega$ or $+j7.69\,\Omega$.
3. $66.7\underline{/-53.1}\,\text{A}$; $82.1\underline{/6.1}\,\text{kV}$.
4. (a) $21.5\underline{/-63.4}\,\text{A}$; $9.57\underline{/53.13}\,\text{A}$; $19.2\underline{/-36.9}\,\text{A}$.
   (b) $5 + j10\,\Omega$; $15 - j20\,\Omega$. (c) $10 + j7.5\,\Omega$.
5. $3.31\,\text{A}$; $110\,\text{W}$; $210\,\text{VAr}$.
6. $1.2\underline{/0}\,\text{A}$; $0.4\underline{/180}\,\text{A}$; $0.28\underline{/135}\,\text{A}$.
7. (a) $237\underline{/-3.16}\,\text{V}$. (b) $250\underline{/-3.8}\,\text{V}$.
8. 96.5%; $6.36\,\text{A}$. Readings on primary: $500\,\text{A}$; $545\,\text{V}$; $26\,\text{kW}$.
9. (a) $8\,\Omega$. (b) $1/4\,\text{A}$; $1/2\,\text{W}$. (c) $2\frac{1}{2}\,\text{W}$; $17\frac{1}{2}\,\text{W}$.
10. $2.1 \times 10^{-8}\,\text{W}$; $2.8 \times 10^{-8}\,\text{W}$; $13 \times 10^{-8}\,\text{F}$.
11. Ratio $= 1:11$; $1.25\,\text{mV}$.
12. (a) $10 - j12\,\Omega$. (b) $19.4\,\text{W}$.
13. $R = 50\,\Omega$; $X = 0\,\Omega$; $1/2\,\text{W}$; $79.1\,\mu\text{H}$; $317\,\text{pF}$.
14. $0.33\,\mu\text{F}\,(Z_1)$, $0.33\,\text{mH}\,(Z_2)$ or $0.33\,\text{mH}\,(Z_1)$, $0.33\,\mu\text{F}\,(Z_2)$. Ratio: 0.33.
15. $11.1\,\Omega$.

Chapter 5
1. (a) $18.6\underline{/3.4}\,\text{A}$; $18.6\underline{/-116.6}\,\text{A}$; $18.6\underline{/123.4}\,\text{A}$.
   (b) $32.1\underline{/-26.6}\,\text{A}$; $32.1\underline{/-146.6}\,\text{A}$; $32.1\underline{/93.4}\,\text{A}$.
   (c) 0.894.

(d) 20.6 kW.
(e) $(20+j10)/3\,\Omega$.
2. 6.97 A.
3. $15.6 \pm j25.6\,\Omega$.
4. (a) 231 A. (b) 0.986. (c) 94.7 kW.
5. 7.19 kW; 5.74 kW.
6. (a) $V_{ab} = 440\underline{/0}$ V; $V_{bc} = 440\underline{/-120}$ V; $V_{ca} = 440\underline{/120}$ V.
   $I_a = 24.7\underline{/-63.6}$ A; $I_b = 11.1\underline{/-36.8}$ A; $I_c = 35\underline{/124.6}$ A.
   $I_1 = 3.04\underline{/-90}$ A; $I_2 = 13.2\underline{/-47.5}$ A; $I_3 = 22\underline{/120}$ A.
   (c) $V_L = 478\underline{/0}$.
7. Power $= W - V^2 R/(R^2 + X^2)$.
8. (a) 0.513 MW/line. (b) 65.4 MW. (c) 139 kV.
9. 25.5 μF.
10. 6.2 kV (line); 97.7%.
11. 551 A; 400 V.

Chapter 6

1. 1.0 s. (a) 0.69 s; 3 s.
2. (a) 12 mA; zero. (b) $i_c = 12 e^{-20000 t}$.
   (c) Zero; 12 V. (d) 360×10^{-8} J. (e) 720×10^{-8} J.
3. $i_1 = \dfrac{V}{R_2} e^{-t/T}$; $i_2 = \dfrac{V}{R_1 + R_2}(1 - e^{-t/T})$; $i = i_1 + i_2$.
   $T = CR_1 R_2/(R_1 + R_2)$; 0.342 mA; 0.25 mA.
4. (a) $i_c = 10 e^{-100 t}$ A; $v_c = 200(1 - e^{-100 t})$ V.
   (b) $i_c = -10 e^{-66.7 t}$ A; $v_c(0.1) = -100$ V.
   (c) $i_c = 15 e^{-100 t}$ A; $v_c = 100(2 - 3 e^{-100 t})$ V.
5. 2.3 ms (closing); 0.6 ms (opening).
6. (a) $i_L = 8.94 \cos(10^3 t - 63.4) - 2 e^{-500 t}$ mA (b) $140°$.
7. $(15 t + 0.3)$ V; 0.3 V.
8. (a) 0.833 mA; 0.833 mA; 0.833 V.
   (b) 0.833 mA; 0.833 mA; 0.833 V; -60 mA/s; -30 mA/s; 1500 V/S.
   (c) 0.083 mA; -0.67 mA; 2.33 V.
9. $R_1 = R_2$; $L = CR_1^2$.
10. $R > 2\sqrt{(L/C)} - r$.
11. $\left[LCD^2 + \left(CR_3 + \dfrac{L}{R} \right) D + \dfrac{R_3}{R} + 1 \right] v_2(t) =$

    $\left[LCD^2 + \left(CR_3 + \dfrac{L}{R_1} \right) D + \dfrac{R_3}{R_1} \right] v_1(t)$
    where $R = R_1 R_2/(R_1 + R_2)$

Answers to problems

$A_1 e^{-\alpha_1 t} + A_2 e^{-\alpha_2 t}$ where α_1 and α_2 are given by:

$$\alpha_1, \alpha_2 = \left(\frac{R_3}{2L} + \frac{1}{2RC}\right) \pm \left[\left(\frac{R_3}{2L} + \frac{1}{2CR}\right)^2 - \frac{1}{LC}\left(1 + \frac{R_3}{R}\right)\right]^{1/2}.$$

12. $\dfrac{E}{2}[e^{-Rt/(L+M)} - e^{-Rt/(L-M)}]$.

13. (a) $v_c(t) = 0.67 + e^{-\alpha t}(0.33 \cos \omega t + 3.35 \sin \omega t)$
    where $\alpha = 0.75$, $\omega = 0.97$.
    (b) $v_c(t) = 0.51 \cos(2t - 130) + e^{-\alpha t}(1.33 \cos \omega t + 4.93 \sin \omega t)$

$$i(t) = \frac{v_c(t)}{2} + \frac{1}{0.5}\frac{d}{dt}v_c(t).$$

14. (a) $0.1\,\text{mH}$; 200. (b) $200\,\text{k}\Omega$.
16. $v_c(t) = 10(8 - 7e^{-0.318t})$;
    $i(t) = 1 + 3.5e^{-0.318t}$.
17. (a) $a = v_2(0)$; $b = 2i_L(0)$; $c = \frac{1}{2}[3v_1(0) + v_2(0)]$.
    (b) $v_2(0) = 1$; $i_L(0) = 1$; $v_1(0) = 1$.
    (c) $v_2(t) = 1 - e^{-t} - \dfrac{2}{\sqrt{3}} e^{-t/2} \sin \dfrac{\sqrt{3}}{2} t$.
18. $V(s) = \dfrac{10}{s^2}(1 - e^{-0.1s}) - \dfrac{1}{s}e^{-0.1s}$

$$i(t) = \left[10C - \frac{10}{\omega}\sqrt{\left(\frac{C}{L}\right)} e^{-\alpha t} \cos(\omega t - \phi_1)\right] u(t)$$

$$-\left[10C - \frac{10}{\omega}\sqrt{\left(\frac{C}{L}\right)} e^{-\alpha(t-0.1)} \cos\{\omega(t-0.1) - \phi_1\}\right.$$

$$\left. + \frac{1}{\omega L} e^{-\alpha(t-0.1)} \sin \omega(t-0.1)\right] u(t-0.1)$$

where $\omega^2 = (1/LC) - (R/2L)^2$; $\phi_1 = \tan^{-1}(\alpha/\omega)$; $\alpha = R/2L$.

20. $V_2(s) = 0.5/[(s+4)(4s^2 + 4s + 2)]$;
    $v_2(t) = (1/100)e^{-4t} + (1/2\sqrt{50})e^{-0.5t}\cos(0.5t - 98.1)$.
21. $v_0(t) = 5[e^{-\alpha_1 t} - e^{-\alpha_2 t}]u(t)$
    $\quad - 5[e^{-\alpha_1(t-a)} - e^{-\alpha_2(t-a)}]u(t-a)$
    where $\alpha_1 \approx 5 \times 10^{-3}$, $\alpha_2 \approx 2 \times 10^6$, $a = 5 \times 10^{-6}$.
23. $H(s) = 1/(s^3 + 2s^2 + 2s + 1)$.
24. $R(\min) = 2\left[\dfrac{L}{C}(1+k)\right]^{1/2}$.
25. $F(s) = \dfrac{\alpha(s+\alpha) - \beta^2}{(s+\alpha)^2 - \beta^2}$.

Chapter 7

1. 10.3 mA; 2.2 V.
2. 10.8 mA; 2.05 V; 19 mW; 0.8 mW.
3. 5.1 mA; 7.3 V; 2.4 V; 4.9 V; $R_{d.c.} = 1430\,\Omega$; $R_{a.c.} = 788\,\Omega$.
4. (a) $0.61\,\text{k}\Omega < R < 2.9\,\text{k}\Omega$. (b) $-0.17\,\text{V}$ or $0.46\,\text{V}$.
5. (a) 7 V. (b) 270 Ω.
6. Slopes: $1, \frac{1}{2}, 0$. Break points 1V, 2V.
7. Breakpoints: (8 V, 0 A) and (16 V, 2 A).
   Slopes (i/v): 1/10, 1/4 and 7/20 corresponding to resistance of 10 Ω, 4 Ω, 2.85 Ω.
8.

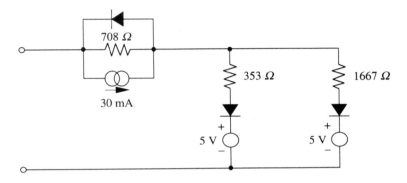

9. $25.9/R$.
10. (a) 0.53 V; 2.3 mA; 1.2 mW. (b) 1.2 mW.

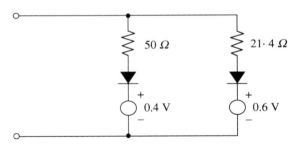

11. $R_1 = 1/3\,\text{k}\Omega$; $R_2 = 2/3\,\text{k}\Omega$; $R_3 = 3\,\text{k}\Omega$; $R_4 = 7/6\,\text{k}\Omega$; $R_5 = 1.46\,\text{k}\Omega$.
12. 0.712 A.
13. (a) 1.67 A; 1.87 V; 1.13 V. (b) 1.16 A; 1.58 V, 0.85 V; 0.58 V.
15. (a) $R_1 = \text{M}\Omega$; $R_2 = 28.6\,\Omega$. (b) 0.1 mA; 31 mA.
16. (c) 10 mA. (d) 1.2 ms.

572   Answers to problems

17. $v_c(t) = V(1 - \cos \omega_0 t)$, $0 < t < \pi/2\omega_0$

    $v_c(t) = V + \omega_0 rCV[e^{-\frac{r}{L}(t - \frac{\pi}{2\omega_0})} - e^{-\frac{1}{rC}(t - \frac{\pi}{2\omega_0})}]$,

    $t > \pi/2\omega_0$.
18. (a) 1.05 A. (b) $C = 0.1$ F.
19. (a) 56.3 mA. (b) 6.3%. (c) 0.73 A. (d) 0.187 A. (e) 3.2%. (f) 0.1%.
20. (a) 360 V; 9.6 V. (b) 262 V; 377 μF.
21. (a) 25 mV. (b) $V_d = \sqrt{2[(V_r^2 + V_s^2 + 2V_rV_s\cos\phi)^{\frac{1}{2}} - (V_r^2 + V_s^2 + 2V_rV_s\cos\phi)^{\frac{1}{2}}]}$
22. 90°; 16.9 kΩ
    $i = 3.24\sin(\omega t - 17.4) - 18e^{-1000t} A(t \geq 2.5 \text{ ms})$.   107°; 197°
23. $f(t) = \dfrac{4A}{\pi} \sum_{n=2k+1}^{\infty} \dfrac{1}{n}\left[\dfrac{\sin(n2\pi p/T)}{n2\pi p/T}\right]\sin\dfrac{n2\pi t}{T}$ $k = 0, 1, 2, \ldots$

    (c) $p = T/6$. (d) $p = T/10$, $T/5$.
24. 60°; 1.15.

Chapter 8

1. $h_{11} = 14/3 \, \Omega$; $h_{12} = 1/3$; $h_{21} = -1/3$; $h_{22} = 1/12$ S.
3. $z_{11} = R_e + R_b$; $z_{12} = R_e$; $z_{21} = R_e - \beta R_d$; $z_{22} = R_d + R_e$.
4. $y_{11} = y_{22} = \dfrac{s(s+3)}{2s+1}$; $y_{12} = y_{21} = \dfrac{s(s+1)}{2s+1}$.
5. (a) $A = -\dfrac{1}{\mu}[1 + Y(Z + R_2)]$; $B = A/R_1$

    $C = -\dfrac{1}{\mu}[2Z + Z^2Y + R_2(1 + ZY)]$; $D = C/R_1$
6. $y_{11} = y_{22} = \dfrac{R^2C^2s^2 + 4RCs + 1}{2R(RCs + 1)}$; $y_{21} = y_{12} = \dfrac{-R^2s^2C^2 + 1}{2R(RCs+1)}$

    Infinite attenuation when $\omega = 1/RC$.
7. (b) $y_{12} = y_{21} = \dfrac{R^2s^3C^3}{2(1+sRC)} = \dfrac{s^3CT^2}{2(1+sT)}$.
8. Iterative impedances: 21.3 Ω and 16.2 Ω.
   Image impedances: 17.1 Ω and 20.5 Ω.
9. $\dfrac{V_{IN}}{V_{OUT}} = \dfrac{1}{sL}[R + sL + \sqrt{(R^2 + 2RsL)}]$; $z_1 = \sqrt{(R^2 + 2RsL)}$

   For given values: $V_{IN}/V_{OUT} = 1.79 - j2.28$; $z_1 = 1270 + j790 \, \Omega$.

Index

ABCD parameters, 511, 515
a.c. bridge, 132–6
active element, 3
active power, 187
admittance, 123–6
  complex, 116, 123
  driving point, 504
  generalized, 348, 361–3
  transfer, 504
admittance matrix, 513
admittance parameters, 508
alternating quantity, 2, 99
ammeter, 251
ampere, 6
amplifier, 78, 510
amplitude of a.c. waveform, 99
amplitude
  response, 367
  spectrum, 474
angular frequency, 99
Argand digram, 106
argument of complex number, 105
asymmetrical network, 87
attenuator, 45, 534
autotransformer, 214–17
auxiliary equation, 288
average power, 187

balanced network, 87, 502
band-pass characteristic, 157, 177
bandwidth, 155
battery model, 43
bias point, 426
bilateral element, 418
Bode diagram, 130
branch, 43

break frequency, 131
break point, 429
bridge circuits, 61, 72, 132–40, 221
Butterworth filter, 372

capacitance, 2, 13, 103
  stored energy in, 24
capacitances
  in parallel, 26
  in series, 24
capacitive circuit
  reactance, 103
  susceptance, 125
capacitors
  loss-angle in, 150
  losses in, 149
  model for, 36
Cartesian coordinates, 106
cascaded networks, 518, 530
chain matrix, 518–25
characteristic impedance, 532
characteristics (of device), 419–25
charge, 5
circle diagram, 129
circuit
  active, 3
  distributed, 3
  dual, 37, 164
  lumped model, 3
  passive, 3, 44
circuit reduction, 84
coefficient of coupling, 33, 175
coil, 26
compensation theorem, 82
complementary function, 285

complex
  algebra, 106
  conjugate, 198
  exponential, 107
  frequency, 348, 369
  impedance, 110
  plane, 106, 108, 361
  power, 198
  quantity, 105
conductance, 19, 69, 124
conduction angle, 20, 452
conductivity, 20
conjugate bridge, 133
conservation
  of charge, 13
  of energy, 9, 13
  of watts and vars, 194
controlled source, 33, 75, 507
convolution, 392–405
  integral, 396
  theorem, 401
copper loss, 21, 213
corner frequency, 131
corresponding ends (of coil), 32, 142
coulomb, 6
coupled coils, 141, 301
coupling coefficient, 33, 175
coupling network, 127
Cramer's rule, 54
critical damping, 294
critical coupling, 174–6
current
  alternating, 2, 98–101
  direct, 2
current convention, 7
current divider, 45, 165
current source, 11, 117

D-impedance, 309
D-operator, 304
damped natural frequency, 294
damped sinusoid, 293
damping, 294, 362–4
damping constant, 294
decibel, 130
delay time, 312
delayed function, 326
delta circuit, 66
delta function, 319
dependent source, 75
determinant, 54
differential operator, 309
differentiating circuit, 316

diode
  ideal, 429
  real, 437
  zener, 437
Dirac function, 319
discontinuous function, 317
discriminant, 293
divider, 45, 350
dot convention, 32, 142
double-energy circuit, 291
doublet (unit), 325
driving point impedance, 143, 505
duality, 37
dual networks, 45, 67, 164
dynamic resistance, 426
  of resonant circuit, 166
dynamometer, 251

effective resistance, 146
effective value of sinusoid, 100
electrical angle, 100
eletrodynamometer, 251
electrokinetic momentum, 347
eletromotive force (e.m.f.), 1, 27
element
  active, 3
  bilateral, 418
  ideal, 36
  lumped, 3
  non-linear, 418
  passive, 3, 36
energy
  initial, 281, 291
  sink, 1, 9
  source, 1, 9
  storage in elements, 2, 24, 28
equivalent circuits
  for capacitor, 149
  for inductor, 147
  series–parallel, 21
  star–delta, 66
Euler's identity, 105
even function, 466
excitation function, 349
exciting current, 202, 205
exponential Fourier series, 471

farad, 23
Faraday's law, 31
field, 1
filters
  for rectifiers, 482–6
  high-pass, 127

Index

low-pass, 372
notch, 140
twin-T, 136, 528
final value theorem, 358
first-order circuit, 291
flux
  density (B), 203
  leakage, 202
  magnetic, 26, 202
  mutual, 31
forced response, 285
form factor, 101
Foster's reactance theorem, 173
Fourier series
  coefficients, 460
  cosine, 466
  exponential, 471
  for rectifier, 469
  sine, 466
frequency
  angular, 99
  complex, 348
  half-power, 155
  natural, 294
  negative, 472
  response, 126
  spectrum, 466
  undamped, 294
frequency changer, 445
function
  network, 349
  rational, 334
  transfer, 126

generalized network function, 349
generator (three-phase), 231
Gibbs phenomena, 465
g-parameters, 507
gyrator, 524

half-power bandwidth, 155
harmonic components, 461
henry, 27
hertz, 99
hybrid parameters, 506
hysteresis loss, 203

ideal source, 11
ideal transformer, 199, 522
image impedance, 531
immittance, 348
impedance
  complex, 110

D-operator, 309
driving point, 143, 504
dynamic, 166
generalized, 348
input, 143
modulus of, 110
transfer, 505
impedance diagram, 115
impedance matrix, 513
impedance transformation, 200
impulse
  function, 317, 319
  response, 320, 351
  train, 392
incremental (slope) resistance, 426
independent parameters, 502
induced e.m.f. 27
inductance, 2, 26, 102
  leakage, 207
  mutual, 31, 35, 141
  self, 26, 27
  stored energy in, 28
inductances
  in parallel, 29
  in series, 30
inductive reactance, 103
inductor
  losses in, 146
  model for, 36
initial conditions, 281, 289
initial value theorem, 358
in-phase component, 192
insertion loss, 536
integrating circuit, 316
inverse matrix, 515
inverse transform, 328
iron loss, 203
iterative impedance, 530

joule, 8
j-operator, 106

Kirchhoff's laws, 13–18, 419
  current, 13
  voltage, 16

ladder method, 87, 372
ladder network, 87
Laplace transform, 328
  of delayed functions, 358
  of derivative, 330
  of integral, 331
lattice network, 72, 521

leakage inductance, 207
linear element, 3, 418
linearity, 57
line spectra, 475
load line, 423
locus diagram, 129
logarithmic decrement, 295
loop, 44
loop equations, 48, 52
loss-angle (of capacitor), 150
low-pass filter, 372
lumped element, 3

magnetizing current, 203
magnetizing force (H), 203
magnitude (of a.c. waveform), 100
mark–space ratio, 376
matching, 201, 218–21
maximum power transfer, 217
Maxwell's cyclic currents, 49
mesh, 44
mesh analysis, 47–50, 52, 71, 545
Millman's theorem, 86
mistuning, 160
modulator, 445
multiple resonance, 169–73
mutual conductance, 69, 70
mutual inductance, 31, 35, 141, 175, 344
mutual resistance, 48, 53

natural response, 285
negative impedance converter, 524
network, 44
network function, 349
nodal analysis, 67–70, 71, 125
nodal equations, 69
node, 13, 43
non-linear elements, 3, 418
Norton's theorem 64
Nyquist diagram, 129

Odd function, 466
ohm, 19
Ohm's law, 18, 112
open-circuit impedance, 505
open-circuit test, 209
operating point, 425
operational amplifier, 78, 510
oscillatory response, 293, 363
output resistance, 43, 80
overdamped circuit, 294, 362

parabola (unit), 325
parallel resonant circuit, 164
Parseval's theorem, 478
partial fractions, 334–9
particular integral, 285
passive circuit, 3
peak inverse voltage, 450
period, 99
periodicity theorem, 382
phase angle, 99
phase response, 239, 367
phase sequence, 246
phase sensitive detector, 447
phase transformation, 270
phasor
  cartesian form, 105
  diagram, 113, 118
  polar form, 106
  rotting, 107
  stationary, 108
pi-network, 516
piecewise-linear circuits, 428–43
poles
  of admittance, 362
  of impedance, 172
  of network function, 360
  repeated, 362
pole-zero methods, 359–72
pole-zero diagram, 361
port, 44
potential difference (p.d.), 7
potentiometer, 46
power
  active and reactive, 187, 190
  apparent, 191
  average, 187–8
  complex, 198
  diagram, 249
  in harmonic components, 478
  instantaneous, 8, 188, 273
  real, 187, 188
  sign convention, 9, 188
  triangle, 191
power factor, 150, 194
power factor correction, 194, 249
practical voltage source, 42
pulse
  repeated, 376
  transform of, 381

quadrature component, 192
quasi-steady state, 377

Index

Q-factor,
  of capacitor, 150
  of inductor, 147
  of resonant circuit, 165, 169, 253, 294

radian, 99
ramp function, 325
rational function, 334
rationalization, 124
reactance, 103, 110
reactive power, 190
reciprocal network, 505
reciprocity theorem, 58, 546
rectifier circuits, 448
  bridge, 453
  full-wave, 453
  half-wave, 448
  ripple in, 482
regulation (of transformer), 211
resistance, 2, 18, 101
resistances
  in parallel, 22
  in series, 21
resistivity, 20
resistor
  model for, 36
  non-linear, 445
resonance, 146, 158
resonance curve, 162
resonant circuits
  inductively coupled, 173
  losses in, 151
  magnification factor in, 157, 165
  parallel, 164
  series, 151
resonant frequency, 153, 158
response function, 349
root mean square (r.m.s.) value, 101, 189, 478
roots (of polynomial), 354
Rosen's theorem, 67

s (complex frequency), 348
  domain, 340
  plane, 361
sampling, 394
Schering bridge, 133
Scott connection, 271
second-order circuit, 291
self inductance, 26
shift theorem, 357

short-circuit test, 210
siemens, 19, 124
sifting property, 394
single-energy circuit, 291
singularity function, 317–27
sink, 9
skin effect, 146
small-signal model, 426
source,
  controlled, 33, 75, 507
  current, 12, 64
  dependent, 33, 75, 507
  ideal, 11
  practical, 42, 64
  transformation, 64
  voltage, 11, 64
spectrum, 474
stagger tuning, 177
star connection, 66, 239
star–delta transformation, 66, 86
steady state, 280
steady-state response, 285, 377
step function, 318
step response, 319
substitution theorem, 81
superposition theorem, 55, 546
susceptance, 124
symmetrical circuits, 87, 502

tee network, 516
Tee-pi transformation, 67, 138, 516
Tellegen's theorem, 194
Thévenin's theorem, 58, 313, 547
Thévenin–Norton transformation, 64, 117
third-harmonic current, 269
three-phase circuit, 233
  balanced load in, 236
  delta connection in, 239
  line voltage in, 235
  neutral point in, 235
  phase sequence in, 239, 246
  phase voltage in, 235
  star connection in, 239
  unbalanced load in, 245
  Y-connection in, 239
thyristor, 455
time constant, 286
time domain, 340
total response, 295
tranfer characteristic, 423
transfer function, 126, 349

transfer impedance, 505
transformer
  auto, 214
  copper loss in, 213
  core loss in, 203, 213
  efficiency in, 213
  equivalent circuit, 261
  ideal, 199
  regulation in, 211
  single phase, 201
  tests, 209, 261
  third harmonic currents in, 269
  three-phase, 258
transformer bridge, 221
transient response, 282, 285
transistor, 423
  model, 426, 427, 509
  parameters, 503
transmission parameters, 511
triac, 459
triangular pulse, 383
tuning, 157
turnover frequency, 131
twin-T network, 136, 528
two-phase voltages, 271
two-port network, 126, 501
two-wattmeter method, 253

unbalanced network, 87

underdamped circuit, 294, 362
universal resonance curve, 161
unsymmetrical network, 503

var, 191
volt, 7
volt-ampere, 191
voltage divider, 44, 350
voltage source
  ideal, 11
  practical, 42, 64

watt, 8
wattmeter, 251
waveform
  rectified sine, 304, 469–71
  sawtooth, 384
  square, 381, 463–6
  trapezoidal, 499
  triangular, 383, 466–8
  TV video, 390
Wheatstone bridge, 61, 72
Wien bridge, 135

Y–Δ transformation, 66, 244
y-parameters, 503

zeros of network function, 172, 360
z-parameters, 505